CHEMICAL
AND
CATALYTIC
REACTION
ENGINEERING

McGRAW-HILL
BOOK COMPANY
New York
St. Louis
San Francisco
Auckland
Düsseldorf
Johannesburg
Kuala Lumpur
London
Mexico
Montreal
New Delhi
Panama
Paris
São Paulo
Singapore
Sydney
Tokyo
Toronto

JAMES J. CARBERRY

Professor of Chemical Engineering
University of Notre Dame

Chemical and Catalytic Reaction Engineering

This book was set in Times Roman.
The editors were B. J. Clark and Michael Gardner;
the production supervisor was Leroy A. Young.
The drawings were done by Oxford Illustrators Limited.
Fairfield Graphics was printer and binder.

Library of Congress Cataloging in Publication Data

Carberry, James J
   Chemical and catalytic reaction engineering.

   (McGraw-Hill chemical engineering series)
   Includes bibliographies.
   1. Chemical processes. 2. Catalysis. I. Title.
TP155.7.C37   660.2'9'9   76–1963
ISBN 0–07–009790–9

**CHEMICAL
AND
CATALYTIC
REACTION
ENGINEERING**

34567890 FGRFGR 78321098

To the patient ones:
my wife, MARGARET VIRGINIA,
my parents,
my daughters, ALISON and MAURA,
and
the Notre Dame students

"O Muse, o alto ingegno, or m'aiutate;
o mente che scrivesti ció ch'io vidi,
qui si parrá la tua nobilitate."

DANTE   INFERNO, II, 7

CONTENTS

PREFACE

Behold, my desire is . . . that mine adversary had written a book
Job, 31: 35

Students of history are familiar with the tale of an ancient king who commanded his wise men to fashion a comprehensive history of their yesteryears. Exasperation seized him when some several dozen volumes were produced. He then recharged his scholars to fashion a somewhat more brief account. Eventually, as is the wont of even modern administrators, the king expressed impatience with a terse one-volume product. In consequence he slaughtered his scribes and then retired to reflect upon the fact that life seems so short and history so long.

Chemical reaction engineering and particularly catalysis and its applications are indeed so complex and the professors' view so primitive that I hasten to plead for merciful understanding before all kings who labor in the real reactor world. My experience in industry's reactor vineyard and subsequent labors in the academy have persuaded me that this text must be viewed as but a commentary, a particular view of what I consider to be a few essential scientific ingredients in an area within which progress is largely realized by art, some science, and a generous portion of serendipity.

A Handbook of Chemical and Catalytic Reaction Engineering would seem to be beyond creation, whether authored by kings and/or philosophers. What is set forth here is designed to stimulate the novice who will build, as do we all, upon these simple elements in order to fashion meaningful solutions to the complex chemical reaction realities which nature visits upon us.

It was Bernard of Chartres of the twelfth century who wisely observed that we sit upon a mountain built by others and thus view the terrain more clearly by reason of those builders, our mentors. The late John Treacy of Notre Dame and that grand gentleman of Yale, the late R. Harding Bliss, patiently nurtured me in this fascinating subject of chemical reaction engineering. Professor Paul Emmett's cosmic course in catalysis at Johns Hopkins and numerous dialogues with the late Sir Hugh Taylor and with my very lively colleagues George Kuczynski and Michel Boudart served admirably to focus my vision upon heterogeneous catalysis and its provocative mysteries.

The mountain is surely composed of the shoulders of many others whose identity will become evident with a study of the body of this work.

This text is so structured that an introductory course may be fashioned with the first five chapters and selected segments of one or more of those which follow. A more advanced course might well commence with Chapter 4. In either case, it is to be noted that Chapter 8 (Catalysis) can be utilized quite independently of the others.

Beyond Chapter 3, the problems are, by design, somewhat unique insofar as some are slightly devoid of necessary data, others overly rich, while there is also provided a reasonable number of open-ended problems, i.e., the student is required to seek out specified literature sources and then is invited to create solutions in the light of his or her (informed) subjective judgments of said data. In accord with design realities, nonunique, yet instructive, solutions should emerge to the benefit of all participants.

I am grateful to have had the opportunity to present portions of the material in a series of lectures at the Shell Department of Chemical Engineering, Cambridge University, as NSF Senior Fellow; at the University of Naples; and at Stanford University.

Mrs. Helen Deranek most admirably transformed the terrors of my hand-written manuscript into typewritten form worthy of human scrutiny while Mrs. W. G. Richardson very patiently performed editorial miracles of revision. I am most grateful to them and to those who generously gave of their time to read the text and render worthy criticisms of it: Professors R. Aris, O. Levenspiel, D. Luss, J. H. Olson, and W. D. Smith. Paul Charles, Joseph Perino and Steve Paspek, class of '76, very carefully freed the manuscript of numerous errors.

JAMES J. CARBERRY

LIST OF SYMBOLS

The scope of chemical and catalytic reaction science and engineering is so great that a simple unambiguous litany of symbols cannot be fashioned. The most frequently assigned meanings of symbols used in the text are given below. Unavoidably different meanings are specified in situ.

a	surface-to-volume ratio
a, b, c	stoichiometric coefficient
$\mathscr{A}$	preexponential factor; reactor aspect ratio
$A, B, C, \ldots$	concentrations of molecular species A, B, C, ...
C_p	heat capacity
d	diameter of particle, bubble, or tube
$\mathscr{D}$	diffusion coefficient (molecular or turbulent)
D	operator; dispersion of catalytic metals
e	void fraction
exp	exponential
E	activation energy
E'	mass-transfer enhancement factor
$\mathscr{E}$	efficiency
f	reduced concentration; a function; ratio of external to total catalyst area
F	molar flow rate
g	gravitational constant; a function
G	mass velocity, ρu
ΔG	free-energy change
h	convective heat-transfer coefficient; height

ΔH reaction-enthalpy change

H bed height, hold-up, HTU

i ith species

I inert concentration

I   inhibitor

j heat-, mass-transfer, factor

J observed, global rate of absorption-reaction

k chemical or chemisorption rate coefficient

k_g, k_{L_0} convective physical mass-transfer coefficients for gas and liquid phases, respectively

k_L reaction-affected liquid-phase mass-transfer coefficient

K ratio of rate coefficients; equilibrium adsorption or kinetic adsorption coefficient; equilibrium constant

$\mathscr{K}$ effective rate coefficient

ℓ length of a pore, chain, or mixing length

L $1/a$, volume-to-external surface area; liquid flow rate; length

ln   natural logarithm

$\mathscr{L}$ Laplace transform

m reaction order; aspect ratio

n number of CSTRs

M molecular weight; Henry's-law coefficient

N number of transfer units (NTU); flux; number of moles

p partial pressure

P or π total pressure; product species

ΔP pressure drop

$q_{g,r}$ heat generation; removal

q recycle flow rate

Q volumetric flow rate

r radial distance; Biot number ratio; chemical reaction rate (intrinsic)

R gas constant; recycle ratio

R_0 particle or tube radius

$\mathscr{R}$ global rate of reaction

s rate coefficient ratio

S selectivity; total surface area of a porous catalyst; sites

$\mathscr{S}$ bubble-cap-tray submergence; selectivity

T temperature

t reduced temperature; T/T_0; time; physiosorbed layer thickness

tanh   hyperbolic tangent

u fluid velocity

U overall heat-transfer coefficient

v molar volume

V volume

x conversion; distance

X mole fraction in liquid phase

y distance; gas-phase mole fraction

Y yield; gas-phase mole fraction

z, Z length coordinate

Greek symbols

α, β, γ reaction orders

α fraction; bulk-to-film volume ratio

β Prater number, internal $\Delta T_a / T_0$

$\bar{\beta}$ External $\Delta T_a / T_0$

γ diffusivity ratio

Γ gamma function

δ film thickness

ε E/RT_0

η effectiveness factor

λ thermal conductivity

μ viscosity

ν kinematic viscosity

ρ density; reduced radius

ϕ Thiele modulus, $L\sqrt{kC_0^{n-1}/\mathscr{D}}$

Φ Wheeler-Weisz observable

θ contact time; surface coverage

Subscripts

A, B, C species

c fluid-reactant core interface

h heat

i ith species; interface

m mass

o initial or bulk condition

s surface condition of particle/pellet

w wall condition

Dimensionless Groups

Bi   Biot number

Da   Damköhler number

Fr   Froude number

Le   Lewis number

Nu   Nusselt number

Pe   Peclet number

Pr   Prandtl number

Re   Reynolds number

Sc   Schmidt number

Sh   Sherwood number

St   Stanton number

CHEMICAL
AND
CATALYTIC
REACTION
ENGINEERING

INTRODUCTION

"Tis ten to one, this play can never please all that are here."
Shakespeare "King Henry VIII"

Given a particular thermodynamically permissible chemical reaction network, the task of the chemical engineer and applied kineticist is essentially that of "engineering" the reaction to achieve a specific goal. That goal, or end, is the transformation of given quantities of particular reactants to particular products. This transformation (reaction) ought to be realized in equipment of reasonable, economical size under tolerable conditions of temperature and pressure. The plant, of which the chemical reactor or reactors are but a part, usually contains preparatory equipment for reactor-feed treatment and additional treating units designed to separate and isolate the reactor products. While the reactor, which is supplied and serviced by auxiliary equipment, might be considered the heart of a chemical or petroleum plant, a view of the overall process must be borne in mind by the reaction engineer; for a serious and complex separation problem may well dictate reactor operation at conversion (fraction of a particular, key reactant which is consumed) and yield levels (fraction of consumed reactant which appears as desired product) which would be considered less than maximum. The reaction, then, must be engineered with a view toward the overall economics of plant design.

For example, it may well be established that a reaction between an olefin, oxygen, and NH_3 provides the highest yield of product only when a vast excess of olefin is fed to the reactor relative to NH_3. However, an analysis of overall plant investment would reveal that the cost of processing the excess unreacted olefin in the reactor effluent is so great that savings would ensue if the lower yield associated with a lower olefin/NH_3 ratio was actually entertained. Yet even in this instance, the reactor engineer must dictate the conditions which guarantee reactor performance at that specified level. Whether the reactor performance is directly cost-determining, or indirectly so, the challenge remains: a specified reactor performance must be achieved. Such an achievement implicitly suggests analysis and prognostication, roles shared by both the chemist engaged in applied research and the chemical engineer explicitly devoted to application.

These problems of reaction analysis and reactor design are best presented in an idealized framework, from which the realities can be appreciated. Ideally, one secures or receives (with prudent apprehension) basic laboratory data relating conversion to the desired product as a function of species concentrations, temperature, etc. Given these raw data or possibly organized data (a chemical-kinetic law or model), the chemical reaction engineer, in principle, organizes this chemical information in concert with physical parameters (heat-transfer coefficients, etc.) and ideally creates a mathematical model.

But why the necessity of a model composed of physicochemical submodels? When a chemist conducts a test-tube experiment, involving, say, a highly exothermic reaction, he need not be concerned with a heat-transfer problem so long as an ice bath is at his elbow. Near isothermality (or at least thermal stability) can generally be realized by the simple expedient of alternately immersing the test tube into the flame and (when things get out of hand) into the ice bath. When, however, the plant or even bench-scale reactor assumes dimensions commensurate with production or semiproduction levels, in situ heat removal is obviously required. For example, a laboratory study of vanadium pentoxide catalysis of SO_2 might involve passing the reactants over a few fine grains of catalyst packed in a tube immersed in a well-agitated heat-transfer medium, e.g., molten salt. A commercial SO_2 oxidizer consists, on the other hand, of a packed bed perhaps several meters in diameter. Heat removal at the plant scale analogous to that employed in the laboratory is hardly feasible. As reaction rate and equilibrium are both highly temperature-dependent, we cannot predict plant-scale SO_2 oxidation behavior unless we have accurate organized information on the modes and rates of heat transport in a packed bed. If such data are available, it is conceivable that a proper combination of these data with laboratory-scale chemical-reaction-rate data will yield a meaningful model and thereby provide the basis for scale-up.

This model, so organized, would represent the physicochemical events and thus permit prediction of reactor performance, either analytically or via computer solution. This ideal situation might be schematized as shown in Fig. 1-1.

The ideal situation suffers when confronted with reality. The platonic archetype cited above departs considerably from the real flowsheet of events which confront the practitioner. It is clear that if the chemical information and/or physical information are indeed less then quite precise, the synthesis process will yield at best ambiguous predictions. Only one step (chemical or physical informa-

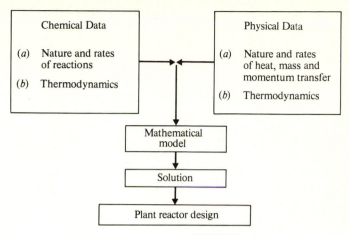

FIGURE 1-1
Ideal reactor-design flowsheet.

tion or synthesis) need be faulty to cast doubt upon the resulting predictions.   In other words, the laboratory chemical data may be inaccurate, our heat-transfer or mass-transfer correlations may be imprecise, or our mathematical model (or mode of solution) may be faulty.

In the past, the practice has been to abandon all hope of rational design. That is, plant design has tended of necessity to be based securely upon results obtained at several levels of reactor production (see Fig. 1-2).   In this classic, cautious mode of investigation and design, the increasing complexities associated with chemical modification due to physical-parameter intervention are observed gradually (and, incidently, at a considerable expense).

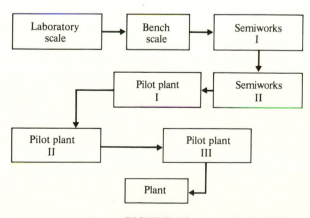

FIGURE 1-2
Traditional reactor-design flowsheet.

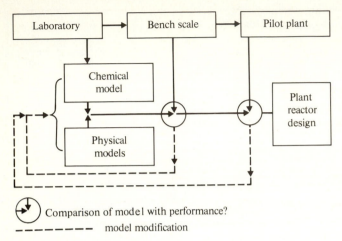

Comparison of model with performance?
— — — — — — —   model modification

FIGURE 1-3
Realistic reactor-design flowsheet.

Between the two extreme modes of inquiry (the dangerous rational-synthesis approach and the cautious but expensive incremental scale-up strategy) some prudent intermediate mode exists. We might schematize this mode as shown in Fig. 1-3.

While Fig. 1-3 may seem more complex than the alternatives, the dangers of the ideal approach and costs of the incremental-scale-up strategy must be considered if, as is not uncommon, we are confronted with a 1000:1 or 10,000:1 scale-up factor between commercial plant and test tube. The intermediate alternative scheme involves *feedback*; i.e., one attempts analysis at laboratory and bench scale, and then a comparison of pilot-plant prediction and performance is made. This confrontation of reality with predictions rooted in physicochemical model prediction allows for model-parameter (physical and chemical) adjustment. Therefore, final design is not an a priori process but one that is informed through feedback.

In sum, the chemical reaction engineer is more likely to be confronted with problems of analysis of existing units (laboratory, bench, pilot, and plant scale) than with the ideal issue of a priori plant-reactor design. It follows that the more intelligent we become in analyzing at any scale of operation, the fewer scales of operation will have to be analyzed. Returning to the SO_2 oxidation example, the more likely route to final plant design will involve (1) combining isothermal laboratory data with physical, heat- or mass-transfer data, (2) comparing predictions based upon a comprehensive physicochemical model with nonisothermal bench or semiworks data, (3) modifying model components to account for discrepancies in step (2), and (4) repeating steps (2) and (3) at larger scales of operation.

Implicit in the above discussion is the notion that with an increase in scale of reactor operation (size) certain physical events intervene to alter, modify, and indeed possibly falsify chemical-kinetic dispositions. In the simple case of a homogeneous reaction, it may well be that heat transport rather than chemical reaction per se determines reactor size and/or its performance.   In a heterogeneous reaction, since, by definition, at least two phases are involved, mass and heat transport between phases as well as with the external environment may be involved. The emphasis in this text lies on heterogeneous reactions, as such reactions are, in the author's opinion, more common and far more challenging.

A discussion of an elementary example should suffice to illustrate the intervention of the physical upon chemical events with respect to the chemical reaction.

1-1 THE REACTION

An important distinction must be made between the intrinsic rate of a chemical reaction and the observed, or global, rate of reaction.   An elementary example will suffice to illustrate this point.   Let us suppose that an inert gas such as He, containing a few mole percent NH_3, is passed over a hot tungsten wire.   The tungsten filament catalyzes the decomposition of NH_3.   The overall reaction is $2NH_3 \xrightarrow{w} N_2 + 3H_2$.

The situation can be schematically set forth as shown in Fig. 1-4.   The intrinsic rate of the chemical reaction will be proportional to the NH_3 concentration *at* the catalyst surface C_s.   Assuming for the sake of illustration that this surface rate is directly proportional to C_s,

$$\text{Intrinsic rate} = k_s C_s \qquad (1\text{-}1)$$

where k_s is the intrinsic chemical-reaction-rate coefficient.   Now in the steady-state circumstance the intrinsic rate must be equal to the rate at which NH_3 is supplied to the surface via gas-film mass transport.   In the traditional manner we designate the mass-transport coefficient $k_g a$.   Thus

$$k_g a(C_0 - C_s) = k_s C_s \qquad (1\text{-}2)$$
$$\text{Mass transport} = \text{surface reaction}$$

Consider next the problem of analyzing the above experiment; i.e., how do we determine the intrinsic coefficient k_s?   As our filament is small, NH_3 conversion

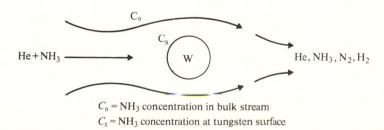

$C_0 = NH_3$ concentration in bulk stream
$C_s = NH_3$ concentration at tungsten surface

FIGURE 1-4
A nonporous-solid–catalyzed reaction-flow network.

will be small indeed, and we can assume an average value of reactant concentration based upon chemical analyses up- and downstream of the tungsten wire. But such measurements provide C_0 not C_s; for the surface concentration is usually unobservable. So our observed rate is given by

$$\text{Observed rate} = k_0 C_0 \qquad (1\text{-}3)$$

Clearly k_0 does not equal k_s, the coefficient sought, unless, as Eq. (1-2) shows, $C_0 = C_s$; that is, $k_g a$ is very large and so provides a supply of NH_3 to the surface at a rate which does not cause a gradient in the gas film. Quantitatively we solve Eq. (1-2) for C_s

$$C_s = \frac{C_0}{1 + k_s/k_g a} \qquad (1\text{-}4)$$

and substitute Eq. (1-4) into (1-1):

$$\text{Intrinsic rate} = k_s C_s = \frac{k_s C_0}{1 + k_s/k_g a} \qquad (1\text{-}5)$$

Bearing in mind that our observed rate must equal the intrinsic rate (note that observed and intrinsic rate *coefficients* are equal only when $C_s = C_0$), we see that

$$k_0 C_0 = k_s C_s$$

or
$$k_0 C_0 = \frac{k_s C_0}{1 + k_s/k_g a} \qquad (1\text{-}6)$$

and so our observed, or *global*, rate coefficient is in general related to the *intrinsic* coefficient (for linear kinetics) by

$$k_0 = \frac{1}{1/k_s + 1/k_g a} \qquad (1\text{-}7)$$

an expression reminiscent of that for the overall heat-transfer coefficient in terms of individual coefficients.

Had our reaction been noncatalytic, say thermal decomposition of NH_3, global and intrinsic rate coefficients would, of course, be identical (assuming the absence of free-radical diffusion intrusions).

What has been noted above concerning mass-diffusional masking of intrinsic kinetics can also be said of the diffusion of heat. That is, one must also anticipate that a bulk-fluid temperature may differ, by reason of film resistance to heat transfer, from that at the reaction site (in the above example, the tungsten surface).

In sum

$$\text{Global rate} = k_0 f(C_0, T_0) \qquad \text{and} \qquad \text{intrinsic rate} = k_s f'(C_s, T_s)$$

In the instance of homogeneous reaction we can generally be confident that $k_0 = k_s$. For a heterogeneous system, one would be ill advised to make such assumptions a priori.

This text treats first the intrinsic events and then explicitly heterogeneous systems in terms of global rates and their physicochemical components.

1-2 THE REACTOR

The device within which the physicochemical transformations are caused can assume various shapes and modes of operation and be operated in a number of possible environments of temperature and pressure.   Given some knowledge of the chemistry and rates of the various reactions involved, the reactor environment which provides the desired quantity of product at an acceptable level of quality must be specified.   Just as keen competence is required in analyzing and formulating quantitative models of the chemical reactions, so that an equally profound awareness of reactor behavior is required of the chemical reaction engineer.

1-3 STRUCTURE OF THIS TEXT

Given the key importance of physicochemical kinetics and reactor behavior, this text is structured to:

1 Review principles and techniques whereby models of intrinsic rates of chemical reactions can be fashioned (Chap. 2)
2 Introduce key concepts of reactor behavior in terms of limiting reactor types and environments (Chap. 3)
3 Set forth governing continuity equations for the nonideal reactor, thereby identifying the nature and magnitudes of real reactor parameters, with design and analyses of simple homogeneous reactor types (Chap. 4)
4 Introduce general concepts underlying global heterogeneous reactions (Chap. 5)
5 Treat specific noncatalytic heterogeneous systems, both fluid-fluid and fluid-solid (Chaps. 6 and 7)
6 Review principles of heterogeneous catalysis and intrinsic rate formulations (Chap. 8)
7 Treat the global rates of heterogeneous catalytic reactions (Chap. 9)
8 Outline principles of design and analysis of common heterogeneous reactor types (Chap. 10)

1-4 THE ANATOMY OF PROCESS DEVELOPMENT

An appropriate framework within which the issues treated in this text can fruitfully be focused is provided by Weekman.[1]   In Fig. 1-5 the flowsheet between exploratory studies and plant operation is indicated.   Note the early intervention of

[1] V. W. Weekman, course notes on heterogeneous catalysts, University of Houston.

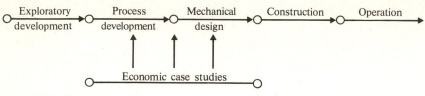

FIGURE 1-5
Scale-up flowsheet.

economic analyses.   In greater detail Fig. 1-6 reveals the role of kinetic-transport experiments, model development, and optimization.

If we suppose that yield-conversion behavior is to be modeled, Fig. 1-7 illustrates such modeling on three self-explanatory levels of sophistication.   Note the hazards of empirical modeling in Fig. 1-7a and b when extrapolation is required.

A scale whereby models can be rated is presented in Fig. 1-8, where the barometer, or index, of fundamentalness is the ratio of the number of fundamental laws invoked to the number of adjustable constants.   Diverse petroleum-refining reactor systems are indicated on this scale.

Prater's principle of "optimum sloppiness" is schematized in Fig. 1-9, where usefulness, cost, and *net* value of the model are plotted against the index of fundamentalness.   This display shows the unreasonableness of seeking a totally fundamental model of a complex reactor.   Such a reactor hosts a complex array of physicochemical rate phenomena, the fundamentals of which are ill understood and thus proper subjects of long-range fundamental research programs, usually defying the time table of process development and plant design.   We may confidently assert, however, that a command of fact and theory will reduce the cost of model creation with evident net value benefit.

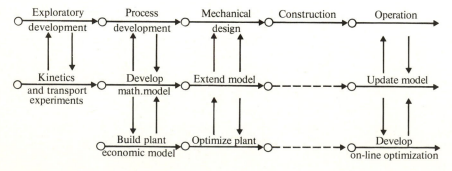

FIGURE 1-6
A somewhat more detailed scale-up flowsheet.

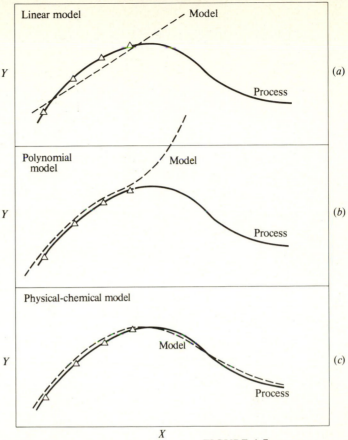

FIGURE 1-7
Diverse extrapolation models.

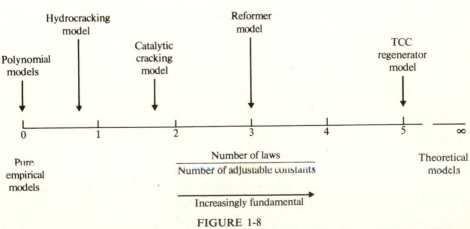

FIGURE 1-8
Model ratings in terms of the index of fundamentalness.

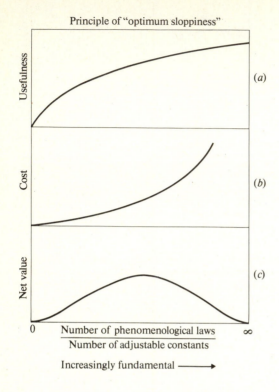

FIGURE 1-9

Prater's principle of optimum sloppiness.

GENERAL REFERENCES

ARIS, RUTHERFORD: *Ind. Eng. Chem.*, **56** (7): 22 (1964).   Provides excellent insight into the nature and structure of reactor analyses.

————: "Elementary Chemical Reactor Analysis," Prentice-Hall, Englewood Cliffs, N.J., 1969.   Theory and applications well set forth.

DENBIGH, K. C., and J. C. R. TURNER: "Chemical Reactor Theory," Cambridge University Press, London, 1971.   A terse, concise, and eloquent exposition.

HOUGEN, O. A., and K. M. WATSON: "Chemical Process Principles," pt. III, "Kinetics and Catalysis," Wiley, New York, 1947.   Remains a fine source of catalytic-model formulation and detailed design illustrations.

LEVENSPIEL, OCTAVE: "Chemical Reaction Engineering," Wiley, New York, 1972.   Clearly written undergraduate text with particular emphasis on noncatalytic reactions and residence-time distribution.

PETERSEN, EUGENE E.: "Chemical Reactor Analysis," Prentice-Hall, Englewood Cliffs, N.J., 1965.   A good treatment of heterogeneous systems.

SMITH, J. M.: "Chemical Engineering Kinetics," 2d ed., McGraw-Hill, New York, 1970.   A complete revision of the classic first edition; this text is rich in illustrative examples, with particular emphasis on heterogeneous catalysis and reactions.

2

BEHAVIOR OF CHEMICAL REACTIONS

"The velocity is delightful . . ." *"The Greville Memoirs," July 18, 1837*

Introduction

In this chapter the experimental bases of chemical-reaction-rate expressions are noted, and definitions are set forth in terms of laboratory observables. Chemical-reaction-rate theories are briefly discussed insofar as such discussion establishes some rational bases in support of laboratory-rooted definitions. A brief comment upon limiting laboratory reactor types is then presented, with some necessary definitions designed to facilitate appreciation of the formal modes of rate analysis. Formal, mathematical descriptions of chemical-reaction rates are presented for both simple and complex reaction schemes, and various means of data analysis are implicitly suggested, with particular emphasis upon yield and selectivity as well as conversion. The chapter concludes with a brief treatment of autocatalytic and chain reactions and some remarks on data procurement and error analyses. This chapter then, is designed as a review of matter usually found in an undergraduate physical chemistry course. An added emphasis is evident in that rate of generation of product (yield) rather than mere disappearance of reactant is given special attention.

Homogeneous, isothermal, constant-volume reaction environments are assumed to prevail in the analyses presented in this chapter. Such an atmosphere must be striven for if meaningful kinetic data are to be extracted from a laboratory reactor. The more complex systems (heterogeneous and/or variable-volume) will be treated in due course. At this juncture, a respect for simple definitions and a familiarity with the general tenets of kinetic analysis seem to take highest priority.

2-1   EXPERIMENTAL BASES OF CHEMICAL KINETICS

Whereas the concern of chemical equilibrium or statics is that of specifying *what* reactions may occur between molecular and/or atomic species, chemical kinetics concerns itself with the *velocity* of reactions between species. Application of chemical-equilibria principles indicates, for example, that in the oxidation of NH_3 reactions may occur to yield N_2, NO, and N_2O as well as H_2O. If interest is in one product, say NO, it is clear that thermodynamics simply assures us that said product can, in principle, be produced. Whether NO can actually be produced and produced to the exclusion of N_2 and N_2O will depend upon the kinetics of the various reactions. If conditions of temperature, pressure, and appropriate catalysis can be found, the rate of NO formation will perhaps be relatively rapid compared with N_2 and N_2O generation. In fact in the presence of a Pt-alloy catalyst at 1000°C and 1 atm pressure, nearly 100 percent yield of NO can be obtained from a mixture of air and about 10 percent NH_3. On the other hand, at temperatures below 600°C, N_2O formation predominates.

Given the obvious importance of chemical kinetics, our concern is now that of defining and specifying how the velocity of a chemical reaction is expressed. Unlike chemical statics, precise chemical-kinetic principles and data do not exist to permit a priori prediction. Chemical kinetics rests largely upon experiment.

In 1850, L. Wilhelmy made the first *quantitative* observations of reaction velocity. In his batch study of sucrose inversion in aqueous solutions of acids,

$$C_{12}H_{22}O_{11} + H_2O \xrightarrow{\text{acid}} C_6H_{12}O_6 + C_6H_{12}O_6 \qquad (2\text{-}1)$$
$$\text{Sucrose} \qquad\qquad\qquad \text{Glucose} \quad\;\; \text{Fructose}$$

Wilhelmy noted that the rate of change of sucrose concentration C with respect to time t is a linear function of the concentration of unconverted sucrose C. That is,

$$-\frac{dC}{dt} = kC \qquad (2\text{-}2)$$

The observed velocity law reveals that reaction (2-1) is irreversible; i.e., the back reaction between glucose and fructose is negligible relative to the forward sucrose-inversion reaction.

The proportionality constant k is known as the *specific rate coefficient*. Actually for reaction (2-1), k is proportional to acid concentration. The acid is clearly a catalyst in the reaction, as it does not appear in the overall reaction statement.

At a fixed temperature and acid concentration, Eq. (2-2) was integrated by Wilhelmy to yield, for an initial concentration C_0,

$$\ln \frac{C}{C_0} = -kt \quad \text{or} \quad \frac{C}{C_0} = \exp[-kt] \tag{2-3}$$

The concentration-vs.-time data agreed nicely with above equation.

In the system studied by Wilhelmy, the rate, being proportional to the first power of concentration, is said to be of *first order* in the concentration of that species. *Order of reaction* with respect to a particular species is the numerical value of the power to which the concentration is raised to faithfully describe the experimental relationship between reaction rate and the concentration of that species; i.e.,

$$\text{Rate} = -\frac{dA}{dt} = kA^{\alpha} \tag{2-4}$$

The overall reaction order is the sum of exponents; i.e., if

$$-\frac{dA}{dt} = kA^{\alpha}B^{\beta}C^{\gamma} \tag{2-5}$$

then the overall order is $\alpha + \beta + \gamma$.

Essentially order is empirical insofar as it is specified on the basis of observed rate-vs.-concentration data. Order need not be a whole number, since it may be zero, fractional, or negative for a specific component.

Thus in the sucrose-inversion reaction, we can say that the rate is first order with respect to sucrose and zero order with respect to the coreactant, water. Zero order with respect to water does not imply that the reaction does not involve water but that there is no apparent *experimental* dependence since water was present in such vast excess that its concentration changed negligibly during reaction. That is,

$$\text{Rate} = k(\text{sucrose})^1(H_2O)^{\beta} = k'(\text{sucrose})^1$$

as $(H_2O)^{\beta}$ is a constant in the experiment. In principle, the rate is βth order in water, and this dependency would be observed experimentally if H_2O were not present in vast excess.

Order and Stoichiometry

Essentially order is determined by the best fit or correspondence between a rate equation and experimental data. It follows that there is no necessary connection between kinetic order and the stoichiometry of the reaction. For example, in the Pd-catalyzed oxidation of CO, the stoichiometry is

$$CO + \tfrac{1}{2}O_2 \longrightarrow CO_2 \tag{2-6}$$

while kinetics suggest *negative* first order in CO. Equation (2-6) is simply an overall statement of the reaction, whereas the process must proceed through a series of more elementary steps. The sum of these elementary processes must yield the overall reaction statement. However, in a kinetic study, one necessarily measures only the slowest step or steps of the sequence. More formally the rate-determining step is measured implicitly, and consequently its concentration dependencies will appear in the rate expression. The resultant rate expression may be complex or simple. In either case, it is not possible to infer a mechanism of reaction (detailed sequence of steps) solely from kinetic rate expressions. For example, let an overall reaction statement be $2A + B \rightleftarrows P$. Suppose the elementary steps (mechanism) to be:

Step 1: $A + B \rightleftharpoons (AB)$ $(AB) = K_1 AB$

Step 2: $(AB) + B \rightleftharpoons (BAB)$ $(BAB) = K_2(AB)B = K_2 K_1 AB^2$

Step 3: $A + (BAB) \rightleftharpoons P + B$

where K_1 and K_2 are equilibrium coefficients.

Overall: $2A + B \rightleftharpoons P$

The following irreversible rate laws can be found:

If step 1 is the slowest, rate $= k_1 AB$.
If step 2 is the slowest, rate $= k_2 K_1 AB^2$.
If step 3 is the slowest, rate $= k_3 K_2 K_1 A^2 B^2$.

Suppose on the other hand, the mechanism is

Step 1: $A + B \rightleftharpoons (AB)$

Step 2: $(AB) + A \rightleftharpoons P$

Overall: $2A + B \rightleftharpoons P$

Then if step 1 controls, rate $= k_1 AB$, while if step 2 controls, rate $= k_2 K_1 A^2 B$, where k_1 and k_2 are rate coefficients. These rate expressions are derived as follows:

In the first illustration, when a step is assumed to be rate-controlling, i.e., the slowest of all steps, all other elementary steps are assumed to be in a state of rapidly established equilibrium, or steady state. Then if step 1 in the first illustration controls, we can write the rate of that reaction as indicated by the stoichiometry of that one elementary step; i.e.,

$$\text{Rate (step 1)} = k_1 AB$$

Should step 2 be the slowest elementary event in the sequence (mechanism), then

$$\text{Rate (step 2)} = k_2(AB)B$$

but by the equilibrium in step 1

$$K_1 = \frac{(AB)}{AB} \qquad \text{or} \qquad (AB) = K_1 AB$$

Hence

$$\text{Rate (step 2)} = k_2(AB)B = k_2 K_1 AB^2$$

When step 3 controls,

$$\text{Rate (step 3)} = k_3 A(BAB)$$

but

$$(BAB) = K_2(AB)B = K_2 K_1 AB^2$$

Therefore

$$\text{Rate (step 3)} = k_3 K_2 K_1 A^2 B^2$$

Some concrete examples regarding order and stoichiometry, model and mechanism can be cited.

Kinetic studies of the decomposition of N_2O_5

$$2N_2O_5 \longrightarrow 2N_2O_4 + O_2 \qquad (2\text{-}7)$$

indicate

$$-\frac{d(N_2O_5)}{dt} = k(N_2O_5)$$

It is incorrect to infer that reaction (2-7), while first order, involves simple (unimolecular) dissociation of N_2O_5 to final products N_2O_4 and O_2. Ogg[1] showed that the mechanism is

$$N_2O_5 \longrightarrow NO_2 + NO_3$$
$$NO_2 + NO_3 \longrightarrow NO + O_2 + NO_2 \qquad (2\text{-}8)$$
$$NO + NO_3 \longrightarrow 2NO_2 \rightleftharpoons N_2O_4$$

Consider the decomposition of acetaldehyde, $CH_3CHO \rightarrow CH_4 + CO$, in the gas phase at 450°C. The rate expression is

$$-\frac{d(CH_3CHO)}{dt} = k(CH_3CHO)^{3/2} \qquad (2\text{-}9)$$

Clearly order and stoichiometry are not synonymous. Further, as molecules react as integers (mono- or bimolecular), the observed rate law cannot represent a simple (elementary) rate-controlling step. In fact, the decomposition is a complex chain reaction.[2]

Molecularity of a reaction refers to the number of molecules of reactants which are involved in the elementary chemical event. Molecularity pertains then to the molecular mechanism, while order pertains to the experimental rate equation. In the Ogg mechanism [Eq. (2-8)] the molecularities are 1, 2, and 2, respectively.

Only if the rate equation reflects the elementary molecular events will order and molecularity be equivalent. In the celebrated reaction

$$2HI \longrightarrow H_2 + I_2 \qquad (2\text{-}10)$$

[1] R. A. Ogg, Jr., *J. Chem. Phys.*, **15**: 337, 613 (1947).
[2] S. W. Benson, " Foundations of Chemical Kinetics," p. 379, McGraw-Hill, New York, 1960.

detailed studies[1] indicate that the mechanism probably involves two HI molecules colliding and rearranging to produce H_2 and I_2. The experimental description of the decomposition (the rate equation) reveals second-order kinetics; i.e.,

$$-\frac{d(HI)}{dt} = k(HI)^2 \qquad (2\text{-}11)$$

In this case order and molecularity coincide. Correspondence between order and overall stoichiometry is occasionally found but must be considered fortuitous, for example, $NO + \frac{1}{2}O_2$.

In sum, the experimentally determined order is not in principle uniquely related to overall stoichiometry, for example, $2A + B \rightarrow P$, so that one cannot expect a rate law of the form

$$Rate = kA^2B$$

As the overall expression or balance can be constructed from several differing elementary sequences (mechanisms), the form of the observed rate law conveys nothing unique concerning the real mechanism.

Example The velocity of the reaction $2NO + O_2 \rightarrow 2NO_2$ was observed by Bodenstein, who found that the irreversible-reaction-rate model is

$$Rate = k(NO)^2(O_2)$$

Here order and stoichiometry coincide; yet later studies indicate that not one but at least two elementary steps are involved, the slowest of the two being bimolecular; i.e.,

Step 1: $NO + O_2 = NO_3$ fast equilibration

Step 2: $NO_3 + NO \longrightarrow 2NO_2$ slow

so that the rate-controlling step is expressed as

$$Rate = k_2(NO_3)(NO)$$

But $NO_3 = K_1(NO)(O_2)$, and therefore

$$Rate = k_2 K_1(NO)^2(O_2) = k(NO)^2(O_2)$$

If we assume an entirely different mechanism, say

Step 1: $NO + NO = (NO)_2$ fast equilibration

Step 2: $(NO)_2 + O_2 \longrightarrow 2NO_2$ slow

then $$Rate = k_2(NO)_2O_2$$

but $(NO)_2 = K_1(NO)^2$, and so

$$Rate = k_2 K_1(NO)^2(O_2)$$

as observed.

[1] C. A. Eckert, and M. Boudart, *Chem. Eng. Sci.*, **18**: 144 (1963).

Therefore it is demonstrated that (1) agreement between the observed rate law and stoichiometry does not prove that the overall reaction balance reflects an elementary (one-step) event and (2) more than one mechanistic sequence can be formulated which, with a shrewd choice of the rate-controlling step, will lead to the observed rate law.

The precise specification of the mechanism requires data other than the kinetic law, e.g., identification of intermediate NO_3 and/or $(NO)_2$, etc. The kinetic law is thus a phenomenological *model* rooted in kinetic observables and transmits to the observer only what apparently occurs as witnessed on the gross laboratory scale. Inferences concerning the detailed molecular events are permissible only when supported by microscopic explorations which complement the kinetic investigation.

The Proportionality or Rate Coefficient

For a reaction $A + B \rightarrow P$ it has been indicated that analysis of laboratory rate data might suggest

$$-\frac{dA}{dt} = k A^{\alpha} B^{\beta} \qquad (2\text{-}12)$$

where α and β are empirically determined. The proportionality coefficient k is a specific rate constant at constant temperature and catalyst concentration. If wall or surface effects are manifest, k will be a function of extent of surface, etc., and (at a constant temperature, catalyst concentration, wall area, etc.) α is determined by observing the rate-versus-A behavior in an excess of B, so that

$$-\frac{dA}{dt} = k' A^{\alpha} \qquad k' = k B^{\beta} \qquad (2\text{-}13)$$

while β is similarly fixed by rate-versus-B behavior in the presence of excess A. The next task is that of specifying the nature of k. If the system is homogeneous (reaction occurs entirely in one phase) and catalyst concentration is fixed or the system is noncatalytic, then, as suggested by Arrhenius, k should depend upon temperature in the following fashion:

$$k = \mathscr{A} \exp\left(-\frac{E}{RT}\right) \qquad (2\text{-}14)$$

while Wilhelmy found

$$k = \mathscr{A} \exp\left(-\frac{b}{T}\right)$$

where $\mathscr{A}$ is called a *preexponential* or *Arrhenius factor* and E is the activation energy. Empirically, chemical-reaction-rate data can be compactly organized in terms of concentration (or pressure) and temperature by

$$-\frac{dA}{dt} = \mathscr{A} A^{\alpha} B^{\beta} \cdots \exp\left(-\frac{E}{RT}\right) \qquad (2\text{-}15)$$

In summary, Eq. (2-15) presents a phenomenological description of reaction velocity in terms of laboratory observables (temperature, concentration of reactants, products, inerts, catalysts, reactor surface area, and its nature). Before looking into specific reaction types and methods of kinetic analysis, it is perhaps worthwhile to touch upon some of the theoretical foundations of chemical kinetics. A study of the theories which give rational support to the laboratory-founded rate expressions [such as Eq. (2-15)] should enhance our understanding of rate processes and consequently lead to a more intelligent use of kinetic data and rate laws.

2-2 THEORETICAL BASES OF CHEMICAL KINETICS

In an attempt to explain the effect of temperature upon the rate of sucrose inversion, Arrhenius[1] suggested that an equilibrium existed between inert and active sucrose molecules. The equilibrium concentration of active molecules would depend upon temperature as dictated by thermodynamics, i.e.,

$$K = (\text{const})\left(\exp \frac{-\Delta H^\circ}{RT}\right)$$

and so $S \rightleftharpoons S^*$ (active sucrose)

$$K = \frac{S^*}{S} = (\text{const})\left(\exp \frac{-\Delta H^\circ}{RT}\right) \qquad (2\text{-}16)$$

Now

$$\text{Rate} = (\text{const})(S^*) = K'S$$

$$\text{Rate} = (\text{const})(S)\exp \frac{-\Delta H^\circ}{RT}$$

or

$$\text{rate} = \mathscr{A}S \exp\left(-\frac{E}{RT}\right) = kS \qquad \text{where } E \equiv \Delta H^\circ$$

E was defined by Arrhenius as the difference in heat content between the active and inert reactant molecules. E has become known as an *energy of activation*,[2] while $\mathscr{A}$ is often termed the *frequency* (of *collision*) *factor*. The relationship between E, ΔH, and $\mathscr{A}$ will be made more explicit below. The Arrhenius equation has been verified by numerous experiments, and so

$$k = \mathscr{A} \exp\left(-\frac{E}{RT}\right) \qquad \text{or} \qquad \ln k = \ln \mathscr{A} - \frac{E}{RT} \qquad (2\text{-}17)$$

A plot of $\ln k$ versus $1/T$ should give a straight line of slope $-E/R$ and intercept $\ln \mathscr{A}$.

While providing a method by which rate-temperature data can be rationally organized, the Arrhenius theory provides no basis for predicting E and $\ln \mathscr{A}$. It is understandable that, given the success of the Arrhenius relation, subsequent

---

[1] S. Arrhenius, *Z. Phys. Chem.*, **4**: 226 (1889).

[2] We shall retain the symbol E and the term activation energy in view of the rather widespread use of the symbol and terminology. Actually activation enthalpy change ΔH is more precise.

efforts would be devoted to more sophisticated interpretation of $\mathscr{A}$ and E. Inspired by developments of the kinetic theory of gases, the collision theory of chemical reaction emerged. The preexponential factor $\mathscr{A}$ was interpreted as a collision frequency Z. This collision frequency is (for at least two reactant molecules or atoms) determined by kinetic theory under limiting conditions.

Detailed treatments can be found in the usual physical chemistry texts. The predictive and instructive powers of collision theory are so limited that further elaboration is not justified. It suffices to note that experimental values of the Arrhenius coefficient $\mathscr{A}$ have been found to depart significantly from values suggested by collision theory. A more powerful model is set forth in the text of Glasstone, Laidler, and Eyring.[1] The essential features of this theory will now be considered.

Transition-State Theory

By *transition-state theory* we mean to emphasize the more telling aspects of *absolute-rate theory*, a term perhaps misleading in view of the very small number of even simple reactions whose absolute rates are subject to a priori prediction.

Somewhat in the spirit of Arrhenius' speculations cited above, transition-state theory asserts that in the reactants' progress along the path to products, an intermediate complex, or transition state, prevails; the transition-state complex exists in equilibrium with reactants; e.g., in the reaction $A + B \rightarrow P$, the theory states

$$A + B \; \rightleftharpoons \; (AB)^{\ddagger} \; \longrightarrow \; P$$

Product (P) appears at a rate governed by the frequency of $(AB)^{\ddagger}$ decomposition in the forward direction, and, as will be shown below, the concentration of $(AB)^{\ddagger}$. Although $(AB)^{\ddagger}$ is not readily measured while A and B are detectable, the reactant-complex equilibrium can be stated in terms of thermodynamic activities of A, B, and $(AB)^{\ddagger}$:

$$K^{\ddagger} = \frac{a_{AB}}{a_A \, a_B} = \frac{\gamma_{AB}}{\gamma_A \gamma_B} \frac{C_{AB}}{C_A C_B} \qquad (2\text{-}18)$$

If it is assumed that the reaction rate is equal to the product of *concentration* of activated complex and the frequency of that complex decomposition, then

$$\text{Rate} = v \frac{\gamma_A \gamma_B}{\gamma_{AB}} K^{\ddagger} C_A C_B \qquad (2\text{-}19)$$

Now $K^{\ddagger} = \exp(-\Delta G^{\ddagger}/RT)$, where $\Delta G^{\ddagger}$ is the free-energy difference between complex and reactants. Absolute-rate theory shows that the decomposition frequency v is $k_B T/h$, where k_B is Boltzmann's constant, T is absolute temperature, and h is Planck's constant. Since $\Delta G^{\ddagger} = \Delta H^{\ddagger} - T\Delta S^{\ddagger}$,

$$\text{Rate} = \frac{k_B T}{h} \frac{\gamma_A \gamma_B}{\gamma_{AB}} C_A C_B \exp \frac{\Delta S_c^{\ddagger}}{R} \exp \frac{-\Delta H^{\ddagger}}{RT} \qquad (2\text{-}20)$$

[1] S. Glasstone, K. J. Laidler, and H. Eyring, "The Theory of Rate Processes," McGraw-Hill, New York, 1941.

but

$$\Delta H^{\ddagger} = E - RT + \Delta(PV)^{\ddagger} = E - RT + \Delta n\, RT = E + (\Delta n - 1)RT \qquad (2\text{-}21)$$

Δn is the change in number of moles between reactants and activated complex:

$$\exp\left(\frac{-\Delta H^{\ddagger}}{RT}\right) = \exp\left(\frac{-E}{RT}\right)\exp(1 - \Delta n)^{\ddagger}$$

For an ideal-gas system, where $\gamma_A = \gamma_B = \gamma_{AB} = 1$,

$$\text{Rate} = \frac{k_B T}{h}\exp\frac{\Delta S_c^{\ddagger}}{R}\exp(1 - \Delta n)^{\ddagger}\exp\left(\frac{-E}{RT}\right)C_A C_B = k_c\, C_A\, C_B$$

Arrhenius suggested $k = \mathscr{A}\exp(-E/RT)$ or

$$\mathscr{A} = \frac{k_B T}{h}\exp\frac{\Delta S_c^{\ddagger}}{R}\exp(1 - \Delta n)^{\ddagger}$$

The a priori calculation of $\Delta S^{\ddagger}$ and $\Delta H^{\ddagger}$ requires knowledge of the nature and structure of the activated complex. Simple systems can be so described using the methods of statistical mechanics to compute $K^{\ddagger}$. Eckert and Boudart[1] illustrate the principles and the technique in the case of HI decomposition, where nonideality requires a priori determination of γ for reactants and the complex.

Relationship between Transition-State Theory and Arrhenius Model[2]

According to transition-state theory, in terms of pressure units,

$$\text{Rate} = \frac{k_B T}{h}K_p^{\ddagger}p_A p_B = k_p p_A p_B$$

Then

$$k_p = \frac{k_B T}{h}K^{\ddagger} = \frac{k_B T}{h}\exp\frac{\Delta S_p^{\ddagger}}{R}\exp\frac{-\Delta H^{\ddagger}}{RT} \qquad (2\text{-}22)$$

Substituting Eq. (2-21) into (2-22) gives

$$k_p = \exp(1 - \Delta n)^{\ddagger}\frac{k_B T}{h}\exp\frac{\Delta S_p^{\ddagger}}{R}\exp\frac{-E}{RT} \qquad (2\text{-}23)$$

Therefore

$$\mathscr{A} = \exp(1 - \Delta n)^{\ddagger}\frac{k_B T}{h}\exp\frac{\Delta S_p^{\ddagger}}{R} \qquad (2\text{-}24)$$

where the rate is expressed in terms of pressure. In terms of concentration, since $k_c = k_p(RT)^{-\Delta n}$, where Δn is equal to the increase in number of molecules associated with creation of the activated complex from reactants,

$$k_c = \exp(1 - \Delta n)^{\ddagger}\frac{k_B T}{h}(RT)^{-\Delta n}\exp\frac{\Delta S_p^{\ddagger}}{R}\exp\frac{-E}{RT} \qquad (2\text{-}25)$$

For example, for $A + B \rightleftharpoons (AB)^{\ddagger}$

$$\Delta n = -1 \qquad \text{and} \qquad (1 - \Delta n)^{\ddagger} = 2 \qquad (2\text{-}26)$$

---

[1] Loc. cit.
[2] $\Delta S_c^{\ddagger} = \Delta S_p^{\ddagger} - \Delta n\, R \ln RT$, and $\exp(\Delta S_c^{\ddagger}/R) = \exp(\Delta S_p^{\ddagger}/R)(RT)^{-\Delta n}$.

we have

$$k_c = e^2 \frac{k_B T}{h} RT \exp \frac{\Delta S^{\ddagger}}{R} \exp \frac{-E}{RT}$$

Activity vs. Concentration

We have indicated that an implicit assumption in the above formulation of transition-state theory is that the rate is proportional to the *concentration* of the activated complex. This leads to the expression for k in the case $A + B \rightarrow (AB)^{\ddagger}$

$$k = \frac{k_B T}{h} \frac{\gamma_A \gamma_B}{\gamma_{AB}} \exp \frac{\Delta S^{\ddagger}}{R} \exp \frac{-\Delta H^{\ddagger}}{RT} \qquad (2\text{-}27)$$

in the instance of nonideality of the reactant phase. If, on the other hand, the rate is proportional to the *activity* of the activated complex, then

$$k = \frac{k_B T}{h} \gamma_A \gamma_B \exp \frac{\Delta S^{\ddagger}}{R} \exp \frac{-\Delta H^{\ddagger}}{RT} \qquad (2\text{-}28)$$

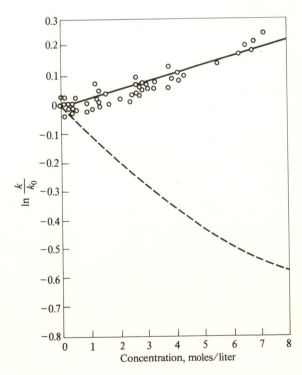

FIGURE 2-1
Variation in rate coefficient with total concentration for HI decomposition at 321.4°C. The solid line represents Eq. (2-27), and the dashed line represents Eq. (2-28). [*C.A. Eckert and M. Boudart, Chem. Eng. Sci.*, **18**:144 (1963).]

Eckert and Boudart[1] analyzed Kistiakowsky's rate data on HI decomposition as a function of concentration. If ideal-gas behavior prevailed, the second-order rate would vary with the square of concentration. Kistiakowsky's measurements revealed a greater effect of pressure than predicted by ideal-gas behavior. Eckert and Boudart's analysis of these data clearly demonstrates that it is the concentration rather than activity of the activated complex which dictates reaction rate [Eq. (2-27)]. Their paper illustrates the essential features of activated-complex formulation as well as the effects of nonideality upon the rate constant. Figure 2-1 shows the experimental and predicted effect of concentration on the specific rate constant.

2-3 SOME DEFINITIONS OF REACTION ENVIRONMENTS

Laboratory Reactor Types

The general types of devices used to obtain raw kinetic data deserve mention at this point, as well as the terminology for describing the extent of reaction and product-generating efficiency.

 Laboratory-reaction studies are conducted in three general fashions:

1 *Batch operation*:

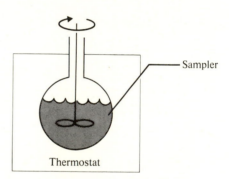

This is a closed system, and progress of reaction is monitored as a function of time t.

2 *Semibatch operation*:

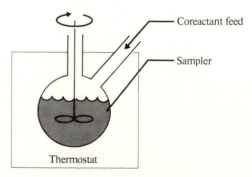

One reactant is charged; the coreactant is added during the course of reaction.

[1] Loc. cit.

3 Continuous-flow operation:

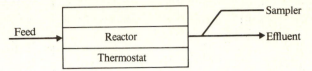

In contrast to batch and semibatch operation, continuous-flow reactors can be operated in steady state; i.e., the composition of effluent remains fixed with time if flow rate, temperature, and feed composition remain invariant. Reaction progress is followed as a function of residence time in the reactor rather than real time. This residence time (reactor volume divided by volumetric feed rate) can assume great importance depending upon the design of the flow reactor. Detailed discussion will be given in the next chapter; however, two limiting situations can be cited:

a Plug-flow reactor (PFR):

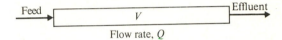

All molecules have same residence time V/Q, and concentrations vary only along the length of the tubular reactor.

b Continuous-flow stirred-tank reactor (CSTR):

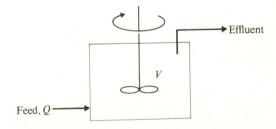

Due to vigorous agitation, the reactor contents are well mixed, so that effluent composition equals that in the tank. As effluent and stirred-tank contents are of uniform composition, progress of reaction is monitored by noting the exit composition vs. average residence-time V/Q behavior in the steady state.

While quantitative discussions of type a and b flow reactors and intermediate types will appear in the next chapter, it is important to note that different information is acquired in securing reaction-progress-vs.-residence-time data from PFR and CSTR. In the PFR, since composition varies along the length of the tube, the effluent composition is an overall, integral composition. On the other hand, as the composition is uniform throughout the steady-state CSTR, the effluent-composition-vs.-average-residence-time data represent point or local values. The PFR is a distributed-parameter unit (composition varies with position), while the CSTR is a lumped-parameter unit (uniform composition at all positions in the reactor). So for the CSTR, a steady-state material balance yields

$$QC_0 = QC + V\mathscr{R}$$

hence

$$\text{Rate, } \mathscr{R} = \frac{C_0 - C}{V/Q}$$

Laboratory Reactor Environments

The composition environment is implicitly noted above. In batch and semibatch operation uniformity throughout the geometric confines of the reaction volume must be realized at every moment during the course of reaction. Hence the necessity of vigorous agitation. Further, in a kinetic study the temperature field is of prime importance. Three general cases can exist:

1 *Isothermal* Heat is exchanged efficiently, so that at all times (or positions in a PFR) the temperature is constant.
2 *Adiabatic* Heat exchange between the reacting mass and the external surroundings is denied, either through design of the experiment, folly, or the inevitable consequence that the reaction rate (and thus heat generation or abstraction) is far too great to allow adequate exchange.
3 *Nonisothermal, nonadiabatic* A case more common than is generally acknowledged and one which is obviously intermediate between cases 1 and 2.

In the treatment of chemical-kinetic analysis which follows, isothermality is assumed in batch operation under conditions such that spatial composition gradients are absent. This means that compositions vary only with time, although the plug-flow case is noted to clarify the rigor which must be observed in expressing the differential reaction rate. Adiabatic operation on a laboratory scale is difficult to realize by design and although it is not as complex to analyze as nonadiabatic, nonisothermal data, it is preferable to follow laboratory procedures which generate isothermal data.

2-4 DEFINITIONS OF EXTENT OF REACTION

Conversion is expressed as the fraction of a key limiting reactant which is consumed to generate all reaction products. It is convenient to define conversion x as

$$x = \frac{\text{moles reactant consumed}}{\text{moles of initial reactant}} \tag{2-29}$$

Yield can be defined in at least two ways: $Y(\text{I})$ is moles of a particular product generated per mole of *initial* key reactant, and $Y(\text{II})$ is moles of a particular product generated per mole of key reactant *consumed*.

Selectivity can be defined as moles of a particular product generated per mole of another product (by-product) generated.   Selectivity is simply the ratio of two yields.

Conversion, yield, and selectivity are further specified as point (local) values and integral (overall) values.

These definitions can be clarified by example.   Consider the oxidation of ethylene (E) in air to produce ethylene oxide (EO) and combustion products (CP) carbon dioxide and water.   Assume the scheme

66435-1

$$
\text{Air} + \text{E} \quad
\begin{array}{c}
\nearrow^{1} \text{EO} \\
\Big\downarrow {\scriptstyle 3} \\
\searrow_{2} \text{CP}
\end{array}
\qquad (2\text{-}30)
$$

In a batch reactor

$$-\frac{d(E)}{dt} = k_1 f(E) + k_2 f'(E) \qquad (2\text{-}31)$$

$$\frac{d(EO)}{dt} = k_1 f(E) - k_3 g(EO) \qquad (2\text{-}32)$$

$$\frac{d(CP)}{dt} = k_2 f'(E) + k_3 g(EO) \qquad (2\text{-}33)$$

where $f(E), f'(E)$, and $g(EO)$ are general complex kinetic functions which need not be specified here.   If E_0 is the initial ethylene concentration, conversion x is simply $(E_0 - E)/E_0$.

The distinction between point and overall conversion is not crucial.   Overall conversion x_0 can be considered the final value at the end of reaction (or exit of a flow reactor), while the point value is clearly the intermediate value.   In virtually all cases, one can expect conversion to increase with extent of exposure, though in an equilibrium reaction, an inadequate heat-removal policy can lead to conversion which passes through a maximum at some intermediate stage, e.g., in a nitrogen oxides–absorption tower.

Point yield of EO is simply the ratio of its rate of net appearance relative to that of reactant disappearance; thus

$$Y_p = \text{point or local yield} = \frac{d(EO)/dt}{d(E)/dt} = \frac{d(EO)}{d(E)}$$

Integral yield of EO is simply the result of integrating the point-value function. This gives either

$$Y_0(\mathrm{I}) = \frac{EO}{E_0} = \text{function of } x, k_1, k_2, k_3, \text{ etc.} \qquad (2\text{-}34)$$

or

$$Y_0(\mathrm{II}) = \frac{EO}{E_0 - E} = \text{function of } x, k_1, k_2, k_3, \text{ etc.} \qquad (2\text{-}35)$$

and so
$$\frac{EO}{E_0 - E} = \frac{EO}{E_0 x} \qquad \text{or} \qquad Y_0(\text{II}) = \frac{Y_0(\text{I})}{x} \qquad (2\text{-}36)$$

Point selectivity, as noted above, is a ratio of yields:
$$\mathscr{S}_p = \frac{Y_p(EO)}{Y_p(CP)} = \frac{d(EO)}{d(CP)}$$

Overall selectivity is secured by integration of $\mathscr{S}_p$, which in this example gives *EO/CP* as some function of conversion, etc. It must be borne in mind that the functions cited regarding integral yield and selectivity can be complex, and indeed in many important cases analytical integration is not possible. In terms of experimental realities, the integrated or overall yield and selectivity are, like conversion, the values found at the end of reaction, while point values are obviously those found at any given moment (or position in a PFR).

The above definitions of conversion, yield, and selectivity are not by any means unique or general.[1] In many quarters "conversion" to a product is accepted terminology for yield. In other instances yield and selectivity are used interchangeably.

2-5 MATHEMATICAL DESCRIPTIONS OF REACTION RATES

We shall now turn our attention to detailed mathematical descriptions of various reaction types. Chemical reactions may be simple, for example, $A \rightarrow B$, or complex

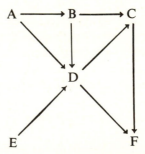

In addition to the inherent complexity illustrated in the latter scheme, mixed orders, reversibility, and volume changes due to stoichiometry conspire to complicate solution of the differential equations describing the system. Our descriptions begin with a definition of the rate of chemical reaction.

Reaction rate is defined as the change in moles n of a component with respect to time per unit of *reaction* volume, i.e.,

$$\text{Rate} = \frac{1}{V}\frac{dn}{dt} = \text{function } (A,\ B,\ C,\ \ldots,\ \text{temperature}) \qquad (2\text{-}37)$$

---

[1] H. H. Voge, Letter to Editor, *Chem. Eng. News*, Feb. 21, 1966.

The expression assumes that a batch system exists, so that time t is the time of retention within a closed vessel (of variable or fixed volume). For fixed volume, batch ($n = CV_0$ = number of moles):

$$\text{Rate} = \frac{1}{V_0}\frac{dn}{dt}$$

$$= \frac{1}{V_0}\frac{dC\,V_0}{dt} = \frac{dC}{dt} \qquad \begin{array}{l} V = V_0,\ C = C_0 \text{ at } t = 0 \\ V = V_0,\ C = C \text{ at } t > 0 \end{array} \qquad (2\text{-}38)$$

and for variable volume, batch ($n = CV$):

$$\text{Rate} = \frac{1}{V}\frac{dn}{dt}$$

$$= \frac{dC}{dt} + \frac{C}{V}\frac{dV}{dt} \qquad \begin{array}{l} V = V_0,\ C = C_0 \text{ at } t = 0 \\ V = V \neq V_0,\ C = C \text{ at } t > 0 \end{array} \qquad (2\text{-}39)$$

Suppose, on the other hand, a simple isothermal tubular reactor (PFR) is being fed at a constant rate F_0, moles per time. A balance over a differential volume dV

$$F - (F + dF) = (\text{rate})\,dV$$

$$\text{gives} \qquad -\frac{dF}{dV} = \text{rate}$$

Two cases may be encountered in such a system: (1) there is no change in moles due to reaction, or (2) a mole change due to reaction exists, in which case the volumetric flow Q will vary with extent of reaction (distance along the PFR). In case 1, at constant temperature and pressure, the volumetric flow rate Q remains constant along the length of the reactor. In case 2, Q changes as the number of moles change due to reaction.

CASE 1 There is no mole change, and $Q = Q_0$:

$$\text{Rate} = \frac{dF}{dV} = \frac{d(QC)}{dV} = Q_0\frac{dC}{dV} = \frac{dC}{d\tau} \qquad \tau = \frac{V}{Q_0} \qquad (2\text{-}40)$$

Comparing this result to the fixed-volume batch case, we see that contact time $\tau = V/Q_0$ in a flow system is equal to holding time t in the fixed-volume batch system *only* if Q_0 refers to reaction temperature and pressure.

CASE 2 There is mole change due to reaction, that is, $Q \neq Q_0$:

$$\text{Rate} = \frac{dF}{dV} = Q\frac{dC}{dV} + C\frac{dQ}{dV} \qquad (2\text{-}41)$$

In this second case, the change in flow rate due to reaction as a function of volume traversed must be specified, just as the change in reaction volume with time must be specified in the batch variable-volume case.

The merits of using conversion x, defined earlier, become evident in formulating the variable-volume reaction scheme. As noted above, for the PFR, in terms of concentration

$$\frac{dF}{dV} = Q\frac{dC}{dV} + C\frac{dQ}{dV} \qquad (2\text{-}42)$$

However, as

$$x = \frac{\text{moles of reactant consumed}}{\text{moles of reactant in feed}}$$

for F_0 = moles reactant fed per time

$$F = F_0 - xF_0 = F_0(1 - x) \qquad \text{and} \qquad F_0 = Q_0 C_0$$

and so

$$\frac{dF}{dV} = -F_0\frac{dx}{dV} = \text{rate} = -C_0\frac{dx}{d\tau_0} \qquad \text{where } \tau_0 = \frac{V}{Q_0} \qquad (2\text{-}43)$$

a far simpler expression than that in terms of concentration.

By similar reasoning, for a batch case, since $n = n_0(1 - x)$,

$$\frac{1}{V}\frac{dn}{dt} = -\frac{n_0}{V}\frac{dx}{dt} = \text{rate} = -C_0\frac{dx}{dt} \qquad (2\text{-}44)$$

Note that in each case (PFR or batch) the rate functionality must be expressed in terms of conversion x to facilitate integration (analytical or otherwise). In this chapter we shall largely confine ourselves to constant-volume batch or flow systems without mole change due to reaction. We justify this restriction at this stage because our intent here is to present integrated forms of rate equations which can be profitably used in interpreting laboratory data usually secured in batch systems of constant volume or in flow systems where reactant dilution with inerts or low conversion is frequently realized to assure a negligible total flow change due to reaction.

2-6 CLASSIFICATION OF REACTIONS

One may categorize reactions by order, reversibility, complexity, whether homogeneous or heterogeneous, and catalyzed or not. For example, ethylene is hydrated with water in the presence of dissolved acid catalysts to produce ethyl alcohol. The reaction is obviously catalytic. At high pressure ethylene gas dissolves and reacts in liquid water (containing H_2SO_4) to yield the alcohol. This is a heterogeneous system (gas-liquid); the reaction occurs in one phase (homogeneous) and is therefore catalyzed in that phase (homogeneous catalysis).

On the other hand, ethylene gas and H_2O vapor will react to give alcohol when passed through a tube packed with solid particles containing acidic properties. This is a heterogeneous system (gas-solid); the reaction occurs at the gas-solid interface (heterogeneous) and is solid-catalyzed (heterogeneous catalysis).

In this chapter we shall deal explicitly with kinetics of reactions in one phase (homogeneous).   Implicitly homogeneous catalysis is manifest in the rate constant k, as we noted earlier in discussing sucrose inversion.   Kinetics of heterogeneous catalysis will be treated in Chap. 8.

For constant-volume batch and tubular-flow reactions with negligible volume change due to reaction, the following topics will be treated:

1   Simple reactions
   a   First, second, third, and nth order, irreversible
   b   Reversible cases for some simple reactions
   c   Relationship between equilibrium and rate constants
2   Complex reactions
   a   Simultaneous reactions
   b   Higher-order simultaneous reactions
   c   Concurrent and consecutive linear reactions
   d   Higher-order consecutive reactions
   e   The general complex linear network
   f   Autocatalysis, chain reactions, explosions, and simple polymerization

2-3   SIMPLE REACTIONS

Preliminary Remarks

In a general case $aA + bB \cdots \rightarrow pP + qQ \cdots$ the reaction rate (at constant volume) is expressed in terms of concentrations A, B, etc., and stoichiometric coefficients as

$$\text{Rate} = -\frac{1}{a}\frac{dA}{dt} = -\frac{1}{b}\frac{dB}{dt} = \frac{1}{p}\frac{dP}{dt} = \frac{1}{q}\frac{dQ}{dt}$$

For example in the gas-phase reaction $2NO + O_2 \rightarrow 2NO_2$

$$\frac{d(NO)}{dt} = 2\frac{d(O_2)}{dt} = -\frac{d(NO_2)}{dt}$$

Since $d(NO)/dt = -k_{NO}(NO)^2(O_2)$, if oxygen disappearance is measured,

$$\frac{d(O_2)}{dt} = -k_{O_2}(NO)^2(O_2) = -\frac{k_{NO}}{2}(NO)^2(O_2)$$

In terms of appearance of product NO_2

$$\frac{d(NO)_2}{dt} = k_{NO_2}(NO)^2(O_2) = k_{NO}(NO)^2(O_2)$$

Note that the rate of NO oxidation is second order in NO, first order in O_2, and zero order in NO_2.   Overall order is third.   The orders found experimentally may suggest to the novice a reaction involving three-body collision (molecularity of 3).   Actually the reaction is complex,[1] as noted above.

---

[1] J. J. Carberry, *Chem. Eng. Sci.*, **9**: 189 (1959).

Irreversible Reactions

Zero-order reactions When the reaction rate is independent of the concentration of a particular substance, the rate is said to be zero order with respect to that species. Zero order can mean two things: (1) the species is not a participant in the reaction, for example, NO_2 in the oxidation of NO at moderate temperatures, or (2) the species is present in such abundant supply that its concentration is virtually constant throughout the course of reaction; i.e., experimentally, its concentration dependency cannot be detected, and *apparent* zero order prevails. Thus in the NO oxidation, in the presence of excess O_2, the rate becomes overall second order (in NO) and apparently zero order in O_2; that is,

$$\frac{d(NO)}{dt} = -k_{NO}(O_2)(NO)^2 \xrightarrow{O_2 \gg NO} -k'(NO)^2$$

where $k' = k_{NO}(O_2) =$ constant, at fixed temperature. In a simple case of overall zero-order reaction

$$-\frac{dA}{dt} = k \qquad (2\text{-}45)$$

or

$$A = -kt + \text{constant} \qquad (2\text{-}46)$$

That is, concentration changes linearly with time. It is difficult to cite an actual homogeneous-reaction case following Eq. (2-45). Such behavior is common in heterogeneous catalytic systems. For a simple surface-catalyzed reaction

$$-\frac{dA}{dt} = k\theta_A$$

where θ_A is the fraction of the catalyst surface covered by A, Langmuir suggested

$$\theta_A = \frac{KA}{1 + KA} \qquad \text{thus} \qquad \frac{dA}{dt} = \frac{-kKA}{1 + KA}$$

If, now $KA \gg 1$, then $dA/dt = -k$ and we have zero-order kinetics. Clearly as A nears complete consumption $KA \ll 1$ and $dA/dt \to kKA$; we then have first-order kinetics. In general, then, zero order with respect to a participating species is a limiting condition which cannot prevail as the species concentration approaches zero. Some examples of zero-order kinetics are NH_3 decomposition over Pt and the decomposition of N_2O in the presence of a Pt-wire catalyst.

The condition of zero order is implicitly imposed in kinetic experiments in order to fix the order of another component, i.e.,

$$\frac{dA}{dt} = -kA^\alpha B^\beta \xrightarrow{B \gg A} -k'A$$

or apparent (imposed) zero order in B.

Note that in analyzing rate data, we may use Eq. (2-45), observing the rate-vs.-concentration behavior, or use the integrated form [Eq. (2-46)], observing the concentration-vs.-time behavior.

Another mode of analysis is based upon determination of the time required to achieve a fixed conversion. Commonly, half-life is used, i.e., the time required to obtain 50 percent conversion. For a zero-order reaction ($A = A_0$ at $t = 0$) Eq. (2-46) is

$$A - A_0 = -k_0 t$$

For $A/A_0 = \frac{1}{2}$,

$$t_{1/2} = \frac{A_0}{2k_0} \qquad (2\text{-}47)$$

For zero-order kinetics of the simple type, half-life varies *linearly* with initial concentration A_0, while fractional conversion $1 - A/A_0$ varies inversely with initial concentration A_0.

First-order reactions   In this instance

$$-\frac{dA}{dt} = k_1 A \qquad (2\text{-}48)$$

On integration, assuming that $A = A_0$ at $t = 0$,

$$\ln \frac{A}{A_0} = -k_1 t \qquad \text{or} \qquad \frac{A}{A_0} = \exp(-k_1 t) \qquad (2\text{-}49)$$

A plot of the natural logarithm of the fraction of reactant remaining A/A_0 versus time yields a straight line of slope $-k$. The half-life for a simple first-order reaction is

$$t_{1/2} = \frac{\ln 2}{k_1} = \frac{0.693}{k_1} \qquad (2\text{-}50)$$

In this case half-life and fractional conversion [Eq. (2-49)] are independent of initial concentration A_0.

Pseudo-first-order reactions   Often an intrinsically higher-order reaction is reduced to an apparent first-order case. For example, $A + B \rightarrow P$. Then

$$\frac{dA}{dt} = -kAB^\beta \qquad \text{if} \quad B \gg A$$

and

$$\frac{dA}{dt} = -k_1 A \qquad \text{where } k_1 = kB^\beta$$

In this, the pseudo-first-order case, half-life, conversion, and rate depend upon initial concentration of the excess component B_0 insofar as the pseudo-first-order rate constant contains B.

Numerous examples of first-order reactions exist in reality as well as in the minds of those seeking analytical solutions to complex reaction-diffusion problems. We have already cited the cases of N_2O_5 decomposition and sucrose inversion. Radioactive decay of unstable nuclei is a classic example of first-order kinetics. Once again we emphasize that whereas all unimolecular reactions are first-order, the converse is not necessarily true.

Second-order reactions Two types (I and II) of second-order reaction require analysis:

Type I: $A + A \rightarrow P$ $\dfrac{-dA}{dt} = kA^2$ (2-51)

Type II: $A + B \rightarrow P$ $\dfrac{-dA}{dt} = kAB$ (2-52)

Obviously type II reduces to type I when $A = B$.

TYPE I SECOND ORDER   Integration of (2-51), where $A = A_0$ at $t = 0$, yields

$$\frac{1}{A} - \frac{1}{A_0} = kt \qquad (2\text{-}53)$$

Half-life ($A/A_0 = \frac{1}{2}$) for type I second order is

$$t_{1/2} = \frac{1}{kA_0} \qquad (2\text{-}54)$$

Note that, in this case, half-life varies inversely with initial reactant concentration while the reciprocal of $1 - x$ varies linearly with initial concentration.

Although comparatively rarer than type II reactions, a number of type I cases can be cited, namely, decomposition of HI, decomposition of NO_2 in the gas phase; liquid-phase ClO^- decomposition, and dimerization of cyclopentadiene in either gas or liquid phase.

Pseudo-second-order reaction follows as a limiting case of intrinsic third-order reaction, e.g., in NO oxidation. As previously noted, this third-order reaction is second order in NO and becomes second order overall in an excess of O_2.

TYPE II SECOND ORDER   In the instance

$$aA + bB \rightarrow P \qquad (2\text{-}55)$$

$$\frac{dA}{dt} = -k_A AB \qquad \text{and} \qquad \frac{dB}{dt} = -k_B AB \qquad (2\text{-}56)$$

where $k_A/a = k_B/b$. If D is the concentration of A reacted at any time ($A = A_0$; $B = B_0$ at $t = 0$), then

$$A = A_0 - D \qquad B = B_0 - \frac{b}{a} D \qquad (2\text{-}57)$$

and
$$\frac{dA}{dt} = -\frac{dD}{dt} = -k_A(A_0 - D)\left(B_0 - \frac{b}{a}D\right) \qquad (2\text{-}58)$$

or
$$\frac{dD}{(A_0 - D)(B_0 - bD/a)} = k_A\,dt \qquad (2\text{-}59)$$

Integration yields

$$\ln\frac{A}{B} = \frac{bA_0 - aB_0}{a}k_A t + \ln\frac{A_0}{B_0} \qquad (2\text{-}60)$$

Types I and II [Eqs. (2-53) and (2-60)] are readily displayed graphically. In type I, a linear relation exists between $1/A$ and time t, while for type II the $\ln(A/B)$-versus-t plot displays linearity, the slope $[(bA_0 - aB_0)k_A]/a$, being positive or negative depending upon stoichiometry and initial concentrations.

Half-life for type II second-order reactions must be specified in terms of either A or B. In terms of species A, for type II

$$t_{1/2} = \frac{a}{k_A(bA_0 - aB_0)} \ln\frac{aB_0}{2aB_0 - bA_0} \qquad (2\text{-}61)$$

Half-life in this case is a complex function of initial species concentrations.

If stoichiometric proportions of reactants are used $[B_0 = (b/a)A_0]$, type II reduces to a type I case; i.e., Eq. (2-58) becomes

$$\frac{dD}{dt} = k_A\frac{b}{a}(A_0 - D)^2 \qquad (2\text{-}62)$$

Type II examples are plentiful, for example, HI formation from gaseous H_2 and I_2, HBr formation from H and Br_2, urea synthesis from $NH_4{}^+$ and CNO^- ions, and organic-ester hydrolysis in nonaqueous media.

It is difficult to visualize pseudo-second-order kinetics of type II (this implies a third-order intrinsic reaction between three different species). One case of intrinsic second order can be derived from an apparent third-order situation. In a study of the gas-phase reactions between NO_2 and alcohols (ROH) to form alkyl nitrites, Fairlie, Carberry, and Treacy[1] found that third-order kinetics prevail

$$-\frac{d(NO_2)}{dt} = k(NO_2)^2(ROH) \qquad (2\text{-}63)$$

However, prompted by the fact that k *increased* with *decreasing* temperature (contrary to the Arrhenius generalization), they analyzed the matter and proposed a mechanism involving.

$$2NO_2 \rightleftharpoons N_2O_4 \qquad \text{rapid}$$
$$N_2O_4 + ROH \longrightarrow RONO + HNO_3 \qquad \text{slow}$$

Then
$$\text{Rate} = k_2(N_2O_4)(ROH) \qquad \text{second order}$$

[1] A. M. Fairlie, J. J. Carberry, and J. Treacy, *J. Am. Chem. Soc.*, **75**: 3786 (1953).

However since $N_2O_4 = K(NO_2)^2$, in terms of NO_2, Eq. (2-63) results, an apparent third-order case yet intrinsically second order in terms of N_2O_4. The second-order rate constant displays normal Arrhenius behavior.

Third-order reactions   Three types exist:

Type I:
$$\frac{dA}{dt} = -kA^3 \qquad (2\text{-}64a)$$

Type II:
$$\frac{dA}{dt} = -kA^2B \qquad (2\text{-}64b)$$

Type III:
$$\frac{dA}{dt} = -kABC \qquad (2\text{-}64c)$$

Third-order reactions of all types are so rare that only a brief treatment is justified.

TYPE I

$$\frac{1}{A^2} - \frac{1}{A_0^2} = 2kt \qquad (2\text{-}65)$$

$$\text{Half-life-}t_{1/2} = \frac{3}{2kA_0^2} \qquad (2\text{-}66)$$

TYPE II   The most common third-order reaction is $aA + bB \to pP$

$$\frac{dA}{dt} = -kA^2B$$

Let

$$B = B_0 - \frac{b}{a}(A_0 - A) \qquad (2\text{-}67)$$

Upon integration

$$\left(\frac{a}{b}B_0 - A_0\right)\left(\frac{1}{A} - \frac{1}{A_0}\right) + \ln\frac{AaB_0}{A_0Bb} = \left(\frac{a}{b}B_0 - A_0\right)^2 kt \qquad (2\text{-}68)$$

As noted, type II cases are more common.   Industrially the crucial step in dictating the rate of HNO_3 formation in the ammonia oxidation process is the gas-phase oxidation of NO, a type II third-order reaction.   As illustrated for NO_2-alcohol kinetics, NO oxidation is also intrinsically second order, involving NO and NO_3; however, the NO_3 concentration is not readily determined; thus the system is treated phenomenologically as a third-order type II case, and when it is so expressed, k exhibits a negative temperature dependency.

Fractional-order reactions   In general the detailed mechanisms of most reactions are complex, consisting, as noted earlier, of a series of elementary steps, which in sum yield the overall reaction statement. It therefore follows that few reactions can be expected to fall within the simple order categories we have cited above. In point of fact, fractional, noninteger orders may be anticipated in analysis of other than the most primitive reaction systems. Consider the general case involving one species

$$-\frac{dA}{dt} = kA^{\alpha} \qquad (2\text{-}69)$$

For $\alpha \neq 1$, integration of (2-69) yields

$$\frac{1}{A^{\alpha-1}} - \frac{1}{A_0^{\alpha-1}} = (\alpha - 1)kt \qquad (2\text{-}70)$$

$$\text{Half-life-}t_{1/2} = \frac{2^{\alpha-1} - 1}{kA_0^{\alpha-1}(\alpha - 1)} \qquad (2\text{-}71)$$

Heterogeneous reactions can readily be described in terms of fractional orders. Phosgene formation from CO (first order) and Cl_2 ($\frac{3}{2}$ order) is an excellent example of an homogeneous fractional-order reaction.

Example and exercises To convey the essential features of reaction velocity models, we have confined our attention to instances where reaction volume remains constant, thus permitting rate formulation in terms of concentration. As noted in Sec. 2-5, when there is a change in the number of moles due to reaction, more subtle modes of rate expression are demanded.

Let us consider derivation of the rate expression for the reaction $A \rightarrow mB$. In a gas-phase system at other than dilute concentrations one has:

1 Batch reactor
    a Constant volume, thus pressure changes with reaction progress.
    b Constant pressure; thus volume changes with reaction progress.
2 Flow reactor
    a Constant volume *and* constant pressure (assuming negligible pressure drop); thus velocity or volumetric flow rate changes with reaction progress.

Let us derive rate expressions of a form convenient for analysis of each case.

CASE 1*a*: BATCH, CONSTANT VOLUME

$$-\frac{1}{V}\frac{da}{dt} = k\left(\frac{a}{V}\right)^{\alpha} \qquad (a)$$

where a is the moles of reactant.

A MOLE BALANCE

$t=0$	$t>0$
a_0	a
b_0	$b_0 + m(a_0 - a)$
c (inert)	c

Total moles $= N = ma_0 + b_0 + c + a(1 - m)$

or

$$a = \frac{N - (b_0 + ma_0 + c)}{1 - m}$$

Since $N_0 = a_0 + b_0 + c$,

$$a = \frac{N - N_0 + (1 + m)a_0}{1 - m}$$

Substituting in Eq. (a) and utilizing the ideal-gas law, we have

$$-\frac{d\pi}{dt} = \frac{k}{(RT)^{\alpha-1}} \frac{\{\pi - \pi_0[1 - (1 - m)y_0]\}}{(1 - m)^{\alpha-1}} \qquad (b)$$

where $y_0 =$ initial mole fraction of A
$\quad \pi =$ pressure at any time
$\quad \pi_0 =$ initial total pressure

As an exercise derive the rate equation in terms of total volume versus t for case 1b.

CASE 2   For plug flow and constant V_0 and π_0

$$-\frac{dF}{dV_0} = k(A)^{\alpha} = k\left(\frac{F}{Q}\right)^{\alpha} \qquad (c)$$

where $A = F/Q$
$\quad F =$ moles of A flowing
$\quad Q =$ total volumetric flow rate
$\quad F_0 = Q_0 A_0$

We have but to express the variable Q as a function of conversion x. Quite simply

$$Q = Q_0[1 - (1 - m)x] \quad \text{and} \quad F = F_0(1 - x) = Q_0 A_0(1 - x)$$

so that Eq. (c) becomes

$$A_0 \frac{dx}{d\tau_0} = k\left(\frac{F_0}{Q_0}\right)^{\alpha}\left[\frac{(1 - x)}{1 - (1 - m)x}\right]^{\alpha} \quad \text{or} \quad \frac{dx}{d\tau_0} = k(A_0)^{\alpha-1}\left[\frac{1 - x}{1 + \epsilon x}\right]^{\alpha} \qquad (d)$$

where $\epsilon = m - 1$ and $\tau_0 = V_0/Q_0$. Note that for the general case

$$aA + bB \rightarrow cC + dD \quad \text{and} \quad m - 1 = c + d - a - b = \epsilon$$

In studies of NH_3 synthesis from N_2 and H_2 over promoted iron catalysts, the differential rate has traditionally been expressed in terms of the mole fraction y of NH_3, that is,

$$\text{Rate} \frac{\text{moles}}{\text{g catalyst time}} = \frac{dy}{(1 + y)^n d(W/F_0)} = f(N_2, H_2, NH_3)$$

where $W =$ weight of catalyst
$\quad F_0 =$ molar feed rate

Unfortunately discord prevailed for some years over the value of the exponent n. Early Russian derivations led to $n = 1$, while workers in the United States derived a result with $n = 3$. The Japanese derivation leads to $n = 2$. Since the system is that of a plug-flow reactor wherein a gas-phase reaction occurs with a change in moles with extent of reaction, the student should be able to demonstrate, on the back of an AIChE membership card, that $n = 2$ and that other values are a consequence of expressing the differential rate in terms of concentration.[1]

Reversible Reactions

We shall consider three classes of reversible reaction: (1) simple first order in each direction. $A \rightleftharpoons B$; (2) second order in each direction, $A + B \rightleftharpoons C + D$; and (3) a mixed case $A \rightleftharpoons B + C$.

Simple reversible reaction   Consider

$$A \underset{k_2}{\overset{k_1}{\rightleftharpoons}} B \qquad (2\text{-}72)$$

Then

$$\frac{dA}{dt} = -k_1 A + k_2 B \qquad (2\text{-}73)$$

A material balance demands that $B = B_0 + A_0 - A$

$$\frac{dA}{dt} = -k_1 A - k_2 A + k_2(B_0 + A_0)$$

$$\frac{dA}{dt} = -(k_1 + k_2)A + k_2(B_0 + A_0) \qquad (2\text{-}74)$$

Now in terms of equilibrium values A_e and B_e

$$K = \frac{B_e}{A_e} = \frac{B_0 + A_0 - A_e}{A_e} \qquad (2\text{-}75)$$

or

$$A_e = \frac{B_0 + A_0}{K + 1} \qquad (2\text{-}76)$$

also

$$k_2 = \frac{k_1 + k_2}{K + 1} \qquad (2\text{-}77)$$

so that

$$\frac{dA}{dt} = -(k_1 + k_2)\left(A - \frac{B_0 + A_0}{K + 1}\right) \qquad (2\text{-}78)$$

In view of (2-76)

$$\frac{dA}{dt} = -(k_1 + k_2)(A - A_e) \qquad (2\text{-}79)$$

or

$$\frac{dA}{dt} = -k(A - A_e)$$

[1] R. B. Anderson and H. L. Toor, *J. Phys. Chem.*, **63**: 1982 (1959).

Let $C = A - A_e$; then

$$\frac{dC}{dt} = -kC \qquad (2\text{-}80)$$

which yields an equivalent first-order irreversible solution

$$\frac{C}{C_0} = \exp(-kt) = \frac{A - A_e}{A_0 - A_e} = \exp[-(k_1 + k_2)t] \qquad (2\text{-}81)$$

The slope of the $\ln[(A - A_e)/(A_0 - A_e)]$-versus-$t$ plot gives $-(k_1 + k_2)$, while A at equilibrium establishes K from Eq. (2-76).

If a reliable value of A_e is not experimentally available, Eq. (2-81) can be employed using trial values of A_e until semilog linearity results. Examples of this simple equilibrium type are gas-phase cis-trans isomerizations and racemization of glucoses.

Higher-order reversible reactions We now consider a case where second-order kinetics characterizes both forward and reverse rates

$$A + B \underset{k_2}{\overset{k_1}{\rightleftharpoons}} C + D \qquad (2\text{-}82)$$

Then

$$\frac{dA}{dt} = -k_1 AB + k_2 CD \qquad (2\text{-}83)$$

Letting $A = A_0 - E$, where E represents the concentration of A *reacted*, we have

$$\frac{dE}{dt} = k_1(A_0 - E)(B_0 - E) - k_2(C_0 + E)(D_0 + E) \qquad (2\text{-}84)$$

or $\quad \dfrac{dE}{dt} = (k_1 A_0 B_0 - k_2 C_0 D_0) - (k_1 A_0 + k_1 B_0 + k_2 C_0 + k_2 D_0)E + (k_1 - k_2)E^2$

$$\frac{dE}{dt} = \alpha + \beta E + \gamma E^2 \qquad (2\text{-}85)$$

Integration yields

$$\ln \frac{E2\gamma/(\beta - \sqrt{q}) + 1}{E2\gamma/(\beta + \sqrt{q}) + 1} = \sqrt{q} \cdot t \qquad \text{where} \qquad q = \beta^2 - 4\alpha\gamma \qquad (2\text{-}86)$$

Equation (2-86) is obviously difficult to use without prior knowledge of k_1 and k_2 or their ratio K. If K is known, a trial-and-error choice of k_1 is required to correlate the data in accord with Eq. (2-86). At low conversion levels ($E \to 0$), k_1 can be determined by treating this early stage as irreversible.

We next consider the mixed-order reversible cases. For

$$A \underset{k_2}{\overset{k_1}{\rightleftharpoons}} B + C \qquad \text{where} \qquad \begin{array}{l} \longrightarrow = \text{first order} \\ \longleftarrow = \text{second order} \end{array} \qquad (2\text{-}87)$$

When we assume that B and $C = 0$ at $t = 0$ and let $A = A_0 - E$,

$$\frac{dE}{dt} = k_1(A_0 - E) - k_2 E^2 \qquad (2\text{-}88)$$

Integrating gives

$$2k_2 \alpha t = \ln \frac{A_0 + E(\beta - \tfrac{1}{2})}{A_0 - E(\beta + \tfrac{1}{2})} \qquad \text{where} \qquad \begin{cases} \alpha = \sqrt{K_1^2/4 + K_1 A_0} \\ \beta = \alpha/K_1 \end{cases} \qquad (2\text{-}89)$$

In this case trial values of β are chosen until the indicated linearity is found. The slope gives $k_2 \alpha$, and equilibrium data provides K_1.   For the case

$$A + B \;\overset{k_1}{\underset{k_2}{\rightleftharpoons}}\; C \qquad \text{where} \qquad \begin{array}{l} \longrightarrow \;= \text{second order} \\ \longleftarrow \;= \text{first order} \end{array}$$

the solution is

$$k_1 \alpha t = \ln \frac{2KA_0 B_0 - E(KA_0 + KB_0 + 1 - \alpha K)}{2KA_0 B_0 - E(KA_0 + KB_0 + 1 + \alpha K)} \qquad (2\text{-}90)$$

where

$$\alpha = \sqrt{(A_0 - B_0)^2 + \frac{2}{K}(A_0 + B_0) + \frac{1}{K^2}}$$

If K is known, α can be computed and then k_1 determined.   When K is not known, trial-and-error procedures are required.

Our brief treatment of equilibrium reactions other than the simple first-order case indicates the complexities inherent in analyzing such systems; however, the skilled experimental strategist might set conditions approaching irreversibility to establish at least one rate constant.

2-8   RELATIONSHIP BETWEEN EQUILIBRIUM AND SPECIFIC RATE CONSTANTS

It is generally assumed in the case of a reversible reaction such as

$$A + B \;\overset{k_1}{\underset{k_{-1}}{\rightleftharpoons}}\; C + D$$

that the equilibrium constant K_{eq} is related uniquely to the ratio of forward to reverse rate constants; thus

$$\frac{k_1}{k_{-1}} = K_{eq}$$

This relationship is by no means generally true, as should be clear from the following argument.

Affinity

Consider a general case of reactants R and products P, that is, $R \rightleftarrows P$.   At equilibrium we have $K_{eq} = P_e/R_e$.   Affinity is defined as the distance from the equilibrium conditions, so that in terms of free energy ΔG affinity is

$$\Delta G_0 \;=\; \Delta G^{\circ} \;-\; \Delta G$$
$$\text{Affinity} = \text{equilibrium} - \text{nonequilibrium state}$$

Since
$$\Delta G = -RT \ln K$$

the nonequilibrium state is simply $-RT \ln(P/R)$, while the equilibrium condition of free energy is

$$-RT \ln \frac{P_e}{R_e} = -RT \ln K_e$$

so that
$$\text{Affinity,}\; -\Delta G_0 = +RT \ln K_e - RT \ln \frac{P}{R} \qquad (2\text{-}91)$$

At equilibrium, of course, affinity is zero, as $P_e/R_e = P/R$.

If a particular reaction is the consequence of several elementary steps we may define the affinity of each step as the difference between the true equilibrium state and the nonequilibrium which prevails.   For example, in the reaction $2A + B \rightleftarrows P$ the elementary steps might be:

Step 1: $\qquad A \;\rightleftharpoons\; A^*$

Step 2: $\qquad A^* + B \;\rightleftharpoons\; X$

Step 3: $\dfrac{A^* + X \;\rightleftharpoons\; P}{2A + B \;\rightleftharpoons\; P}$

To produce 1 mol of P, step 1 must occur twice and steps 2 and 3 once each. Overall reaction affinity, expressed in terms of the affinities of each elementary step, then becomes

$$\Delta G_0 = 2\Delta G_1 + \Delta G_2 + \Delta G_3$$

or, in general,

$$\Delta G_0 = \sum v_i \, \Delta G_i$$

Now consider the ratio of reaction rates at nonequilibrium as governed by step r:

$$\frac{\vec{r}}{\overleftarrow{r}} = \exp \frac{-\Delta G_r}{RT}$$

Of course, when overall equilibrium is realized, the affinity ΔG_r is zero and forward $\vec{r}$ and reverse $\overleftarrow{r}$ rates are equal. Under nonequilibrium conditions, the net rate of reaction $\vec{r} - \overleftarrow{r}$ is given by

$$r_{net} = \vec{r} - \overleftarrow{r} = \vec{r}\left(1 - \exp \frac{\Delta G_r}{RT}\right) \qquad (2.93)$$

where ΔG_r is the affinity of the rate-controlling step in a sequence of the several steps which compose the overall reaction.

Now let it be supposed that when an overall reaction is the sum of several elementary steps, all save one are at equilibrium (affinity is zero in all but one of the elementary steps); then since

$$\Delta G_0 = \sum v_i \, \Delta G_i = v_r \, \Delta G_r$$

we have

$$\Delta G_r = \frac{\Delta G_0}{v_r}$$

and so

$$\vec{r} - \overleftarrow{r} = \vec{r}\left(1 - \exp \frac{\Delta G_0}{v_r RT}\right)$$

Since

$$-\Delta G_0 = +RT \ln K_e - RT \ln \frac{P}{R}$$

we have

$$\exp \frac{\Delta G_0}{RT} = \frac{P}{K_e R}$$

and so

$$\vec{r} - \overleftarrow{r} = r_{\text{net}} = \vec{r}\left[1 - \left(\frac{P}{K_e R}\right)^{1/v_r}\right] \qquad (2\text{-}94)$$

Since k_1/k_2 must be independent of concentration,

$$\frac{\overleftarrow{r}}{\vec{r}} = \left(\frac{P}{K_e R}\right)^{-1/v_r} \qquad \text{and} \qquad \frac{k_1}{k_{-1}} = (K_e)^{1/v_r}$$

The ratio of forward to reverse reaction-rate expression $\vec{r}/\overleftarrow{r}$ found from kinetic studies will depend on the mechanism (rate-controlling step or steps), and such a ratio is clearly not equal to $P/K_e R$ but depends on v_r, the stoichiometric number (the number of times the rate-controlling step must occur to produce the overall quantity of product P). Consider the following example.

The important reaction of dissolved N_2O_4 with water to produce HNO_3 and NO has been studied by Denbigh and Prince,[1] with the result that the net rate of reaction at constant water and acid concentration is given by

$$r_{\text{net}} = k_1(N_2O_4) - k_{-1}(N_2O_4)^{1/4}(NO)^{1/2}$$

The overall reaction stoichiometry may be written

$$1.5N_2O_4 + H_2O \rightleftharpoons 2HNO_3 + NO$$

and so

$$K_e = \frac{(HNO_3)^2(NO)}{(N_2O_4)^{1.5}(H_2O)}$$

Equating the measured values of $\vec{r}/\overleftarrow{r}$ to $(P/K_e R)^{-1/v}$ in accord with Eqs. (2-94), we find, at constant H_2O concentration and thus constant HNO_3 concentration,

$$\frac{\overleftarrow{r}}{\vec{r}} = \frac{k_{-1}}{k_1} \frac{(N_2O_4)^{1/4}(NO)^{1/2}}{N_2O_4} = \left(\frac{P}{K_e R}\right)^{1/v_r} = K_e^{-1/v_r}\left[\frac{(HNO_3)^2(NO)}{(N_2O_4)^{1.5}(H_2O)}\right]^{1/v_r}$$

[1] K. G. Denbigh and A. J. Prince, *J. Chem. Soc.*, 1947: 790.

or

$$\frac{k_{-1}}{k_1}\frac{(HNO_3)^{\alpha}(NO)^{1/2}}{(H_2O)^{\beta}(N_2O_4)^{3/4}} = \frac{1}{K_e^{1/v_r}}\left[\frac{(HNO_3)^2}{H_2O}\right]^{1/v_r}\left[\frac{NO}{(N_2O_4)^{1.5}}\right]^{1/v_r}$$

Therefore

$$v_r = 2 \qquad \alpha = 1 \qquad \beta = \tfrac{1}{2}$$

and

$$\frac{k_1}{k_{-1}} = (K_e)^{1/v_r} \qquad (2\text{-}95)$$

The reaction is obviously complex, involving some elementary steps, one of these being rate-controlling and characterized by a stoichiometric number of 2. More will be said of this issue in a later chapter; however, at this juncture it need only be stated that the relationship between the ratio of measured rate coefficients and the equilibrium constant expressed in terms of overall reaction is not necessarily a simple one, as suggested by Guldburg and Waage. It can be generally asserted that the relationship between net rate, forward rate, K_e, and the overall reactant-product equilibrium stoichiometry for any reaction is

$$r_{net} = \bar{r}\left[1 - \left(\frac{products}{K_{eq} \times reactants}\right)^{1/v_r}\right] \qquad (2\text{-}96)$$

2-9   COMPLEX REACTIONS

We define a complex reaction as one which is capable of proceeding by more than one path. Reactions treated up to this point are simple insofar as only one path is involved. True, equilibrium may limit conversion to that product, and in irreversible cases conversion may be low due to intrinsic rate limitations, e.g., temperature. In any case, steps can be taken to enhance conversion and thus the yield in simple systems. In complex systems, on the other hand, it is selectivity and yield which are important. Since more than one product can be generated by diverse routes, our concern is with those factors which dictate conversion to one of these products exclusively (yield). For example, in the complex scheme

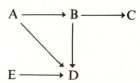

it may matter little if A or E can be totally converted. Far more important is the issue of conversion of A to B, let us say, or to C or D. If B is the desired product, it is not dA/dt which is of sole interest but dB/dt, and then the yield of B as a function of conversion of A. Given methods whereby complex-reaction-rate data can be analyzed to determine the reaction-path network and respective rate constants, the system may possibly be engineered or optimized to commercial advantage.[1]

---

[1] When a large number of components constitute the reactant feed, lumping may be employed. See for example, J. Wei and J. C. W. Kuo, *Ind. Eng. Chem. Fundam.*, **8**: 114, 124 (1968); D. Luss and P. Hutchinson, *Chem. Eng. J.*, **2**: 172 (1971).

Simultaneous Reactions

For a reaction that is first order in each step

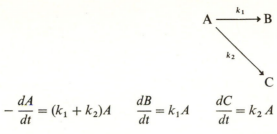

$$-\frac{dA}{dt} = (k_1 + k_2)A \qquad \frac{dB}{dt} = k_1 A \qquad \frac{dC}{dt} = k_2 A$$

Then $A/A_0 = \exp[-(k_1 + k_2)t]$. Substituting for A and integrating ($B = B_0$, $C = C_0$ at $t = 0$) gives the overall yields

$$\frac{B}{A_0} = \frac{B_0}{A_0} + \frac{k_1}{k_1 + k_2}\{1 - \exp[-(k_1 + k_2)t]\}$$

$$\frac{C}{A_0} = \frac{C_0}{A_0} + \frac{k_2}{k_1 + k_2}\{1 - \exp[-(k_1 + k_2)t]\}$$

$$(2\text{-}97)$$

Dividing $B - B_0$ by $C - C_0$, we obtain $(B - B_0)/(C - C_0) = k_1/k_2$, the selectivity, a result which is found immediately from $(dB/dt)/(dC/dt)$. Selectivity in this linear reaction system depends only on the rate-constant ratio. This is true so long as all reactions are of the same order. If the orders differ,

$$\frac{dB}{dC} = \frac{k_1 A^\alpha}{k_2 A^\beta} = \frac{k_1}{k_2} A^{\alpha - \beta}$$

To facilitate integration in this case, it is preferable to divide the product-formation rate by that of the reactant disappearance; thus the point yield is

$$\frac{dB}{dA} = -\frac{k_1 A^\alpha}{k_1 A^\alpha + k_2 A^\beta} = -\frac{1}{1 + (k_2/k_1)(A)^{\beta - \alpha}} \qquad (2\text{-}98)$$

Specific treatment of these mixed non-first-order cases is given later. For identical order, since $dB/dC = k_1/k_2$, the slope of the B-versus-C curve yields k_1/k_2, while $k_1 + k_2$ is found by simple analysis of the reactant A-versus-time data.

The point yield in the identical-order case is, of course,

$$-\frac{dB}{dA} = \frac{k_1}{k_1 + k_2} \qquad -\frac{dC}{dA} = \frac{k_2}{k_1 + k_2}$$

which upon integration gives the yield of each product as a function of conversion of reactant A. It is extremely important to note that where both simultaneous reactions are of identical order, B/C, the selectivity, is totally independent of time or extent of reaction, being solely determined by rate-constant ratio. Yield B/A_0 depends upon the rate coefficients and conversion.

Our treatment of simultaneous reaction schemes illustrates that certain advantages lie in eliminating time by dividing one rate expression by another, particularly division by the reactant disappearance rate. In this fashion, the appearance of a particular product is related upon integration to disappearance (conversion) of the reactant. The result expresses yield or selectivity, the prime objective in analyzing complex-reaction networks.

Higher-Order Simultaneous Reactions

Consider a mixed-order case

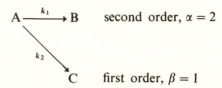

which is best handled on a time-free basis. dB/dA is given by Eq. (2-98) ($\alpha = 2$, $\beta = 1$) as

$$\frac{dB}{dA} = -\frac{k_1 A^2}{k_1 A^2 + k_2 A} = -\frac{k_1 A}{k_1 A + k_2}$$

Integration gives ($B = B_0$, $A = A_0$) for B

$$\frac{B}{A_0} = \frac{B_0}{A_0} + \left(1 - \frac{A}{A_0}\right) + \frac{k_2}{k_1 A_0} \ln \frac{A/A_0 + k_2/k_1 A_0}{1 + k_2/k_1 A_0} \qquad (2\text{-}99)$$

and for C, by the same reasoning,

$$\frac{C}{A_0} = \frac{C_0}{A_0} + \frac{k_2}{k_1 A_0} \ln \frac{1 + k_2/k_1 A_0}{A/A_0 + k_2/k_1 A_0} \qquad (2\text{-}100)$$

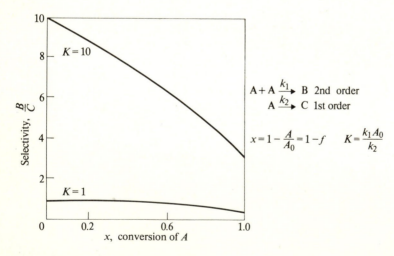

FIGURE 2-2
Yield of B relative to that of C versus conversion of A for the reaction
A + A $\xrightarrow{\quad 1 \quad}$ (second order) and A $\xrightarrow{\quad 2 \quad}$ C (first order).

When C_0 and $B_0 = 0$, the ratio of B to C, overall selectivity, is

$$\frac{B}{C} = \frac{1 - A/A_0}{\left(\dfrac{k_2}{k_1 A_0}\right) \ln \dfrac{1 + k_2/k_1 A_0}{A/A_0 + k_2/k_1 A_0}} - 1 \qquad (2\text{-}101)$$

Typical selectivity profiles are shown in Fig. 2-2.

Parallel or Concurrent Reactions

$$A \xrightarrow{\;k_1\;} B$$

$$R \xrightarrow{\;k_2\;} Y \qquad (2\text{-}102)$$

$$-\frac{dA}{dt} = \frac{dB}{dt} = k_1 A \qquad \text{and} \qquad -\frac{dR}{dt} = \frac{dY}{dt} = k_2 R \qquad (2\text{-}103)$$

$$\frac{dA}{dR} = \frac{dB}{dY} = \frac{k_1 A}{k_2 R} \qquad (2\text{-}104)$$

$$\ln \frac{A}{A_0} = \frac{k_1}{k_2} \ln \frac{R}{R_0}$$

or

$$\frac{A}{A_0} = \left(\frac{R}{R_0}\right)^{k_1/k_2} \qquad (2\text{-}105)$$

or

$$\frac{B}{A_0} = 1 - \left(\frac{R}{R_0}\right)^{k_1/k_2} \qquad \text{for } B_0 = 0 \qquad (2\text{-}106)$$

Equation (2-105) states that a log-log plot of A/A_0 versus R/R_0 has a slope equal to k_1/k_2, while k_1 is determined by the $\ln(A/A_0)$-versus-time relation. Yield and selectivity in this parallel network do depend upon time as well as rate constants since, for $B_0 = 0$,

$$\frac{B}{A_0} = 1 - [\exp(-k_2 t)]^{k_1/k_2} \qquad (2\text{-}107)$$

or

$$\frac{B}{A_0} = 1 - \exp(-k_1 t)$$

Similarly for Y, $Y/R_0 = 1 - \exp(-k_2 t)$. Therefore, selectivity is

$$\frac{B}{Y} = \frac{A_0}{R_0} \frac{1 - \exp(-k_1 t)}{1 - \exp(-k_2 t)} \qquad (2\text{-}108)$$

Parallel reactions are essentially independent; in consequence higher-order cases are treated by the methods outlined under Simple Reactions, as the parallel system is merely a multiple, independent, network of simple reactions.

Consecutive Linear Reactions

The classic first-order sequence is

$$A \xrightarrow{k_1} B \xrightarrow{k_2} C$$

$$-\frac{dA}{dt} = k_1 A \quad \text{and} \quad \frac{dB}{dt} = k_1 A - k_2 B \qquad (2\text{-}109)$$

C is determined by material balance. In terms of time t since $A/A_0 = \exp(-k_1 t)$,

$$\frac{dB}{dt} = k_1 A_0 \exp(-k_1 t) - k_2 B \qquad (2\text{-}110)$$

This linear, first-order differential equation is of the form

$$\frac{dy}{dx} + Py = Q$$

the general solution being

$$y = \exp\left(-\int P\, dx\right)\left[\int Q \exp\left(\int P\, dx\right) dx + \text{const}\right]$$

Therefore, for $B = B_0$ at $t = 0$

$$\frac{B}{A_0} = \frac{k_1}{k_2 - k_1}\left[\exp(-k_1 t) - \exp(-k_2 t)\right] + \frac{B_0}{A_0}\exp(-k_2 t) \qquad (2\text{-}111)$$

The time at which the maximum in B occurs is obtained by differentiation of (2-111); for $B_0 = 0$

$$t_{\max B} = \frac{\ln(k_2/k_1)}{k_2 - k_1} \qquad (2\text{-}112)$$

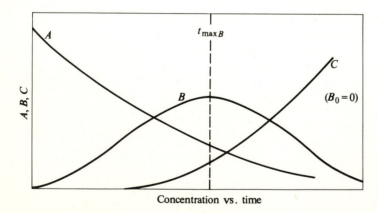

Concentration vs. time

FIGURE 2-3
Typical consecutive-reaction profile: A → B → C.

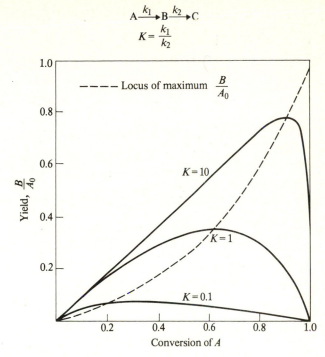

FIGURE 2-4
Yield of B versus conversion of A for diverse values of
$K = k_1/k_2$.

while the maximum concentration is

$$\frac{B_{max}}{A_0} = \left(\frac{k_1}{k_2}\right)^{k_2/(k_2-k_1)} \tag{2-113}$$

As k_1 is readily determined from $\ln(A/A_0)$-versus-time data, the maximum time or concentration datum provides the means whereby k_2 is established.   Figure 2-3 schematically displays a typical consecutive-reaction profile, from which it is clear that the yield of B or C is crucially time-dependent for given rate-coefficient values.

The first-order consecutive reaction scheme is nicely handled on a time-free basis:

$$\frac{dB}{dA} = -1 + \frac{k_2}{k_1}\frac{B}{A} \tag{2-114}$$

which is readily integrated to

$$\frac{B}{A_0} = \frac{1}{1 - k_2/k_1}\left[\left(\frac{A}{A_0}\right)^{k_2/k_1} - \frac{A}{A_0}\right] + \frac{B_0}{A_0}\left(\frac{A}{A_0}\right)^{k_2/k_1} \tag{2-115}$$

Yield of B is then uniquely related to conversion of A and the rate constant ratio k_1/k_2, as shown in Fig. 2-4.

When $k_1 = k_2$, application of L'Hospital's rule gives, on a time-free basis,

$$\frac{B}{A_0} = \frac{A}{A_0}\left(\frac{B_0}{A_0} - \ln\frac{A}{A_0}\right) \qquad (2\text{-}116)$$

The behavior of B with respect to time is complicated when $B_0 \neq 0$. In that case

$$t_{\max B} = \frac{1}{k_2 - k_1}\left[\ln\frac{k_2}{k_1}\left(1 + \frac{B_0}{A_0} - \frac{k_2}{k_1}\frac{B_0}{A_0}\right)\right] \qquad (2\text{-}117)$$

That a maximum in B at $t > 0$ does not necessarily exist when $B_0 \neq 0$ is evident in terms of initial rate:

$$\left(\frac{dB}{dt}\right)_{t=0} = k_1 A_0 - k_2 B_0 \qquad (2\text{-}118)$$

If $k_1 A_0 < k_2 B_0$, the initial slope is negative and no maximum can exist.

Higher-Order Consecutive Reactions

Suppose we have a mixed-order consecutive scheme

$$A \longrightarrow B \qquad \text{first order}$$
$$A + B \longrightarrow C \qquad \text{second order} \qquad (2\text{-}119)$$

Then

$$\frac{dA}{dt} = -k_1 A - k_2 AB \qquad (2\text{-}120)$$

and

$$\frac{dB}{dt} = k_1 A - k_2 AB \qquad (2\text{-}121)$$

This reaction network illustrates the merit of analysis on a time-free basis. For example, as Benson shows,[1] if we eliminate A by differentiating (2-121) and substitute (2-120) in the result, we find

$$(k_1 - k_2 B)\frac{d^2 B}{dt^2} + k_2\left(\frac{dB}{dt}\right)^2 + (k_1^2 - k_2^2 B^2)\frac{dB}{dt} = 0$$

which is a nonlinear differential equation. If, instead we divide (2-121) by (2-120) to eliminate time, we obtain

$$\frac{dB}{dA} = \frac{k_1 A - k_2 AB}{-k_1 A - k_2 AB} = \frac{B - k_1/k_2}{B + k_1/k_2} \qquad (2\text{-}122)$$

which can be integrated for the condition $B = B_0$ at $A = A_0$, where $K = k_1/k_2$:

$$\frac{B}{A_0} + \frac{K+1}{A_0}\ln\frac{K-B}{K-B_0} = \frac{B_0}{A_0} - \left(1 - \frac{A}{A_0}\right) \qquad (2\text{-}123)$$

[1] " Foundations of Chemical Kinetics."

While the result is not explicit in B/A_0, the behavior of B as a function of conversion of A can be graphically displayed[1] for various values of K/A_0, where $B_0 = 0$ and $K = k_1/k_2$.† Other networks are nicely handled on a time-free basis.

If both steps are second order (type I)

$$A + A \longrightarrow B \qquad \text{second order}$$
$$B + B \longrightarrow C \qquad \text{second order} \qquad (2\text{-}124)$$

then

$$\frac{dA}{dt} = -k_1 A^2 \qquad \frac{dB}{dt} = \frac{k_1}{2} A^2 - k_2 B^2 \qquad (2\text{-}125)$$

Eliminating time gives

$$\frac{dB}{dA} = -\frac{1}{2} + \frac{k_2}{k_1} \left(\frac{B}{A}\right)^2$$

Let $B = Ay$ and $K = k_2/k_1$; then

$$\frac{dy}{d \ln A} = Ky^2 - y - \tfrac{1}{2} \qquad (2\text{-}126)$$

Integrating for the case where $B_0 = 0$ gives

$$\ln \frac{A}{A_0} = \frac{1}{\sqrt{1 + 2K}} \ln \left[\frac{2Ky - (1 + \sqrt{1 + 2K})}{2Ky - (1 - \sqrt{1 + 2K})} \frac{1 - \sqrt{1 + 2K}}{1 + \sqrt{1 + 2K}}\right]$$

or, letting $\sqrt{1 + 2K} = \alpha$, we have

$$\ln \frac{A}{A_0} = \frac{1}{\alpha} \ln \frac{2Ky/(1 + \alpha) - 1}{2Ky/(1 - \alpha) - 1}$$

Simplifying gives

$$\frac{y(1 - \alpha) + 1}{y(1 + \alpha) + 1} = \left(\frac{A}{A_0}\right)^\alpha \qquad (2\text{-}127)$$

or since $y = B/A$,

$$\frac{B}{A_0} = \frac{(A/A_0)[1 - (A/A_0)^\alpha]}{(A/A_0)^\alpha (1 + \alpha) - (1 - \alpha)} \qquad (2\text{-}128)$$

We next treat a consecutive case in which the intermediate is attacked by reactant via second-order kinetics, i.e.,

$$A + B \longrightarrow C \qquad \text{second order}$$
$$B + C \longrightarrow D \qquad \text{second order} \qquad (2\text{-}129)$$

$$\frac{dA}{dt} = -k_1 AB \qquad \frac{dB}{dt} = -k_1 AB - k_2 BC \qquad \frac{dC}{dt} = k_1 AB - k_2 BC$$

---

[1] Ibid., pp. 43–44.
† Graphical displays of yield or conversion for systems analyzed in this section are set forth in Chap. 3 for various levels of backmixing.

Again eliminating time, we have

$$\frac{dC}{dA} = -1 + \frac{k_2}{k_1}\frac{C}{A} \qquad (2\text{-}130)$$

Note that (2-130) is identical to (2-114) for the reaction $A \rightarrow B \rightarrow C$, a consequence of the fact that the reactant B is common to both consecutive steps.   Hence

$$\frac{C}{A_0} = \frac{1}{1 - k_2/k_1}\left[\left(\frac{A}{A_0}\right)^{k_2/k_1} - \frac{A}{A_0}\right] + \frac{C_0}{A_0}\left(\frac{A}{A_0}\right)^{k_2/k_1} \qquad (2\text{-}131)$$

When $k_1 = k_2$, we have, as in the first-order consecutive case

$$\frac{C}{A_0} = \frac{A}{A_0}\left(\frac{C_0}{A_0} - \ln\frac{A}{A_0}\right) \qquad (2\text{-}132)$$

General Complex Linear Reactions

Consider the following system of *first-order reactions*:

$$A \xrightarrow{k_1} B \xrightarrow{k_4} D$$

with k_2 and k_3 branches leading to

$$E \xrightarrow{k_5} C \qquad (2\text{-}133)$$

Our task is to derive and integrate the equations which define the yield of B, C, and D as a function of the conversion of A.   This system is of interest since it involves simultaneous, consecutive, and parallel reaction types.

For the scheme shown (2-133), we have

$$\frac{dA}{dt} = -(k_1 + k_2)A \qquad (2\text{-}134a)$$

$$\frac{dB}{dt} = k_1 A - (k_3 + k_4)B \qquad (2\text{-}134b)$$

$$\frac{dC}{dt} = k_2 A + k_3 B + k_5 E \qquad (2\text{-}134c)$$

$$\frac{dD}{dt} = k_4 B \qquad (2\text{-}134d)$$

$$\frac{dE}{dt} = -k_5 E \qquad (2\text{-}134e)$$

Eliminating time by dividing by dA/dt gives

$$\frac{dB}{dA} = -\frac{k_1}{k_1 + k_2} + \frac{k_3 + k_4}{k_1 + k_2}\frac{B}{A} \qquad (2\text{-}135a)$$

$$\frac{dC}{dA} = -\frac{k_2}{k_1 + k_2} - \frac{k_3}{k_1 + k_2}\frac{B}{A} - \frac{k_5}{k_1 + k_2}\frac{E}{A} \qquad (2\text{-}135b)$$

$$\frac{dD}{dA} = -\frac{k_4}{k_1 + k_2}\frac{B}{A} \qquad (2\text{-}135c)$$

$$\frac{dE}{dA} = \frac{k_5}{k_1 + k_2}\frac{E}{A} \qquad (2\text{-}135d)$$

Inspection indicates that integration of (2-135a) to give B/A and (2-135d) to obtain E/A is all that is required to solve for the yields of B, C, and D. Equation (2-135a) is a linear differential equation of the form

$$\frac{dy}{dx} + P(x)y = Q(x)$$

the general solution to which is

$$B = A^{K_1}\left(-K_2 \int \frac{dA}{A^{K_1}} + \text{const}\right) \qquad (2\text{-}136)$$

in terms of this problem, where

$$K_1 = \frac{k_3 + k_4}{k_1 + k_2} \qquad \text{and} \qquad K_2 = \frac{k_1}{k_1 + k_2}$$

When $B = B_0$ at $A = A_0$, (2-136) becomes

$$\frac{B}{A_0} = \left(\frac{A}{A_0}\right)^{K_1}\left\{\frac{B_0}{A_0} + \frac{K_2}{K_1 - 1}\left[\left(\frac{A}{A_0}\right)^{1-K_1} - 1\right]\right\} \qquad (2\text{-}137)$$

In order to solve for C/A_0, E/A must be obtained by solving (2-135d):

$$\frac{dE}{dA} = \left(\frac{k_5}{k_1 + k_2}\right)\frac{E}{A} = K_6\frac{E}{A} \qquad (2\text{-}135d)$$

Integrating gives

$$\frac{E}{A} = \frac{E_0}{A_0}\left(\frac{A}{A_0}\right)^{K_6-1} \qquad (2\text{-}138)$$

Substituting for E/A and B/A in (2-135b) and integrating, we obtain

$$\frac{C}{A_0} = \frac{C_0}{A_0} - \frac{K_2 K_3 + K_1 K_5 - K_5}{1 - K_1}\left(1 - \frac{A}{A_0}\right) + \left[\frac{K_3 B_0}{K_1 A_0} + \frac{K_2 K_3}{K_1/(1 - K_1)}\right]\left[1 - \left(\frac{A}{A_0}\right)^{K_1}\right]$$

$$+ \frac{E_0}{A_0}\left[1 - \left(\frac{A}{A_0}\right)^{K_6}\right] \qquad (2\text{-}139)$$

where

$$K_3 = \frac{k_3}{k_1 + k_2} \qquad \text{and} \qquad K_5 = \frac{k_2}{k_1 + k_2}$$

The yield of D is also readily obtained by integrating (2-135c), letting $K_4 = k_4/(k_1 + k_2)$:

$$\frac{D}{A_0} = \frac{D_0}{A_0} - \frac{K_2 K_4}{1 - K_1}\left(1 - \frac{A}{A_0}\right) + \left[\frac{K_4 B_0}{K_1 A_0} + \frac{K_2 K_4}{K_1/(1 - K_1)}\right]\left[1 - \left(\frac{A}{A_0}\right)^{K_1}\right] \qquad (2\text{-}140)$$

In the case where $k_3 + k_4 = k_1 + k_2$, that is, $K_1 = 1$, application of L'Hospital's rule to Eqs. (2-137), (2-139), and (2-140) yields, for B,

$$\frac{B}{A_0} = \frac{A}{A_0}\left(\frac{B_0}{A_0} - K_2 \ln \frac{A}{A_0}\right) \qquad (2\text{-}137a)$$

for C,

$$\frac{C}{A_0} = \frac{C_0}{A_0} + \left(1 - \frac{A}{A_0}\right)\left(K_5 + K_2 K_3 + K_3 \frac{B_0}{A_0}\right)$$

$$+ K_2 K_3 \frac{A}{A_0} \ln \frac{A}{A_0} + \frac{E_0}{A_0}\left[1 - \left(\frac{A}{A_0}\right)^{K_6}\right] \qquad (2\text{-}139a)$$

and for D,

$$\frac{D}{A_0} = \frac{D_0}{A_0} + \left(1 + \frac{A}{A_0}\right)\left(K_2 K_4 + K_4 \frac{B_0}{A_0}\right) + K_2 K_4 \frac{A}{A_0} \ln \frac{A}{A_0} \qquad (2\text{-}140a)$$

As indeed they must, these general solutions reduce to various simple cases. For example, the simple simultaneous-reaction solution follows by setting k_3, k_4, and k_5 equal to 0, while the consecutive case results when k_2, k_3, and $k_5 = 0$. When k_2, k_3, and $k_4 = 0$, the parallel-reaction solution results.

2-10 AUTOCATALYTIC REACTIONS (HOMOGENEOUS)

An autocatalytic reaction is one in which a product C of the reaction catalyzes or promotes further reaction of reactants, $A + C \rightarrow 2C + P \cdots$

$$-\frac{dA}{dt} = kAC \qquad (2\text{-}141)$$

Let $M_0 = C_0 + A_0$. Now $C = C_0 + A_0 - A$, or $C = M_0 - A$, and so

$$-\frac{dA}{dt} = kA(M_0 - A) \qquad (2\text{-}142)$$

Equation (2-142) is of the integral form $\int \dfrac{dx}{x(ax + b)}$, where $b = M_0$ and $a = -1$, and so

$$\left. {}_{A_0} \!\!\left[\frac{1}{M_0} \ln \frac{A}{M_0 - A}\right] \right.^{A} = -kt \qquad (2\text{-}143)$$

$$\ln \left(\frac{A}{A_0}\frac{M_0 - A_0}{M_0 - A}\right) = -M_0 kt \qquad (2\text{-}144)$$

When we solve for A/A_0, the reactant concentration-vs.-time behavior is given by

$$\frac{A}{A_0} = \frac{(M_0/C_0) \exp(-M_0 kt)}{1 + (A_0/C_0) \exp(-M_0 kt)} \qquad (2\text{-}145)$$

In terms of reactant conversion $1 - A/A_0$, we find

$$x = 1 - \frac{A}{A_0} = \frac{1 - \exp(-M_0 kt)}{1 + (A_0/C_0) \exp(-M_0 kt)} \qquad (2\text{-}146)$$

Figure 2-5 illustrates typical conversion-time behavior for an autocatalytic reaction. Note the inflection point, characteristic of autocatalysis. The rate-vs.-concentration behavior is interesting: Eq. (2-142) states that

$$\text{Rate} = \frac{dA}{dt} = -kA(M_0 - A)$$

when $\qquad\qquad A = A_0 \ (t = 0) \qquad \left.\frac{dA}{dt}\right|_{t=0} = -kA_0 C_0 \qquad (2\text{-}147)$

since $M_0 = C_0 + A_0$. Now C_0, the initial catalyst concentration, will often be small since catalyst is created in situ due to reaction. The value of C_0 need only be sufficient to initiate reaction. In consequence, if $A_0 \gg C_0$, then $M_0 \to A_0$. In terms of Eq. (2-142),

$$-\frac{dA}{dt} \approx kA(A_0 - A) \qquad (2\text{-}148)$$

which indicates that dA/dt becomes a maximum at A/A_0 of $\frac{1}{2}$. In general, differentiation of (2-142) will precisely define the position of rate maximum with respect to A:

$$-\frac{d(\text{rate})}{dA} = k[(M_0 - A) - A] = 0$$

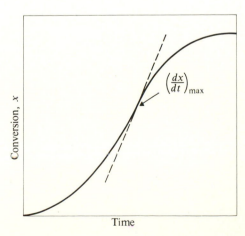

FIGURE 2-5
Characteristic conversion-vs.-time behavior
for a simple autocatalytic reaction.

or

$$\left(\frac{A}{A_0}\right)_{\text{max rate}} = \frac{M_0}{2A_0} \qquad (2\text{-}149)$$

The time at which this maximum rate occurs is obtained by substituting (2-149) into (2-144)

$$t_{\text{max rate}} = \frac{1}{M_0 k} \ln \frac{A_0}{M_0 - A_0} = \frac{1}{M_0 k} \ln \frac{A_0}{C_0} \qquad (2\text{-}150)$$

If, then, the time of rate maximum can be determined, k can be found as M_0 and A_0 are known. The rate-vs.-conversion $(1 - A/A_0)$ curve will be as shown in Fig. 2-6 since, by Eqs. (2-150) and (2-145), we find

$$x_{\text{max rate}} = \frac{1 - C_0/A_0}{2} \qquad (2\text{-}151)$$

From the conversion-vs.-time data (Fig. 2-5) t_{max} (at the inflection, maximum-rate, point) is obtained, and thence k is found by Eq. (2-150).

Note that in Fig. 2-6, for autocatalysis, where C_0 is not large compared with A_0, the rate *increases* with conversion up to a point governed by the C_0/A_0 ratio. This type of reaction system is termed *abnormal*, in contrast with normal reactions (in which the rate *decreases* with conversion, or extent of reaction under isothermal conditions). As C_0 becomes greater than A_0 in autocatalysis, the conversion level at which the rate becomes a maximum shifts to the left in Fig. 2-6, until at $C_0 \gg A_0$ autocatalytic character is masked by the zero-order behavior of C_0 and *normal* reaction-rate–conversion character emerges.

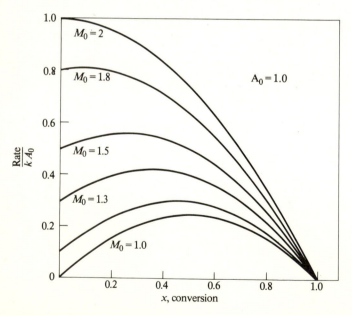

FIGURE 2-6
Autocatalytic rate-vs.-conversion behavior with the parameter $M_0 = (A_0 + C_0)/A_0$.

2-11 CHAIN REACTIONS

A chain reaction is essentially a consecutive one involving three stages:

Initiation: $A + A \longrightarrow M$ intermediate

Propagation: $M + A \longrightarrow B$ intermediate

Termination: $B + A \longrightarrow C$ stable product

> The *initiation reaction* produces the intermediate M, which then generates subsequent reactions.
>
> The *propagation reactions* are subsequent steps in which initiated intermediates M react to produce other intermediates B; in the process reactants may be consumed.
>
> *Termination reactions* are steps which cause annihilation of intermediates. Termination may result via intermediate consumption to form a stable product, or an active intermediate may become deactivated via collision with the reactor wall.

The classic example of a chain reaction is $H_2 + Br_2 \rightarrow 2HBr$. Bodenstein and Lind[1] studied the kinetics of this reaction between 200 and 300°C and found that the rate data were adequately defined by the expression

$$\frac{d(HBr)}{dt} = \frac{kH_2(Br_2)^{1/2}}{1 + k'HBr/Br_2} \qquad (2\text{-}152)$$

where $k' = \frac{1}{10}$ and is virtually temperature-independent and

$$k = \mathscr{A} \exp \frac{-40,200}{RT}$$

About a decade later other workers suggested the mechanism

Initiation: $Br_2 \xrightarrow{\ k_1\ } 2Br$

Propagation: $Br + H_2 \xrightarrow{\ k_2\ } HBr + H$

$\qquad\qquad\quad H + Br_2 \xrightarrow{\ k_3\ } HBr + Br$ $\qquad\qquad\qquad (2\text{-}153)$

$\qquad\qquad\quad H + HBr \xrightarrow{\ k_4\ } H_2 + Br$

Termination: $2Br \xrightarrow{\ k_5\ } Br_2$

The rate of HBr formation is[2]

$$\frac{d(HBr)}{dt} = k_2(Br)H_2 + k_3H(Br_2) - k_4H(HBr) \qquad (2\text{-}154a)$$

---

[1] M. Bodenstein and S. C. Lind, *Z. Phys. Chem.*, **57**: 168 (1907).
[2] Note that since each step is assumed to be elementary, the order and stoichiometry of each step are identical.

The rates of chain-carrier (H and Br) disappearance are, for Br

$$\frac{d(Br)}{dt} = 2k_1 Br_2 - k_2(Br)H_2 + k_3 H(Br_2) + k_4 H(HBr) - 2k_5(Br)^2 \quad (2\text{-}154b)$$

and for H

$$\frac{d(H)}{dt} = k_2 H_2(Br) - k_3 H(Br_2) - k_4 H(HBr) \quad (2\text{-}154c)$$

In principle, Eqs. (2-154a) to (2-154c) could be solved simultaneously. However, the solution can be simplified by the rather realistic assumption that the concentrations of chain carriers H and Br are very low and after a brief induction period during which these concentrations are developed, a steady state ensues, that is, $d(H)/dt$ and $d(Br)/dt = 0$. Therefore (2-154b) and (2-154c) equal zero. Thus from (2-154c)

$$k_2 H_2(Br) - k_4 H(HBr) = k_3 H(Br_2)$$

Equation (2-154a) then becomes

$$\frac{d(HBr)}{dt} = 2k_3 H(Br_2) \quad (2\text{-}155)$$

The concentration of H can be expressed by first adding (2-154b) and (2-154c), which gives

$$Br = \left(\frac{k_1}{k_5} Br_2\right)^{1/2}$$

Substituting for Br in (2-154c) and solving for H gives

$$H = \frac{k_2 H_2(Br)}{k_3 Br_2 + k_4(HBr)}$$

or

$$H = \frac{k_2 (k_1/k_5)^{1/2} H_2(Br_2)^{1/2}}{k_3 Br_2 + k_4(HBr)}$$

Substituting for H in (2-155) gives

$$\frac{d(HBr)}{dt} = \frac{2k_2 (k_1/k_5)^{1/2} H_2(Br_2)^{1/2}}{1 + (k_4/k_3)HBr/Br_2}$$

Comparison with the empirical rate equation (2-152) indicates

$$k = 2k_2 \left(\frac{k_1}{k_5}\right)^{1/2} \quad \text{and} \quad k' = \frac{k_4}{k_3} = \frac{1}{10}$$

The negligible temperature dependency of k' is thus rationalized, for if the activation energies E_4 and E_3 are nearly equal, the ratio of these rate constants will be virtually temperature-independent. The reader is referred to Frost and Pearson[1] for detailed comments on the H_2-Br_2 reaction.

[1] A. A. Frost and R. G. Pearson, "Kinetics and Mechanism," Wiley, New York, 1961.

2-12 GENERAL TREATMENT OF CHAIN REACTIONS[1]

Consider the general scheme of two carriers R and S and second-order termination:

Initiation: Reactants $\longrightarrow$ $mR + nS$ rate $= mr_1 + nr_1$

Propagation: $R\cdots \longrightarrow S\cdots$ rate $= r_{p_1} R$

         $S\cdots \longrightarrow R$ rate $= r_{p_2} S$

Termination: $R + R \longrightarrow \cdots$ rate $= r_t R^2$

R and S are chain carriers. Termination occurs by second-order kinetics. The rates of each step are mr_1, initiation; $r_{p_1} R$ and $r_{p_2} S$, propagation of each carrier, respectively; and $r_t R^2$, termination. The r's include rate constants and all concentration factors other than carrier concentration. Steady state requires for R

$$mr_1 - r_{p_1} R + r_{p_2} S - 2r_t R^2 = 0$$

and for S
$$nr_1 + r_{p_1} R - r_{p_2} S = 0$$

(2-156)

We can solve for R and S to obtain

$$R = \left[(m + n)\frac{r_1}{2r_t}\right]^{1/2} \qquad S = \frac{nr_1 + r_{p_1}[(m + n)r_1/2r_t]^{1/2}}{r_{p_2}} \qquad (2\text{-}157)$$

If the product species is formed during propagation, say $R \cdots \rightarrow S + P$, the rate of product formation is

$$\frac{dP}{dt} = r_{p_1}\left[(m + n)\frac{r_1}{2r}\right]^{1/2} \qquad (2\text{-}158)$$

If $S \cdots \rightarrow P + R$,

$$\frac{dP}{dt} = r_{p_1}\left[(m + n)\frac{r_1}{2r_t}\right]^{1/2} + nr_1 \qquad (2\text{-}159)$$

If, on the other hand, product results in the termination step,

$$\frac{dP}{dt} = r_t\left[\frac{(m + n)r_1}{2r_t}\right] \qquad (2\text{-}160)$$

In general, the propagation rate for αth-order termination is proportional to

$$r_p\left(\frac{r_1}{r_t}\right)^{1/\alpha} \qquad (2\text{-}161)$$

Chain length is defined as the propagation rate divided by the initiation rate, i.e.,

$$\ell = \frac{r_{p_1} R}{r_1(m + n)} \qquad (2\text{-}162)$$

[1] This treatment follows Frost and Pearson in part.

For two carriers and second-order termination $r_1(m + n) = 2r_t R^2$ [Eq. (2-156)], so that

$$\ell = \frac{r_{p_1}}{2r_t R} = \frac{r_{p_1}}{2r_t[(m + n)r_1/2r_t]^{1/2}}$$

or

$$\ell = \frac{r_{p_1}}{[2(m + n)r_1 r_t]^{1/2}} \qquad (2\text{-}163)$$

For a single carrier $n = 0$, while for first-order termination, where $n = 0$,

$$\ell = \frac{r_p}{r_t}$$

In general ℓ is proportional to

$$\frac{r_p}{r_1}\left(\frac{r_1}{r_t}\right)^{1/\alpha}$$

for α-order termination.

Activation Energy of Chain Reactions

In view of Eq. (2-161), applying the Arrhenius relation, we obtain the rate of propagation as proportional to

$$\exp\frac{-[E_p + (1/\alpha)(E_1 - E_t)]}{RT} \qquad (2\text{-}164)$$

Thus the apparent activation energy E_a is

$$E_a = E_p + \frac{1}{\alpha}(E_1 - E_t)$$

which shows that the experimental activation energy need not equal that of initiation E_1, as α may be 2 and $E_t > 0$.

Branching-Chain Reactions (Explosions)

If a chain carrier R generates two other chain carriers in the propagation process, we have

Initiation:	$\cdots$	$\longrightarrow$	mR	rate $= r_1 m$
Propagation and	R $+ \cdots$	$\longrightarrow$	R $+ \cdots$	rate $= r_p R$
branching	R $+ \cdots$	$\longrightarrow$	2R $+ \cdots$	rate $= r_b R$
Termination:	R $+$	$\longrightarrow$		rate $= r_t R$

In steady state

$$mr_1 + r_b R - r_t R = 0 \qquad (2\text{-}165)$$

Therefore

$$R = \frac{mr_1}{r_t - r_b}$$

The propagation rate is then

$$r_p R = \frac{m r_1 r_p}{r_t - r_b} \qquad (2\text{-}166)$$

Note that the branching rate r_b is subtracted from that of termination r_t. Branching, as Frost and Pearson note, is a sort of negative termination.

   If now conditions of temperature, pressure, and composition are such that branching and termination rates become nearly equal, $r_t - r_b \to 0$ and the propagation rate approaches infinity. This is an explosion. If termination is due to wall collision, pressure and surface area per unit volume can be manipulated to enhance r_t and thus change the explosive limit.

2-13 POLYMERIZATION

The kinetics of polymerization is complex; however, some light can be shed on the primitive aspects of the process by noting the similarity between chain and polymerization reactions.   For a monomer M we have consecutive creation of polymer P_r of varying extent of polymerization:

$$
\begin{array}{llll}
\text{Initiation:} & M & \longrightarrow & P_1 \\
\text{Propagation:} & M + P_1 & \longrightarrow & P_2 \\
 & M + P_2 & \longrightarrow & P_3 \\
 & M + P_r & \longrightarrow & P_{r+1} \\
\text{Termination:} & P_{r+1} + P_r & \longrightarrow & \text{product}
\end{array}
\qquad (2\text{-}167)
$$

Clearly the above reaction sequence does not correspond to a true chain reaction in that the original chain carrier P_1 is not regenerated.   Essentially we are identifying P_1 with $P_r \cdots$ and may then treat the system as we have true chain reactions. Applying Eq. (2-161), where

$$r_p = k_p M \qquad r_1 = k_1 \text{ (initiator)} \qquad r_t = k_t$$

the polymerization rate becomes proportional to

$$k_p \left(\frac{k_1}{k_t}\right)^{1/2} \text{(initiator)}^{1/2} M \qquad (2\text{-}168)$$

a rate law commonly found in polymerization studies.   For example, the radical polymerization of styrene (initiated by azoisobutyronitrile) conforms to the above expression.   For more sophisticated treatments of polymerization kinetics we suggest a study of the work of Zeman and Amundson.[1]

[1] R. Zeman and N. R. Amundson, *AIChE J.*, **9:** 208 (1963).

2-14 DATA PROCUREMENT

Broadly speaking, laboratory inquiries might provide us with data of (1) reactant or product concentration vs. real or contact time or (2) reaction rate vs. reactant-product concentration, and (3) either or both types of information as a function of temperature and/or catalyst concentration. Raw data obtained in batch or PFR experiments (x versus t or θ) are usually in *integral* form. The kinetic model or rate equation is developed from x-versus-t or θ data by (1) fitting the integrated form of one of the candidate rate equations to the raw data or (2) differentiating the raw x-versus-time data to secure rate-vs.-concentration data. The differentiation may be performed graphically or analytically.

Raw data obtained in differential or well-mixed flow-reactor experiments are obtained by operating a PFR at small conversion levels, in which case

$$\text{Rate} = \frac{1}{V}\frac{dn}{d\theta} = \frac{1}{V}\overline{\frac{n_0 - n}{\theta}} \qquad \text{as } \overline{n_0 - n} \to 0$$

In a CSTR

$$\text{Rate} = \frac{C_{in} - C_{out}}{\theta}$$

Table 2-1   REACTION-KINETICS DATA PROCUREMENT

c = concentration, t = time, θ = holding time

Reactor type	Data procured	Treatment of data	Comment
Batch	c versus t	Fit c-versus-t data to integrated model or differentiate c-versus-t data	Simple but limited to slow reactions; temperature control a problem
Semibatch	c versus t at fixed rate of coreactant feed	Fit c-versus-t data to integrated model or differentiate c-versus-t data	Rate and temperature controlled by addition rate; mathematical analysis complex for nonlinear kinetics
Plug flow, integral	c versus θ	Fit c-versus-θ data to integrated model or differentiate c-versus-θ data	Useful for homo- and heterogeneous reactions but often non-isothermal
Differential	Rate versus c	Material balance gives rate; log-log plot of rate versus c gives order and k	Small conversion can cause analytical problems and loss of precision
Perfectly mixed flow (CSTR)	Rate versus c	Material balance gives rate; log-log plot of rate versus c gives order and k	Large conversions give greater precision; useful for homogeneous reactions; catalytic CSTRs have been designed

One notes that point rates are neatly obtained in a CSTR or open-loop recycle reactor; in either case a simple material balance establishes the rate.[1]

Table 2-1 summarizes the chemical-kinetic-data procurement picture.[2]

The Wei-Prater Analysis

A signal advance in the resolution and analysis of complex-reaction kinetic networks for both homogeneous and heterogeneous reactions was set forth by Wei and Prater.[3] In essence their analysis permits one to uncouple complex reaction networks, thus permitting the analyst to extract rate coefficients and to specify the reaction paths. Details of the rationale as well as examples are presented in the Wei and Prater classic exposition, and further discussion is provided in Boudart's text (see the Additional References).

2-15 ANALYSIS OF ERRORS IN KINETIC DATA

Some brief remarks on errors and their magnitude associated with reduction of rate data are in order. The fractional error in a dependent variable d as a function of independent variables v_i is given by

$$\left(\frac{\Delta d}{d}\right)^2 = \sum_{i=1}^{n} \left(\frac{\partial \ln d}{\partial \ln v_i}\right)^2 \left(\frac{\Delta v_i}{v_i}\right)^2 \qquad (2\text{-}169)$$

where $\Delta v_i / v_i$ is the fractional random error in v_i. Let it be supposed that reaction-rate data are being secured in a continuously fed, well-stirred reactor (CSTR), so that

$$r = \text{rate} = -\frac{dC}{d\theta} \equiv \frac{C_0 - C}{V/Q} = \frac{Fy_i}{V} \qquad \text{where } F = QC$$

where F = molar flow rate

$\quad y_i$ = mole fraction of product in effluent

$\quad V$ = reaction volume

We assume there is no product in the feed. Applying Eq. (2-169) for the independent variables F, y, and V, we obtain

$$\frac{\Delta r}{r} = \left[\left(\frac{\Delta F}{F}\right)^2 + \left(\frac{\Delta y_i}{y}\right)^2 + \left(\frac{\Delta V}{V}\right)^2\right]^{1/2} \qquad (2\text{-}170)$$

and so the squares of the individual errors are additive. Of practical concern is the question: How accurate must one's measurements of species concentration, flow

---

[1] L. P. Hammett, *J. Am. Chem. Soc.*, **70:** 3444 (1948), **78:** 521 (1956), **72:** 280, 283, 287 (1950), **80:** 2415 (1958).

K. G. Denbigh, *Trans. Faraday Soc.*, **40:** 352 (1944), **44:** 479 (1948).

[2] The treatment of data using sophisticated statistical techniques is set forth in D. L. Marquardt, *J. Soc. Ind. Appl. Math.*, **2:** 431 (1963); L. Lapidus and T. I. Peterson, *Chem. Eng. Sci.*, **21:** 655 (1966); R. Mezaki and J. R. Kittrell, *AIChE. J.*, **13:** 176 (1967); R. Bellman et al., Quasilinearization and Estimation of Chemical Rate Constants from Kinetic Data, *Rand Corp. Mem.* RM-4721-NIH, August 1965.

[3] James Wei and C. D. Prater, *Adv. Catal.*, **13:** 203 (1962).

rate, etc., be for the measured rate to be within a desired range of certainty? If, for example, we want $\Delta r/r$ to be 0.05 and our reaction-volume and flow-rate errors are no greater than 0.5 and 3 percent, respectively, how great a relative error in in product mole fraction can be tolerated? By Eq. (2-170)

$$\frac{\Delta y}{y} = [(0.05)^2 - (0.005)^2 - (0.03)^2]^{1/2}$$

Then $\Delta y/y = 0.04$.

Our chemical analysis must therefore be accurate within 4 percent or less if the derived rate is to be in error no more than 5 percent. Note that if absolute errors ΔF, ΔV, and Δy are fixed, the relative error in rate is reduced by operation characterized by large values of F, V, and y. The disadvantage of the differential reactor becomes obvious: Δy is, by definition, small in differential-reactor operation. CSTR operation is to be preferred since a large product mole fraction can be tolerated without sacrificing the merits of point-rate determination.

It should be obvious that for the rate coefficient k, its determination and precision will critically depend upon relative errors in compositions and rate, per se. As Benson notes, when k is computed from concentrations (or partial pressures) measured at two time intervals, the error in k is reduced when relative time and concentration intervals are large. This point is nicely illustrated with respect to errors in k and activation-energy determination from the $(\ln k)$-versus-$1/T$ correlation. From the Arrhenius relation

$$d \ln k = \frac{E}{RT^2} dT$$

or
$$\frac{dk}{k} = \frac{E}{RT} \frac{dT}{T} \qquad (2\text{-}171)$$

So the expected error in T (in kelvins) produces a consequent error in k, the magnitude of which depends upon activation energy E. Detailed analyses reveal that the error in E increases as the temperature interval over which k is measured is reduced.

Summary

Benson[1] provides a convenient summary to guide those concerned with the procurement of precise rate parameters. To obtain the rate coefficient with an estimated error of $\pm\epsilon$, it is necessary:

1 To measure concentrations with an accuracy of

$$\pm \frac{\text{Change in concentration}}{\text{Largest concentration}} \frac{\epsilon}{1.4} \%$$

---

[1] " Foundations of Chemical Kinetics."

2 To measure time (or holding time) with an accuracy of

$$\pm \frac{\text{Time interval}}{\text{Largest time}} \frac{\epsilon}{1.4} \%$$

3 To measure temperature with an accuracy of

$$\pm \frac{\epsilon}{35} \% = \pm \frac{\epsilon T}{3500} \quad K$$

The consequent accuracy of E for two measurements of k, each of accuracy $\pm \epsilon$ percent, will be

$$\overline{\Delta^2 E^{1/2}} = \pm \frac{\text{highest temperature}}{\text{temperature interval}} \frac{\epsilon}{18}$$

Benson declares these to be rough but reliable rules, accurate for reactions where E/RT is about 35.

We conclude this chapter with a simple illustration of kinetic-data treatment involving two methods of data procurement and diverse modes of data reduction to obtain a chemical-reaction-rate model.

Example

Let it be supposed that we are to develop a kinetic model for the liquid-phase esterification of teraphthalic acid (TPA) with ethylene glycol (EG) as catalyzed by an acid, the exact nature of which is known only to the legal department of our institution.   The product is a building-block monomer for a synthetic-fiber polymer of incredible virtues, not the least of which is its selling price (of little interest to our scientists, of prudent interest to our engineers, and of consuming interest to our stockholders).   The overall reaction can be stated as

$$\text{TPA} + \text{EG} \quad \xrightarrow{\ 1\ } \quad \text{M} \quad \xrightarrow{\ 2\ } \quad \text{P} \quad \xrightarrow{\ 3\ } \quad \$$$

Step 3 is instantaneous according to our MBA wizards, while step 2 shall be the concern of the Islandic Division of our firm.   So alerted, we may treat the reaction as

$$\text{TPA} + \text{EG} \quad \xrightarrow{\ k\ } \quad \text{M} \cdots$$

The TPA, ordinarily a solid, is dissolved in a powerful solvent and is to be reacted in excess EG.   The chemists in residence have TPA concentration-vs.-time data from their favorite laboratory reactor, a batch kettle.   The modernists (chemical engineers) have provided data obtained in a laboratory CSTR.   Earlier tests with the CSTR prove that effluent concentration does indeed equal that uniform concentration throughout the baffled CSTR at a speed above 2000 r/min.   Both sets of data were obtained at 120°C.

These data, expressed in IOUs (Intentionally Obscure Units) are as follows:

	Batch data		CSTR data		
Run	TPA concentration A	Time	Feed concentration A_0	Effluent concentration A	Holding time θ
1	9	0			
	7	1			
	5	3			
	3	6			
	0	16			
2			10	9	0.57
3			10	4	10
4			14	9	3.5
5			14	8	4.8
6			14	2	36
7			7	1	40

Data reduction   We now must treat these data to secure a reaction-rate model; i.e., the order with respect to TPA and the rate coefficient k at 120°C are to be extracted.

Batch-data treatments   The rate-concentration dependency is obtained from these integral data by determining the rate from the A-versus-t data.   This is done by (1) differentiating the A-versus-t curve at various values of A or (2) assuming an

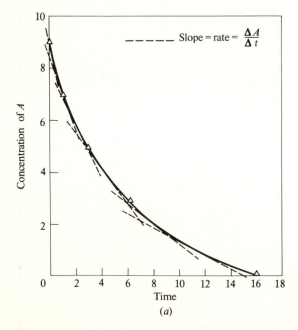

FIGURE 2-7a
Batch-concentration-vs.-time data.

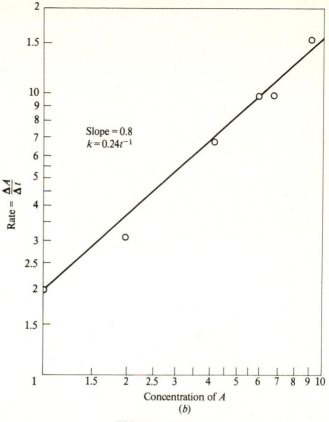

FIGURE 2-7*b*
Log-log plot of rate vs. concentration of A.

order and comparing the A-versus-t data with the integrated form of the appropriate rate equation.

Differentiation of concentration-vs.-time data Differentiation of the A-versus-t data can be carried out analytically if one first fits these data with a polynomial function, or (with considerable saving of time) the A-versus-t curve can be differentiated graphically. The raw data, plotted in Fig. 2-7*a*, are graphically differentiated to yield

Rate $\Delta A/\Delta t$	1.6	1.1	1	0.67	0.29	0.2
A	9	7	6	4	2	1

In Fig. 2-7*b* these data are plotted on log-log coordinates. A least-squares treatment establishes a straight line through these derived data. Since the slope (order) is found to be 0.8,

$$\text{Rate} = k(A^{0.8}) \qquad (a)$$

and the coefficient $k = 0.24t^{-1}$.

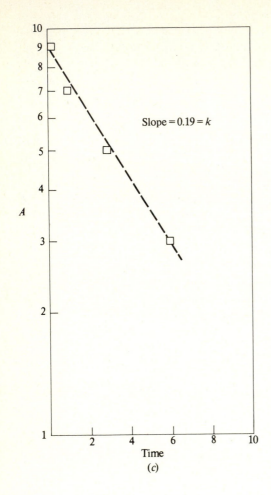

Slope $= 0.19 = k$

A

Time

(c)

FIGURE 2-7c
Integral first-order semilog of Av vs. time.

Integral treatment of A-versus-t data If the reaction is first order,

$$\frac{A}{A_0} = \exp\left(-kt\right) \qquad (b)$$

The reduced concentration data are plotted in semilog coordinates in Fig. 2-7c. A straight-line fit passing through A/A_0 of unity is shown. The slope provides the rate coefficient k. For the *assumed* first-order model $k = 0.19$. In view of the findings using the differentiation technique we could treat the A-versus-t data in accord with 0.8-order kinetics, i.e., Eq. (2-70). For 0.8 order

$$A^{0.2} - A_0^{0.2} = -0.2kt$$

or

$$\frac{[(A/A_0)^{0.2} - 1]A_0^{0.2}}{-0.2t} = k_{0.8} \qquad (c)$$

   In this case we simply calculate k for the available values of A/A_0 and t. If the order is indeed 0.8, a constant value of k should be found.   The same could have been done using Eq. (b), of course.   By Eq. (c)

A/A_0	t	$k_{0.8}$	k_1
0.775	1	0.374	0.253
0.557	3	0.32	0.196
0.333	6	0.253	0.182

The k values reveal a trend inconsistent with the notion of a rate constant.   Point values of k assuming linear kinetics [Eq. (b)] can be computed and are shown in the above tabulation.   Again inconstancy is evident.

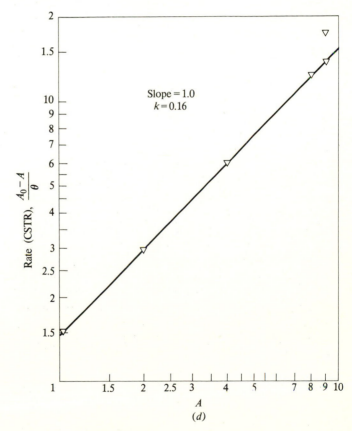

FIGURE 2-7d
Log-log plot of rate vs. concentration for CSTR data.

CSTR data treatment   A simple material balance about the CSTR establishes the point rate at the effluent (and therefore reactor) concentration

$$\text{Rate} = \frac{A_0 - A}{\theta} = k_c A^\alpha$$

The rate data are

Rate $\frac{A_0 - A}{\theta}$	A	$k_c\ (\alpha = 1)$
1.75	9	0.194
0.6	4	0.15
1.4	9	0.155
1.25	8	0.156
0.3	2	0.15
0.15	1	0.15

The data are plotted on log-log coordinates in Fig. 2-7d.   A linear fit indicates a slope of unity and $k_c = 0.16$.   The consistency of k_c is checked by a point computation of k_c assuming linear kinetics, since

$$\text{Rate} = \frac{A_0 - A}{\theta} = k_c A$$

Point values of k_c are tabulated above, from which it is apparent that (1) direct measurement of reaction rate at a uniform point concentration of reactant yields more consistent rate data than extracting the rates by differentiation of conversion-vs.-time data or by fitting such data to an integrated equation for an assumed order and (2) referring to CSTR runs 2 and 4, CSTR differential conversion data (run 2) are of intrinsically poorer reliability than integral data secured in a CSTR or its equivalent.

ADDITIONAL REFERENCES

The text of M. Boudart, "Kinetics of Chemical Processes," Prentice-Hall, Englewood Cliffs, N.J., 1968, provides unique insights into the diverse aspects of the topic, while an invaluable reference pertaining to the procurement of rate data and their interpretation is to be found in S. L. Friess, E. S. Lewis, and A. Weissberger (eds.), "Techniques of Organic Chemistry," vol. VIII, pt. I, Interscience, New York, 1961. The excellent text W. Moore, "Physical Chemistry," Prentice-Hall, Englewood Cliffs, N.J., 1962, is recommended as a primer in that key subject.   K. Laidler, "Chemical Kinetics," McGraw-Hill, New York, 1965, is also a valuable reference.

As Boudart notes, the first chapter of S. Glasstone, K. Laidler, and H. Eyring, "The Theory of Rate Processes," McGraw-Hill, New York, 1941, constitutes a fine exposition of transition-state theory.   Classic insights into chemical kinetics are

set forth in Boudart's translation of N. N. Semenov, " Problems in Chemical Kinetics and Reactivity," vols. I and II, Princeton University Press, Princeton, N.J.,
1958.
In anticipation of topics to be subsequently treated in this text, the student will find invaluable the classic D. A. Frank-Kamenetskii, " Diffusion and Heat Exchange in Chemical Kinetics," trans. J. P. Appleton, Plenum, New York, 1969.

PROBLEMS

2-1 NH_3 is synthesized from N_2 and H_2 over a promoted iron catalyst at 450°C and 200
atm.   The reaction is $N_2 + 3H_2 \rightleftharpoons 2NH_3$.
   (*a*)   What is the effect of increasing total pressure upon the equilibrium yield of
      NH_3?   Upon the rate of reaction?
   (*b*)   Discuss the consequences of operation at 550°C upon reaction rate and equilib-
      rium yield.
   (*c*)   What is the reaction order with respect to each species?

2-2 SO_2 is catalytically oxidized to SO_3 over a supported Pt catalyst or a supported
V_2O_5 catalyst.   At a total pressure of 1 atm how is the equilibrium yield of SO_3
affected by an increase in temperature for each catalyst?

$$ SO_2 + \tfrac{1}{2}O_2 \quad \rightleftharpoons \quad SO_3 $$

2-3 How does the rate coefficient of a chemical reaction change with a 10°C increase in
temperature if the base temperature is 100°C and (*a*) $E = 10$ kcal, (*b*) 20 kcal, (*c*) 30
kcal?   Repeat for a base temperature of 500°C.

2-4 An irreversible dimerization $2M \rightarrow D$ is studied in an isothermal batch reactor at
130°C.   Initial concentration of M is 1 g mol/liter, and after 1 h its value is 0.1 g
mol/liter.   Estimate its concentration after 30 min of reaction if the reaction is (*a*)
first order, (*b*) second order.   The reaction occurs in the liquid phase.

2-5 From the following data, determine the activation energy for the formation of
methyl ethyl ether in alcoholic solution

T, °C	0	6	18	30
$k \times 10^5$ liters/(g mol)(s)	5.6	11.8	48.8	208

2-6 Acetic acid is employed as a solvent in the homogeneous catalyzed oxidation of
toluene to its acid.   At 200°C the acetic acid concentration falls from 1 to 0.64
mol/liter in 10 min and to 0.445 mol/liter in 20 min.   What is the order of the
acetic acid decomposition reaction?

2-7 Toluic acid is esterified with methanol in the liquid phase with H_2SO_4 as catalyst.
At 80°C the reaction is reversible, and half-life is independent of initial concentrations in both directions.   In excess methanol and initial acid concentration of 1 g
mol/liter it requires 10 min for the acid concentration to drop to 0.75 mol/liter and
another 10 min for its value to reach 0.6 mol/liter.   Estimate the value of the
equilibrium constant at 80°C.

2-8 What are the units of an αth-order rate coefficient (*a*) in terms of concentration and
(*b*) in terms of partial pressure?

2-9   Leyes and Othmer [*Ind. Eng. Chem.*, **37**: 968 (1946)] present the following batch homogeneous data for the esterification reaction of butanol (B) and acetic acid (A).

Temperature = 100°C   Moles $A_0 = 0.2332$ Moles $B_0 = 1.16$

Time, h	A converted, mol	Time, h	A converted, mol	Time, h	A converted, mol
0	0	3	0.03662	6	0.06086
1	0.01636	4	0.04525	7	0.06833
2	0.02732	5	0.05405	8	0.07398

Determine a suitable rate equation and establish the value of the constants.

2-10   The catalytic chlorination of an aromatic (A) by Cl_2 (B) yields the monochloride (C) and degradation by-product (D): $A + B \xrightarrow{k_1} C + D$.   Further chlorination produces the dichloride (E) and D: $C + B \xrightarrow{k_2} E + D$.   Assuming first-order reactions, derive an expression from which k_2/k_1 can be ascertained from experimental values of C and E versus A.   Initial values of C and E are zero.

2-11   Given the following data on the decomposition of furfural in 1% H_2SO_4 at 160°C find a reasonable interpretation of these data and estimate the extent of reaction at 3½ h.

Time, h	Furfural remaining, g/100 ml
0	2
0.5	1.95
1	1.87
1.5	1.79
2	1.74
2.5	1.69

Comment critically upon this experiment.

2-12   A dimer (B) is catalytically produced by second-order reaction of A, and the dimer reacts further to yield an isomer (C): $2A \xrightarrow{k_1} B \xrightarrow{k_2} C$.   Assuming order and stoichiometry conform, determine the values of k_1 and k_2 from the following data:

t_1,h	Liquid-phase concentration, mol/liter		
	A	B	C
0	1	0	0
0.03	0.76	0.098	0.02
0.06	0.63	0.132	0.06
0.1	0.51	0.14	0.1
0.15	0.39	0.12	0.17
0.2	0.33	0.10	0.24
0.3	0.25	0.05	0.32

2-13   Troupe and Kobe [*Ind. Eng. Chem.*, **42**: 801 (1950)] report data on the esterification of 85% lactic acid (*A*) with methanol (M) and interpret their results by a second-order irreversible kinetic model.   However, the two sets of data set forth below do not conform to the proposed model   Propose an alternative interpretation.

Run	Time, h	Acid converted, mol/100 g	
1	0	0	Run 1 $T = 100°$; $0.1\%H_2SO_4$;
	0.25	0.216	$A_0 = 0.745/100g$ mixture
	0.5	0.313	$M_0 = 0.621/100g$ mixture
	1	0.384	
	2	0.422	
	4	0.42	
2†	0	0	
	0.25	0.219	
	0.50	0.308	
	1	0.378	
	2	0.415	
	3	0.421	
	4	0.427	
	6	0.431	
	8	0.436	

† Same as run 1 but $A_0 = 0.595$ and $M_0 = 1.073$.

2-14   An olefin (B) produced catalytically by dehydrogenation of a saturated hydrocarbon (A) undergoes isomerization to produce by product (C).   From the following data determine the appropriate rate coefficients, assuming linear kinetics.

	Concentration, mol/liter		
Time, min	A	B	C
0	100	0	0
0.2	90.5	8.5	1.0
0.4	81.9	14.6	3.5
0.6	74.1	19.3	6.6
0.8	67.0	23.0	10.0
1	60.7	26.0	13.3
1.5	47.2	31.8	21.0
2.0	36.8	35.9	27.3
2.5	28.7	39.7	31.6
3.0	22.3	41.9	35.8
4.0	13.5	44.8	41.7
5.0	8.2	46.9	44.9
6.0	5.0	48.1	46.9
7.0	3.0	48.6	48.4
8.0	1.8	49.3	49.1
10.0	0.7	49.7	49.6

2-15   Normal butanol is an effective solvent in the homogeneous catalyzed oxidation of *p*-xylene; however, thermal decomposition of the alcohol is suspected, and laboratory CSTR data are taken to ascertain the severity and kinetics of the alcohol's destruction.   In a well-stirred reactor of 1 liter reactant-phase capacity, the liquid reactant is fed and product removed at constant rate.   Conversion is measured at steady state for three feed rates with the following results for feed of 100% alcohol:

Feed rate, ml/min	100	200	400
Alcohol in effluent, mole %	20	40	60

Density is constant at 1 g/ml.   Determine reaction order and the rate coefficient.

2-16 The hydrolysis of methyl acetate is catalyzed by reaction product acetic acid. When initial acetate concentration is 0.5 and the acid 0.05 mol/liter, 60 percent conversion of acetate occurs in 1 h. At what time does the rate become a maximum? What is its value and the conversion at maximum rate?

3

BEHAVIOR OF CHEMICAL REACTORS

" For he, by geometric scale, could take the size of pots of ale."
S. Butler " Hudibras "

Introduction

Chapter 2 set forth directly and by referral the principles and techniques whereby a rate model of a chemical reaction network can be constructed from laboratory data. It was emphasized that experiments which provide the raw data for subsequent kinetic analysis should be characterized by a well-defined concentration–residence-time–temperature environment.

Assuming that the prerequisites for valid laboratory kinetic experimentation have been realized, we now consider the behavior of reaction networks in environments which depart from those usually governing laboratory inquiries. To focus attention upon the principles involved, limiting reactor types and environments will be considered in this chapter, thus permitting some generalizations which will prove fruitful in analyses of complex reactor or reaction networks.

Limiting reactor types considered are the continuously fed stirred-tank reactor (CSTR) and the plug-flow reactor (PFR), both under isothermal conditions. Simple intermediate mixing levels between the two will also be explored. The influence of *isothermal* backmixing upon conversion, yield, and selectivity will be demonstrated. The influence of radial gradients will be set forth by a consideration of the laminar-flow reactor (LFR).

For limiting reactor types, limiting temperature environments will then be discussed, namely, isothermal, implicitly treated, and then its extreme opposite, adiabatic operation. The behavior indexes are, of course, conversion, yield, and selectivity.

While the concern of Chap. 2 was with reaction per se, in this chapter the reactor (type and environment) is of prime concern. A one-dimensional model is largely invoked at this juncture, i.e., concentration, temperature distributions, and their alteration are considered *only* in the direction of flow. For we are here concerned with the influence upon reactor behavior of backmixing of both heat and mass, and to facilitate isolation of these signal effects, gradients perpendicular to flow are assumed to be negligible, except in our brief treatment of laminar flow, which of course involves radial gradients.

3-1 ISOTHERMAL BACKMIXED AND PLUG-FLOW REACTORS

It was noted in Chap. 2 that the extent of isothermal reaction depends upon the magnitude of the rate constant, concentration, its initial value (for non-first-order kinetics), and contact or residence time. For a fixed temperature and first-order reaction, for example, conversion is accurately predicted (k is known), or k can be determined for a known conversion level *if* residence or contact time is known with precision. In a batch system, holding time can be rather accurately determined *if* provision for rapid heat-up and quenching is realized. Suppose, however, that a portion of the reactant fluid is held under reaction conditions for a time less than t and that other portions are held longer than t. When k is known for a first-order reaction, it is *not* possible to predict conversion from the simple relation

$$x = 1 - \frac{C}{C_0} = 1 - \exp(-kt)$$

If the contact- or residence-time distribution is precisely known, prediction may be possible. The residence-time function is simply a statement of the fraction of molecules in a system found to have a contact or residence time in an apparatus between time t and $t + dt$. Let us consider two extreme cases: (1) all molecules enjoy the same residence time; (2) an exponential distribution prevails.

Case 1 corresponds to piston or plug flow in a continuously fed PFR, while case 2 corresponds to a perfectly backmixed flow reactor (CSTR). Therefore, in the PFR we have zero backmixing, while in the CSTR there is infinite (perfect) backmixing.

Suppose we add to a feed stream in steady flow a pulse of inert tracer and monitor the effluent for both the PFR and CSTR. Figure 3-1a illustrates these two cases.

In case 1, there being no backmixing, all tracer molecules emerge at $t = \theta$, the average residence time for the reactor, equal to V/Q. In case 2, since there is infinite backmixing, a fraction of tracer emerges immediately, the balance appearing at various times greater and less than θ. Between these two extreme distributions a number of intermediate distributions exist. The general instance

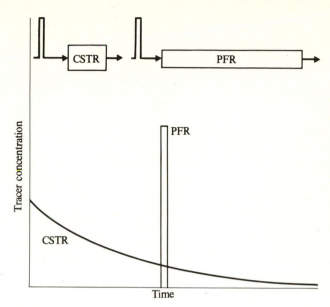

FIGURE 3-1*a*
Response of limiting reactor types (PFR and CSTR) to an inert-tracer pulse input.

will be dealt with later. Here we shall limit ourselves to the two ideals, the PFR and the CSTR, both in steady state.

Definitions

The PFR is one in which all entering molecules enjoy the same residence or contact time, which is the average determined by reactor volume divided by volumetric feed rate evaluated at reactor temperature and pressure.

The CSTR is one in which there is an exponential distribution of residence times. An operational definition of a CSTR is that the steady-state effluent concentration is equal to the uniform concentration within the vessel at any time.

Levenspiel[1] employs a useful illustration to demonstrate the difference in performance between plug-flow devices (Fig. 3-1*b*) and perfectly mixed flow devices (Fig. 3-1*c*). He considers a tubular heat exchanger (PF) and a perfectly mixed (stirred-tank) unit (CST). For identical flow rates, the driving force in case 1, is the log mean, $(\Delta T)_{lm} = 21.4°C$, while in case 2 since effluent temperature equals the average value in the perfectly mixed vessel, the driving force is 1°C, that is, $\Delta T = 1 - 0$. In consequence for the same heat-removal duty, the perfectly mixed vessel requires $21.4/1 = 21.4$ times the heat-transfer capacity of the plug-flow unit.

[1] O. Levenspiel, seminar, University of Notre Dame, April 1965.

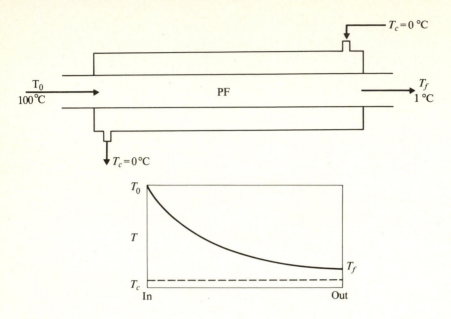

FIGURE 3-1b
Plug-flow heat exchanger.

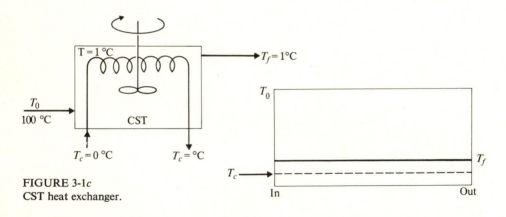

FIGURE 3-1c
CST heat exchanger.

In heat exchange this means that $(UA)_{\text{CST}} = 21.4(UA)_{\text{PF}}$. Now we can extend Levenspiel's demonstration. Given the analogy

$$\text{Heat transfer} = \frac{\text{calories}}{\text{time}} = \frac{dq}{d\theta} = UA\,\Delta T$$

$$\text{Chemical reaction} = \frac{\text{moles}}{\text{time}} = \frac{dn}{d\theta} = kVC^{\alpha}$$

the effects of mixing upon normal chemical-reaction rate became apparent. Note, however, that in the heat-exchange example, the product UA (coefficient times area) is involved, so that two alternatives remain to compensate for perfect mixing (reduced driving force): (1) increase A and/or (2) increase U. By analogy then, in a reactor network, volume V and/or the coefficient k can be increased to compensate for backmixing. In the heat-transfer parable, it is unlikely that U can be significantly increased in the CST exchanger; however, when we recall the principles of Chap. 2, it is clear that since the chemical rate coefficient k depends exponentially upon temperature, it can indeed be readily increased, possibly to assume a magnitude which more than compensates for the reduced reaction driving force imposed by backmixing. Furthermore, while the heat-transfer rate is linear in ΔT (log mean or otherwise), the reaction rate, moles per time, is not necessarily linear in species concentration ($\alpha \geqslant 1.0$). Therefore for a given reactor type, mixing effects in a chemical reactor will depend upon reaction order and the temperature field. The integrated result, i.e., conversion, yield, or selectivity, will then be a function of (1) temperature, (2) concentration dependencies and distribution, and (3) contact- or residence-time distribution.

3-2 THE CSTR AND PFR RESIDENCE-TIME DISTRIBUTION

Step Input

We consider a well-stirred vessel initially free of nonreacting tracer operating in steady state. At $t = 0$ a step input of inert tracer of concentration C_0 is imposed on the system. A material balance yields

$$QC_0 = QC + V\frac{dC}{dt} \qquad (3\text{-}1)$$

Note that the CSTR condition implies that effluent concentration C is equal to the average value within the vessel. The boundary condition for a step input is that at $t = 0$ effluent tracer concentration is 0; that is, at $t = 0$, $C = 0$. Let $\theta = V/Q$; then

$$\int_0^C \frac{dC}{C_0 - C} = \frac{1}{\theta}\int_0^t dt \qquad (3\text{-}2)$$

or

$$\frac{C}{C_0} = 1 - \exp\left(-\frac{t}{\theta}\right) \qquad (3\text{-}3)$$

as shown in Fig. 3-1d. By definition the PFR response is identical to the input signal (Fig. 3-1e). In general, a number of intermediate cases can be found (Fig. 3-1f).

We must now show how the C/C_0-versus-time curve for a step input is related to the residence-time distribution function $J(t)$. The assumption of perfect mixing need not be retained; therefore, consider the step response curve in Fig. 3-2.[1]

At any time t after tracer of concentration C_0 is introduced in the feed, let the concentration in the effluent be C. Divide (segregate) this effluent stream into

---

[1] This demonstration is due to J. M. Smith.

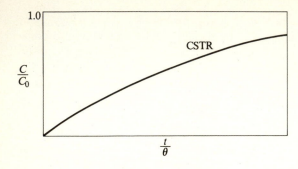

FIGURE 3-1d
CSTR response to a step input.

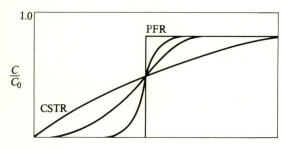

FIGURE 3-1e
PFR response to a step input.

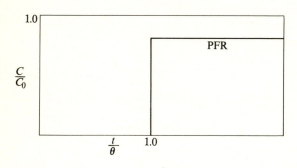

FIGURE 3-1f
Response to step input of PFR, CSTR, and reactors of intermediate mixing levels.

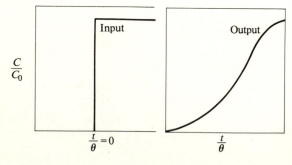

FIGURE 3-2
General response of a reactor to step input of tracer.

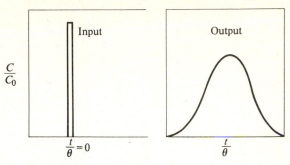

FIGURE 3-3
General response of a reactor to pulse input of tracer.

two volume fractions; the first consists of fluid which has been in the reactor for a time less than t. This defines $J(t)$. The second fraction consists of fluid which has a residence time in the vessel greater than t, and this fraction equals $1 - J(t)$. The fraction $1 - J(t)$ might have no tracer in it, while $J(t)$ will have a concentration C_0. The mass rate of tracer flow in the effluent stream will be

$$Q[1 - J(t)]0 + Q[J(t)]C_0 = Q[J(t)]C_0 = QC$$

Therefore
$$J(t) = \frac{C}{C_0} \qquad (3\text{-}4)$$

In consequence the measured response to a step (C/C_0 vs. t/θ) is identical to a residence-time-distribution function.

Pulse-Function Input

Suppose an instantaneous pulse, or spike, of tracer is imposed on the system (Fig. 3-1). Actually our pulse would approximate a square wave of concentration C_0 in a time interval Δt. As shown in Fig. 3-3, the effluent would show a maximum; i.e., a smeared pulse would emerge.

When the effluent is analyzed during a time interval Δt, the mass of tracer is $QC\,\Delta t$. As $J(t)$ is the fraction of effluent with a residence time less than t, ΔJ is the fraction in the residence-time increment Δt. Again assuming our incremental sample to consist of a fraction at C_0 and another of zero concentration, the mass of tracer in the effluent in time interval Δt is given in terms of C_0 by

$$QC_0\,\Delta J\,\Delta t = QC\,\Delta t \qquad (3\text{-}5)$$

or
$$\frac{C}{C_0\,\Delta t} = \frac{\Delta J}{\Delta t} = \frac{dJ}{dt}\bigg|_{\Delta t \to 0}$$

Now the total mass introduced in the pulse is $m_p = QC_0\,\Delta t$; thus

$$\frac{CQ}{m_p} = \frac{dJ}{dt} \qquad (3\text{-}6)$$

The pulse-response curve is therefore equal to the derivative of the step-response curve. (A true mathematical pulse is injected in a time interval of $\Delta t \to 0$ and since $C_0 \Delta t$ is finite, C_0 must approach infinity as Δt approaches zero.)

Figure 3-3 now becomes physically meaningful. The pulse data for output C/C_0 vs. time tell us that the largest fraction of emerging molecules has a residence time equal to t/θ at the C/C_0 maximum, a few molecules have small residence times, and a few very long retention times. If dJ can be expressed analytically, combining this information with the kinetic expression should, in principle, allow us to calculate conversion. In general it is true that

$$\bar{C} = \int_0^1 C \, dJ \qquad (3\text{-}7)$$

That is, the average emerging reactant concentration $\bar{C}$ must be the sum of all effluent streams, each of which may have a unique residence time. To secure a direct analytical solution to Eq. (3-7), C must be expressed as a function of residence time t and dJ expressed as a function of that residence-time distribution. In Chap. 2, integrated expressions related C to t for various reaction types, for example, $C = C_0 \exp(-kt)$ for first-order kinetics. In this chapter, Eqs. (3-2) to (3-4) yield, for example, the $J = C/C_0$-vs.-t relation for a CSTR, that is,

$$J = \frac{C}{C_0} = 1 - \exp\left(-\frac{t}{\theta}\right) \qquad (3\text{-}8)$$

The fraction of molecules in the reduced residence-time interval $d(t/\theta)$ for a CSTR is

$$dJ = d\frac{t}{\theta} \exp\left(-\frac{t}{\theta}\right) \qquad (3\text{-}9)$$

Substituting for C and dJ in Eq. (3-7) in the case of a first-order reaction in a CSTR (note that at $J = 0$, $t/\theta = 0$ and at $J = 1$, $t/\theta = \infty$), we have

$$\bar{C} = C_0 \int_0^\infty \exp(-kt) \exp\left(-\frac{t}{\theta}\right) d\frac{t}{\theta} \qquad (3\text{-}10)$$

$$\bar{C} = \frac{C_0}{\theta} \frac{1}{k + 1/\theta} = \frac{C_0}{1 + k\theta}$$

for a CSTR, first-order reaction. For a PFR, $C = C_0 \exp(-k\theta)$; but since $k\theta$ is a constant,

$$\bar{C} = C_0 \int_0^1 \exp(-k\theta) \, dJ = C_0 \exp(-k\theta) \int_0^1 dJ \qquad (3\text{-}11)$$

or

$$\bar{C} = C_0 \exp(-k\theta)$$

Segregated-Flow Model (Residence-Time-Distribution Model)

The implicit assumption expressed by Eq. (3-7) is that the effluent stream can be considered as comprising various fluid elements each of which has its own unique residence time. For a CSTR an exponential distribution of t is obtained, while for a PFR no distribution exists, that is, $t = \theta = $ constant for all elements. Such a model is segregated in that each element is essentially considered an independent entity which travels through the reactor in a particular time t. This is termed *segregated flow*. Implicit in the notion of segregated flow is the assumption that mixing or intermingling of one element with others *does not* occur in the reactor but only afterward (macromixing). Such a system is equivalent to a bank of tubular reactors set in parallel, each tube of the assembly being of different length or volume so that the reactant residence time is different in each unit with the net result (for a CSTR) that an exponential residence-time distribution results.

Nonsegregated-Flow Model (Material-Balance Model)

We may, however, view matters in terms of intimacy of mixing in the reactor (micromixing). Consider a CSTR. By definition the perfectly backmixed continuous-reactor effluent concentration must equal the average concentration within the vessel. With flow and chemical reaction in a steady state, the assumption of perfect mixing permits us to express the situation by a simple material balance

$$QC_0 = Q\bar{C} + Vk\bar{C}^\alpha \qquad (3\text{-}12)$$

For first-order kinetics, $\alpha = 1$ and $\theta = V/Q$; then

$$\frac{\bar{C}}{C_0} = \frac{1}{1 + k\theta} \qquad (3\text{-}13)$$

a result identical to that found by the residence-time-distribution (segregated) approach [Eq. (3-10)]. For first-order kinetics it makes no difference in terms of conversion whether one assumes nonsegregated or segregated flow. Time alone and not the degree of micromixing determines conversion. This is not surprising since a first-order process depends not upon the interaction of molecules but only upon the time of exposure of the reacting molecule. It follows that when interaction of species determines rate behavior, assumptions concerning the degree of mixing or segregation (micro-macromixedness) will affect the prediction of conversion. This will now be illustrated for one-half- and second-order kinetics.

3-3 ORDER AND SEGREGATION

Orders Less than Unity

To illustrate the effects of segregation for orders less than unity, consider a one-half-order reaction, $kC^{1/2}$. In Chap. 2 we found, for $\alpha = \frac{1}{2}$,

$$\sqrt{C} - \sqrt{C_0} = -\frac{kt}{2} \qquad (3\text{-}14)$$

for a batch or PFR system. Rearranging (3-14) in terms of C/C_0 gives

$$\frac{C}{C_0} = \left(1 - \frac{kt}{2\sqrt{C_0}}\right)^2 \qquad (3\text{-}15)$$

Substituting in Eq. (3-7) we have for a CSTR with an exponential (CSTR) residence-time distribution

$$\bar{C} = \frac{C_0}{\theta} \int_0^\infty \left(1 - \frac{kt}{2\sqrt{C_0}}\right)^2 \exp\left(-\frac{t}{\theta}\right) dt \qquad (3\text{-}16)$$

Expanding and integrating

$$\frac{\bar{C}}{C_0} = 1 + 2\left(\frac{k\theta}{2\sqrt{C_0}}\right)^2 - 2\frac{k\theta}{2\sqrt{C_0}} \qquad (3\text{-}17)$$

Consider now the result, for $\alpha = \frac{1}{2}$, assuming complete (micro) mixing (material-balance approach) or nonsegregated flow. For a CSTR,

$$QC_0 = Q\bar{C} + Vk\sqrt{\bar{C}}$$

Solving for $\bar{C}/C_0$, we obtain

$$\frac{\bar{C}}{C_0} = 1 + 2\left(\frac{k\theta}{2\sqrt{C_0}}\right)^2 - 2\frac{k\theta}{2\sqrt{C_0}}\sqrt{1 + \left(\frac{k\theta}{2\sqrt{C_0}}\right)^2} \qquad (3\text{-}18)$$

In Fig. 3-4, $\bar{C}/C_0$ is plotted vs. $k\theta/2\sqrt{C_0}$ for segregated (macro) and perfectly (micro) mixed CSTR models. A difference in predicted conversion clearly exists between the two models of the CSTR. The assumption of perfect intimacy of mixing leads to a higher predicted conversion (lower $\bar{C}/C_0$) for a given value of $k\theta/2\sqrt{C_0}$ than for the CSTR residence-time model (segregated flow).

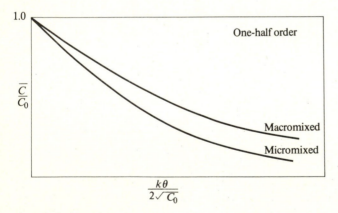

FIGURE 3-4
Reactant remaining vs. dimensionless extent of reaction time for one-half-order reaction for macromixed (segregated) and micro-mixed (nonsegregated) CSTR.

Orders Greater than Unity: $2A \rightarrow P$

For a PFR (see Chap. 2) of second-order type I,

$$\frac{C}{C_0} = \frac{1}{1 + kC_0\theta} \qquad (3\text{-}19)$$

For a CSTR, by a material balance (nonsegregated)

$$QC_0 = Q\bar{C} + Vk\bar{C}^2 \qquad (3\text{-}20)$$

which yields for the nonsegregated, micromixed CSTR

$$\frac{\bar{C}}{C_0} = \left[\sqrt{\frac{\beta}{2}\left(\frac{\beta}{2} + 2\right)} - \frac{\beta}{2} \right] \qquad \beta = \frac{1}{kC_0\theta} \qquad (3\text{-}21)$$

For small values of $k\theta C_0$ (large β), the binomial expansion gives

$$\frac{\bar{C}}{C_0} = 1 - \frac{1}{\beta} + \frac{2}{\beta^2} - \frac{3}{\beta^3} + \cdots \qquad (3\text{-}22)$$

Consider the evaluation of $\bar{C}/C_0$ using the residence-time-distribution approach for a segregated CSTR. Substituting (3-19) into (3-7) gives

$$\bar{C} = \int_0^1 C \, dJ = \frac{C_0}{\theta} \int_0^\infty \frac{\exp(-t/\theta)}{1 + kC_0 t} \, dt \qquad (3\text{-}23)$$

If we let

$$\beta = \frac{1}{kC_0\theta} \qquad \text{and} \qquad 1 + kC_0 t = kC_0\theta u = \frac{u}{\beta} \qquad (3\text{-}24)$$

then

$$\frac{t}{\theta} = u - \beta$$

and

$$\bar{C} = C_0 \beta \int_\beta^\infty \frac{\exp[-(u - \beta)]}{u} \, du = C_0 \beta \exp(\beta) \int_\beta^\infty \frac{\exp(-u)}{u} \, du$$

$$\bar{C} = C_0 \beta \exp(\beta) E_i(\beta) \qquad (3\text{-}25)$$

where E_i is the exponential integral. For small values of $kC_0\theta$ (large β)

$$\frac{\bar{C}}{C_0} = 1 - \frac{1!}{\beta} + \frac{2!}{\beta^2} - \frac{3!}{\beta^3} + \cdots \qquad (3\text{-}26)$$

For small β (large $k\theta C_0$) the tabulated values of $E_i(\beta)$ are used.

In Fig. 3-5 performance of the nonsegregated and segregated CSTR are compared for second-order reaction. At a value of $kC_0\theta = 1$ the material balance predicts a value of reactant remaining of 0.3 while the residence-time solution predicts about 0.2.

The residence-time solution, as mentioned above, assumes segregated packets of reactant, each of which enjoys a unique contact time. In essence this analysis

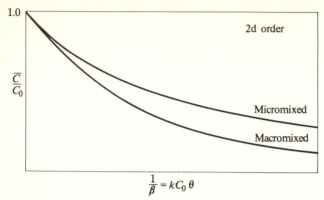

FIGURE 3-5
Reactant remaining vs. dimensionless extent of reaction time for
second-order reaction for macro- and micromixed CSTR.

treats the system as an assembly of small batch reactors, moving without mutual interaction through the CSTR with a specific and unique residence time, the sum of which reveals a CSTR distribution. This is essentially a segregated system. It might describe, for example, conversion of a dispersed noncoalescing phase in which droplet residence time follows the CSTR (exponential) distribution.

On the other hand, the material-balance (nonsegregated) approach simply assumes complete microscale mixing with a resulting average composition $\bar{C}$. In essence this model assumes no segregation and in consequence predicts lower conversion for orders greater than 1 for a given β than does the segregated model. Since first-order reaction depends not upon intimacy of mixing or interaction but only upon time for a given rate constant k, both segregated and nonsegregated models yield identical results. For orders less than unity, nonsegregated conversion is greater than segregated.

Example An elementary thought experiment[1] lucidly demonstrates this point. Consider equal volumes of a reactant C. In one vessel the concentration is 1; in the other 3. What is the velocity of reaction (rate × volume) if (1) we react each independently and then mix the result (macromix) or (2) we first mix (micromix) and then react? Let k be unity in each case. See Table 3-1.

Patently, segregation makes no difference, nor does the order of mixing or reaction affect first-order kinetics; however, in second order (or any higher than first order) premixing (micromixing) is detrimental. For orders less than unity premixing (micromixing) is advantageous.

[1] A. B. Metzner and R. L. Pigford, "Scale-up in Practice," Reinhold, New York, 1958.

Selectivity in Segregated and Nonsegregated Reactors

For a selectivity-sensitive network such as

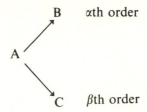

where reaction orders differ, the overall selectivity B/C will surely be sensitive to the ratio of macromixing to micromixing, and while conversion taxations are easily remedied by contact-time or temperature adjustments, a selectivity deficit is not so readily cured. Yet it often develops that a modest selectivity improvement significantly improves process economics. The consequences of early vs. late mixing with respect to selectivity should not be overlooked. We shall consider this topic when attention is devoted to series reactor networks involving CSTR-PFR combinations in Sec. 3-6.

3-4 REACTION ORDER AND NONSEGREGATED BACKMIXING

It is pertinent at this point to ascertain how reaction order influences the extent to which conversion is retarded by backmixing. Since the mixing process essentially alters the concentration field within the reactor, intuition suggests that the higher the reaction order the greater the influence of backmixing upon reaction rate (and therefore conversion for a given contact time).

Intuition is readily supported by quantitative treatment in this instance. Consider cases of zero, one-half, and second-order kinetics. The extent to which backmixing influences reaction rate in each case is reflected in terms of the required ratio of contact time in a CSTR θ_c relative to that in a PFR θ_p for a given level of conversion and temperature. *Nonsegregated* flow will be assumed in the CSTR

Table 3.1			Concentration, mol/unit volume			Velocity of reaction, mol/unit time		
Case	C		First order	Second order	One-half order	First order	Second order	One-half order
I, macromixed (segregated)	1 $\xrightarrow{\text{reaction}}$ 3 $\xrightarrow{\text{reaction}}$		1 3	1 9	1 1.73 $\Big\}$ $\xrightarrow{\text{mix}}$	4	10	2.73
II, micromixed (unsegregated)	1 3 $\Big\}$ $\xrightarrow{\text{mix}}$ 2				$\xrightarrow{\text{reaction}}$	4	8	2.83

analysis for simplicity.   While differences were demonstrated for nonlinear kinetics between segregated and nonsegregated CSTR models, these are secondary in comparison with differences in PFR and CSTR performance.   In what follows, a steady state is assumed.

Normal Reaction Kinetics

Zero-order kinetics   From Eq. (2-45) the PFR or batch rate is

$$-\frac{dC}{dt} = k$$

On integration with $C = C_0$, $t = 0$, and $C = C$, $t = \theta_p$

$$C_0 = C + k\theta_p$$

For a CSTR, by material balance, in = out + accumulation

$$QC_0 = QC + Vk$$

or
$$C_0 = C + k\theta_c \qquad \text{where } \theta_c = \frac{V}{Q} \qquad (3\text{-}27)$$

Therefore for zero-order kinetics

$$\frac{\text{CSTR contact time}}{\text{PFR contact time}} = \frac{\theta_c}{\theta_p} = 1$$

for all values of conversion.   Since rate is independent of concentration, rate is independent of the concentration distribution and consequently of extent of backmixing.

First-order reaction   For one CSTR relative to the PFR, the contact-time ratio is

$$\frac{\theta_c}{\theta_p} = \frac{1 - C/C_0}{(C/C_0)\ln(C_0/C)} = \frac{1 - f}{f \ln(1/f)} \qquad \text{where } f = \frac{C}{C_0} \qquad (3\text{-}28)$$

Here the contact-time ratio is dependent upon concentration f.

One-half-order reaction   For batch reactor or PFR Eq. (2-70) becomes, for $\alpha = \frac{1}{2}$,

$$\sqrt{C} - \sqrt{C_0} = -\frac{k}{2}\theta_p$$

Solving for θ_p, where $f = C/C_0$ and $K = k/\sqrt{C_0}$,

$$\theta_p = \frac{1 - \sqrt{f}}{K/2}$$

For a CSTR, a material balance yields

$$C_0 = C + \theta_c k\sqrt{C} \quad \text{or} \quad \theta_c = \frac{1-f}{K\sqrt{f}}$$

The ratio of CSTR to PFR contact time is

$$\frac{\theta_c}{\theta_p} = \frac{f-1}{2(\sqrt{f}-f)} \tag{3-29}$$

Second-order kinetics For a PFR, rearranging (2-53), where $t = \theta_p$ and $C/C_0 = f$, we obtain

$$\theta_p = \frac{1-f}{kC_0 f}$$

For a CSTR (nonsegregated)

$$C_0 = C + \theta_c kC^2 \quad \text{or} \quad \theta_c = \frac{1-f}{kC_0 f^2}$$

Thus

$$\frac{\theta_c}{\theta_p} = \frac{1}{f} \tag{3-30}$$

Abnormal Reaction Kinetics

Simple autocatalysis The simple cases considered so far involve a rate-conversion relationship which decreases for positive orders, i.e., normal kinetics. Autocatalysis is characterized by a rate-conversion functionality which first increases (abnormal kinetics), passes through a maximum, and then decreases (see Fig. 2-6). Consider the simple autocatalytic reaction $A + C \rightarrow 2C + P$, where C is the autocatalytic agent. Qualitatively we can appreciate the fact that backmixing involves the "return" of an agent C, which enhances the rate. Therefore, we can anticipate unusual θ_c/θ_p-versus-conversion behavior in autocatalysis.

As an exercise, compute θ_c/θ_p versus conversion of A for the simple autocatalytic step $A + C \rightarrow 2C + P$. As this step is assumed to be an elementary one, the order is defined. Assume that $C_0 \ll A_0$. As another exercise, consider a simple first-order exothermic reaction $A \xrightarrow{k} P + \text{heat}$. Discuss the influence of backmixing upon conversion in an adiabatic system, bearing in mind the "autocatalytic" influence of heat generation upon k.

Orders less than zero A rate equation of the form

$$-\frac{dC}{dt} = \frac{k}{C}$$

is occasionally found in catalytic reaction systems; e.g., in excess air, Pd-, Pt-, or Ru-metal–catalyzed oxidation of CO to CO_2 is found to be of negative first order

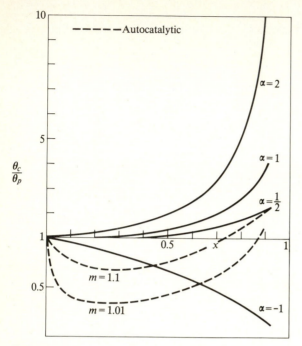

FIGURE 3-6
CSTR and PFR holding-time ratio vs. conversion for diverse
reaction orders (micromixed CSTR model).

in CO.[1]   Obviously a rate equation of this form is a limiting one, inapplicable at
near zero value of reactant CO.   Detailed discussion of this point is set forth in
Chap. 9.

In effect a negative order with respect to the reactant suggests inhibition, so
that the rate increases with conversion.   In consequence backmixing will be advantageous.   Proceeding as we did in the common cases of positive-order reactions,
θ_c/θ_p can be computed for the negative-order case ($\alpha = -1$).   The student should
verify the result, which is, for $\alpha = -1$,

$$\frac{\theta_c}{\theta_p} = \frac{f - f^2}{1 - f^2} = \frac{2f}{1 + f}$$

In Fig. 3-6 the ratio θ_c/θ_p as a function of conversion, $x = 1 - f$, is shown
for orders greater than, equal to, and less than zero, as well as for autocatalysis.
Numerical values of θ_c/θ_p are given in Table 3-2 for various orders and second-
order autocatalysis, where $m_0 = (A_0 + C_0)/A_0$.

For other than zero- and first-order kinetics these ratios will differ if segregated flow is assumed in the CSTR model.   Specifically θ_c/θ_p will be greater for

[1] D. Tajbl, J. Simons, and J. J. Carberry, *Ind. Eng. Chem. Fundam.*, **5**: 171 (1966).

positive orders less than unity and less than those shown for orders greater than unity if segregated flow is assumed. However, the trend indicated in the above analysis remains qualitatively correct. *The higher the nonautocatalytic reaction order, the more taxing the effect of mixing upon reaction rate.* This generalization will prove to be of signal importance in the analysis of selectivity or yield in complex schemes.

Summary

Under *isothermal* conditions, the backmixing influence upon conversion can be generalized as follows:

1 Isothermal backmixing is detrimental to simple normal reaction conversion to the extent that concentration affects reaction rate; so zero-order reactions are unaffected by backmixing and the detrimental influence becomes more telling as reaction order increases. Autocatalysis defies this trend.

2 The magnitude of the backmixing influence is readily assessed by a comparison of CSTR and PFR performance.

3 However, a quantitative comparison of backmixing based upon residence-time data and kinetics is possible only for linear kinetics.

4 In nonlinear rate models, the nature of mixing (segregated or nonsegregated or a combination of both) must be known for a *precise* assessment of backmixing effects upon reaction.

5 Segregated backmixing favors conversion for reaction orders greater than unity, while the converse is true for positive reaction orders less than unity.

6 Hence late micromixing (CSTR following a PFR) is desirable for reactions of order greater than unity, while early micromixing (CSTR preceding a PFR) is advantageous for reactions of order less than unity. For a first-order reaction, the order of mixing is obviously of no consequence.

7 Abnormal reactions (autocatalytic and negative-order reactions) defy these generalizations.

While the relation between conversion and backmixing for nonlinear kinetics is a function of the extent of segregation, for totally backmixed *homogeneous* reaction environments, the extent of segregation will doubtless be limited, with

Table 3.2

						Order		
	-1	0	$\frac{1}{2}$	1	2	2 autocatalytic		
$1-f$						$m_0 = 1.0$	$m_0 = 1.1$	$m_0 = 2$
0.1	0.945	1	1	1.03	1.11	0.408	0.772	1.03
0.3	0.825	1	1.09	1.18	1.43	0.37	0.675	1.065
0.5	0.667	1	1.2	1.45	2	0.427	0.737	1.22
0.7	0.463	1	1.41	1.94	3.33	0.608	0.98	1.59
0.9	0.182	1	2.08	3.9	10	1.47	2.15	3.23

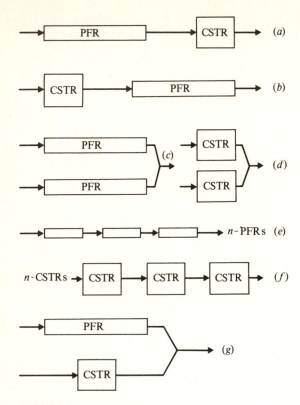

FIGURE 3-7
Diverse CSTR-PRF combinations.

the consequence that the assumption of zero segregation (nonsegregation or micro-mixedness) is justified in analyses. If, on the other hand, we are confronted with an immiscible system, e.g., mixed acid nitration of aromatics, the assumption of total segregation might be more realistic.

We now turn our attention to multiple-reactor systems.

3-5 ISOTHERMAL CSTR-PFR COMBINATIONS

Various combinations of the CSTR and PFR are possible (see Fig. 3-7). It is obvious that systems (c) and (d) simply lead to double productivity relative to the single unit. For a given residence time θ in a single unit, the installation of another parallel unit simply allows twice the feed rate and therefore twice the productivity. It is not reasonable to operate parallel units at anything but equal conversion levels; otherwise, dilution by effluent from the less efficient unit will nullify the performance of the parallel mate. A seventh combination is shown in Fig. 3-7*g*.

In view of our remarks concerning the desirability of equal conversion in each parallel unit, however, system (g) offers no advantage over other combinations and is, indeed, an unreasonable combination.

System (e) is equivalent to a PFR of length or volume equal to n times the single unit. Total volume is merely the sum of each PFR volume, and so for a fixed feed Q, the total holding time θ is the sum of each unit's contact time.

Example

Consider systems (a) and (b). Let $k = 1$, $C_0 = 1$, and $\theta = 1$ in each reactor.

FIRST-ORDER REACTION:

System (a): $\qquad C = 1 \exp(-1) = 0.367 = \text{feed to CSTR}$

$$C_{\text{final}} = \frac{C}{1 + k\theta} = \frac{0.376}{2} = 0.183$$

System (b): $\qquad C_1 = \frac{1}{1 + k\theta} = 0.5 = \text{feed to PFR}$

$$C_{\text{final}} = 0.5 \exp(-1) = 0.183$$

SECOND-ORDER REACTION:

System (a): $\qquad C_1 = \frac{1}{1 + k\theta C_0} = 0.5 = \text{feed to CSTR}$

$$C_{\text{final}} = \frac{\sqrt{1 + 4k\theta C_1} - 1}{2k\theta} = 0.365$$

System (b): $\qquad C_1 = \frac{\sqrt{1 + 4k\theta C_0} - 1}{2k\theta} = 0.615 = \text{feed to PFR}$

$$C_{\text{final}} = \frac{0.615}{1 + 0.615} = 0.38$$

This simple illustration demonstrates, once again, that linear (first-order) kinetic systems depend only upon holding time for a given k and are independent of the mechanism or device which provides that residence-time function. Nonlinear systems, on the other hand, are sensitive to the manner and order of mixing.

As an exercise, repeat the above example for $\alpha = -1$ and $k\theta C_0^{-2} = 0.1$ and 0.2 Any problems?

3-6 INTERMEDIATE BACKMIXING

Often neither a single-CSTR nor a PFR residence-time distribution prevails; instead backmixing is of an intensity intermediate between the limiting extremes. Simulation of such cases is of obvious importance.

CSTRs in Series

Combination (f) consists of n perfectly mixed reactors arranged in tandem so that the effluent of one vessel is the feed to the next. We focus first, upon two CSTRs in series, each of equal volume. The residence time or response to a step input of inert tracer will be developed before considering reaction in such a system. For the first vessel the effluent tracer concentration C is

$$C_1 = C_0 \left[1 - \exp\left(-\frac{2t}{\theta_t} \right) \right] \qquad \text{where} \qquad \begin{aligned} \theta_t &= (V_1 + V_2)/Q \\ \theta_1 &= \theta_t/2 \end{aligned} \qquad (3\text{-}31)$$

For the second CSTR, a material balance gives

$$QC_1 = QC_2 + V_2 \frac{dC_2}{dt}$$

$$\frac{dC_2}{dt} = \frac{2}{\theta_t}(C_1 - C_2) = \frac{2}{\theta_t} \left\{ C_0 \left[1 - \exp\left(-\frac{2t}{\theta_t} \right) \right] - C_2 \right\} \qquad (3\text{-}32)$$

or

$$\frac{dC_2}{dt} + \frac{2C_2}{\theta_t} = \frac{2C_0}{\theta_t} \left[1 - \exp\left(-\frac{2t}{\theta_t} \right) \right]$$

which is a first-order linear differential equation of the form

$$\frac{dy}{dx} + yP(x) = Q(x)$$

The general solution is

$$y = \left[\exp\left(-\int P \, dx \right) \right] \left\{ \left[\int Q \exp\left(\int P \, dx \right) \right] dx + \text{const} \right\}$$

For a step input $C = 0$, $t = 0$,

$$C_2 = \left[\exp\left(-\frac{2t}{\theta_t} \right) \right] \left(\left\{ K_1 \int \left[1 - \exp\left(-\frac{2t}{\theta_t} \right) \right] \left(\exp\frac{2t}{\theta_t} \right) dt \right\} + K_2 \right)$$

Therefore

$$J(t) = \frac{C_2}{C_0} = 1 - \left(1 + \frac{2t}{\theta_t} \right) \exp\left(-\frac{2t}{\theta_t} \right)$$

This procedure can be extended to any number of equal-volume CSTRs in series. The resulting responses for $n = 1$, 2, 10, 20 and an infinite number of CSTRs in series are shown in Fig. 3-8. It is apparent that as the number of CSTRs in series is increased, the residence-time function approaches that of a PFR ($n \to \infty$). The corresponding pulse responses are shown in Fig. 3-9, representing the derivatives of the $J(t)$ curves.

The series network of CSTRs provides an illustration of residence-time distributions intermediate between plug flow and the uniform environment characteristic of a CSTR with its exponential distribution of residence times. The useful-

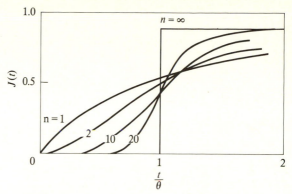

FIGURE 3-8
Step-input response for n CSTRs in series ($n = 1$ to ∞).

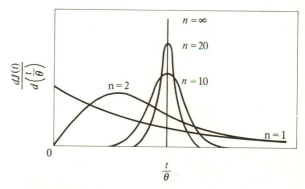

FIGURE 3-9
Pulse-input response for n CSTRs in series ($n = 1$ to ∞).

ness of the CSTR series network is that a reactor which reveals neither a PFR nor a single-CSTR response may be described in terms of n CSTRs. If an n-CSTR residence-time distribution does in fact exist, reactor performance is capable of being predicted *for first-order kinetics*. Once again it must be emphasized that in other than first-order cases, more must be known about the nature of mixing (segregated and/or nonsegregated) before a *precise* prediction of performance is feasible.

It is worth noting at this point that complex-reactor geometries, while revealing residence-time functions intermediate between PFR and single-CSTR character, are often not easily described by n CSTRs in series. The fluidized-bed reactor is typical of a unit displaying a J curve which does not conform to the n-CSTR pattern.

Reaction in the Series CSTR Isothermal Network

In a series of n CSTRs, each of equal volume V, in which reaction occurs in steady state, we have for the first vessel, for first-order kinetics,

$$QC_0 = QC_1 + VkC_1 \quad \text{or} \quad C_1 = \frac{C_0}{1 + k\theta_1}$$

For the second and nth vessels, respectively,

$$C_2 = \frac{C_1}{1 + k\theta_2} \quad \text{and} \quad C_n = \frac{C_{n-1}}{1 + k\theta_n}$$

At constant flow rate for equal-volume vessels, $\theta_1 = \theta_2 = \theta_n$, and thus by successive substitutions

$$\frac{C_n}{C_0} = \frac{1}{(1 + k\theta_1)^n} = \frac{1}{(1 + k\theta_t/n)^n} \quad (3\text{-}33)$$

where θ_t is total holding time for the system.

A useful and instructive comparison between PFR and n-CSTR performance can now be undertaken. Conversion $(1 - C/C_0)$ in a PFR is given by

$$x_{\text{PFR}} = 1 - \frac{C}{C_0} = 1 - \exp(-k\theta_p) \quad (3\text{-}34)$$

compared with that in an n-CSTR network

$$x_{n\text{-CSTR}} = 1 - \frac{1}{(1 + k\theta_1)^n} \quad (3\text{-}35)$$

Solving for $n\theta_1$ in (3-35) and θ_p in (3-34) gives

$$\theta_t = n\theta_1 = \left[\frac{(1 - x)^{-1/n} - 1}{k} \right] n \quad n\text{-CSTRs} \quad (3\text{-}36)$$

and

$$\theta_p = - \frac{\ln(1 - x)}{k} \quad \text{PFR} \quad (3\text{-}37)$$

Then $\theta_p/n\theta_1$ represents the respective holding time in a PFR (or batch reactor) and n-CSTR network for a given conversion level x:

$$\frac{\theta_p}{n\theta_1} = \frac{-\ln(1 - x)}{[(1 - x)^{-1/n} - 1]n} \quad (3\text{-}38)$$

As an exercise show at what values of n, $\theta_p/n\theta_1$ becomes 90 and 99 percent of unity. This exercise should demonstrate that as n increases in the n-CSTR series network, PFR (or batch) performance is realized. This must, of course, follow logically from our previous demonstration that the residence-time behavior of n CSTRs approaches PFR character with increasing n.

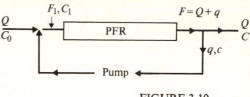

FIGURE 3-10
Recycle reactor model.

The Recycle Model of Intermediate Backmixing

There exists an alternative model to the n-CSTR series whereby the range of mixing between CSTR and plug flow can be represented, namely, the PFR with recycle (see Fig. 3-10). For a simple first-order isothermal reaction occurring in the PFR,

$$\frac{dC}{d\tau} = -kC \qquad d\tau = \frac{dV}{F}$$

or

$$-\tau = -\int \frac{dV}{F} = \frac{1}{k} \int_{C_1}^{C} \frac{dC}{C} \tag{3-39}$$

By a material balance

$$FC_1 = QC_0 + Cq$$

and

$$C_1 = \frac{RC + C_0}{R + 1} \qquad \text{where } R = \frac{q}{Q} = \text{recycle ratio}$$

When we let $C/C_0 = f$,

$$k\tau = -\ln \frac{C}{C_1} = -\ln \frac{f(R + 1)}{Rf + 1} \tag{3-40}$$

For large values of the recycle ratio R, the term $[f(R + 1)]/(Rf + 1)$ approaches unity, and for this condition we can expand the log term, retaining only the first term of the series, that is, $\ln x = (x - 1)/x$, giving

$$-k\tau = \frac{f(R + 1)/(Rf + 1) - 1}{f(R + 1)/(Rf + 1)} = \frac{1 - f}{f(R + 1)}$$

Solving for f

$$f = \frac{1}{1 + (R + 1)\tau k} \tag{3-41}$$

Recall that

$$\tau = \frac{V}{F} = \frac{V}{Q + q} = \frac{\theta}{R + 1} \qquad \text{so} \qquad \theta = \frac{V}{Q}$$

and therefore

$$f = \frac{1}{1 + k\theta} = 1 \text{ CSTR} \qquad \text{at large } R$$

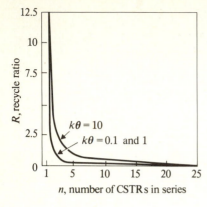

FIGURE 3-11
Recycle ratio R versus n-CSTR model
for linear kinetics. [*B. G. Gillespie and
J. J. Carberry, Ind. Eng. Chem. Fundam.,
5:164 (1966).*]

Thus a PFR operating at a large recycle ratio R behaves as a single well-stirred, perfectly mixed vessel. When $R = 0$, we have simple PFR performance. Hence mixing behavior intermediate between plug flow (no mixing) and perfect mixing (one CSTR) can be characterized by finite values of R.

Evidently a relationship must exist between the PFR with recycle and n CSTRs in series [(Eqs. (3-40) and (3-33)]

$$f_n = \frac{1}{(1 + k\theta_t/n)^n} = f_R = \frac{\exp\left[-k\theta/(R+1)\right]}{1 + R\{1 - \exp\left[-k\theta/(R+1)\right]\}} \qquad (3\text{-}42)$$

Figure 3-11 provides the relationship between n and R.

The Axial-Dispersion Model of Intermediate Backmixing

We can view modest backmixing as dispersion or macroscopic diffusion superimposed upon plug flow. In Chap. 2 the mass-continuity equation for an isothermal PFR [Eq. (2-43)] is derived

$$F = F + dF + \mathscr{R} \, dV$$

Input = output + accumulation

or
$$-\frac{dF}{dV} = \mathscr{R} \qquad (3\text{-}43)$$

As $F = QC$ and $Q = uA$, for constant velocity u and cross-sectional area A, when it is recognized that $dV = A \, dz$, the relation between concentration C and reactor length is

$$-u \frac{dC}{dz} = \mathscr{R}$$

Let us suppose that superimposed upon convective flow in the z direction there exists a diffusion/dispersion process (backmixing) which *might* be described, by analogy to molecular diffusion, by Fick's law

$$\text{Dispersion} = \mathscr{D}_a \frac{d^2 C}{dz^2}$$

Adding this dispersion rate to that of convective transport, we have

$$\mathscr{D}_a \frac{d^2 C}{dz^2} - u \frac{dC}{dz} = \mathscr{R}$$

$\mathscr{D}_a$ is an apparent, experimentally determined axial diffusion or dispersion coefficient which accounts for the departure from plug flow of the residence-time distribution. Letting $f = C/C_0$ and $Z = z/L$ for linear kinetics, we find

$$\frac{\mathscr{D}_a}{Lu} \frac{d^2 f}{dZ^2} - \frac{df}{dZ} = \frac{\mathscr{R}\theta}{C_0} \qquad (3\text{-}44)$$

where $\theta = L/u$, the average holding time. The dimensionless group $Lu/\mathscr{D}_a$ is the axial Peclet number based upon reactor length L.

Inspection of Eq. (3-44) indicates that as the Peclet value becomes large, the dispersion rate approaches zero and PFR behavior prevails, while, of course, deviation from such behavior is manifest with decreasing values of the axial Peclet number. Physically the axial Peclet number represents

$$\text{Pe}_a = \frac{Lu}{\mathscr{D}_a} = \frac{\text{convective flow}}{\text{diffusive flow}} \text{ in axial direction}$$

When dispersion predominates, total backmixing (CSTR) is found.

Given a value of Pe_a, backmixing effects for a specific velocity and reactor length are readily assessed by analytical solution of Eq. (3-44) for isothermal kinetics or by numerical solution for nonisothermality and/or nonlinear kinetics. In nonlinear and/or nonisothermal circumstances numerical solution poses stability difficulties which are remedied in the literature.[1]

Backmixing surely affects all save zero-order reactions, and within the general scope of these influences subtle differences prevail between microscale and macroscale modes of backmixing for all except zero and linear systems.

Of greater importance than conversion alteration, due to gradient alteration by backmixing, is that of yield or selectivity modification by virtue of backmixing. When one encounters such complex reaction networks, an assessment of yield or selectivity taxation is most readily realized if nonsegregation (micromixedness) is assumed in contrasting PFR and CSTR performance. The general problem will be addressed in such terms. However, an illustration of selectivity alteration due to segregation is justified, so that the compromises implicit in later, simpler analyses are now made clear.

---

[1] J. J. Carberry and M. M. Wendel, *AIChE J.*, **9**: 129 (1963).

Selectivity in Series CSTR-PFR Combinations[1]

We now consider order-sensitive selectivity in terms of macromixing and micro-mixing, i.e., late and early mixing, respectively. In the first case we have a PFR-CSTR series network and in the second a CSTR-PFR series network. For the complex reaction, where B is the desired product,

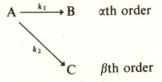

$$A \xrightarrow{k_1} B \qquad \alpha\text{th order}$$
$$\searrow k_2$$
$$\qquad\qquad C \qquad \beta\text{th order}$$

the overall selectivity B/C will depend upon the rate-coefficient ratio $K = k_1/k_2$ and reaction orders (recall that when $\alpha = \beta$, selectivity depends solely upon K). We consider instances in which $\alpha \neq \beta$. Specifically for K equal to 1 and 2, we dwell upon

$$
\begin{array}{cccc}
\alpha = 1 & \alpha = 2 & \alpha = 1 & \alpha = \frac{1}{2} \\
\beta = 2 & \beta = 1 & \beta = \frac{1}{2} & \beta = 1
\end{array}
$$

In all cases it is assumed that final conversion of A is 98 percent, while that between the CSTR and PFR is 49 percent. The student should verify the computed results set forth in Table 3-3.

It will usually prove difficult to determine the extent of segregation; however, the results in Table 3-3 suggest that some telling selectivity improvements can be realized by a wise use of CSTR-PFR series combinations.

Table 3.3 SELECTIVITY B/C AS A FUNCTION OF MICROMIXING OR MACROMIXING FOR VARIOUS ORDERS AND RATE-COEFFICIENT RATIOS IN THE REACTION

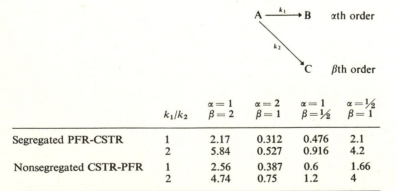

$$A \xrightarrow{k_1} B \qquad \alpha\text{th order}$$
$$\searrow k_2$$
$$\qquad\qquad C \qquad \beta\text{th order}$$

	k_1/k_2	$\alpha = 1$ $\beta = 2$	$\alpha = 2$ $\beta = 1$	$\alpha = 1$ $\beta = \frac{1}{2}$	$\alpha = \frac{1}{2}$ $\beta = 1$
Segregated PFR-CSTR	1	2.17	0.312	0.476	2.1
	2	5.84	0.527	0.916	4.2
Nonsegregated CSTR-PFR	1	2.56	0.387	0.6	1.66
	2	4.74	0.75	1.2	4

[1] P. Costa and J. J. Carberry, *Chem. Eng. Sci.*, **28**: 2257 (1973).

3-7   EFFECT OF NONSEGREGATED MIXING UPON ISOTHERMAL SELECTIVITY

The following reaction networks will be treated herein or by referral:

Simultaneous reactions:

$$A \overset{k_1}{\nearrow} B \qquad \overset{k_2}{\searrow} C$$

both same order   (*a*)

$$A \overset{k_1}{\nearrow} B \qquad \overset{k_2}{\searrow} C$$

second order

first order   (*b*)

Consecutive reactions: $A \xrightarrow{k_1} B \xrightarrow{k_2} C$ all first order   (*a*)

$$\begin{array}{ll} A \longrightarrow B & \text{first order} \\ A + B \longrightarrow C & \text{second order} \end{array}$$ (*b*)

$$\begin{array}{ll} A + A \longrightarrow B & \text{second order} \\ B + B \longrightarrow C & \text{second order} \end{array}$$ (*c*)

$$\begin{array}{ll} A + B \longrightarrow C & \text{second order} \\ B + C \longrightarrow D & \text{second order} \end{array}$$ (*d*)

General network:

$$A \xrightarrow{k_1} B \xrightarrow{k_4} D$$
$$\downarrow{k_2} \qquad \downarrow{k_3}$$
$$E \xrightarrow{k_5} C$$

all first order

Simultaneous and Cocurrent Reactions

Batch or PFR relationships for the networks under consideration were developed and presented in Chap. 2.   In the general case of simultaneous reaction

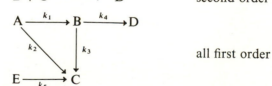

$$A \xrightarrow{k_1} B$$
$$\searrow{k_2}$$
$$C$$

we have

$$\frac{dB}{dt} = k_1 A^\alpha \qquad \text{and} \qquad \frac{dC}{dt} = k_2 A^\beta$$

Then

$$\frac{dB}{dC} = \frac{k_1}{k_2} A^{\alpha - \beta}$$

or

$$\frac{dB}{dA} = -\frac{1}{1 + (k_2/k_1)A^{\beta - \alpha}} \qquad (2\text{-}98)$$

For the PFR or batch reaction, Eq. (2-98) must be integrated to provide the B-versus-A relationship. For the single CSTR, on the other hand, (2-98) becomes a difference equation since A and B are constant throughout the reactor (perfect mixing) and equal to the effluent concentrations ($K = k_2/k_1$):

$$\frac{dB}{dA} = \frac{B - B_0}{A - A_0} = \frac{-1}{1 + KA^{\beta - \alpha}} \qquad (3\text{-}44)$$

When the reaction orders are identical ($\alpha = \beta$), for the CSTR Eq. (3-44) is

$$\frac{B - B_0}{A - A_0} = -\frac{1}{1 + K} = \frac{-k_1}{k_1 + k_2} \qquad (3\text{-}45)$$

which is identical to the integrated form of Eq. (2-98) for batch or PFR operation. Recall that since the ratio of rates dB/dA is independent of concentration [(Eq. (3-45)], mixing cannot influence the ratio and therefore selectivity (not conversion) is not sensitive to mixing. *For simultaneous reactions of identical orders, yield or selectivity is the same in a CSTR and PFR. Selectivity is zero order in this network when $\alpha = \beta$.*

When $\alpha \neq \beta$, mixing will influence selectivity as illustrated in the scheme

$$A \xrightarrow{\;k_1\;} B \qquad \text{second order, } \alpha = 2$$
$$\searrow^{k_2}$$
$$C \qquad \text{first order, } \beta = 1$$

In Chap. 2 we found [Eq. (2-99)]

$$\frac{B}{A_0} = \frac{B_0}{A_0} + \left(1 - \frac{A}{A_0}\right) + \frac{K}{A_0} \ln \frac{A/A_0 + K/A_0}{1 + K/A_0}$$

and, from Eq. (2-100),

$$\frac{C}{A_0} = \frac{C_0}{A_0} + \frac{K}{A_0} \ln \frac{1 + K/A_0}{A/A_0 + K/A_0}$$

for PFR or batch systems.

For a CSTR, Eq. (3-44) becomes for $\alpha = 2$, $\beta = 1$

$$\frac{B}{A_0} = \frac{1 - A/A_0}{1 + \dfrac{K}{A_0} \dfrac{A_0}{A}} + \frac{B_0}{A_0} \qquad (3\text{-}46)$$

For the first-order reaction product C (in a CSTR)

$$\frac{dC}{dA} = \frac{C - C_0}{A - C_0} = \frac{-1}{1 + A/K}$$

or
$$\frac{C}{A_0} = \frac{C_0}{A_0} + \frac{1 - A/A_0}{1 + \dfrac{A}{A_0} \dfrac{A_0}{K}} \qquad (3\text{-}47)$$

For the condition where product is absent from the feed stream, selectivity B/C is

$$\left(\frac{B}{C}\right)_{CSTR} = \frac{1 + \dfrac{A}{A_0}\dfrac{A_0}{K}}{1 + \dfrac{K}{A_0}\dfrac{A_0}{A}} \qquad (3\text{-}48)$$

From Chap. 2, for PFR, this ratio, from Eq. (2-101), is

$$\left(\frac{B}{C}\right)_{PFR} = \frac{1 - A/A_0}{\dfrac{K}{A_0}\ln\dfrac{1 + K/A_0}{A/A_0 + K/A_0}} - 1$$

The effects of mixing upon yield for this system can be gainfully displayed by calculating the B/C ratio for PFR relative to CSTR as a function of conversion $(1 - A/A_0)$.

Calculate the B/C ratio as an exercise.

As noted earlier, isothermal backmixing taxes the conversion step of higher order more severely than the first-order competing step. In consequence, if B is the desired product, selectivity is markedly reduced by mixing. If, on the other hand, C is the desired product, mixing will be of benefit *in terms of yield* as conversion to undesired product B is retarded.

That the reaction step of lowest order is less adversely affected by mixing should be evident from our earlier discussion with one-half, first, and second-order reactions. If, for example, our simultaneous system is

$$A \xrightarrow{k_1} B \qquad \alpha = \text{one-half order}$$
$$ \searrow^{k_2}$$
$$ C \qquad \beta = \text{first order}$$

then it will be the yield of C which is adversely affected in a CSTR. Should B be the desired product, a yield advantage would result in a reactor designed to maximize mixing. A numerical comparison of B/C values for a CSTR is given in Table 3-4 for two cases discussed for $k_2/k_1 A_0 = 1$.

While we made it clear in discussing the effects of mixing upon conversion in simple systems that these effects are lessened as reaction order decreases, e.g., no

Table 3.4

$\dfrac{A}{A_0}$	Case I: $\alpha = 2$ $\beta = 1$	Case II: $\alpha = \frac{1}{2}$ $\beta = 1$
0.8	0.8	1.11
0.4	0.4	1.58

$(B/C)_{CSTR}$

effect of mixing for zero-order reactions, the concept of order with respect to selectivity and the effect of mixing upon selectivity now need emphasis. For the general simultaneous network, the product ratio B/C is

$$\left(\frac{B}{C}\right)_{CSTR} = \frac{1 + A^{\alpha - \beta}/K}{1 + KA^{\beta - \alpha}} \qquad (3\text{-}49)$$

When $\alpha = \beta$, the selectivity ratio B/C is concentration-independent; i.e., *selectivity is zero order.* Mixing essentially alters the spatial distribution of the concentration within the reactor; therefore, if yield is concentration-independent for $\alpha = \beta$, it follows that mixing cannot influence selectivity when $\alpha = \beta$. The greater the difference in α and β, the greater the order with respect to selectivity and consequently the greater the influence of mixing upon it.

In any simultaneous competitive-reaction situation, the step characterized by the lowest order will be least influenced by mixing, while higher-order steps will be most influenced. Yield advantage due to mixing then resides with the lowest-order step. This generalization applies to parallel as well as simultaneous reaction networks:

$$A \xrightarrow{k_1} B \qquad \text{second order}$$

$$X \xrightarrow{k_2} Y \qquad \text{first order}$$

For this case mixing will favor the yield of Y.

We now turn our attention to another important selectivity scheme, that of consecutive reactions involving creation of a species which may suffer further reaction to produce another product.

Consecutive Reactions

For the simple sequence $A \xrightarrow{k_1} B \xrightarrow{k_2} C$, where B is the desired product, the B/A_0 ratio for a PFR is given as a function of conversion by Eq. (2-115), where $f = A/A_0$ and $K = k_2/k_1$:

$$\frac{B}{A_0} = \frac{1}{1 - K}(f^K - f) + \frac{B_0}{A_0} f^K$$

For a CSTR, the differential equation (2-114) become a difference equation, and B/A_0 is readily found

$$\frac{dB}{dA} = \frac{B - B_0}{A - A_0} = -1 + K\frac{B}{A}$$

or
$$\frac{B}{A_0} = \frac{1 - f}{1 + K[(1 - f)/f]} \qquad \text{for } B_0 = 0 \qquad (3\text{-}50)$$

As an exercise, for $K = 0.1$ and 0.9 calculate the yield B/A_0 at $x = 0.2$, 0.5, 0.7, and 0.9 for a PFR and CSTR.

Backmixing is clearly detrimental to the yield of B in this isothermal reaction sequence. When it is recalled that backmixing adversely affects the conversion of A, there appears to be little virtue in promoting backmixing for consecutive reactions if the intermediate is the desired product, as both conversion and yield suffer. Obviously if C is the desired product, mixing is beneficial, e.g., in catalytic oxidation of automotive-exhaust species.

Summary

Extremes of residence-time distribution (PFR vs. CSTR) as they affect yield or selectivity have been set forth above for simultaneous, cocurrent (or parallel), and consecutive reactions of a simple nature. Isothermal backmixing extremes (zero and infinite) have been demonstrated for simple yield- or selectivity-sensitive reaction networks. In accord with earlier findings, reactions of higher order are more severely arrested relative to lower-order events, which allows us to generalize yield or selectivity effects for simultaneous and parallel networks. Overexposure of intermediate in a consecutive network proves deleterious. Given these simple generalizations for simple networks, we now consider somewhat more complex systems under intermediate conditions of backmixing, i.e., isothermal reactor environments intermediate between the PFR and CSTR limits.

3-8 YIELD AT INTERMEDIATE LEVELS OF MIXING

Earlier it was shown [Eqs. (3-39) to (3-42)] that degrees of mixing intermediate in character between CSTR and PFR can be simulated in terms of a PFR with recycle (Fig. 3-10). The recycle model will be developed for some typical selectivity schemes which we have already considered in terms of the CSTR and PFR extremes.

Consider the first-order consecutive sequence

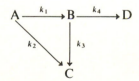

where we found for a PFR [Eq. (2-137)]

$$\frac{B}{A_0} = \frac{K_2}{1 - K_1}(f^{K_1} - f) + \frac{B_0}{A_0}f^{K_1} \qquad (3\text{-}51)$$

where
$$K_1 - \frac{k_3 + k_4}{k_1 + k_2} \quad \text{and} \quad K_2 = \frac{k_1}{k_1 + k_2}$$

For a PFR with recycle of a portion q, containing A, B, and C (effluent values), the inlet stream to the PFR (not the system) contains A and B, designated A_1 and B_1. That is, in (3-51), $A_0 = A_1$ and $B_0 = B_1$ must be expressed in terms of

R ($=q/Q$), which determines the PFR inlet concentrations. A simple material balance, assuming $B = 0$ *in the feed to the system*, gives

$$A_1 = \frac{RA + A_0}{R + 1} \qquad B_1 = \frac{RB}{R + 1} \qquad R = \frac{q}{Q} \qquad f = \frac{A}{A_1} = \frac{A(R + 1)}{RA + A_0} \tag{3-52}$$

Then

$$\frac{B}{A_1} = \frac{B(R + 1)}{RA + A_0} = \frac{B/A_0(R + 1)}{Rf + 1} \tag{3-53}$$

and

$$\frac{B_1}{A_1} = \frac{RB/A_0}{Rf + 1} \tag{3-54}$$

Substituting in Eq. (3-51) and solving for B/A_0, we have

$$\frac{B}{A_0} = \frac{\dfrac{fK_2}{1 - K_1}\left\{\left[\dfrac{f(R + 1)}{Rf + 1}\right]^{K_1 - 1} - 1\right\}}{1 - \dfrac{Rf}{Rf + 1}\left[\dfrac{f(R + 1)}{Rf + 1}\right]^{K_1 - 1}} \tag{3-55}$$

The results for various values of R are displayed in Fig. 3-12 for the limiting case in which k_2 and $k_4 = 0$.

As an exercise compute the yield of B versus conversion of A for (*a*) $K_1 = 0.1$ and $K_2 = 0.1$, (*b*) $K_1 = 0.01$ and $K_2 = 1$ for diverse values of R.

Recycle Model for Nonlinear Kinetics

For mixed orders

$$A + A \xrightarrow{\;k_1\;} B \qquad \text{second order}$$
$$\searrow^{k_2}$$
$$C \qquad \text{first order}$$

we found that the ratio of B to C for a PFR, where $K = k_2/k_1 A_1$ and $f = A/A_1$, is, from Eq. (2-101),

$$\frac{B - B_1}{C - C_1} = \frac{1 - f}{K \ln\left[(1 + K)/(f + K)\right]} - 1$$

In terms of recycle ratio, where $C_0 = B_0 = 0$,

$$\frac{B - \dfrac{RB}{R + 1}}{C - \dfrac{RC}{R + 1}} = \frac{1 - \dfrac{f(R + 1)}{Rf + 1}}{\dfrac{K(R + 1)}{Rf + 1} \ln \dfrac{Rf + 1 + K(R + 1)}{Rf + f + K(R + 1)}} - 1 \tag{3-56}$$

or

$$\frac{B}{C} = \frac{1 - f}{K(R + 1) \ln \dfrac{Rf + 1 + K(R + 1)}{Rf + f + K(R + 1)}} - 1 \tag{3-57}$$

where $f = A/A_0$ and $K = k_2/k_1 A_0$. In Fig. 3-13 B/C is plotted versus $1 - f$ for $K = 0.1, 1$ and 10. Other illustrations will be found in the literature.[1]

[1] B. G. Gillespie and J. J. Carberry, *Ind. Eng. Chem. Fundam.*, **5**: 164 (1966).

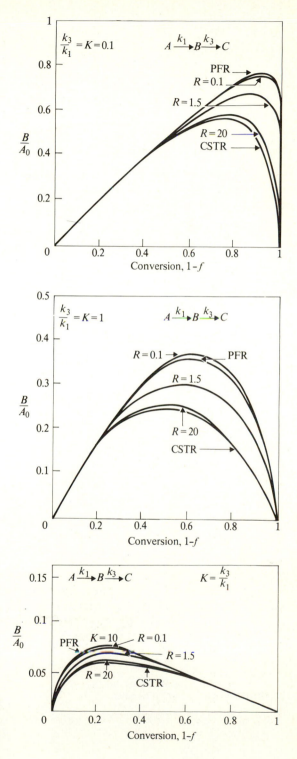

FIGURE 3-12
Yield of B versus conversion for linear reaction $A \xrightarrow{k_1} B \xrightarrow{k_3} C$ for various values of $k_3/k_1 = K$ and levels of mixing.   [*B. G. Gillespie and J. J. Carberry*, **5** : *164* (*1966*).]

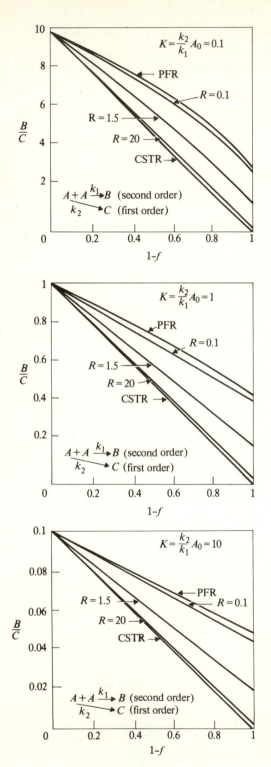

FIGURE 3-13
Yield of B relative to C for various levels of backmixing. [*B. G. Gillespie and J. J. Carberry, Ind. Eng. Chem. Fundam.*, **5**:*164 (1966).*]

Our treatment of selectivity-sensitive kinetic schemes in terms of PFR, CSTR, and intermediate degrees of mixing (recycle PFR) allows us to generalize concerning the influence of mixing upon isothermal selectivity. Thus, excluding autocatalysis:

1 For an isothermal consecutive-reaction system mixing favors the formation of the ultimate product; i.e., yield of intermediate is lower than in the absence of mixing for a given conversion (Fig. 3-12).

2 For isothermal simultaneous reactions, mixing favors the reaction of lowest order (Fig. 3-13). If each reaction is of the same order, the product ratio is unaffected by mixing.

We shall now discuss a system which poses a problem of design in that the yield is a function of both consecutive and simultaneous steps, the simultaneous one being second order. The system is

$$A \xrightarrow{k_1} B \xrightarrow{k_2} C \qquad A + A \xrightarrow{k_3} D$$

where B is a desired product and C and D undesired by-products. This system, first cited by van de Vusse[1] poses a dilemma in the light of conclusions 1 and 2 above, for according to conclusion 1, the yield of B is favored in a PFR, yet the by-product, second-order reaction, which produces undesired D, is best suppressed in a CSTR. Van de Vusse treated this system for PFR and CSTR extremes as a function of the rate-constant ratios

$$K_1 = \frac{k_3 A_0}{k_1} \qquad \text{and} \qquad K_2 = \frac{k_2}{k_1}$$

On a time-free basis, the rate of formation of B ($b = B/A_0$) to disappearance of A ($f = A/A_0$) is

$$\frac{db}{df} = \frac{K_2 b}{f(1 + K_1 f)} - \frac{1}{1 + K_1 f} \qquad (3\text{-}58)$$

Then

$$b = \exp\left(K_2 \ln \frac{f}{1 + K_1 f}\right)\left[\int \frac{-\exp\left(-K_2 \ln \frac{f}{1 + K_1 f}\right)}{1 + K_1 f} df + \text{const}\right]$$

or

$$b = \left(\frac{f}{1 + K_1 f}\right)^{K_2}\left[\int \frac{-[f/(1 + K_1 f)]^{-K_2}}{1 + K_1 f} df + \text{const}\right] \qquad (3\text{-}59)$$

The integral in Eq. (3-59) can be solved analytically only for specific values of K_2, for example, $\frac{1}{16}$, $\frac{1}{8}$, $\frac{1}{2}$, 1, 2. Van de Vusse presents solutions of (3-59) for a range of K_2 values, where $B_0 = 0$ for a PFR. The CSTR result is, of course, obtained directly from Eq. (3-58).

[1] J. G. van de Vusse, *Chem. Eng. Sci.*, **91**: 994 (1964).

As an exercise, derive the CSTR yield B/A_0 for the specific rate-coefficient ratios cited here.   When $B = B_0$, the PFR results are

$$K_2 = \frac{k_2}{k_1} = \frac{1}{2} \qquad b_0 = \frac{B_0}{A_0}$$

$$b = -\left[\frac{f}{K_1(1 + K_1 f)}\right]^{1/2}\left[\ln \frac{2K_1 f + 1 + 2\sqrt{K_1 f(1 + K_1 f)} - b_0\sqrt{K_1(1 + K_1)}}{2K_1 + 1 + 2\sqrt{K_1(1 + K_1)}}\right]$$

$$(3\text{-}60)$$

$$K_2 = 1: \qquad b = \frac{-f}{1 + K_1 f}\left[\ln \frac{1}{f} - b_0(1 + K_1)\right] \qquad (3\text{-}61)$$

$$K_2 = 2: \qquad b = \left(\frac{f}{1 + K_1 f}\right)^2\left[\frac{1}{f} - K_1 \ln f - 1 + b_0(1 + K)^2\right] \qquad (3\text{-}62)$$

While van de Vusse explored the two mixing extremes (PFR and CSTR) for various values of K_1 and K_2, an investigation of intermediate degrees of mixing seems attractive.[1]   Here the merits of the recycle model became apparent as the mixing parameter (recycle ratio R) enters the problem as a boundary condition. Direct substitution of feed concentrations in terms of R into Eqs. (3-60) to (3-62) yields B/A_0 as a function of extent of mixing as characterized by R.

After some algebraic manipulations we obtain, noting that $K_1 f$ is independent of R,

$$K_2 = \tfrac{1}{2}: \qquad K_1' = K_1 \frac{Rf + 1}{R + 1}$$

$$\frac{B}{A_0} = b = -\frac{\left[\dfrac{f}{K_1(1 + K_1 f)}\right]^{1/2}\left[\ln \dfrac{2K_1 f + 1 + 2\sqrt{K_1 f(1 + K_1 f)}}{2K_1' + 1 + 2\sqrt{K_1'(1 + K_1')}}\right]}{1 - \left[\dfrac{f}{K_1(1 + K_1 f)}\right]^{1/2}\dfrac{R}{Rf + 1}\sqrt{K_1'(1 + K_1')}} \qquad (3\text{-}63)$$

$$K_2 = 1: \qquad b = \frac{-\dfrac{f}{1 + K_1 f}\ln \dfrac{Rf + 1}{f(R + 1)}}{1 - \dfrac{f}{1 + K_1 f}\dfrac{R}{Rf + 1}(1 + K_1')} \qquad (3\text{-}64)$$

$$K_2 = 2: \qquad b = \frac{\left(\dfrac{f}{1 + K_1 f}\right)^2\left[\dfrac{1}{f} - K_1 \ln \dfrac{f(R + 1)}{Rf + 1} - \dfrac{R + 1}{Rf + 1}\right]}{1 - \left(\dfrac{f}{1 + K_1 f}\right)^2\dfrac{R + 1}{Rf + 1}\dfrac{R}{Rf + 1}(1 + K_1')^2} \qquad (3\text{-}65)$$

Results are shown in Fig. 3-14, which indicates that for the isothermal consecutive kinetic system involving a simultaneous second-order step, neither the plug-flow

B. G. Gillespie and J. J. Carberry, *Chem. Eng. Sci.*, **21**: 472 (1966).

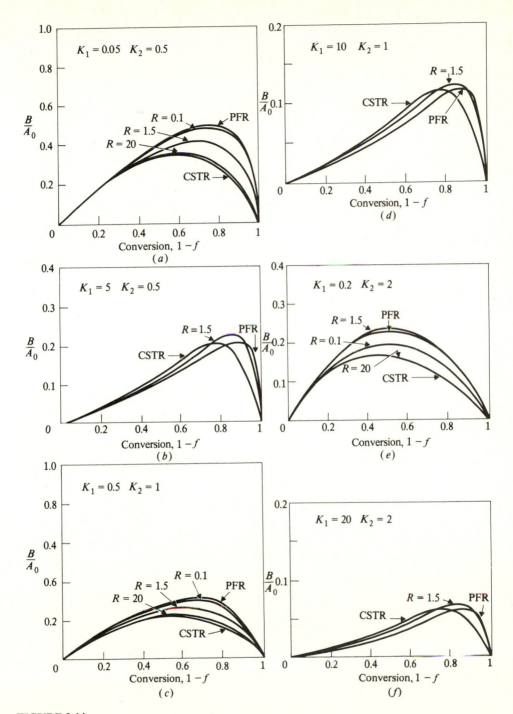

FIGURE 3-14
Yield in van de Vusse network. [*B. G. Gillespie and J. J. Carberry, Chem. Eng. Sci.*, **21**:*472* (1966).]

nor perfect-mixing ideal necessarily leads to a maximum of B.   Instead an intermediate level of mixing gives the best yield of B for certain values of the rate-constant ratios.

Meaning of recycle model   The recycle model is basically a nonsegregated flow model.   For the first-order systems this fact is irrelevant, as noted earlier.   However, when nonlinear kinetics is involved, a difference in predicted performance results between the assumption of segregated and nonsegregated mixing.   Our analysis of the van de Vusse system is quantitatively sound only if nonsegregation actually prevails.   Qualitatively, the analysis is valid insofar as trends are indicated; e.g., an intermediate degree of mixing may give the best yield of B.

Other models might be offered to simulate intermediate levels of mixing. Mention has already been made of the series of n CSTRs which simulate an intermediate level of residence-time distribution and therefore a level of mixing residing between plug and perfectly mixed flow.   Cholette and associates[1] have made effective use of PFR-CSTR series combinations to describe intermediate mixing effects for isothermal, exothermal, and endothermal adiabatic reactor operations. As has been noted in this chapter, the backmixing process is also described in terms of plug flow with axial diffusion.

For isothermal operation, Cholette and his coworkers conclude, in agreement with out thought experiment and associated calculations, that for a PFR followed by a CSTR conversions are higher if reaction order is greater than unity and less if the order is less than unity.   For a CSTR followed by a PFR the opposite holds.

One may well ask: Which of the diverse total and partial backmixing models best reflects reality?   There is a paucity of data on reaction in partially back-mixed systems, particularly as regards selectivity alteration.   The choice of mixing model becomes a matter of convenience and ease of application.   From this point of view, it is evident that plug flow with axial dispersion is the least convenient method of describing intermediate levels of backmixing.   Even for a simple isothermal first-order reaction, one must solve the equation (Fig. 3-15)

$$\frac{1}{\text{Pe}_a}\frac{d^2f}{dz^2} - \frac{df}{dz} = kf\theta \qquad (3\text{-}66)$$

while for the series of n CSTRs, one has

$$f = \frac{1}{(1 + k\theta_n)^n} \qquad (3\text{-}67)$$

or, in terms of the recycle model,

$$f = \frac{\exp\{-[k\theta/(R+1)]\}}{1 + R\{1 - \exp[-k\theta/(R+1)]\}} \qquad (3\text{-}68)$$

We have shown that nonlinearity of kinetics presents no particular difficulty in formulating the series-CSTR or the recycle model.   For nonlinear kinetics,

[1] A. Cholette, et. al., *Can. J. Chem. Eng.*, **37**: 105 (1959), **38**: 192 (1961).

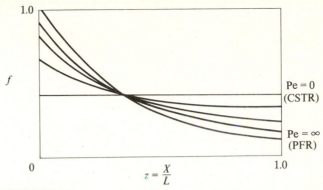

FIGURE 3-15
Isothermal first-order-concentration–reactor-length profiles for plug
flow with axial mixing ($Pe_a = 0$ to ∞; CSTR to PFR).

Eq. (3-66) must be resolved numerically and not by routine techniques.

  With complex reaction networks the comparative ease of application of the
series-CSTR and recycle models compared with that of axial diffusion and plug
flow becomes quite evident.   For isothermal reaction networks, we can recommend
that simulation of partial backmixing be undertaken in terms of the series-CSTR
or recycle models.

  For partial mixing in a nonisothermal reactor, e.g., a tubular reactor with
heat exchange so limited that an axial temperature profile is established, it is evident
that one must resort to simulation via the plug-flow–dispersion model.   But in this
instance the nonlinearity due to the Arrhenius functionality requires numerical
solution of two nonlinear differential equations, so that it is idle to contemplate
simple analogies.

3-9 THE LAMINAR FLOW REACTOR (LFR)

In dealing with the isothermal CSTR and PFR we are treating the issue of residence-time-distribution effects generated by mixing or its absence in the direction
of bulk flow.   Implicit in our creation of the PFR model is the notion that velocity
and/or concentration gradients *do not* exist in the direction perpendicular to flow,
i.e., in the radial direction for the rather usual instance of cylindrical, tubular
reactor utilization.   Indeed the term piston or plug flow derives from the assumption of radial uniformity of temperature, concentrations, and velocity.

  However, fluid mechanics teaches us that if the Reynolds number of fluid
flowing in a tube is below a value of 2100, laminar flow prevails, with a consequent
parabolic velocity distribution, given by

$$u_r = u_{max}\left[1 - \left(\frac{r}{R}\right)^2\right] \qquad (3\text{-}69)$$

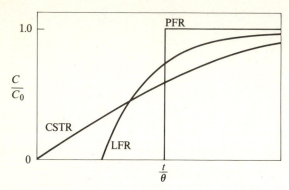

FIGURE 3-16
Step response for a PFR, CSTR, and LFR.

Hence residence time of any fluid element at a particular radial position r is

$$\theta_r = \frac{L}{u_r} \qquad (3\text{-}70)$$

from which we note that molecules in the centerline of the laminar-flow tubular reactor experience a residence time which is one-half the average. The residence-time distribution for laminar flow is shown in Fig. 3-16 in contrast with that for a PFR and CSTR.

As an exercise, derive the residence-time distribution for laminar flow (Fig. 3-16).

We observe that the residence-time distribution for the LFR is one intermediate between the PFR and CSTR, and thus we can anticipate that conversion and yield behavior of the LFR will lie between those established for the PFR and CSTR.

In processing of normal newtonian reacting fluids, laminar flow conditions are hardly sought. Conditions of turbulent flow, so necessary for effective heat and mass transport, are design objectives. However, when dealing with fluids of very high viscosity, as in polymer-reaction processing equipment, the condition of turbulent flow (Re > 2100) may not be readily realized at tolerable levels of momentum loss (pressure drop). For such (usually nonnewtonian) fluids, laminar flow must be tolerated, in which case analyses of the LFR become meaningful.

Furthermore, even with turbulent flow, if a radial temperature gradient exists, there must also exist a radial distribution in local reaction rate with a consequent radial concentration gradient. Usually, this condition occurs more commonly in fixed-bed reactors. The point is that whether one deals with a true LFR or a turbulent-flow unit within which a radial concentration distribution is induced by an extrafluid mechanical factor (temperature distribution), an assessment of the influence of radial concentration distribution upon conversion and selectivity is in order.

Seminal treatments of the isothermal LFR have been provided for the conversion problem for zero-, first-, and second-order kinetics.[1] Johnson[2] has neatly organized a general treatment of the isothermal LFR not only for conversion but yield in a linear consecutive-reaction network, an extension of practical import since polymerizations involve consecutive reactions. The essence of Johnson's analyses follows.

Conversion in an LFR

Fluid velocity as a function of reduced radius r/R in laminar flow is given by Eq. (3-69); hence contact time is

$$\theta_r = \frac{L}{u_{max}[1 - (r/R)^2]} \qquad (3\text{-}71)$$

Assuming that the fluid in any annular element behaves as it does in a constant-volume isothermal reactor, the differential molar flow rate F is

$$dF = Au2\pi r\, dr \qquad (3\text{-}72)$$

where A is species concentration. For simple kinetics we assume

$$-\frac{dA}{d\theta} = kA^\alpha$$

where, for $\alpha \neq 1$, integration gives

$$A = A_0[1 + k(\alpha - 1)A_0^{\alpha-1}\theta_r]^{1/(1-\alpha)} \qquad (3\text{-}73)$$

We substitute Eqs. (3-71) and (3-73) into (3-72) and define

$$V = 1 - \left(\frac{r}{R}\right)^2 \qquad \text{and} \qquad \beta = \frac{k(\alpha - 1)A_0^{\alpha-1}L}{u_{max}} \qquad (3\text{-}74)$$

to obtain

$$F = \pi R^2 u_{max} A_0 \int_0^1 \left(1 + \frac{\beta}{V}\right)^{1/(1-\alpha)} V\, dV \qquad (3\text{-}75)$$

For second-order reaction, $\alpha = 2$, integration yields conversion x

$$x = 1 - \frac{F}{F_0} = 2\beta - 2\beta^2 \ln\left(1 + \frac{1}{\beta}\right) \qquad (3\text{-}76)$$

For a PFR and $\alpha = 2$, we have

$$x = \frac{2\beta}{1 + 2\beta} \qquad (3\text{-}77)$$

[1] R. C. L. Bosworth, *Phil. Mag.*, **39**: 847 (1948); K. G. Denbigh, *J. Appl. Chem.*, **1**: 227 (1951); F. A. Cleland and R. H. Wilhelm, *AIChE J.*, **2**: 489 (1956).
[2] M. M. Johnson, *Ind. Eng. Chem. Fundam.*, **9**(4): 681 (1970).

As an exercise, verify Eq. (3-77).   For β values of 0.2, 0.5, 1, and 4 compute the ratio of conversion in a PFR to that in a LFR.

For the first-order reaction $A = A_0 \exp(-k\theta)$, substituting into Eq. (3-72), as before, we find

$$\frac{F}{F_0} = 2 \int_0^1 \exp\left(-\frac{a}{V}\right) V \, dV \qquad \text{where } a = \frac{kL}{u_{max}} \qquad (3\text{-}78)$$

Letting $Z = 1/V$ leads to

$$\frac{F}{F_0} = 2 \int_1^\infty \frac{\exp(-aZ)}{Z^3} \, dZ$$

or the third-order exponential integral

$$\frac{F}{F_0} = 2E_3(a) \qquad (3\text{-}79)$$

Example   For $\alpha = 1$ and $a = 1.1$ compute the conversion in a LFR and in a PFR.   In the "Handbook of Mathematical Functions" we find $E_3(1.1) = 0.1$; thus

$$x_{LFR} = 1 - 2E_3(1.1) = 0.8 = 1 - \frac{F}{F_0}$$

For the PFR

$$x = 1 - \exp(-2a)$$

since

$$\frac{kL}{u_{max}} = \frac{kL}{2u_{av}} = \frac{k\theta}{2}$$

That is, we assume average contact time L/u_{av} to be the same in each reactor.   For $a = 1.1$ in plug flow,

$$x = 1 - \exp(-2.2) = 0.87$$

Selectivity in an LFR

For the linear consecutive network $A \xrightarrow{1} B \xrightarrow{2} C$ we seek the yield of B in laminar flow, which, by arguments similar to those above, is

$$\frac{N_B}{2N_{A_0}} = \frac{k_1}{k_2 - k_1} \int_0^1 \left[\exp\left(-\frac{a}{v}\right) - \exp\left(-\frac{a'}{v}\right) \right] V \, dV \qquad (3\text{-}80)$$

where

$$a = \frac{k_1 L}{u_{max}} \qquad \text{and} \qquad a' = \frac{k_2 L}{u_{max}}$$

Therefore

$$\frac{N_B}{2N_{A_0}} = \frac{k_1}{k_2 - k_1} [E_3(a) - E_3(a')] \qquad (3\text{-}81)$$

As an exercise, derive Eq. (3-81).

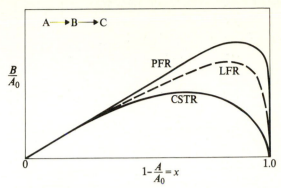

FIGURE 3-17
Yield of B in $A \overset{k_1}{\to} B \overset{k_2}{\to} C$ for LFR, PFR, and CSTR.

The LFR yield of B is schematically shown as a function of conversion of A in Fig. 3-17, in contrast with the PFR and CSTR. Should pressure-drop considerations force one to deal with laminar flow with consequent conversion and yield taxation, the static pipeline mixer (a flow inverter) may remedy matters by reducing the spread of residence-time distribution.[1]

3-10 NONISOTHERMAL REACTOR PERFORMANCE

Since heat release or absorption occurs with chemical reaction, the isothermal conditions assumed to prevail in our earlier discussions may not exist in real reactor systems unless the heat effect is negligible and/or suitable design permits removal or addition of heat at a rate which exactly balances the release or abstraction of thermal energy at each point in time or space.

Reactions accompanied by negligible enthalpy change are rare indeed. For a given enthalpy change (heat of reaction) the heat capacity of the total reactant-bearing stream may be high enough to yield a tolerably small temperature change with extent of reaction. Thus liquid-phase reactions are more easily conducted isothermally, particularly in aqueous solvent systems.

In general, rapid reaction assures us smaller reactor-volume requirements; yet the very rapidity of reaction leads, quite logically, to rapid rates of heat release or abstraction. Heat capacity of the solvent or gaseous carrier is limited, so that in most cases the realization of isothermal conditions requires some ingenuity in the design of heat-transfer devices within the reactor. Some detailed discussion of the heat-transfer problem will be undertaken later. At this point, we consider the problem of conversion and selectivity in a simple case of nonisothermality, namely, the nonisothermal CSTR. In such a system, part of the heat released in, say, an exothermic reaction is absorbed by the reacting stream, thus raising its temperature

[1] T. Bor, *Br. Chem. Eng.*, **16**(7): 610 (1971).

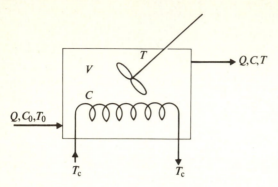

FIGURE 3-18
Nonisothermal CSTR; that is, $T \neq T_0$.

to an extent which depends upon the heat of reaction, conversion, and the thermal capacity of the stream and the extent to which heat is transferred to the coolant.

We consider nonisothermal behavior for two reasons: (1) real reactors are seldom operable under isothermal conditions, and (2), as a consequence, what has been said about conversion, selectivity, and mixing effects for isothermal systems in the earlier sections of this book may not necessarily be true under nonisothermal circumstances.   Indeed, as we shall see, for certain reaction networks, nonisothermal operation may yield dividends in selectivity as well as conversion.

Nonisothermal Homogeneous Reaction in a CSTR[1] and a PFR

A CSTR is operated nonisothermally; i.e., the heat generated (or absorbed) due to reaction is only partially removed (or added) with the consequence that the feed stream absorbs (or yields) the untransferred heat, thus raising (or lowering) the reactant-stream temperature.   With operation in steady state, referring to Fig. 3-18, we invoke both a mass and heat balance in accord with

$$\text{Input} = \text{output} + \text{generation} + \text{loss}$$

For mass

$$QC_0 = QC + \mathcal{R}V$$

or

$$\frac{C}{C_0} = 1 - \frac{\mathcal{R}\theta}{C_0} \qquad (3\text{-}82)$$

Note that our dimensionless rate (for any single reaction order) is

$$\frac{\mathcal{R}\theta}{C_0} = 1 - \frac{C}{C_0} \qquad (3\text{-}83)$$

For heat, assuming physical properties to be sensibly temperature-independent,

$$\rho C_p Q T_0 = \rho C_p Q T - \mathcal{R}(-\Delta H)V + U_0 A(T - T_c) \qquad (3\text{-}84)$$

---

[1] Obviously a nonisothermal CSTR is indeed isothermal with respect to spatial coordinates. For the CSTR, nonisothermality refers to reaction temperature relative to feed temperature.

Rearranging Eq. (3-84) gives

$$T\left(1 + \frac{U_0 A}{\rho C_p Q}\right) = T_0 + \frac{\mathscr{R}(-\Delta H)\theta}{\rho C_p} + \frac{U_0 A}{\rho C_p Q} T_c \qquad (3\text{-}85)$$

When we eliminate $\mathscr{R}\theta$ from Eq. (3-85), note that $(C_0 - C)/C_0$ can be defined as reactant conversion x, and then let

$$t = \frac{T}{T_0} \qquad \bar{U} = \frac{U_0 A}{\rho C_p Q} \qquad \text{and} \qquad \gamma = \frac{T_c}{T_0}$$

we have

$$t = \frac{T}{T_0} = \frac{1 + \bar{U}\gamma}{1 + \bar{U}} + \frac{-\Delta H \, C_0 \, x}{\rho C_p T_0(1 + \bar{U})} \qquad (3\text{-}86)$$

Note that $-\Delta H$ is positive for exothermal reaction and negative for endothermal reaction. Examination of Eq. (3-86) indicates the following points:

1 As $U_0 A$ increases $T \rightarrow T_0$; that is, the rate of heat removal (or addition) approaches a value guaranteeing isothermality relative to feed temperature.

2 $T \rightarrow T_0$ as $Q\rho C_p$ increases; i.e., as the thermal capacity of the stream increases, the temperature change is less for a given rate of generation and removal.

3 When $U_0 A \rightarrow 0$ (no heat transferred between coolant and reacting mass), adiabatic conditions prevail; i.e., all heat liberated (or absorbed) is adsorbed (or removed) by the reacting stream.

For the nonisothermal CSTR under discussion, Eq. (3-86) indicates that conversion x is simply related to the temperature of reaction T for a specified value of reaction enthalpy change, heat capacity and heat-transfer coefficient, transfer area, and volumetric feed rate.

The situation is far more complex in a PFR since x is a function of distance along the reactor, the rate of heat transfer between coolant and reacting fluid varies (for constant $U_0 A$) along the length of the reactor, and T is a function of distance along the reactor, as shown in Fig. 3-19.[1] If $UA = 0$ for the PFR, we

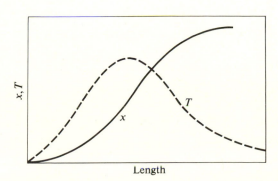

Length

FIGURE 3-19
Exothermic conversion-temperature profiles in a PFR (nonisothermal and nonadiabatic).

---

[1] It is herein assumed that T is not a function of radial position.

have the adiabatic PFR.   Then T is uniquely related to x via the heat-mass balance. For a simple PFR we have, in terms of conversion of a single reaction,

$$\frac{dx}{d\theta} = k(1 - x) = \mathcal{R} \qquad (3\text{-}87a)$$

For heat, the balance gives

$$\frac{dt}{d\theta} = \frac{-\Delta H\, C_0}{\rho C_p T_0} k(1 - x) = \beta\mathcal{R} \qquad (3\text{-}87b)$$

Eliminating $\mathcal{R}$ between Eqs. (3-87a) and (3-87b) gives

$$\frac{dt}{d\theta} = \beta\frac{dx}{d\theta}$$

which upon integration yields

$$t = \frac{T}{T_0} = 1 + \bar{\beta}x \qquad (3\text{-}88)$$

where

$$\bar{\beta} = \frac{-\Delta H C_0}{\rho C_p T_0}$$

The differential equation describing reactant temperature versus θ (or distance along the PFR), becomes

$$\frac{dt}{d\theta} = \beta(1 - x)k_0 \exp\left[-\epsilon\left(\frac{1}{1 + \bar{\beta}x} - 1\right)\right] \qquad (3\text{-}89)$$

where $\epsilon = E/RT_0$, while for mass

$$\frac{dx}{d\theta} = k_0(1 - x) \exp\left[-\epsilon\left(\frac{1}{1 + \bar{\beta}x} - 1\right)\right] \qquad (3\text{-}90)$$

In view of the unique relationship between t and x for adiabatic operation [Eq. (3-88)] Eq. (3-90) is readily solved graphically to yield both x-versus-θ and t-versus-θ

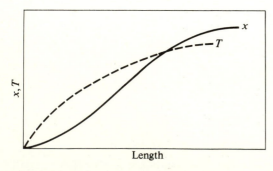

FIGURE 3-20
Exothermic conversion temperature in an adiabatic
PFR.

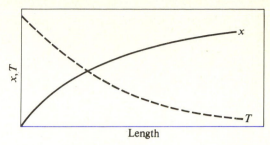

FIGURE 3-21
Adiabatic PFR (endothermic reaction).

profiles, as shown in Fig. 3-20 (exothermal reaction, $\bar{\beta} = +$) and Fig. 3-21 (endo-thermal reaction, $\bar{\beta} = -$).

Conversion in an Adiabatic CSTR

To dramatize the relative influence of backmixing (thermal as opposed to mass) we now consider the adiabatic CSTR. As discussed in Chap. 2, the specific rate constant varies with temperature according to the Arrhenius expression

$$k = \mathscr{A} \exp\left(-\frac{E}{RT}\right)$$

At some specified temperature (conveniently the reactor-feed temperature T_0)

$$k_0 = \mathscr{A}_0 \exp\left(-\frac{E}{RT_0}\right)$$

Assuming $\mathscr{A}$ to be relatively temperature-independent, k at T is

$$\frac{k}{k_0} = \exp\left[-\epsilon\left(\frac{1}{t} - 1\right)\right] \qquad (3\text{-}91)$$

where $t = T/T_0$. For the adiabatic CSTR, t is

$$t = 1 + \frac{-\Delta H\, C_0\, x}{\rho C_p\, T_0} = 1 + \bar{\beta}x$$

For the reaction $A \rightarrow B$, in a CSTR

$$\frac{C}{C_0} = f = 1 - x = \frac{1}{1 + k\theta}$$

or in terms of k as a function of $t = 1 + \bar{\beta}x$

$$1 - x = \frac{1}{1 + k_0\,\theta \exp\left[\dfrac{-E}{RT_0}\left(\dfrac{1}{1 + \bar{\beta}x} - 1\right)\right]} \qquad (3\text{-}92)$$

which establishes the relation between conversion x, $1 - f$, and $k_0\,\theta$ (reactor volume) as a function of E/RT_0 and net heat generation $\bar{\beta}$.

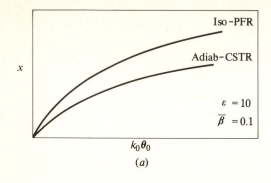

(a)

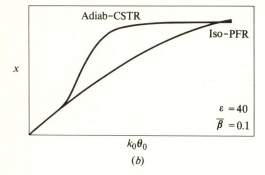

(b)

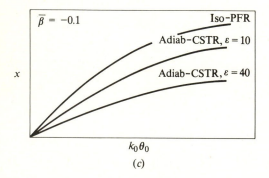

(c)

FIGURE 3-22
Adiabatic CSTR conversion versus $k_0\theta_0$ contrasted with isothermal PFR performance. (a) Small ϵ, $\bar{\beta}=0.1$; (b) large ϵ, $\bar{\beta}=0.1$; (c) endothermic reaction, $\bar{\beta}=-0.1$. [*J. M. Douglas, Chem. Eng. Prog. Symp. Ser.*, **60** (*48*): *1 (1969)*.]

Consider two exothermic adiabatic cases where $\bar{\beta}=0.1$ in each but $E/RT_0 =10$ in case 1 and 40 in case 2. We now compare the x-versus-$k_0\theta$ relation in each case against the isothermal PFR performance, where, of course, $1-x = \exp[-k_0\theta]$.

By trial and error, the CSTR adiabatic solution is realized with the results shown in Fig. 3-22.

In Fig. 3-22*a*, although the temperature of the CSTR is higher than that of the PFR (as $x \to 1$, $t_{PFR} = 1$ and $t_{CSTR} = 1.1$) the CSTR temperature benefit does not cause a significant rate-coefficient enhancement due to a low activation energy ($E/RT_0 = 10$). For a given conversion, a large $k_0\theta$ (reactor volume) is required due to the backmixing in the CSTR. In case 1, the detrimental influence of perfect mixing upon rate is more significant and is not compensated for by the higher CSTR operating temperature.

In case 2, the ultimate 10 percent increase in temperature as $x \rightarrow 1$ in the CSTR significantly increases k, since $E/RT_0 = 40$, with the result that the rate-coefficient enhancement more than compensates for the reactant dilution due to mixing in the CSTR. For endothermal cases, the CSTR is always less efficient (Fig. 3-22c).

Yield in an Adiabatic CSTR

A useful basis for comparing nonisothermal selectivity effects is its comparison with yield in the isothermal CSTR. For the case $A \xrightarrow{k_1} B \xrightarrow{k_2} C$ heat generation is, of course, due to formation of B and C. For the nonisothermal CSTR, assuming for purposes of illustration that $\bar{U} = 0$, the temperature T/T_0 is

$$t = 1 + \frac{-\Delta H_1 A_0(x - y)}{\rho C_p T_0} + \frac{[-\Delta H_1 + (-\Delta H_2)]A_0 y}{\rho C_p T_0} \tag{3-93}$$

Let

$$\bar{\beta}_1 = \frac{-\Delta H_1 A_0}{\rho C_p T_0} \quad \text{and} \quad \bar{\beta}_2 = \frac{-\Delta H_2 A_0}{\rho C_p T_0}$$

then

$$t = 1 + \bar{\beta}_1 x + \bar{\beta}_2 y \tag{3-94}$$

where $x =$ moles A converted per mole A_0
   $y =$ moles C formed per mole A_0
   $x - y =$ mole B formed $= B/A_0$ per mole A_0

Then for a CSTR,

$$\frac{B}{A_0} = x - y = \frac{x}{1 + K[x/(1 - x)]} \quad \text{where } K = \frac{k_2}{k_1}$$

From the definition of k [Eq. (3-91)]

$$K = \frac{k_2}{k_1} = K_0 \exp\left[\frac{E_1 - E_2}{RT_0}\left(\frac{1}{t} - 1\right)\right] \quad K_0 = \left(\frac{k_2}{k_1}\right)_0 \tag{3-95}$$

The nonisothermal CSTR yield of B relative to the isothermal CSTR case is

$$\frac{(x - y)_{ad}}{(x - y)_{iso}} = \frac{1 + K_0 \dfrac{x}{1 - x}}{1 + K_0 \dfrac{x}{1 - x} \exp\left[\Delta\epsilon\left(\dfrac{1}{t} - 1\right)\right]} \quad \text{where } \Delta\epsilon = \frac{E_1 - E_2}{RT_0} \tag{3-96}$$

and t is given by Eq. (3-94). The relative yield will depend, then, upon $\Delta\epsilon$, $\bar{\beta}_1$, $\bar{\beta}_2$, and x for a given value of K_0.

Before presenting some quantitative results, a qualitative insight is to be gained by considering an Arrhenius plot of $(\ln k)$-versus-$1/T$ data for a consecutive reaction coefficient (see Fig. 3-23). Consider the case where $E_1 > E_2$.

Assume that $K_0 > 1$, that is, the isothermal rate-constant ratio is determined at a temperature corresponding to $1/T_0 = a$. If the reaction is net exothermic,

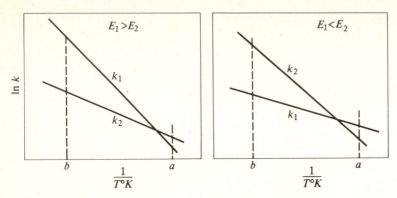

FIGURE 3-23
Arrhenius plot of k_1 and k_2 versus $1/T$ for $E_1 < E_2$ and $E_1 > E_2$.

$1/T$ decreases, reducing k_2/k_1 and thereby enhancing the yield. For an endothermic process k_2/k_1 becomes larger than K_0, and yield is less than the isothermal value. The same reasoning and conclusions apply if K_0 corresponds to $1/T_0 = b$.

When on the other hand, $E_2 > E_1$ (Fig. 3-23), the isothermal event is determined by K_0 corresponding to $1/T_0 = a$. An exothermic condition increases $K = k_2/k_1$, thus decreasing yield, while an endothermic condition enhances yield. These conclusions can be generalized by observing the behavior of $(k_1/k_2)_T/(k_1/k_2)_0$ as a function of $t \; (\gtrless 1.0)$ and $\Delta\epsilon \; [=(E_1 - E_2)/RT_0]$, as set forth in Fig. 3-24.

Yield in Mixed Endothermal and Exothermal Systems

One is often confronted with a complex reaction network in which one step is exothermic and the other endothermic. For example, hydrogenation (exothermic) may occur with cracking (endothermic), either simultaneously or consecutively. Hutchings[1] has computed yield-conversion profiles for such systems by comparing adiabatic CSTR yield relative to that in a PFR (isothermal), for various values of activation-energy differences and enthalpies ($\pm$). His results are reproduced in Fig. 3-25 a to d.

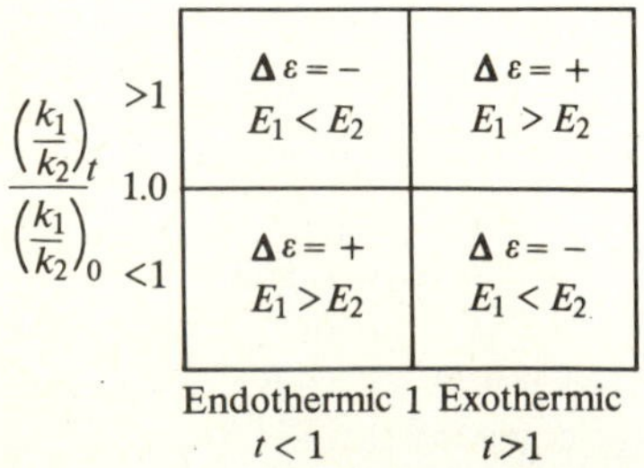

	$\Delta\epsilon = -$	$\Delta\epsilon = +$
$\left(\dfrac{k_1}{k_2}\right)_t > 1$	$E_1 < E_2$	$E_1 > E_2$
$\dfrac{\left(\dfrac{k_1}{k_2}\right)_t}{\left(\dfrac{k_1}{k_2}\right)_0}$ 1.0		
$\left(\dfrac{k_1}{k_2}\right)_0 < 1$	$\Delta\epsilon = +$ $E_1 > E_2$	$\Delta\epsilon = -$ $E_1 < E_2$

Endothermic 1 Exothermic
$t < 1$ $\qquad$ $t > 1$

FIGURE 3-24
Quadrant displaying the influence of exothermicity and endothermicity upon the ratio of rate coefficients.

[1] J. Hutchings, Ph.D. dissertation, University of Notre Dame, Notre Dame, Ind., 1968.

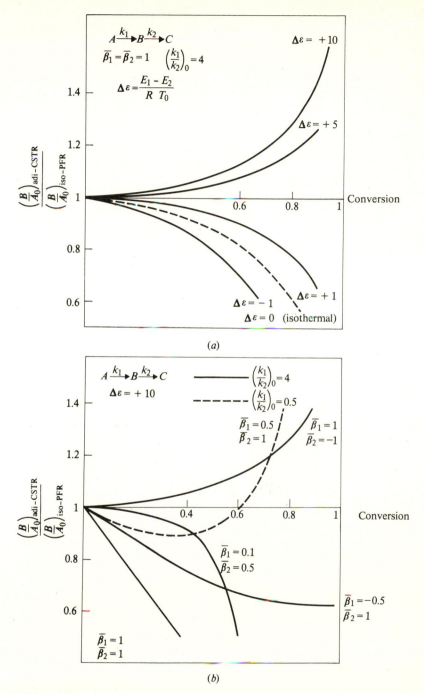

FIGURE 3-25 *a, b, c, d*
Yield or selectivity in an adiabatic CSTR relative to that in an isothermal PFR vs. conversion for consecutive and simultaneous reaction networks.   [*J. J. Carberry, Ind. Eng. Chem.*, **58**:40 (1966).]

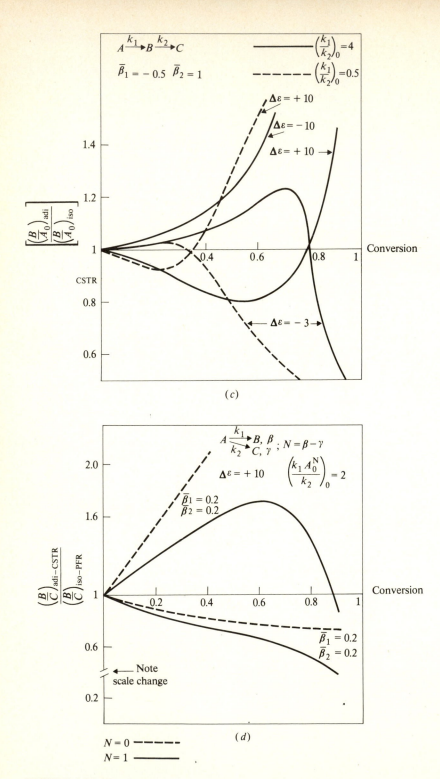

FIGURE 3-25 (*Continued*)

We deem that sufficient evidence has now been set forth to persuade the chemical reaction engineer of the validity of the following theses:

Isothermal environments

1 For normal reaction kinetics (rate decreases with conversion), backmixing is detrimental.

2 For abnormal reaction kinetics (rate increases with conversion), as found in the early stages of autocatalysis and in instances of negative-order kinetics, backmixing is of benefit.

3 The penalty associated with backmixing is proportional to reaction order.

4 In complex, multipathed reaction networks, the steps of highest order are most severely damped by backmixing.

5 In consecutive-reaction networks, overexposure of intermediate by virtue of backmixing prompts its destruction; hence isothermal backmixing is detrimental to yield of intermediate in consecutive networks.

In these isothermal networks, the CSTR-PFR comparison is essentially a comparison of dilution effects sponsored by backmixing (or recycling).

Non isothermal environments With a net release (or abstraction) of heat, besides concentration modification there is temperature modification of the reaction-reactor environment.

1 For normal reaction kinetics, backmixing of heat in an exothermic reaction may well totally compensate for reactant dilution effects noted in item 1 above, with the consequence that nonisothermal backmixing can enhance reactor performance relative to the isothermal or even nonisothermal PFR. Obviously in endothermic reaction, nonisothermality compounds the taxation due to mass backmixing.

2 For abnormal kinetics, exothermic heat generation magnifies the backmixing benefit noted in item 2 above.

3 The penalty or benefit associated with thermal backmixing is a function of combined endothermicity and exothermicity $(\pm \bar{\beta})$ and the sensitivity of the rate coefficient to temperature as revealed by $\epsilon = E/RT_0$.

4 In complex, multipathed, nonisothermal reaction circumstances, yield will patently depend not only on relative reaction orders (as noted in item 4 above) but also upon relative activation energies; see, for example, Fig. 3-25.

In these nonisothermal, backmixed networks, the CSTR-PFR comparison is essentially a comparison of competing (exothermic) and enhancing (endothermic) influences of heat and mass recycling.

Insofar as real reactor environments are rarely isothermal, it is of signal importance that we anticipate that backmixing of heat can be far more influential upon conversion and yield than backmixing or dispersion of mass. This is hardly a surprise in view of the fact that the rate of a simple reaction is exponentially dependent upon temperature and, at most, about second order in concentration.

The dispersion of heat will generally prove dominant relative to that of mass, save in unusual cases of small heat release ($\bar{\beta} \to 0$) or a low sensitivity of k to temperature ($\epsilon \to 0$).

3-11 UNIQUENESS OF THE STEADY STATE

In deriving the model of nonisothermal CSTR [Eqs. (3-82) to (3-92)], we have implicitly assumed that the derived temperature t is unique. Obviously in the CSTR, the steady-state temperature is that established by a just balance of heat generation ($\pm$) with heat removal ($\pm$). Again we shall assume constancy of physical properties and reaction enthalpy with respect to temperature.[1]

Equating heat generation (exothermic) in a nonisothermal CSTR to its removal by flow and intrinsic heat exchange provides a graphic display of possible steady states.

For a simple reaction A $\to$ B heat generation is given by

$$q_{\text{gen}} = -\Delta H \, kC \qquad \text{for linear kinetics} \qquad (3\text{-}97)$$

Recall that in a CSTR, C is given by

$$\frac{C}{C_0} = \frac{1}{1 + k\theta} \qquad (3\text{-}98)$$

where

$$k = \mathscr{A} \exp\left(-\frac{E}{RT}\right) \qquad (3\text{-}99)$$

Heat removal is provided by the enthalpy balance about the CSTR, which is the sum of heat removed by the effluent stream and that by internal heat exchange. Referring to Fig. 3-18, we have

$$q_{\text{rem}} = \frac{\rho C_p Q}{V}(T - T_0) + \frac{U_0 A}{V}(T - T_c)$$

or

$$q_{\text{rem}} V = (\rho C_p Q + U_0 A)T - \rho C_p Q T_0 - U_0 A T_c \qquad (3\text{-}100)$$

Substituting Eq. (3-99) into (3-98) and the result into Eq. (3-97) gives us the q_{gen}-versus-reaction-temperature function. Equation (3-100) provides the q_{rem}-versus-temperature relation, which is linear in view of our assumption that $\rho C_p Q$ and $U_0 A$ are temperature-independent. Figure 3-26 reveals the heat-generation–heat-removal functions vs. temperature phrased in reduced terms, established as follows. We express both Eqs. (3-97) and (3-100) in dimensionless form:

$$\bar{q}_{\text{gen}} = \frac{q_{\text{gen}}\,\theta}{\rho C_p T_0} = \frac{-\Delta H \,\theta C_0 \, k(1-x)}{\rho C_p T_0} = \bar{\beta} k_0 \,\theta(1-x) \exp\left[-\epsilon\left(\frac{1}{t}-1\right)\right] \qquad (3\text{-}101)$$

---

[1] Relaxation of such assumptions leads to somewhat more complex relations, the detailed anatomy of which can be found in the usual sources, e.g., R. B. Bird, W. Stewart, and E. N. Lightfoot, "Transport Phenomena," Wiley, New York, 1960.

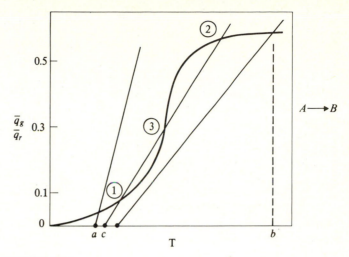

FIGURE 3-26
Heat-generation–heat-removal functions for an exothermic simple reaction in a CSTR.

Recall that for linear kinetics in a CSTR $1 - x$ is

$$1 - x = \frac{1}{1 + k_0 \theta \exp\{-\epsilon[(1/t) - 1]\}} \qquad (3\text{-}102)$$

with the result that dimensionless heat generation is given as a function of reduced temperature $t = T/T_0$ as

$$\bar{q}_{gen} = \frac{\bar{\beta}(k_0 \theta) \exp\{-\epsilon[(1/t) - 1]\}}{1 + k_0 \theta \exp\{-\epsilon[(1/t) - 1]\}} \qquad (3\text{-}103)$$

The heat-removal function can also be rephrased in dimensionless form to yield

$$\bar{q}_{rem} = \frac{q_{rem}\,V}{\rho C_p T_0\, Q} = (1 + \bar{U})t - (1 + \bar{U}\gamma)$$

$$(3\text{-}104)$$

where $\qquad \bar{U} = \dfrac{U_0 A}{\rho C_p Q} \qquad \text{and} \qquad \gamma = \dfrac{T_c}{T_0}$

We note that $\bar{q}_{gen}$ is governed by $-\Delta H$ and the Arrhenius parameters $\mathscr{A}$ and E/RT_0, while the heat-removal relationship, being linear, is dictated by the slope $1 + U$ and intercept $-(1 + \bar{U}\gamma)$. The various possibilities for a given heat-generation function are displayed in Fig. 3-26.

We observe that under certain conditions there exist three intersections in $\bar{q}_{gen}$ and $\bar{q}_{rem}$. By varying feed temperature and/or coolant temperature, the heat-removal line is moved left or right. When it is moved to the left, e.g., reduction in feed temperature, steady state at a low, possibly negligible, rate of generation (reaction) is realized (point *a*), a condition of reaction extinction or quenching. As it is moved to the right, the more vigorous steady state (point *b*) is realized.   At a feed and coolant temperature corresponding to point *c*, there are three intersections; at this condition, states 1 and 2 are stable, but state 3 is unstable because the rate of generation of heat with respect to T, $d\bar{q}_{gen}/dt$, is greater than that of removal, $d\bar{q}_{rem}/dt$.   (This argument is not totally rigorous, as we shall see in our subsequent consideration of transient stability.   Its physical meaning, however, is intuitively persuasive.)

If at point *c* we increase the slope of the $\bar{q}_{rem}$-versus-*t* relation by increasing say Q or U_0, extinction is realized (point *a*); on the other hand, decreasing the slope, say by adiabatic operation ($U_0 = 0$), ignition is achieved (point *b*).

In summary, we note that at the stable points of (steady-state) operation, points 1 and 2

$$\frac{d\bar{q}_{rem}}{dt} > \frac{d\bar{q}_{gen}}{dt} \qquad (3\text{-}105)$$

while at the unstable point, 3,

$$\frac{d\bar{q}_{rem}}{dt} < \frac{d\bar{q}_{gen}}{dt} \qquad (3\text{-}106)$$

In physical terms, Eq. (3-105) signifies that our ability to remove heat is greater than the reaction's ability to generate heat.   When the contrary condition prevails [Eq. (3-106)], operation at that point is impossible and the system will move to either point 1 or 2, depending upon the direction of *t* (whether to the left or right of point 3).

Uniqueness in Complex Reaction Networks

For a simple reaction $A \rightarrow B$ we have learned that two steady states may exist while one unstable state may be manifest.   Let it be supposed that the reaction of interest is complex

$$A \xrightarrow{k_1} B \xrightarrow{k_2} C$$
$$\underset{k_3}{\searrow}$$
$$D \qquad (3\text{-}107)$$

The heat generated for linear kinetics in a CSTR is

$$q_{gen} = [-\Delta H_1\, k_1 + (-\Delta H_3)k_3]A + (-\Delta H_2)k_2\, B \qquad (3\text{-}108)$$

A steady-state material balance provides the conversion of A, x, and y to C:

$$A = \frac{A_0}{1 + (k_1 + k_3)\theta}$$

$$B = \frac{A_0 k_1 \theta}{(1 + k_2\,\theta)[1 + (k_1 + k_3)\theta]} \qquad (3\text{-}109)$$

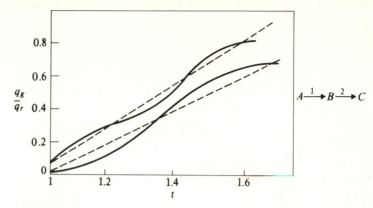

FIGURE 3-27
Heat-generation–heat-removal functions for an exothermic consecutive reaction in a CSTR.

Hence our dimensionless heat-generation function is

$$\frac{q_{\text{gen}}\theta}{\rho C_p T_0} = \frac{\bar{\beta}_1 k_1 \theta + \bar{\beta}_3 k_3 \theta}{1 + (k_1 + k_3)\theta} + \frac{\bar{\beta}_2 k_1 k_2 \theta^2}{(1 + k_2 \theta)[1 + (k_1 + k_3)\theta]} \qquad (3\text{-}110)$$

where

$$\bar{\beta}_i = \frac{-\Delta H_i A_0}{\rho C_p T_0} \qquad \epsilon_i = \frac{E_i}{R T_0}$$

and

$$(k\theta)_i = (k_0 \theta)_i \exp\left[-\epsilon_i\left(\frac{1}{t} - 1\right)\right]$$

Obviously Eq. (3-110) reduces to the simple consecutive and simultaneous cases.

Figure 3-27 gives the heat-generation curves for the consecutive network (A → B → C) for diverse values of $\bar{\beta}$ and ϵ. Note that under certain circumstances more than three solutions of the generation-removal balance are indicated, two of which are unstable, as $dq_{\text{gen}}/dT > dq_{\text{rem}}/dT$ at two points. Behavior in instances of mixed endothermal and exothermal consecutive reactions is shown in Fig. 3-28, and examples for the mixed consecutive and simultaneous reaction network are shown in Fig. 3-29.

Isothermal Uniqueness

In this instance we consider the possibility of multiplicity of the steady state at *one* temperature. An excellent example is found in reality, the oxidation of CO over supported Pt catalyst, where the reaction-rate model in excess O_2 is

$$\mathscr{R} = \frac{k(CO)}{[1 + K(CO)]^2} \qquad (3\text{-}111)$$

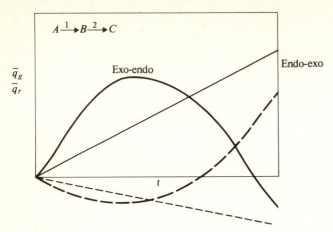

FIGURE 3-28
q_{gen}-versus-q_{rem} functions for consecutive reactions in a CSTR for mixed endothermicity and exothermicity.

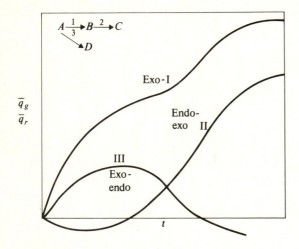

FIGURE 3-29
q_{gen}-versus-q_{rem} for diverse mixed consecutive and simultaneous reactions in a CSTR.

We assume this reaction to occur in an isothermal CSTR. In steady state, where $C \equiv CO$,

$$Q(C_0 - C) = \frac{VkC}{(1 + KC)^2} \quad (3\text{-}112)$$

which when solved for C yields a cubic equation. Three real roots can exist. This situation is nicely displayed in Fig. 3-30, where Eq. (3-111) is plotted versus CO.

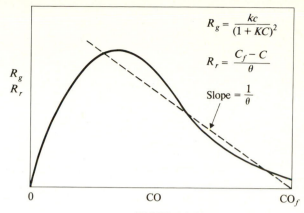

FIGURE 3-30
Isothermal multiplicity in a CSTR.

Note that the rate changes from first order to negative order as CO concentration increases. The operating line for a CSTR is simply

$$\mathscr{R} = \frac{(CO)_f - CO}{\theta} \qquad (3\text{-}113)$$

As indicated in Fig. 3-30, for a particular feed composition $(CO)_f$ and holding time θ, the generation function [Eq. (3-111)] intersects the removal function [Eq. (3-113)] at three points, one of which, the middle point, is unstable.

Recalling the mixed normal and abnormal kinetic behavior displayed in autocatalysis, in which a rate maximum at conversion points greater than zero can be found, one must anticipate the possibility of isothermal steady-state multiplicity in such systems.

We observe that nonuniqueness requires no heat generation. In fact Luss[1] proved that two simultaneous endothermic reactions can yield nonunique solutions.[1]

3-12 REACTOR STABILITY[2]

Although we have demonstrated that at a point in a distributed-parameter system or in a CSTR (a point in terms of C and T) two steady states may prevail while another solution is unstable, it is imperative that a more formal discussion of instability be fashioned. This we do by considering a simple first-order reaction in a CSTR. Transient operation is assumed; by employing the Taylor expansion to describe small departures from the steady state and then subtracting the steady-

---

[1] D. Luss and G. T. Chen, *Chem. Eng. Sci.*, **30**: 1483 (1975).
[2] In spirit this development follows that of S. Carrá and L. Forni, "Aspetti Cinetici della Teoria del Reattore-Chimico," Tamburini, Milan, 1974.

state relation from the Taylor expansion of the transient, we secure an equation which will be analyzed for its stability.

We seek, then, the answer to the question: If the reactor is subjected to a small temperature disturbance, will it return to its stable steady state or runaway? For first-order reaction in a CSTR, in steady state, we have

$$\frac{C_0 - C_s}{C_0} = \frac{\mathscr{R}_s \theta}{C_0} \qquad (3\text{-}114)$$

while the heat balance is

$$\frac{T_0 - T_s}{T_0} = \frac{U_0 a(T_s - T_c)\theta}{\rho C_p T_0} - \frac{\mathscr{R}_s \theta(-\Delta H)}{\rho C_p T_0} \qquad \text{where } \mathscr{R}_s = k_s C_s \qquad (3\text{-}115)$$

In transient, we find

$$\frac{C_0 - C}{C_0} = \frac{\mathscr{R}\theta}{C_0} + \theta \frac{d(C/C_0)}{d\tau} \qquad (3\text{-}116)$$

and

$$\mathscr{R} = kC = \mathscr{A}_0 C \exp\left(-\frac{E}{RT}\right) \qquad (3\text{-}117)$$

The transient heat balance is

$$\frac{T_0 - T}{T_0} = \frac{U_0 a\theta}{\rho C_p T_0}(T - T_c) - \frac{\mathscr{R}\theta(-\Delta H)}{\rho C_p T_0} + \theta \frac{d(T/T_0)}{d\tau} \qquad (3\text{-}118)$$

For small departures from the steady state, we can expand Eq. (3-117) by the Taylor series to obtain

$$\mathscr{R} = k_s C_s + (C - C_s)\left(\frac{\partial \mathscr{R}}{\partial C}\right)_s + (T - T_s)\left(\frac{\partial \mathscr{R}}{\partial T}\right)_s \qquad (3\text{-}119)$$

now

$$\frac{\partial \mathscr{R}}{\partial C} = k_s \qquad \text{and} \qquad \left(\frac{\partial \mathscr{R}}{\partial T}\right)_s = \frac{k_s C_s}{T_s} \frac{E}{RT_s} \qquad (3\text{-}120)$$

Hence

$$\mathscr{R} = k_s C_s + (C - C_s)k_s + \frac{k_s C_s}{T_s} \frac{E}{RT_s}(T - T_s) \qquad (3\text{-}121)$$

We substitute this equation into the transient equations (3-116) and (3-118) and then subtract the steady-state relations from the linearized transients to obtain, in dimensionless form,

$$\frac{-d[(C - C_s)/C_s]}{d(k_s \tau)} = \left(1 + \frac{1}{k_s \theta}\right)\frac{C - C_s}{C_s} + \frac{E}{RT_s}\frac{T - T_s}{T_s} \qquad (3\text{-}122)$$

and

$$\frac{-d[(T - T_s)/T_s]}{d(k_s \tau)} = -\frac{(-\Delta H) C_s}{\rho C_p T_s}\frac{C - C_s}{C_s} + \left(\frac{1}{k_s \theta} + \frac{U_0 a}{\rho C_p k_s} - \frac{E}{RT_s}\frac{-\Delta H}{\rho C_p T_s}\frac{C_s}{T_s}\right)\frac{T - T_s}{T_s}$$

$$(3\text{-}123)$$

Recalling that

$$\frac{-\Delta H\, C_s}{\rho C_p\, T_s} = \bar\beta = \text{dimensionless adiabatic } \Delta T$$

$$\frac{E}{RT_s} = \epsilon = \text{Arrhenius number}$$

$$k_s\,\theta = \text{Da} = \text{Damköhler number}$$

and

$$\frac{U_0\, a}{\rho C_p\, k_s} = \bar U = \text{time-constant ratio } \frac{\text{reaction}}{\text{heat transfer}}$$

let

$$\frac{C - C_s}{C_s} = y \qquad \frac{T - T_s}{T_s} = t \qquad k_s\,\tau = \phi$$

Then Eqs. (3-122) and (3-123) become

$$-\frac{dy}{d\phi} = A_{11} y + A_{12} t \qquad (3\text{-}124)$$

$$-\frac{dt}{d\phi} = A_{21} y + A_{22} t \qquad (3\text{-}125)$$

$$A_{11} = 1 + \frac{1}{\text{Da}} \qquad A_{12} = \epsilon \qquad A_{21} = -\bar\beta \qquad A_{22} = \frac{1}{\text{Da}} + \bar U - \epsilon\bar\beta$$

If we now differentiate (3-125) and substitute for both y and its derivative, we obtain

$$\frac{d^2 t}{d\phi^2} + \frac{dt}{d\theta}(A_{11} + A_{22}) + (A_{11}A_{22} - A_{21}A_{12})t = 0 \qquad (3\text{-}126)$$

Setting $t = \exp S\phi$ and substituting into Eq. (3-126) gives

$$S^2 + S(A_{11} + A_{22}) + (A_{11}A_{22} - A_{21}A_{12}) = 0 \qquad (3\text{-}127)$$

the roots of which are S_1 and S_2; the general solutions of Eqs. (3-124) and (3-125) are

$$y = A_1 \exp S_1\phi + A_2 \exp S_2\phi$$
$$t = A_3 \exp S_1\phi + A_4 \exp S_2\phi \qquad (3\text{-}128)$$

If the reactor is to return to its stable, steady state after a slight disturbance, S_1 and S_2 must both be negative. The roots of Eq. (3-127) are

$$S_1, S_2 = \frac{-(A_{11} + A_{22}) \pm \sqrt{(A_{11} + A_{22})^2 - 4(A_{11}A_{22} - A_{12}A_{21})}}{2}$$

or

$$A_{11} + A_{22} > 0 \qquad A_{11}A_{22} - A_{12}A_{21} > 0$$

or

$$A_{11}A_{22} > A_{12}A_{21}$$

In terms of our reactor-reaction parameter, stability is assured if

$$1 + \frac{2}{Da} + \bar{U} - \epsilon\bar{\beta} > 0 \qquad (3\text{-}129)$$

and

$$\left(1 + \frac{1}{Da}\right)\left(\frac{1}{Da} + \bar{U} - \epsilon\bar{\beta}\right) + \epsilon\bar{\beta} > 0 \qquad (3\text{-}130)$$

As an exercise, discuss the physicochemical significance of the above criteria.[1]

3-13 OPTIMIZATION

Optimization is essentially the manipulation by design and/or operation of key system parameters to achieve a maximum in the value of some critical index of that system's efficiency. In chemical processing it is safe to assert that overall profitability is a sound objective function of optimization. As remarked in Chap. 1, cases may be found in which neither minimum reactor volume nor maximum product yield but complex separation costs will be profit-determining in terms of overall plant economy. Such situations are beyond the scope of this book. Instead, the reactor per se will be considered the focal point of optimization, and for it one readily perceives diverse optimization objective functions:

1 If conversion is of prime concern, one seeks the maximum in moles converted per unit reactor volume per unit time or the maximum conversion in the smallest reactor of minimum contact or holding time. We term this space-time conversion (STC)

$$STC = \frac{x}{Vt}$$

Evidently, the maximum STC is realized at the highest temperatures commensurate with the integrity of materials of reactor construction *and* product survival. For example, in the homogeneous combustion of polluting organic vapors, the maximum STC is governed only by durability of the reaction chamber to temperature.

2 More commonly yield of a desired intermediate in a complex reaction network is of prime concern, in which case a maximum in space-time yield (STY) might be our objective function:

$$STY = \frac{Y}{Vt}$$

Consequently, a maximum in STY may not necessarily be found at V and t values corresponding to the maximum in STC. Indeed in the partial oxidation of an organic, for example, it is evident that the maximum tolerable temperature, while maximizing STC, could well minimize STY.

---

[1] Note that in the adiabatic case, $\bar{U} = 0$, condition (3-129) is sufficient for stability. Further Eq. (3-129) is the slope condition, Eq. (3-105).

Temperature is not the only negotiable parameter in the pursuit of optimization. Reactor type and mode of operation are engineering variables. Reconsider the van de Vusse reaction network

$$A \xrightarrow{k_1} B \xrightarrow{k_2} C$$

$$A + A \xrightarrow{k_3} D$$

where B is the desired product. Assuming that E values are unaffected by diffusion:

1 If $E_1 > E_2$ and E_3, STY is increased with increasing temperature.
2 If $E_3 < E_1 < E_2$, temperature should be increased and then decreased to enhance STY.
3 If $k_1 > k_2 > k_3$, a PFR increases STY.
'4 If k_3 is significant relative to k_1 and k_2, a CSTR or a series of CSTRs would give the best STY.
5 If condition 4 prevails; further STY improvements can be realized by distributing the reactant A along the reaction path, as such a policy (sidestream, or interstage, addition) will maintain a low point concentration of A to the detriment of second-order by-product generation, step 3.

In sum, for a given complex reaction-reactor network there exists, in principle, an optimum temperature-coreactant profile which sponsors a maximum STY.

A Simple Optimization Problem

A rich literature on optimization exists, thus freeing us to treat a simple problem which illustrates the specification of a temperature sequence to approach an optimum STY. The reversible exothermic reaction A $\underset{k_{-1}}{\overset{k_1}{\rightleftarrows}}$ B poses an STY issue since an increase in temperature, while increasing the conversion of A, decreases the equilibrium yield of B.

The catalytic oxidation of SO_2 over V_2O_5 or Pt ($SO_2 + \frac{1}{2}O_2 \underset{k_{-1}}{\overset{k_1}{\rightleftarrows}} SO_3$) is a perfect example of our model reaction. As no threat of by-product formation exists and this exothermic reaction is self-arresting due to equilibrium limitations, it is safely conducted adiabatically. At an SO_2-air feed temperature of 450°C, conversion in a single adiabatic reactor is but 80 percent, corresponding to 98 percent of the equilibrium conversion at the exit temperature of about 650°C. To achieve near total conversion of SO_2 one might (1) abandon adiabaticity and employ a near-isothermal multitubular reactor whereby heat exchange provides a near-constant modest temperature commensurate with high reaction rate and high equilibrium conversion or (2) retain the far less expensive adiabatic reactor but cool between adiabatic stages, which effectively shifts the equilibrium conversion point to the larger desired value as the process stream moves from the first to nth adiabatic stage.

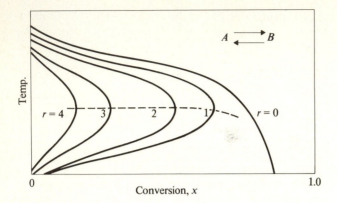

FIGURE 3-31
Reaction temperature vs. conversion at various values of the
rate for an exothermic reversible reaction.

Since, in fact, SO_2 is commercially oxidized in a multistaged adiabatic catalytic-reactor network with interstage heat removal, our model reaction is of more
than academic interest. For an exothermic reversible reaction, one can plot the
reaction temperature vs. conversion for a given value of reaction rate as a parameter,
as schematically set forth in Fig. 3-31, which is constructed as follows. Assuming
that order and stoichiometry conform, in terms of conversion x

$$\frac{\text{Net rate}}{\pi/RT} = k_0(1-x)\exp\left[-\epsilon_1\left(\frac{1}{t}-1\right)\right] - k_{-0}x\exp\left[-\epsilon_{-1}\left(\frac{1}{t}-1\right)\right] \qquad (3\text{-}131)$$

or

$$\frac{r}{k_0\pi/RT_0} = \frac{1}{t}\left\{(1-x)\exp\left[-\epsilon_1\left(\frac{1}{t}-1\right)\right] - \frac{x}{K_0}\exp\left[-\epsilon_{-1}\left(\frac{1}{t}-1\right)\right]\right\}$$

where $\qquad t = \dfrac{T}{T_0} \qquad \epsilon_1 = \dfrac{E_1}{RT_0} \qquad \epsilon_{-1} = \dfrac{E_{-1}}{RT_0}$

and k_0 and K_0 are the forward rate coefficient and equilibrium constant at the
reference temperature T_0. At constant values of x, the net rate is computed versus
t for given values of the base rate coefficients. Results are set forth in Fig. 3-32.
A cross-plot with net rate as a parameter yields Fig. 3-31.

The dashed curve in Fig. 3-32 represents the loci of points of maximum
conversion for a given rate and therefore the optimum temperature sequence.
Let us suppose that the feed condition to the first adiabatic stage to be x_0 and T_0.
Referring to Fig. 3-33, assume a three-stage adiabatic system with indirect interstage heat removal. A simple adiabatic heat balance provides conversion-

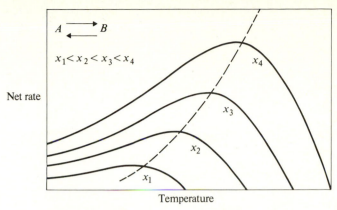

FIGURE 3-32
Net rate vs. temperature with conversion as a parameter.

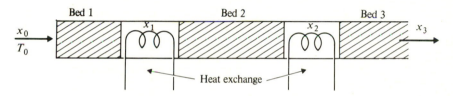

FIGURE 3-33
Schematic of an adiabatic reactor train with indirect interstage heat removal.

temperature relationship per stage, and so for the first stage the effluent temperature
is

$$T_1 = T_0 + \frac{-\Delta H \, A_0}{\rho C_p} (x_1 - x_0) \qquad (3\text{-}132)$$

The point (T_1, x_1) in Fig. 3-34 is situated on the curve of constant rate r_3.
Indirect interstage cooling then reduces the temperature to T_2 [point (x_1, T_2)], the
feed condition for stage 2. Repeating for the next two stages, we arrive at a final
conversion x_3 at T_5.

Two points become apparent in terms of design policy: (1) a single-stage
reactor would at best provide an exit conversion far less than x_3, specifically x_e, at
best; (2) an infinite number of stages corresponds to progression along the optimum
temperature profile, and, in principle, an equilibrium conversion corresponding to
x_{max} could be realized. This limiting case is equivalent to a flow reactor operated
with an imposed optimum temperature profile.

Design (1), a single-stage adiabatic reactor, fails to give the desired conversion.
Design (2) while providing the highest conversion, involves a heat-exchange policy

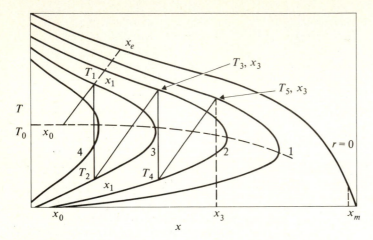

FIGURE 3-34
Graphic display of an approach to optimization of the system set forth in
Fig. 3-33.

which is implemented on a plant scale with severe difficulty and considerable
expense. The optimum is obviously a compromise, a three- or four-stage adiabatic
reactor train.

Summary

While we have deemed it appropriate to explore the behavior of an ideal reactor, the
compromise with reality is not really a taxing one. For although real reactors exhibit
temperature, concentration, and (occasionally) residence-time distributions which do not
conform to one of the ideals set forth in this chapter, the limiting ideals should nevertheless
be studied so that the analyst can become aware of the benefit or loss which accompanies
backmixing, nonisothermality, and reactant allocation with respect to the specific reactor
function.

Real reactor environments are commonly complex, and within them enormously
complicated, often ill-understood, reactions take place. Since the analyst-designer
rarely has a precise kinetic model of a well-defined reaction network at his disposal,
he does not enjoy the luxury of unambiguous specification of a unique plant-scale reactor
type and environment. As indicated in the discussion of optimization, one can, at best,
approach the optimum. Realization of true optimization is to be enjoyed only in the most
primitive, trivial reaction-reactor networks.

Command of the principles set forth in this chapter should permit the student to
fashion meaningful judgments regarding:

1 The influence of isothermal backmixing upon conversion and yield for both normal
and abnormal kinetics
2 The influence of thermal backmixing upon conversion and yield for all kinetic situations

3   The importance of micromixing and macromixing upon conversion and yield
4   The role of radial concentration gradients upon conversion and yield as exemplified by the LFR
5   The existence of both isothermal and nonisothermal steady-state multiplicity
6   The factors which govern instability as a consequence of small departures from the steady state
7   The realities of optimization as regards reactor type, feed allocation, and temperature distribution

ADDITIONAL REFERENCES

The residence-time-distribution problem and analyses thereof are discussed in depth by K. Bischoff and O. Levenspiel, *Ind. Eng. Chem.*, **51**: 1431 (1959); **53**: 313 (1961), while micro-macromixedness and yield are analyzed by K. Rietema, *Chem. Eng. Sci.*, **8**: 103 (1958).

F. A. Cleland and R. H. Wilhelm, *AIChE J.*, **2**: 489 (1956), present both experimental and theoretical aspects of reaction diffusion in laminar flow; see also R. C. L. Bosworth, *Phil. Mag.*, **39**: 847 (1948). Mixing-reaction analyses for endothermic and/or exothermic reaction under micro-macromixed conditions are tersely summarized by J. M. Douglas, *Chem. Eng. Prog. Symp. Ser.*, **60** (48): 1 (1964), whereas short-circuiting in tubular reactors is discussed by W. Hennel, *Br. Chem. Eng.*, **10** (6): 386 (1965).

The work of G. R. Worrell and L. C. Eagleton, *Can. J. Chem. Eng.*, **42**: 254 (1964) is an experimental study of mixing and segregation in a CSTR, while the influence of product recycle and temperature upon autocatalytic reactions is analyzed by Y. Ahn, L. Fan, and L. E. Erickson, *AIChE J.*, **12**: 534 (1966). An analysis of the complex residence-time distribution of a catalytic cracking-regenerator unit is given by T. B. Metcalfe, *Chem. Eng. Prog.*, **60**: 71 (1964). The residence-time-distribution problem and its relevance to biological problems is treated in depth by R. Aris, "Intracellular Transport," p. 167, Academic, New York, 1966.

The issue of steady-state multiplicity and stability criteria is set forth in the pioneering work of R. Aris and N. R. Amundson, *Chem. Eng. Sci.*, **7**: 121, 132 (1958). D. E. Boynton, W. B. Nichols, and H. M. Spurlin, *Ind. Eng. Chem.*, **51** (4): 489 (1959), analyze the stability of highly exothermic, potentially explosive chemical reactions. R. Aris, "The Optimal Design of Chemical Reactors," Academic, New York ,1961, remains a signal source of mathematic insight into the optimization topic.

PROBLEMS

3-1   In the heat-transfer parable (Sec. 3-1), for the conditions stated we found UA for the CST to be 21.4 times that of the plug-flow exchanger. Compute this ratio for the same overall conditions for two, three, and five CST exchangers when the heat removal is evenly distributed among the CSTs.

3-2   For a first-order reaction of 20 kcal activation energy and a feed temperature of 150°C, at what temperature must a CSTR be operated to be as efficient as an isothermal PFR at conversions of (*a*) 50 percent, (*b*) 90 percent, and (*c*) 99 percent? Repeat for second, one-half, and negative first orders.

3-3   Execute the Metzner-Pigford thought experiment for (a) zero-order and (b) negative-first-order reaction.

3-4   In the noble-metal–catalyzed oxidation of CO, the order in CO is negative first order at a conversion up to 99.9 percent CO and positive first order from that point to total conversion. Devise a PFR-CSTR combination which will provide minimum total contact time.

3-5   For the simple linear reversible reaction $A \underset{k_{-1}}{\overset{k_1}{\rightleftharpoons}} B$ develop the relationship between conversion and total contact time for (a) one, (b) three, and (c) five CSTRs. Repeat for an open-loop recycle reactor as a function of recycle ratio R.

3-6   For purposes of an order-of-magnitude economic analysis, the MBA wizards at your establishment want an estimate of STY (moles B per time $\times$ total volume) for the homogeneously catalyzed oxidation of propylene (A) to propylene oxide (B), the sequence being

where C is CO_2 plus H_2O. Assuming that linear kinetics prevail, develop the isothermal reaction rate and yield relationships for (a) PFR, (b) n CSTRs, (c) a recycle reactor, (d) a PFR with axial mixing. Comment upon the relative ease of solution for these diverse models.

3-7   An environmental engineer and his law firm suggest that the issue raised in Prob. 3-6 might also be profitably analyzed in terms of a single CSTR with recycle. Validate Dickens' generalization with respect to the law and, incidently, Carberry's extension of it, to the level of universality.

3-8   Returning to reality, let us consider Prob. 3-6 under nonisothermal conditions. Compare the adiabatic PFR, n-CSTR, recycle, and axial-dispersion models, again in terms of relative ease of resolution. Specifically if

$$k_1 = k_{0_1} \exp\left[-20\left(\frac{1}{t}-1\right)\right] \qquad k_2 = k_{0_2} \exp\left[-10\left(\frac{1}{t}-1\right)\right]$$

$$k_3 = k_{0_3} \exp\left[-30\left(\frac{1}{t}-1\right)\right]$$

where k_0 refers to the feed temperature values, choose the intermediate-mixing model which gives an answer without recourse to a computer. Slide rules and pocket computers are acceptable.

$$k_{0_1} = 0.1 \text{ s}^{-1} \qquad k_{0_2} = 0.05 \text{ s}^{-1} \qquad k_{0_3} = 0.03 \text{ s}^{-1}$$

$$T_0 = 600°C \qquad \bar{\beta}_1 = 0.3 \qquad \bar{\beta}_2 = 0.8 \qquad \bar{\beta}_3 = 0.5$$

3-9   For second-order reaction in a series of two CSTRs, what is the optimum ratio of holding times θ_2/θ_1?

3-10   For an adiabatic CSTR, derive the mass- and heat-continuity equations for the reaction $A \overset{1}{\rightarrow} B \overset{2}{\rightarrow} C$. Now express both conversion of A and yield of B as a function of temperature.

3-11   Utilizing the data given in Prob. 3-8, for the following values of $\bar{\beta}$ explore the temperature sequence in a three-stage CSTR network which provides the best yield of B, that is, B/A versus conversion $1 - A/A_0$.

Case	β_1	β_2	β_3
1	0.2	0.5	0.1
2	−0.1	0.1	0.5
3	0·5	−0.3	0.5
4	0.5	0.3	−0.4

3-12   At a feed temperature of 175°C the rate coefficient for an irreversible linear reaction is $k_0 = 0.1$ s^{-1}, $-\Delta H = 35{,}000$ cal/mol; ρC_p is that of an aqueous solution. Holding time can be varied between 1 and 100 min. Reactor volume is 100 liters. UA is fixed at 10, where heat-transfer area is 1 m². Activation energy is 45 kcal. Explore the issue of steady-state multiplicity under (*a*) adiabatic and (*b*) nonisothermal conditions. $C_0 = 10^{-2}$ mol/ml.

3-13   Consider the reaction $A \xrightarrow{1} B \xrightarrow{2} C \xrightarrow{3} D$ as conducted in an adiabatic CSTR. In terms of heat generation per se, explore a range of dimensionless generation parameters (mixed endothermal and exothermal) to determine how many stable and unstable states are possible.

3-14   For the exothermic V$_2$O$_5$-catalyzed reaction $SO_2 + \frac{1}{2}O_2 \rightleftharpoons SO_3$ qualitatively sketch the heat-generation-vs.-temperature profile. Do the same for an endothermic reversible reaction. Cite a case other than A, B, C, etc.

3-15   In principle, hydrocarbons and CO can be totally oxidized in a high-temperature afterburner (downstream of the automobile engine). Let us suppose that said reactor is a CSTR of contact time of 10 s operating adiabatically. For CO oxidation, k at 1000°C is 0.1 s^{-1}; the activation energy is 70 kcal; the average CO mole fraction is 1.0 percent, and properties of air can be assumed. Is the reactor stable at (*a*) 1000°C, (*b*) 800°C? At what temperature and/or CO mole fraction is stability promised?

3-16   Discuss what optimization strategy you might employ to optimize STC and/or STY in the following situations (B is the desired product).

(*a*) Isothermal:

  (*i*) (*a*)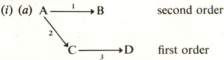

  (*ii*)   Same as (*i*), but orders reversed

  (*iii*)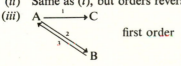

  (*iv*) $A \xrightarrow{1} C \xrightarrow{2} B$ first order

  (*v*) $A \xrightarrow{1} C \underset{3}{\overset{2}{\rightleftharpoons}} B$ first order

  (*vi*) $A \xrightarrow{1} B \xrightarrow{2} C$

    $A + B \xrightarrow{3} D$ first order

    $A + C \xrightarrow{4} D$

(b) Nonisothermal: as in (a) but

Case	ϵ_1	ϵ_2	ϵ_3	ϵ_4	β_1	β_2	β_3	β_4
1	30	20	30	10	0.1	0.2	0.1	0.4
2	30	20	30	10	0.1	−0.2	−0.1	0.4
3	10	10	30	40	0.1	−0.2	−0.1	0.4
4	10	30	20	50	0.1	0.2	0.1	0.1

3-17 A component reaction of reforming is endothermic dehydrogenation, e.g., cyclo-hexane ⇌ benzene. Assuming that $\epsilon_1 = 20$ and determining the enthalpy change, compute profiles in the spirit of Figs. 3-31 and 3-32 for cyclohexane dehydrogenation. $T_0 = 200°C$

3-18 Let it be supposed that the only reactor in hand is a partially baffled cylindrical tank, equipped with cooling-heating coils. Pulse-testing experiments indicate that two pulse-response maxima exist, compounded by a rather long tail in the issuing signal. You are required to utilize this reactor to esterify an aromatic acid, the rate data having been obtained as in the example in Sec. 2-15; that is, that kinetics applies. What do you do?

3-19 Fashion a question related to the issues treated in this chapter and pose it to your colleagues and professor.

3-20 Resketch Fig. 3-34 for the case of direct (cold-shot) interstage cooling by reactant.

4

CONSERVATION EQUATIONS FOR REACTORS

" We first survey the plot, then draw the model."
Shakespeare   King Henry IV, Part Two

Introduction

In the preceding chapters we discussed the conduct of chemical reaction in three ideal reactor types, the CSTR, LFR, and PFR, and in instances where fluid mixing is at levels intermediate between the PFR (no mixing) and the CSTR (perfect mixing).   For the PFR, by definition, concentration and temperature profiles exist *only* in the direction of flow, i.e., axially.   In the CSTR, of course, total uniformity of temperature and concentrations exists, by definition.   Now in principle, both T and C could be functions of three-dimensional space; e.g., in a stirred-tank reactor with poor agitation, both T and C might vary with height, radial distance, and angular displacement.

The CSTR is the stirred-tank ideal, and since T and C are uniformly distributed (in practice, by vigorous agitation), one terms the CSTR a lumped system with respect to spatial coordinates.   The PFR, with and without finite mixing, is a one-dimensional reactor (in the steady state) in that C and T are functions only of length or axial position.   This is a partially lumped system, since we have assumed that C and T are uniform at all radial positions at a given axial location.

In this chapter, attention will be devoted to the development of the conservation equations expressing fluid velocity, concentration, and temperature in

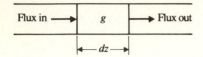

FIGURE 4-1
Differential flux balance for a PFR.

two spatial dimensions and with respect to time for a cylindrical reactor. These relations will then be reduced to various limiting models, e.g., PFR and CSTR. In the course of this reduction process, the assumptions implicit in the ideal models developed earlier will be made explicit.

4-1 TRANSPORT PROCESSES

In deriving the appropriate conservation equations, one simply deals with differences in fluxes which result by reason of generation, withdrawal, or addition of mass, energy, and momentum within a differential element. Consider a simple one-dimensional case, e.g., a tubular reactor in which mass is uniformly distributed across the tube (no radial gradient) (Fig. 4-1). The difference in the flux in the z direction is equal to generation g in the steady state (no accumulation of matter in the differential segment dz).

Definition of Flux

For a species i the flux is defined as transport per unit area perpendicular to flow; thus in mole units, N is expressed as moles per area per unit time. Transport of matter (or energy or momentum) in a given direction is due to bulk flow *and* diffusion, i.e., in the z direction:

$$\text{Total flux} = N_z = u_z C + J_z = \text{flow} + \text{diffusion}$$

where N_z = flux in z direction
  u_z = velocity in z direction
  J_z = diffusive flux in z direction
  C = mole concentration of species under consideration
Let A_z be the area perpendicular to the flux. Now

$$J_z = -\mathscr{D}_z \frac{dC}{dz}$$

where $\mathscr{D}_z$ is a diffusion coefficient, not necessarily molecular in nature.
  With reference to Fig. 4-1, we can write

Rate of change of i in z direction = generation or $-$consumption

$$\frac{d(N_z A_z)}{dz} dz = [-kf(C)]A_z \, dz \qquad (4\text{-}1)$$

or
$$\frac{d(N_z A_z)}{A_z \, dz} = \frac{d(N_z A_z)}{dV} = -kf(C) \qquad (4\text{-}2)$$

Since $N_z = u_z C - \mathcal{D}_z \, dC/dz$, we have

$$\frac{dA_z u_z C}{dV} - \frac{dA_z}{dV}\left(\mathcal{D}_z \frac{dC}{dz}\right) = -kf(C) \qquad (4\text{-}3)$$

Assuming that A_z is independent of z and that $\mathcal{D}_z$ is constant, and noting that AuC is the molar flow F, we have

$$\frac{dF}{A_z \, dz} - \mathcal{D}_z \frac{d^2 C}{dz^2} = -kf(C) \qquad (4\text{-}4)$$

When $\mathcal{D}_z = 0$, we obtain the PFR relation

$$\frac{dF}{A_z \, dz} = \frac{dF}{dV} = -kf(C) \qquad (4\text{-}5)$$

If Au, the volumetric flow rate, is constant, then, since $F = QC$,

$$Q \frac{dC}{dV} = \frac{dC}{d\tau} = -kf(C) \qquad (4\text{-}6)$$

which is the simple PFR result when Q is a constant, as is the case when no mole change due to reaction occurs to cause a variation in Q with extent of reaction. If a mole change due to reaction is significant, then

$$\frac{dF}{dV} = C \frac{dQ}{dV} + Q \frac{dC}{dV} = -kf(C) \qquad (4\text{-}7)$$

However, the use of conversion avoids the form like that given above. Define conversion x as the moles of reactant consumed per mole of feed F_0, or

$$F = F_0 - F_0 x = F_0(1 - x)$$

so that
$$\frac{dF}{dV} = F_0 \frac{dx}{dV} = kf(C) \qquad (4\text{-}8)$$

Then merely expressing $f(C)$ in terms of x permits integration (analytical or otherwise).

The above development simply illustrates the fact that when we derived the PFR relation in preceding chapters, we were implicitly (1) assuming radial uniformity, (2) ignoring axial transport due to diffusion, and (3) assuming, in some instances, a volumetric flow rate independent of extent of reaction. As the following section will show, a number of other assumptions are implicit in the PFR model.

In the treatment which follows, we urge the student to refer to standard

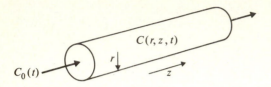

FIGURE 4-2
Schematic of a cylindrical reactor.

transport-phenomena texts[1] for detailed discussions, particularly of momentum and thermal-energy transport.

The Mass, Momentum, and Energy Equations

We consider a particular species i being consumed by chemical reaction in a cylindrical reactor of radial coordinate r and length z, as illustrated in Fig. 4-2. This is a distributed system in that the key species concentration is a function of r, z, and t (time), assuming transient operation.   Further, as chemical reaction involves either generation or abstraction of heat, the temperature (and therefore reaction rate) will be a function of r, z, and t for a given rate of heat addition or removal through the reactor wall.   While angular symmetry may be assumed, provision for reactant addition or leakage (a form of nonchemical generation or consumption) can be assumed.   Thus the continuity equation (of species i) for an annular ring of volume $2\pi r\,dr\,dz$ is readily derived in terms of the balance (see Fig. 4-3):

Rate of accumulation of moles of i = rate of change in z direction due to flow and diffusion + rate of change in the r direction due to flow and diffusion + generation of i due to reaction g_c + generation of i due to interphase addition g_a

The molar flux due to flow and diffusion is $N_j = u_j C_i + J_j$ in a given direction. Since $A_z = 2\pi r\,dr$ and $A_r = 2\pi r\,dz$ we have

$$\frac{\partial VC}{\partial t} + \frac{\partial(N_z A_z)}{\partial z}\,dz + \frac{\partial(N_r A_r)}{\partial r}\,dr = g_c + g_a \qquad (4\text{-}9)$$

Dividing by $V = 2\pi r\,dr\,dz$ and carrying out the indicated differentiation, we obtain, assuming that $\mathcal{D}_z$ and $\mathcal{D}_r$ are constant,

$$\frac{1}{V}\frac{\partial VC}{\partial t} + \frac{\partial A_z uC}{A_z\,\partial z} - \mathcal{D}_z\frac{\partial^2 C}{\partial z^2} - \frac{\mathcal{D}_r}{r}\frac{\partial}{\partial r}\left(r\frac{\partial C}{\partial r}\right) = \frac{g_c}{V} + \frac{g_a}{V} \qquad (4\text{-}10)$$

where the radial velocity u_r is zero and radial $\mathcal{D}_r$ and axial $\mathcal{D}_z$ diffusion coefficients are assumed independent of position.   We note that $A_z\,dz = dV$ and that $Q =$

[1] R. B. Bird, W. E. Stewart, and E. N. Lightfoot, "Transport Phenomena," Wiley, New York, 1960; C. O. Bennett and J. Myers, "Momentum, Heat, and Mass Transfer," 2d ed., McGraw-Hill, New York, 1974.

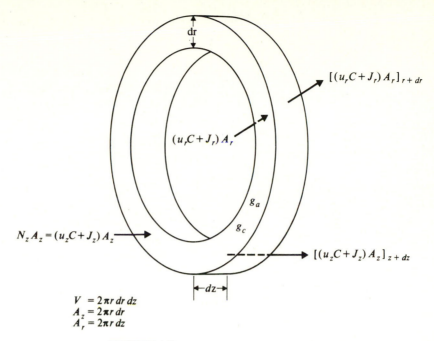

$$V = 2\pi r \, dr \, dz$$
$$A_z = 2\pi r \, dr$$
$$A_r = 2\pi r \, dz$$

FIGURE 4-3
Cylindrical differential ring for derivation of continuity equations.

$A_z u_z$, $n = VC$, and $\mathcal{R} = g/V$; then

$$\underbrace{\frac{1}{V}\frac{\partial n}{\partial t}}_{(1)} + \underbrace{\frac{\partial Q\,C}{\partial V}}_{(2)} - \underbrace{\mathcal{D}_z \frac{\partial^2 C}{\partial^2 z}}_{(3)} - \underbrace{\frac{\mathcal{D}_r}{r}\frac{\partial}{\partial r}\left(r\frac{\partial C}{\partial r}\right)}_{(4)} = \underbrace{\mathcal{R}_c + \mathcal{R}_a}_{(5)} \qquad (4\text{-}11)$$

The continuity equation for species i then expresses (1) the change in moles of i with time in a volume V in terms of (2) the change with volume traversed; (3) the axial diffusion or dispersion of i; (4) the radial dispersion; and (5) generation via reaction and addition from another phase.

By an analogous procedure, an equation for thermal-energy concentration, $\rho C_p T \equiv C$, is derived,[1] where K is the thermal diffusivity:

$$\frac{\partial T}{\partial t} + u\frac{\partial T}{\partial z} - K_z\frac{\partial^2 T}{\partial z^2} - \frac{K_r}{r}\frac{\partial}{\partial r}\left(r\frac{\partial T}{\partial r}\right) = \frac{q_c}{\rho C_p} + \frac{q_a}{\rho C_p} \qquad (4\text{-}12)$$

and q is heat liberated or abstracted via reaction ($= -\Delta H \mathcal{R}$) and addition or removal.

---

[1] See Bird, Stewart, and Lightfoot for a discussion of the implicit assumptions.

Anticipating that fluid velocity may be a function of radial and axial distance, we find the equation of motion in terms of the time-averaged velocity ($u\rho \equiv C$):

$$\frac{\partial \rho u}{\partial t} + u \frac{\partial \rho u}{\partial z} = -\frac{\partial P}{\partial z} - \frac{1}{r}\frac{\partial r\tau}{\partial r}: \qquad \tau = \tau^l + \tau^\tau \qquad (4\text{-}13)$$

In laminar flow under a pressure gradient, the equation of motion reduces for steady state to

$$0 = -\frac{\partial P}{\partial z} - \frac{1}{r}\frac{\partial r\tau^l}{\partial r} \qquad (4\text{-}14)$$

which yields the parabolic velocity distribution

$$u = \frac{\Delta P R^2}{4\mu L}\left[1 - \left(\frac{r}{R}\right)^2\right] = 2u_{av}\left[1 - \left(\frac{r}{R}\right)^2\right] \qquad (4\text{-}15)$$

In turbulent flow, a more common circumstance, solutions of the equation of motion are complex, and it is often more convenient to express the radial velocity distribution in the form

$$u = u_{av}\left[1 - \left(\frac{r}{R}\right)^n\right] \qquad n \to \infty \qquad (4\text{-}16)$$

for plug flow.

In addition to describing radial velocity distribution, the momentum equation relates pressure drop through the reactor in terms of fluid properties and reactor dimensions.

4-2 NATURE OF TRANSPORT COEFFICIENTS

In Eq. (4-11), two diffusion coefficients appear, $\mathscr{D}_z$ and $\mathscr{D}_r$, which govern axial and radial diffusive transport in the reactor. Their thermal analogs, K_z and K_r, appear in Eq. (4-12). These diffusivities are more accurately termed *dispersion coefficients*, as they are generally not molecular in nature, being determined in most cases by turbulence within the reactor and by velocity variations. Often these coefficients are called mixing coefficients since their magnitude governs the degree of mixing which prevails in the axial and/or radial direction. To illustrate, consider Eq. (4-4), where the volumetric flow rate is constant; hence for first-order isothermal reaction

$$\mathscr{D}_z \frac{d^2 C}{dz^2} - u \frac{dC}{dz} = kC \qquad (4\text{-}4a)$$

which describes a PFR with axial mixing or dispersion. Let $f = C/C_0$ and $Z = z/L$, where L is reactor length and C_0 initial reactant concentration; thus

$$\frac{1}{Pe}\frac{d^2 f}{dZ^2} - \frac{df}{dZ} = k\frac{L}{u}f = k\theta f \qquad (4\text{-}17)$$

$Pe = Lu/\mathscr{D}_z$ is termed the *axial Peclet number based upon reactor length.* To solve this differential equation, the appropriate boundary conditions must be specified. In general at $Z = 0$ and 1, the flux entering a boundary must equal that passing through the boundary. At $Z = 0$ (reactor entrance)

$$\text{Flux}\bigg|_{Z=-0} = \text{flux}\bigg|_{Z=+0}$$

$$\left(uC_0 - \mathscr{D}_z'\frac{dC}{dz}\right)_{-0} = \left(uC - \mathscr{D}_z\frac{dC}{dz}\right)_{+0} \qquad \text{at } Z = 0$$

$\mathscr{D}_z'$ refers to the dispersion coefficient upstream and is not necessarily equal to that in the reactor or downstream. When $\mathscr{D}_z' = \mathscr{D}_z \cdots = 0$, then at $Z = 0$, $C = C_0$ or $f = 1$, the usual PFR inlet boundary condition. At $Z = 1$, when $\mathscr{D}$'s are zero, $C = C$, that is, the final concentration; this boundary condition is unnecessary for a PFR without dispersion, i.e., Eq. (4-5). For finite $\mathscr{D}_z$ in the reactor, at $Z = 0$ assuming $\mathscr{D}_z' = 0$, we have

$$uC_0 = uC - \mathscr{D}_z\frac{dC}{dz}\bigg|_0 \qquad (4\text{-}18)$$

or in reduced terms

$$f = 1 + \frac{1}{\text{Pe}}\frac{df}{dZ}\bigg|_{Z=0} \qquad (4\text{-}19)$$

Since $Pe = Lu/\mathscr{D}_z$, a long reactor presents a large Pe value for a given velocity and dispersion coefficient, with the result that $f \to 1$ for large values of Pe. For a given velocity L is fixed by the kinetics through $\theta = L/u$, and we are not free to manipulate L in the reactor section as we are in the upstream section. Since df/dZ is inherently negative, axial dispersion reduces f at $Z = 0$ from the PFR value of unity to a value below unity. Physically, this means that axial mixing causes backflow of reacted material upstream and forward mixing of reactant.

At $Z = 1$, since $(uC)_- = (uC)_+$, we have

$$\frac{1}{\text{Pe}}\frac{df}{dZ}\bigg|_{Z-} = \frac{1}{\text{Pe}'}\frac{df}{dZ}\bigg|_{Z+}$$

where Pe' is the downstream Peclet number. As (by definition) no reaction occurs there, $(df/dZ)_+ = 0$ and

$$\frac{df}{dZ}\bigg|_{Z=1} = 0$$

These boundary conditions, set forth and clarified by Wehner and Wilhelm,[1] permit solution to Eq. (4-17) with results displayed graphically in Fig. 4-4. With an infinite Peclet number ($\mathscr{D}_z \to 0$) the usual PFR first-order decay profile results,

[1] J. F. Wehner and R. H. Wilhelm, *Chem. Eng. Sci.*, **6**: 89 (1956).

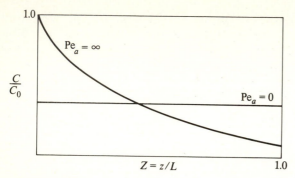

FIGURE 4-4
Reduced reactant concentration vs. length for $Pe_a = 0$ and ∞.

while as $\mathscr{D}_z \to \infty$ or $Pe \to 0$, a uniform concentration exists throughout the reactor; i.e., CSTR character is evident.   The *axial dispersion coefficient is simply another index of mixing in a reactor.*   The larger the value of $\mathscr{D}_z$ for a given L and u, the more drastically the axial concentration (and temperature) profile will be damped. In the limit, uniformity of concentration and temperature exists, and we have the perfectly mixed reactor (CSTR).   For a given kinetic scheme if axial mixing proves to be detrimental to conversion and/or selectivity, the design must be such that Pe is as large as possible.   In the simple homogeneous case discussed here, a long reactor operating at high velocity is required.

Radial mixing, governed by the coefficient $\mathscr{D}_r$, is always desirable, since we wish to avoid variation in conversion and selectivity, which results if molecules at the centerline of the reactor enjoy a different retention time from those at the wall. In an isothermal reactor, a radial concentration gradient is the result of a radial velocity gradient.   In a nonisothermal reactor, the radial concentration gradient is usually established because a temperature gradient (and hence reaction-rate gradient) exists across the tube in the radial direction.   Figure 4-5 schematizes this condition for an exothermic reaction.

We see that for a thermally sensitive product, for example, degradation is likely at the tube axis, where T is high, and/or insufficient conversion may occur at the tube wall, where T is low.   However, since the fluid velocity must approach zero at the wall, some modification of Fig. 4-5 is required.   Considering radial-velocity and temperature gradients, the concentration profile may resemble that

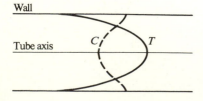

FIGURE 4-5
Radial-temperature–reactant-concentra-
tion profile at a particular axial position
in a cylindrical reactor (exothermic).

FIGURE 4-6
Radial variation in fluid velocity, reactant concentration, and temperature
profiles in a cylindrical reactor (exothermic).

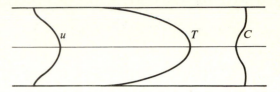

in Fig. 4-6. Ideally, for true PFR performance, a minimum radial velocity, temperature, and concentration gradient should exist with a maximum axial gradient. In other words, $\mathscr{D}_z \to 0$ and $\mathscr{D}_r \to \infty$. When radial gradients are negligible, the system is said to be a lumped one in the radial direction. The CSTR is obviously a lumped system with respect to all directions.

The factors which dictate the magnitudes of $\mathscr{D}_z$ and $\mathscr{D}_r$ depend upon the nature of the system, i.e., whether it is a homogeneous tubular reactor, with or without baffles, or a tube packed with particles or catalyst pellets, either fixed or fluidized. For example, in a simple tank reactor, a high intensity of agitation plus baffling effectively creates almost infinitely large values of $\mathscr{D}_z$, $\mathscr{D}_r$, and $\mathscr{D}_\theta$, thus creating uniformity of concentration and temperature. Before discussing these coefficients and their correlation in detail, the reduction of the general equations to simpler situations will be undertaken.

4-3 REDUCTION OF MORE GENERAL EQUATIONS TO LIMITING REACTOR TYPES

Isothermal Reactors

These can be analyzed by consideration of the continuity equation (4-11) and that of momentum (4-13). Consider Eq. (4-11) and each of its terms (1) to (5) in the light of various reactor types.

1 Batch reactor As there is no flow into or out of the batch reactor, term (2) is zero, as is the external (addition or withdrawal) component $\mathscr{R}_a$. If intensive agitation is provided to ensure uniformity of composition, terms (3) and (4), which describe spatial gradients, are zero, so that

$$\frac{1}{V}\frac{dn}{dt} = \mathscr{R}_c \qquad (4\text{-}20)$$

Equation (4-20) applies to a batch reactor *with* volume change due to reaction. In the *absence* of volume change since $n = VC$,

$$\frac{1}{V}\frac{dn}{dt} = \frac{dC}{dt} = \mathscr{R}_c \qquad (4\text{-}21)$$

2 *Semibatch reactor*   By definition, semibatch operation involves at least one addition or withdrawal stream

$$\frac{1}{V}\frac{dn}{dt} = \mathcal{R}_c + \mathcal{R}_a \qquad (4\text{-}22)$$

assuming uniformity of composition at any time.

3 *PFR*   Radial gradients are absent $[(4) = 0]$, there is negligible mixing in the axial direction $[(3) = 0]$, hence $(1) + (2) = 5$

$$\frac{1}{V}\frac{\partial n}{\partial t} + \frac{\partial Q\,C}{\partial V} = \mathcal{R}_c + \mathcal{R}_a \qquad (4\text{-}23)$$

or assuming $Q = $ constant and $\mathcal{R}_a = 0$

$$\frac{\partial C}{\partial t} + u\frac{\partial C}{\partial z} = \mathcal{R}_c \qquad (4\text{-}24)$$

In steady state, with $u = $ constant,

$$u\frac{dC}{dz} = \mathcal{R}_c \qquad (4\text{-}25)$$

For varying Q (or u) due to mole change resulting from reaction,

$$\frac{dQ\,C}{dV} = \frac{dF}{dV} = -F_0\frac{dx}{dV} = \mathcal{R}_c \qquad (4\text{-}26)$$

4 *Steady-state laminar flow with negligible axial mixing* $[(1)$ and $(3) = 0$, and $\mathcal{R}_a = 0]$

$$2\left[1 - \left(\frac{r}{R}\right)^2\right]\frac{\partial u_{av}\,C}{\partial z} - \mathcal{D}_r\left(\frac{\partial^2 C}{\partial r^2} + \frac{1}{r}\frac{\partial C}{\partial r}\right) = \mathcal{R}_c \qquad (4\text{-}27)$$

For any velocity profile

$$u = u_{av}\left[1 - \left(\frac{r}{R}\right)^n\right]$$

5 *Perfectly mixed reactor (CSTR)* $[(3)$ and $(4) = 0]$

$$\frac{1}{V}\frac{\partial n}{\partial t} + \frac{\Delta Q\,C}{\Delta V} = \mathcal{R}_c \qquad (4\text{-}28)$$

Now

$$\frac{\Delta Q\,C}{\Delta V} = \frac{Q_0\,C_0 - QC}{V} = \frac{C_0 - \gamma C}{V/Q_0}$$

$\gamma = Q/Q_0$ a function of stoichiometry, pressure, and temperature.   Since the CSTR is usually a constant-volume reactor,

$$\theta_0\frac{dC}{dt} = \gamma C - C_0 + \mathcal{R}_c\theta_0 \qquad (4\text{-}29)$$

In steady state, where $\mathscr{R}_c = -kC$ and $\gamma = 1$,

$$\frac{C}{C_0} = \frac{1}{1 + k\theta_0} \qquad (4\text{-}30)$$

In these cases, single-phase reaction has been assumed, in that $\mathscr{R}_c$ is expressed in terms of fluid-phase concentration C. If, for example, we are dealing with a PFR containing pellets which catalyze the reaction, $\mathscr{R}_c$ while expressed in terms of C, contains whatever other coefficients are necessary to link C, the fluid-phase concentration, with the actual surface concentration of reactant and product species.

Strictly speaking in a solid catalytic reaction, where no homogeneous reaction occurs, the continuity equation for the fluid phase is rigorously written with $\mathscr{R}_c = 0$ (no homogeneous reaction) while $\mathscr{R}_a$ expresses the rate of generation of species (or removal) by another phase (the solid catalyst). The point might seem academic, yet in the solid-catalyzed oxidation of NO to NO_2, $\mathscr{R}_c$ represents the homogeneous kinetics and $\mathscr{R}_a$ that due to reaction occurring in the catalyst phase.

Nonisothermal Reactors

In this instance, since concentration of reactant is contained in q [Eq. (4-12)] and $\mathscr{R}_c$ and hence q are exponential functions of temperature, the continuity and thermal-energy equations are coupled and simultaneous solution is required. The limiting forms of the continuity equation cited above remain valid. However, the limiting forms of Eq. (4-12) require specifications. We find then:

6 *Batch (single reaction) adiabatic*

$$\frac{dT}{dt} = \frac{-\Delta H}{\rho C_p} \mathscr{R}_c \qquad (4\text{-}31)$$

Batch, nonadiabatic, well stirred

$$\frac{dT}{dt} = \frac{-\Delta H \mathscr{R}_c}{\rho C_p} - \frac{Ua}{\rho C_p}(T - T_c) \qquad \text{(see item 9)} \qquad (4\text{-}31a)$$

where U = overall heat-transfer coefficient
a = area per unit volume
T_c = coolant temperature

7 *Adiabatic PFR ($\rho u C_p$ = constant), steady state*

$$\frac{dT}{dz} = \frac{q}{u\rho C_p} \qquad (4\text{-}32)$$

8 *Nonadiabatic, nonisothermal PFR (steady state), $K_z \to 0$*

$$\frac{\partial T}{\partial z} - \frac{K_r}{u}\left(\frac{\partial^2 T}{\partial r^2} + \frac{1}{r}\frac{\partial T}{\partial r}\right) = \frac{q}{u\rho C_p} \qquad (4\text{-}33)$$

9   *Nonisothermal PFR, negligible radial gradient*

$$\frac{dT}{dz} + \frac{Ua(T - T_c)}{u\rho C_p} = \frac{q}{u\rho C_p} \qquad (4\text{-}34)$$

where T = average bulk temperature at any position z
      T_c = coolant or heat-source temperature at wall
This system is lumped (see also item 6) radially, as an overall coefficient based on bulk-temperature–wall-temperature difference is invoked to describe radial heat transport in lieu of the distributed radial term in Eq. (4-12) [and (4-33)], i.e., term (4).

10   *Nonisothermal, steady-state CSTR* ($a = A/V$)

$$\frac{dQ\,\rho C_p T}{dV} \left(= \frac{\Delta Q\,\rho C_p T}{\Delta V} \right) + Ua(T - T_c) = q \qquad (4\text{-}35)$$

or

$$T - T_0 = \frac{qV}{Q\rho C_p} - \frac{UA}{Q\rho C_p}(T - T_c)$$

Rearranging, where $\alpha = T_c/T_0$, gives

$$\frac{T}{T_0} = \frac{1 + UA\alpha/Q\rho C_p}{1 + UA/Q\rho C_p} + \frac{qV}{Q\rho C_p T_0(1 + UA/Q\rho C_p)} \qquad (4\text{-}36)$$

As $q = -\Delta H\,\mathscr{R}_c$, $\mathscr{R}_c$ can be eliminated via the continuity equation for a CSTR to yield the reduced temperature T/T_0 as a function of conversion, thermal properties, and transfer coefficient, as illustrated in Chap. 3 for the adiabatic CSTR ($UA = 0$).

In the general nonisothermal, nonadiabatic PFR model, transport coefficients other than those already defined are required to specify the system, namely, the thermal boundary conditions at the wall through which heat is either supplied or removed.   In general, the flux condition is

$$-K_r \frac{dT}{dr}\bigg|_{r=R} = \frac{U_w}{\rho C_p}(T - T_c) \qquad (4\text{-}37)$$

where T_c can be constant or vary with z.   The adiabatic condition is, of course,

$$-K_r \frac{dT}{dr}\bigg|_{r=R} = 0$$

A constant flux condition may exist if $U_w(T - T_c)$ remains constant as a function of reactor length. U_w is an overall wall coefficient which includes the heat-transfer coefficient at the wall on the process side, wall conductivity, and the coolant-side transfer coefficient.
In sum, then, data must be available or a valid means of prediction must be

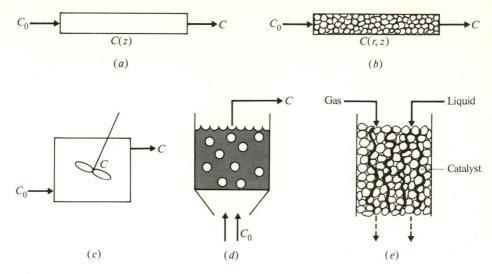

FIGURE 4-7
Diverse real reactor types: (a) tubular; (b) packed bed; (c) CSTR; (d) fluidized bed; (e) trickle bed.

at hand whereby the designer can determine the gross transport coefficients which describe:

1   Axial dispersion of heat and mass
2   Radial dispersion of heat and mass
3   Heat-transport coefficients at reactor wall on the process and coolant side
4   Pressure drop and velocity profiles in the reactor

The general features of some techniques by which the dispersion coefficients are experimentally determined will now be outlined and a summary of experimental results and correlations will be given.   Two very common distributed-reactor types will be featured: (a) the tubular reactor, within which homogeneous reactions are often conducted, and (b) the packed-bed reactor, often used to conduct heterogeneous catalyzed reactions.

Some other important reactor types include (c) the stirred vessel, operated batchwise, semibatchwise, or continuously (CSTR), (d) the fluidized-bed reactor, and (e) the trickle-bed reactor.   These are schematized in Fig. 4-7.

It is not our purpose in this chapter to present comprehensive correlations of all pertinent transport coefficients for common reactor types.   Detailed information of this kind will be set forth in subsequent illustrations.   At this juncture, we wish to simply introduce methods of transport-coefficient measurement and correlation and then illustrate their usefulness.

This purpose is best served by confining attention to mass dispersion in the radial and axial directions.   For homogeneous systems (unpacked tube), the mass-heat analogy provides a basis for extending the mass data to heat transport; however, in packed beds, the modes of heat transfer include, for example, particle conduction as well as fluid transport. Hence no analogy exists between total mass and heat dispersion in a packed bed, for clearly no mass can be conducted through the particles in the axial and/or radial direction. This more complex heat-dispersion problem will be treated in Chap. 10 in dealing with packed-bed, nonisothermal reactors.

4-4 DETERMINATION OF DISPERSION COEFFICIENTS

Radial Dispersion of Mass

In general, the dispersion coefficients are measured under nonreacting circumstances. An inert tracer is used and its behavior noted under various circumstances of flow and reactor geometry. In radial dispersion or mixing, the experiment is easily visualized. With a fluid passing steadily through a tube (packed or unpacked), a dye or some other easily detected tracer is introduced at the tube axis at a velocity equal to that *of the mainstream* flow (Fig. 4-8).

As shown in Fig. 4-8, the steadily injected tracer will fan out, or disperse, radially in a fashion dictated by the nature of the system. As flow rate approaches zero, of course, molecular diffusivity of the tracer will govern radial diffusion. However, interest most often lies in higher, turbulent flow rates at which radial diffusion is governed by turbulence level of the mainstream as well as whatever obstruction, e.g., packing, exists within the tube.

By sampling the dispersing tracer at one or more points downstream of injection, at several radial positions, a radial concentration profile is secured. As the experiment is a steady-state one, without reaction, if we ignore axial mixing (justified if $\partial c/\partial z$ and hence $\partial^2 c/\partial z^2$† is small) the continuity equation (4-11) in cylindrical coordinates becomes

$$\frac{u\,\partial C}{\partial z} - \mathscr{D}_r\left(\frac{\partial^2 C}{\partial r^2} + \frac{1}{r}\frac{\partial C}{\partial r}\right) = 0 \qquad (4\text{-}38)$$

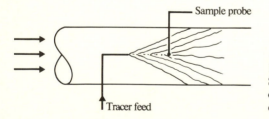

FIGURE 4-8
Schematic of an experimental network designed to ascertain radial-dispersion data.

† If $\partial c/\partial z$ is constant, $\partial^2 c/\partial z^2 = 0$.

Pulse of
tracer

Effluent pulse

FIGURE 4-9
Schematic of experimental methodology for the determination of axial
dispersion.

which for appropriate boundary conditions[1] can be solved to yield a reduced
tracer concentration as a function of r at a given z in terms of $\mathscr{D}_r$. In the experiment all is known except $\mathscr{D}_r$. By fitting the solution of Eq. (4-38) to experimental
radial tracer profiles $\mathscr{D}_r$ is obtained. Note that $\mathscr{D}_r$ and u are assumed independent
of r in this case. By conducting these experiments at various Reynolds numbers,
reactor geometries, etc., $\mathscr{D}_r$ expressed in dimensionless form is correlated in terms
of key dimensionless variables.

Axial Dispersion

The simple approach to the determination of $\mathscr{D}_z$ involves assuming that u is
constant and $\mathscr{D}_r$ is zero. A steady-state reaction-free approach is clearly fruitless
since no axial gradient is established in this case. In a transient, however, we have,
from Eq. (4-11),

$$\frac{\partial C}{\partial t} = \mathscr{D}_z \frac{\partial^2 C}{\partial z^2} - u \frac{\partial C}{\partial z} \qquad (4\text{-}39)$$

By imposing a transient, using a detectable tracer, the experimental data can be
compared with the appropriate solution of Eq. (4-39) for determination of $\mathscr{D}_z$.
Experimentally we might, for example, inject a pulse of tracer into the steadily
flowing bulk stream passing through the reactor (Fig. 4-9).

    If $\mathscr{D}_z$ is finite, the injected pulse will spread due to axial dispersion, and this
measured spread can be expressed in terms of $\mathscr{D}_z$. As with radial dispersion, $\mathscr{D}_z$
values can then be correlated in terms of the usual fluid-mechanical and geometric
variables.

Experimental Results

Bischoff and Levenspiel[2] have thoroughly analyzed available axial and radial
mixing data for both packed and unpacked tubular vessels, expressing the results
as $\mathscr{D}_z/\nu$ and $\mathscr{D}_r/\nu$ versus the Reynolds number, where ν is the kinematic viscosity.

[1] R. A. Bernard and R. H. Wilhelm, *Chem. Eng. Prog.*, **46**: 233 (1950); O. Levenspiel and K. B. Bischoff,
Adv. Chem. Eng., **4**: 95 (1963).
[2] K. B. Bischoff and O. Levenspiel, *Chem. Eng. Sci.*, **17**: 245 (1962).

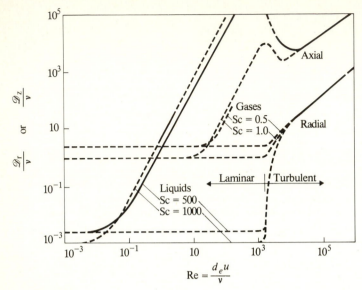

FIGURE 4-10
Radial and axial dispersion correlations for unpacked tubes. [*K. B. Bischoff and O. Levenspiel, Chem. Eng. Sci.*, **17**: 257 (1962).]

For unpacked tubes, the general picture is displayed in Fig. 4-10, where

$$\text{Re} = \frac{d_e u}{v} \qquad d_e = \frac{4 \text{ (free vol of fluid)}}{\text{wetted area}} \qquad \text{Sc} = \frac{v}{\mathscr{D}_{molec}} = \text{Schmidt number}$$

For packed beds, the results are shown in Fig. 4-11 on the same basis.

The dashed regions of Figs. 4-10 and 4-11 indicate regions devoid of experimental data. For both packed and unpacked tubes, it is apparent that $\mathscr{D}_z$ is larger than $\mathscr{D}_r$ at a given value of the Reynolds number, defined in terms of the hydraulic diameter d_e as recommended by Wilhelm.[1] For unpacked tubes, $d_e = d_t$, the tube diameter. For a packed bed of void fraction ε and particle diameter d_p

$$d_e = \frac{d_t}{\tfrac{3}{2}(d_t/d_p)(1 - \varepsilon) + 1} \qquad (4\text{-}40)$$

Since $\mathscr{D}/v$ is a reciprocal Schmidt number, it logically approaches the molecular value at low (laminar) flow rates. In the turbulent regime $\mathscr{D}/v$ becomes independent of the diffusing species and varies linearly with Reynolds number. Since

$$\text{Re} \cdot \text{Sc} = \frac{d_e u}{\mathscr{D}} = \text{Peclet number}$$

[1] R. H. Wilhelm, *Chem. Eng. Prog.*, **49**: 150 (1953).

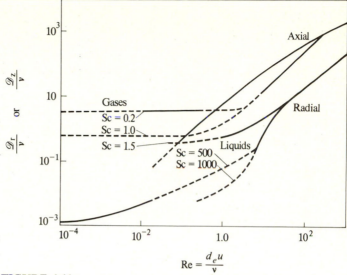

FIGURE 4-11
Radial and axial dispersion correlations for packed beds. [*K. B. Bischoff and O. Levenspiel, Chem. Eng. Sci.,* **17**: *257 (1962).*]

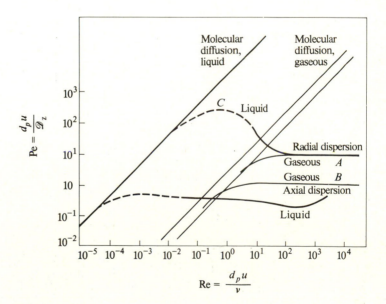

FIGURE 4-12
Wilhelm's display of radial- and axial-dispersion Peclet numbers vs. Reynolds number for laminar and tubulent flow in packed beds. [*R. H. Wilhelm, Pure Appl. Chem.,* **5**: *403 (1962).*]

defined on a diameter basis, we see that in turbulent flow the Peclet number for both packed and unpacked tubes tends to become a constant, independent of Re. This point and its implications can be illustrated nicely by confining attention to packed beds.

Actual data for radial and axial mass dispersion in packed beds are shown in a comprehensive fashion in Fig. 4-12.

In the turbulent regime (Re > about 40), some features are worth noting:

1 *Radial dispersion*   For both gases and liquids, the radial Peclet number assumes a constant value of about 10.
2 *Axial dispersion*   For gases, the axial Peclet number is found to be about 2. For liquids, lower values prevail (between 0.3 and 1), though at Re approaching 1000, the liquid axial Peclet number appears to approach that value of 2 found for gas dispersion.

At low Reynolds numbers, molecular diffusion manifests itself as indeed it must with the suppression of turbulence and mixing.

4-5   INTERPRETATION OF PACKED-BED DISPERSION DATA

The dispersion and mixing phenomena in packed beds doubtless reflect complex underlying events.   In the turbulent regime, which is of prime interest, mixing, stream splitting, velocity variations (local and average), fluid accelerations, and decelerations all contribute to both radial and axial dispersion.   Thus the phenomenological coefficients $\mathscr{D}_r$ and $\mathscr{D}_z$ have lumped into them all manner of subtle, complex functions.

It is possible, however, to fashion models of a simple nature which permit estimates of the radial and axial Peclet number for turbulent flow in packed beds. These models are not offered as explanations of dispersion but simply serve to illustrate the essential features of these events in fixed beds.

Implicit in the dispersion-coefficient approach is the notion that axial and radial turbulent mixing is a diffusive process, analogous to the molecular event. As a basis for model development, we shall derive first the Einstein relation for diffusion in a field of mobile particles, as displayed in Fig. 4-13.   By Fick's law, the diffusive flux is

$$N = -\mathscr{D}\frac{dC}{dX} \qquad \text{mol/(area)(time)}$$

Referring to Fig. 4-13, the transport of particles from left to right per unit area is, assuming an equal chance of particles moving in either direction,

$$\bar{n}_a = \tfrac{1}{2}C_a(X_2 - X_1) \qquad \text{mol/area} \qquad (4\text{-}41)$$

and from right to left we have

$$n_b = \tfrac{1}{2}C_b(X_2 - X_1) \qquad (4\text{-}42)$$

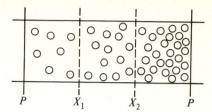

FIGURE 4-13
Einstein's diffusion model. [*J. J. Car-berry, AIChE J.,* **4:** *13M (1958).*]

Then net transport is

$$\bar{n}_b - \bar{n}_a = \tfrac{1}{2}(X_2 - X_1)(C_b - C_a) \qquad (4\text{-}43)$$

If $X_2 - X_1$ is small, Fick's law can be written

$$N = \mathscr{D}\,\frac{C_b - C_a}{X_2 - X_1} \qquad (4\text{-}44)$$

or in a time interval θ_D

$$N\theta_D = \frac{\mathscr{D}(C_b - C_a)\theta_D}{X_2 - X_1} \qquad \text{mol/area} \qquad (4\text{-}45)$$

which is also equal to $\bar{n}_b - \bar{n}_a$. Then

$$\tfrac{1}{2}(X_2 - X_1)(C_b - C_a) = \frac{\mathscr{D}(C_b - C_a)\theta_D}{X_2 - X_1} \qquad (4\text{-}46)$$

which yields

$$\mathscr{D} = \frac{(X_2 - X_1)^2}{2\theta_D} = \frac{\ell^2}{2\theta_D} \qquad (4\text{-}47)$$

the well-known Einstein relationship, where ℓ is a mixing or diffusion length and θ_D the diffusion time.

Radial Diffusion (Dispersion)

Baron[1] and Ranz[2] fruitfully viewed radial dispersion as the result of stream splitting and sidestepping. That is, as shown in Fig. 4-14, a stream of fluid at a particular radial position strikes a piece of packing in its axial journey and is split in two by the collision, and, on the average, one-half the stream moves laterally to the right, the other to the left. This event occurs repeatedly, with the result that the original (tagged) single stream is laterally dispersed, or fans out, toward the wall.

In terms of the Einstein relation (4-47), we can say that upon splitting the stream moves a *diffusive* distance ℓ equal to about one-half a particle diameter,

[1] T. Baron, *Chem. Eng. Prog.,* **48:** 118 (1952).
[2] W. Ranz, *Chem. Eng. Prog.,* **48:** 247 (1952).

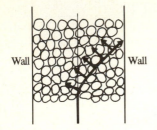

FIGURE 4-14
Idealization of radial dispersion (stream splitting) in turbulent flow through a fixed bed.

that is, $\ell \approx d_p/2$. The time for a jump or laterally dispersing split to occur is of the order of the time required for the fluid to traverse one axial layer of packing, or $\theta_D \approx d_p/u$. Substituting into (4-47), we obtain

$$\mathscr{D}_r = \frac{d_p u}{8} \qquad \text{or} \qquad \frac{d_p u}{\mathscr{D}_r} = \text{Pe}_r \approx 8 \qquad (4\text{-}48)$$

which agrees nicely with the average experimental value of 10 shown in Fig. 4-12. While the model is a simple one, it captures the essential feature of radial dispersion in a packed bed. Wall effects are ignored, and detailed data[1] indicate that Pe_r is a function of radial position within a packed bed.

Axial Dispersion

If we view a packed bed as consisting of an array of voids into which fluid flows at a high velocity from small-area ports created by reason of close particle-particle contact, it can be visualized that as a result of acceleration in the ports and deceleration upon entering the voids, mixing occurs. In the limit, if perfect mixing occurs in each void, the bed may be viewed as a series of perfectly mixed vessels interconnected by ports consisting of closely packed regions. This model is analogous to mixing in a tank caused by a jet feed nozzle which creates high turbulence (mixing) in the tank.

In terms of Eq. (4-47), the distance ℓ over which the fluid journeys (diffuses) before void-cell mixing is, on the average, d_p. Setting $\ell \approx d_p$ and $\theta_D = d_p/u$, we obtain

$$\mathscr{D}_z = \frac{d_p u}{2} \qquad \text{or} \qquad \frac{d_p u}{\mathscr{D}_z} = \text{Pe}_a \approx 2 \qquad (4\text{-}49)$$

in excellent agreement with the axial mixing data for *gases* at Reynolds numbers (based upon superficial velocity u_0) greater than about 10.†

This result [Eq. (4-49)] is also predicted by other arguments.[2] Packed-bed

[1] R. Fahien and J. M. Smith, *AIChE J.*, 1: 28 (1955).
† K. W. McHenry and R. H. Wilhelm, *AIChE J*, 3: 83 (1957).
[2] R. Aris and N. R. Amundson, *AIChE J.*, 3: 280 (1957); H. Kramers and G. Alberda, *Chem. Eng. Sci.*, 2: 173 (1953).

axial-dispersion data for liquids present a problem, the Peclet numbers being severalfold lower than the predicted and experimental value for gases. One can invoke the notion of imperfect void-cell mixing to rationalize the liquid dispersion data;[1] however, the resulting void-cell mixing "efficiency" provides no insight into the events which govern the apparently anomalous behavior. The arguments of Prausnitz,[2] based as they are on fluid-mechanical parameters, seem more promising. In any event, liquid axial-dispersion data for packed beds do not appear to obey the simple diffusion law too often invoked in experimental work. For it has been confirmed[3] experimentally that in pulse-injection experiments designed to determine dispersion or diffusion coefficients for liquid flow through fixed beds, pulse amplitude does not decrease with the square root of bed length, as required by the usual model of diffusion with plug flow [Eq. (4-39)]. Indeed the discrepancy increases with decreasing particle size, suggesting a capacitance, or holdback, effect, which would increase, per unit volume of bed, with decreasing packing size. Thus the tracer becomes "lost" in dead or stagnant zones at particle-particle contact points, to be released by diffusional processes largely molecular in nature. Given the fact that gaseous molecular diffusivities are about 10^5 times greater than those for liquids, it is not surprising that capacitance becomes more evident in liquid systems. As Turner[4] and Aris[5] have shown analytically, a bed capacitance generates lower axial Peclet numbers, in qualitative agreement with laboratory data.

Implications of Fixed-Bed Data

By Eq. (4-47), we find that the time constant for a diffusive event is

$$\theta_D = \frac{\ell^2}{2\mathscr{D}}$$

and in a packed bed $\theta_D \approx \ell/u$. If we now define an axial mixing length ℓ in a packed bed as the total bed length divided by the number of perfect mixers n, we have for $\ell = L/n$

$$\theta_D \neq \frac{d_p}{u} \quad \text{but} \quad \frac{\ell}{u} = \frac{L/n}{u} = \frac{(L/n)^2}{2\mathscr{D}_z}$$

or

$$n = \frac{Lu}{2\mathscr{D}_z} \qquad (4\text{-}50)$$

More completely, since $n \rightarrow 1$ as $\mathscr{D}_z \rightarrow \infty$, for large $\mathscr{D}_z$

$$n = \frac{Lu}{2\mathscr{D}_z} + 1 \qquad (4\text{-}51)$$

[1] J. J. Carberry, *AIChE J.*, **4**: 13M (1958).
[2] J. Prausnitz, *AIChE J.*, **4**: 14M (1958).
[3] J. J. Carberry and R. H. Bretton, *AIChE J.*, **4**: 367 (1958).
[4] G. A. Turner, *Chem. Eng. Sci.*, **1**: 156 (1958).
[5] R. Aris, *Chem. Eng. Sci.*, **11**: 194 (1959).

Equation (4-50) or (4-51) relates the dispersion coefficient to n perfectly mixed vessels set in series. This relation is useful in drawing inferences concerning the influence of, say, axial mixing upon reaction rate. In packed-bed processes, for a reacting gas at Reynolds numbers above about 20, where $Pe_a = 2$, by Eq. (4-50)

$$n = \frac{L}{2d_p} \frac{d_p u}{\mathcal{D}_z} = \frac{L}{2d_p} Pe_a \qquad (4\text{-}52)$$

which for $Pe_a = 2$, indicates that the bed may be treated as a series of n CSTRs, where, by (4-52), for $Pe_a = 2$

$$n = \frac{L}{d_p} \qquad (4\text{-}53)$$

From demonstrations set forth in Chap. 3 it is apparent that if L/d_p is very large (about 20 or greater), the series-CSTR system behaves exactly like a PFR.[1] This point is tellingly illustrated by evaluating, for first-order isothermal reaction, the fraction of reactant remaining in a fixed bed of $n (= L/d_p)$ relative to the case where $n = \infty$. In Fig. 4-15 this ratio is plotted versus $k\theta_0$ for various values of $n = L/2d_p Pe_a$. Except for instances characterized by small L/d_p and/or Pe, axial dispersion exerts a negligible influence upon steady-state conversion.

As shown in Chap. 3, the mixing effect is more severe with higher-order

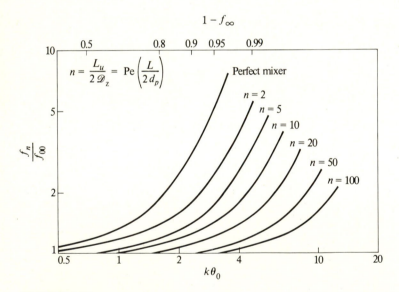

FIGURE 4-15
Relative influence of axial mass dispersion upon first-order processes.
[*J. J. Carberry, Can. J. Chem. Eng.,* **36**: 207 (1958).]

[1] Axial thermal Peclet values of about 0.6 are found in fixed beds; hence $n \approx 150$ for nonisothermal plug flow.

kinetics and less telling in the instance of orders less than unity. Nevertheless, with rare exceptions, a fixed-bed reactor is generally characterized by L/d_p values of the order of at least 50 to several hundred. In consequence, axial dispersion or mixing can generally be ignored in the design of fixed-bed reactors. On the other hand, its influence cannot be dismissed a priori in laboratory studies often characterized by Reynolds numbers or L/d_p ratios so low that Pe_a becomes less than 2 and/or n assumes a value which should cause concern. Under certain circumstances, axial-dispersion corrections can be made, as outlined by Epstein.[1] Such a correction procedure is well suited to heat- or mass-transport studies in which a log mean driving force (which implicitly assumes no axial dispersion) can be corrected by a factor which accounts for driving-force damping due to axial mixing.

Noting that the dispersion Peclet number expresses the transport by bulk flow $(d_p u)$ relative to that by dispersion ($\mathscr{D}_z$ or $\mathscr{D}_r$), it is clear that since the radial Peclet number is 5 to 6 times greater than that of axial dispersion, the radial dispersion coefficient $\mathscr{D}_r$ is of much lower magnitude than the axial counterpart. In consequence, radial mass gradients (established by radial velocity and/or temperature gradients) are less easily damped than the axial profiles. If we use Eq. (4-47) to define the radial mixing-time constant as

$$(\theta_D)_r = \frac{\ell_r}{u} = \frac{\ell_r^{\,2}}{2\mathscr{D}_r}$$

then if $\qquad\qquad \ell_r = \dfrac{d_t}{m} \qquad$ where $m =$ number of *radial* mixers

on solving for m we obtain

$$m = \frac{d_t u}{2\mathscr{D}_r} = \frac{d_t}{2d_p}\left(\frac{d_p u}{\mathscr{D}_r}\right)_r \qquad (4\text{-}54)$$

where d_t is tube diameter. As we have noted, for turbulent flow in a packed bed $(d_p u_0/v >$ about 20) $Pe_r = 10$; thus

$$m = 5\frac{d_t}{d_p} \qquad (4\text{-}55)$$

By Eq. (4-55), within the limits imposed by the simplifications which support the development, we see that (as with axial mixing) radial mixing decreases (m increases) with increasing values of the aspect ratio (d_t/d_p for radial dispersion; L/d_p for axial dispersion) (see Fig. 4-15).

To effect a minimum radial gradient, m should be minimized; however, its value cannot approach unity, as in the axial-mixing case ($n \to 1$, $L/d_p \to 1$ for $Pe_a = 2$). For with radial dispersion as $d_t/d_p \to 1$, $m \to 5$. This is not a valid extrapolation, however, since radial Peclet numbers are extracted from data assuming that the bed is a continuum. Clearly as d_t/d_p decreases to, say, values

[1] N. Epstein, *Can. J. Chem. Eng.*, **36**: 210 (1958).

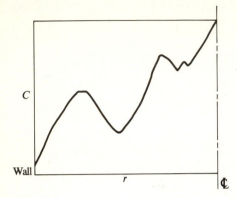

C

Wall

r

₵

FIGURE 4-16
Schematic of radial concentration (or
temperature) distribution at small values
of tube-to-particle diameter ratio in a
fixed bed.

of 5 or less, two new situations can be anticipated: (1) As mentioned above, the
system is no longer a continuum but statistical, as schematized in Fig. 4-16, and
(2) fluid bypassing increases as d_t/d_p decreases.

Furthermore, since $\mathscr{D}_r$ is a function of radial position in a packed bed,
Eq. (4-55) is only qualitatively correct for large d_t/d_p values, say values greater
than 5 to 10. With this restriction, it can be seen that total radial mixing is
achieved with difficulty. For we are limited to several radial mixers at best, and
this situation is far from the desired condition of perfect mixing in the radial
direction.

To summarize packed-bed mass-dispersion data, we can state that (1) axial
mixing effects can be neglected in steady-state isothermal operation at Re > 20 so
long as L/d_p is greater than about 50 and (2) since radial mixing is at best incomplete, design and analysis must account for radial gradients. Clearly in the
absence of a radial temperature or velocity gradient, no radial mass gradient can
exist unless, of course, a reaction is occurring at the tube wall. The far more complex problem of radial dispersion of heat will be discussed in Chap. 10.

Unpacked Tubes

Dispersion of a homogeneous fluid flowing through an unpacked tube will be
governed by turbulence level. In laminar flow, it can be anticipated that molecular
processes dictate both axial and radial diffusion.

From Fig. 4-10 it will be noted that the radial Peclet number for turbulent
flow in unpacked tubes is of the order of 500 to 1000; that is,

$$\text{Pe}_r \approx 10^3 \qquad \text{unpacked tube} \qquad (4\text{-}56)$$

while the axial Peclet number is

$$\text{Pe}_a \approx 10 \qquad \text{where } \text{Pe}_a = \frac{d_t u}{\mathscr{D}_z} \qquad (4\text{-}57)$$

Mass and temperature profiles in turbulent flow can be expected to obey the heat, mass, and momentum analogy. In consequence, radial gradients will be confined to regions adjacent to the tube wall, as is the case with the velocity gradient in turbulent flow.

The concept of a series of mixers used with advantage in the case of packed beds cannot be validly extended to unpacked tubes, as it becomes somewhat difficult to specify a mixing length ℓ and time θ_D.

For unpacked tubes,

$$n = \frac{Lu}{2\mathscr{D}_z} = \frac{L}{2d_t}\frac{d_t u}{\mathscr{D}_z} \qquad (4\text{-}58)$$

By (4-57), for turbulent flow in an unpacked tube, the number of equivalent mixers is

$$n = \frac{L}{2d_t}(\text{Pe} \approx 10)_a \approx 5\frac{L}{d_t} \qquad (4\text{-}59)$$

For radial diffusion, Eq. (4-54) states that

$$m = \frac{d_t u}{2\mathscr{D}_r}$$

Then, by (4-56), for radial dispersion in turbulent flow, the number of radial mixers is of the order

$$m = \frac{10^3}{2} \approx 400 \text{ to } 500 \qquad (4\text{-}60)$$

These results [Eqs. (4-59) and (4-60)] are extremely rough, qualitative estimates, which, however, indicate the greater importance of radial transport limitations relative to axial mixing. Bear in mind that radial and axial transport ideals are diametrically different: we want a maximum of radial mixing (small m) and a minimum of axial mixing (large n). Even for so unfavorable an unpacked-tube-aspect ratio L/d_t of, say, 2, the system (in turbulent flow) behaves like 10 CSTRs (nearly plug flow), while there appears to exist an even smaller possibility of radial mixing ($m \approx 400$), in the unpacked tubular reactor. This assumes, of course, that there exists a cause which establishes, potentially, a radial gradient, e.g., wall reaction; heat transfer through the wall; or radial velocity gradient.[1]

We shall now devote some attention to design and analysis of common reactor types where single-phase, homogeneous reaction(s), take place. While, in principle, the governing continuity equations for these elementary reaction environments have been deduced earlier, it may prove of benefit to set forth a few illustrations, and a comment on scale-up via dimensionless analysis seems in order.

---

[1] Turbulence theory suggests that $m \approx 3$.

4-6 SCALE-UP OF CHEMICAL REACTORS

Consider the batch conduct of an isothermal first-order reaction. Assume that the reaction model has been established in a laboratory spherical flask of diameter D_L. The integrated reaction-rate model is, of course,

$$\frac{A}{A_0} = \exp\left(-kt\right)$$

where kt is a dimensionless group, a Damköhler number. In principle, knowledge of kt permits prediction of A/A_0 in a plant-scale batch reactor of effective diameter D_P. However, this assumes that the corresponding continuity equation for energy is unchanged on scale-up. In the laboratory study dT/dt in Eq. (4-32) is zero since heat exchange matches heat generation ($\pm$). But let us assume that this heat exchange is effected through an exchange area per unit volume $6/D_L$. Exact reactor similarity cannot be maintained on scale-up since at the plant level heat-exchange area per unit volume is $6/D_P$. Obviously one need merely employ immersion heat-exchange coils to solve the problem, but we utilize this primitive example to illustrate that in chemical reaction systems, reactor scale-up via dimensional analysis, commonly used with success in the unit operations, is not possible except in the most trivial reaction systems, e.g., linear kinetics with negligible enthalpy of reaction.

As our reactor model becomes more complex, the number of nondimensional groups increases in both the continuity equations and boundary conditions, and we shall generally find that these real rather than ideal reactor models are governed by nondimensional groups, not all of which can be maintained constant upon scale-up.

For example, in a tubular-flow reactor, even assuming isothermality, a desired conversion demands a specific contact time θ, consistent with constancy of $k\theta$ for the linear case. Now heat transfer and pressure drop are governed by the fluid Reynolds number. Thus if $\theta = L/u$ is maintained constant and productivity $Q = u \times$ area is increased by the desired scale-up factor, the Reynolds number cannot remain constant and thus fluid-mechanical similarity is lost. Should scale-up be critically dependent upon fluid-mechanically governed processes (transport of heat, mass, and/or momentum), similitude cannot be achieved on the two scales.

When nonisothermality is introduced and then heterogeneity of the reaction and/or reactor environment, mere intuition suggests the impossibility of maintaining similitude on differing scales of operation, a point persuasively advanced by Bosworth for both homogeneous and heterogeneous reaction-reactor systems.[1]

Therefore design and analysis rest upon soundly formulating both the reaction and reactor model. Nature provides the reaction network to be modeled, while the reaction engineer enjoys freedom to impose diverse reactor models upon the reactions involved to try to negotiate an effective reactor network which profitably grants product.

---

[1] R. C. L. Bosworth, "Transport Processes in Applied Chemistry," Wiley, New York, 1956.

4-7 HOMOGENEOUS REACTOR DESIGN (QUALITATIVE)

Armed with a model of the dynamics of the reaction network (Chap. 2) and a model of the reactor, one can, in principle, resolve the governing equations either analytically or numerically to establish plant-reactor dimensions and mode of operation. However, in fact, our reaction model is too often incomplete and imprecise while our reactor model may represent a compromise with reality.

Reactor design is rarely an a priori accomplishment; instead as noted in Chap. 1, model predictions and actual performance data must be contrasted at several reactor scales for the model to be refined to a point where it can confidently be used to predict large-scale behavior.

With the above qualification, we may comment on several reactor systems commonly used in the conduct of homogeneous reactions.

Batch Reactors

Batch reactors, processing liquid reactants, are commonly used in laboratory studies and at plant scale when production does not justify continuous operation. Pharmaceuticals are often manufactured in batch reactors by reason of long contact-time requirements and relatively low annual production. Equations (4-20) and (4-31) or (4-31a) govern nonisothermal batch-reactor behavior.

In the general instance of a complex reaction conducted in a nonisothermal, nonadiabatic batch reactor, numerical solution of Eqs. (4-20) and (4-31a) is required. We shall illustrate the procedure for the equivalent problem, the non-isothermal, nonadiabatic PFR [Eqs. (4-25) and (4-34)] later in this chapter.

Semibatch Reactors

In this reactor, a reactant is fed to the vessel containing a batch of a coreactant. Gas-liquid reactions are often realized in this fashion. For the homogeneous reaction, we shall deal with two miscible liquid reactants. As such, the volume of the reacting mass increases with time. Equation (4-22) is the governing mass-continuity equation. Semibatch reactors are of a distinct advantage since (1) control of temperature can be realized by controlling the rate of reactant addition and (2) control of product distribution is achieved by manipulation of rate of reactant addition.

Continuous Reactors

Consistency of productivity, both qualitatively and quantitatively, is realized by using continuous reactors. Further, labor costs involved in charging and discharging the reactor, necessities in batch and semibatch processing, are absent except during shutdowns for maintenance, repairs, etc.

As noted in Chap. 3, the continuously fed reactor assumes diverse modes of behavior with respect to conversion and yield, depending upon fluid residence-time distribution (CSTR, PFR, and intermediate cases). The choice of continuous reactor type is governed by conversion-yield factors cited in Chap. 3. Other obvious factors intervene; e.g., a homogeneous gas-phase reaction is not easily visualized as occurring in a mechanically agitated CSTR. Should backmixing be desired, the recycle reactor would meet that demand in a gas-phase homogeneous system. The use of stirred tanks on a production scale is exclusively limited to reaction systems involving a liquid phase.

The tubular reactor usually operates in plug flow as long as length-to-tube diameter is much greater than unity (say $L/d_t > 20$) and the flow is turbulent. Both homogeneous and solid-catalyzed reactions are commonly conducted in tubular reactors. In the packed- or fixed-bed catalytic reactor, the criterion for negligible backmixing (plug flow) is phrased, as noted earlier, in terms of bed length to catalyst-pellet diameter $L/d_p > 50$ (isothermal) and $L/d_p > 150$ (non-isothermal).

Tubular reactors may assume the form of a simple U tube (laboratory scale)

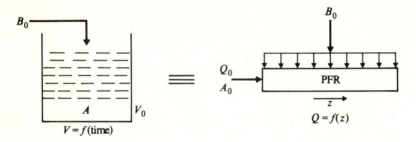

FIGURE 4-17
Variable-volume semibatch reactor and PFR with side-stream addition.

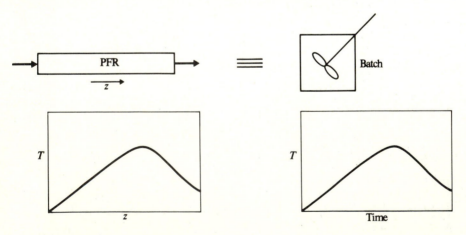

FIGURE 4-18
Nonisothermal PFR and batch reactor.

or on pilot and plant scales a coil or an array of parallel straight tubes immersed in a shell containing a heat sink or source.

   Illustrative analyses will now be presented for (1) an isothermal, variable-volume, semibatch reactor and (2) a nonisothermal, nonadiabatic plug-flow tubular reactor.   These situations represent analogs:

   1 The semibatch variable-volume reactor is equivalent to a plug-flow tubular reactor with an evenly distributed side-stream addition, as shown in Fig. 4-17.
   2 The nonisothermal, nonadiabatic PFR is equivalent to a batch reactor in which temperature varies with time, as shown in Fig. 4-18.

   Adiabatic operation of a PFR will also be illustrated, as will a design procedure when a detailed kinetic model is not at hand.   Finally, transient behavior of a series network of CSTRs will be analyzed, as will be an example of transient polymerization.

4-8 SEMIBATCH REACTOR ANALYSIS AND DESIGN

Consider the reaction $A + B \underset{2}{\overset{1}{\rightleftharpoons}} P$, where a liquid-phase reactant A is pumped into the autoclave to occupy an initial volume V_0. At $t = 0$, B, a miscible liquid coreactant, is pumped into the reactor at a constant rate F (moles per time). Molar density ρ is assumed constant for reactants and mixtures thereof, as is reaction temperature, a condition readily realized since heat release now depends not only upon reaction rate per se but on the negotiable rate of addition.

   We shall first express the continuity equation in terms of concentrations, which results in a differential equation of some complexity.   Rederivation in terms of moles yields a much simpler differential equation.

   In terms of concentrations, assuming $A \gg B$ at any time

$$\frac{1}{V}\frac{dB\,V}{dt} = -k_1 B + k_2 P + \frac{F}{V} \qquad (4\text{-}61)$$

Now

$$V = V_0 + \frac{Ft}{\rho} \qquad (4\text{-}62)$$

For product P,

$$\frac{1}{V}\frac{dP\,V}{dt} = k_1 B - k_2 P \qquad (4\text{-}63)$$

Also

$$\frac{1}{V}\frac{dB\,V}{dt} = \frac{dB}{dt} + \frac{B}{V}\frac{dV}{dt}$$

By Eq. (4-62), where $\bar{F} = F/\rho$,

$$\frac{1}{V}\frac{dB\,V}{dt} = \frac{dB}{dt} + \frac{B\bar{F}}{V_0 + \bar{F}t}$$

and

$$\frac{1}{V}\frac{dP\,V}{dt} = \frac{dP}{dt} + \frac{P\bar{F}}{V_0 + \bar{F}t}$$

Therefore

$$\frac{dB}{dt} + \frac{B\bar{F}}{V_0 + Ft} = -k_1 B + k_2 P + \frac{F}{V} \qquad (4\text{-}64)$$

and

$$\frac{dP}{dt} + \frac{P\bar{F}}{V_0 + \bar{F}t} = k_1 B - k_2 P \qquad (4\text{-}65)$$

Now we solve Eq. (4-65) for B:

$$B = \frac{1}{k_1}\frac{dP}{dt} + \frac{P}{k_1(m+t)} + \frac{k_2}{k_1}P \qquad (4\text{-}66)$$

where $m = V_0/\bar{F}$. Differentiating B and substituting B and its derivative into Eq. (4-64), we obtain

$$\frac{d^2P}{dt^2} + \frac{dP}{dt}\left(\frac{2}{m+t} + k_1 + k_2\right) + \frac{P}{m+t}(k_1 + k_2) = \frac{k_1\rho}{m+t} \qquad (4\text{-}67)$$

an equation not readily susceptible to analytical resolution unless a convenient change of variable can be derived.

As an exercise find a convenient change of variable which will facilitate solution of Eq. (4-67).

We now phrase the problem in terms of moles:

$$\frac{db}{dt} = -k_1 b + k_2 p + F \qquad (4\text{-}68)$$

$$\frac{dp}{dt} = k_1 b - k_2 p \qquad (4\text{-}69)$$

As before, we solve for b and substitute it and its derivative into Eq. (4-68) to obtain

$$\frac{d^2p}{dt^2} + \frac{dp}{dt}(k_1 + k_2) - Fk_1 = 0 \qquad (4\text{-}70)$$

Let $y = dp/dt$; then

$$\frac{dy}{dt} + y(k_1 + k_2) - Fk_1 = 0$$

Integrating between y and zero, that is $dp/dt = 0$ at $t = 0$, we have

$$y = \frac{dp}{dt} = \frac{Fk_1}{k_1 + k_2}\{1 - \exp[-(k_1 + k_2)t]\} \qquad (4\text{-}71)$$

Upon a second integration, and recalling that

$$V = \frac{F}{\rho}(m+t)$$

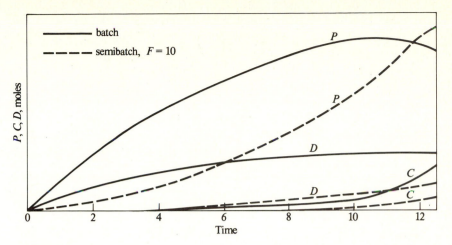

FIGURE 4-19
Profiles of number of moles of each species vs. time for the isothermal semibatch and batch reaction

$$A + B \xrightarrow{\;1\;} P \xrightarrow{\;2\;} C \qquad k_1 = 0.1$$
$$B + B \xrightarrow{\;3\;} D \qquad\qquad k_2 = 0.01$$
$$\qquad\qquad\qquad\qquad\qquad k_3 = 0.05$$

there results the product concentration P-versus-time functionality

$$P = \frac{\rho p}{F(m+t)} = \frac{\rho k_1}{(m+t)(k_1+k_2)} \left[t + \frac{\exp -(k_1+k_2)t}{k_1+k_2} \right]_0^t \qquad (4\text{-}72)$$

Equation (4-72) reduces, of course, to the simple irreversible case when $k_2 = 0$.

Table 4-1   TERMINAL SEMIBATCH SELECTIVITY   S   AND YIELD   Y AS A FUNC-TION OF FEED RATE F[†]

$$S = \frac{P}{C+D} \qquad Y = \frac{P}{A_0 X}$$

F, moles per unit time	S	Y, %
0 (batch)	2.5	72
1	17	95
10	5.8	85
50	3.38	77

† Period of operation is 10 time units in each case.   Initial volume $V_0 = 100$

$$A + B \xrightarrow{\;1\;} P \xrightarrow{\;2\;} C$$

$$B + B \xrightarrow{\;3\;} D$$

Both reactant B and product P profiles can be predicted as a function of time and F for specific values of the rate constants, a matter of importance if, for example, species B and/or P must be maintained beneath particular concentration levels because of threatened concentration-dependent side reactions.

For example, in the nitric acid–sponsored oxidation of p-xylene to product teraphthalic acid, the acid concentration (B) at any time in this variable-volume batch reaction must be maintained below a specific value to minimize by-product nitration reactions.

Semibatch operation, or its equivalent, plug flow with side-stream addition, is of merit when, as in the van de Vusse reaction network

$$A + B \xrightarrow{\ 1\ } P \xrightarrow{\ 2\ } C$$

$$B + B \xrightarrow{\ 3\ } D$$

yield of desired species P is taxed by a second-order simultaneous reaction (step 3).

In this case B is fed to a "heel" of A at such a rate that the formation of D is minimized due to a low concentration of B.   As demonstrated by Lund and Seagrave[1] for the van de Vusse network, yields superior to batch (or PFR) and CSTR operation can be realized by semibatch operation, a finding reminiscent of that of Gillespie and Carberry[2] for the partially recycled PFR.

For the van de Vusse scheme, the variable-volume isothermal batch-reactor equations are

$$\frac{db}{dt} = -k_1 b - \frac{k_3 b^2}{V} + F \qquad (4\text{-}73)$$

$$\frac{dp}{dt} = k_1 b - k_2 p \qquad (4\text{-}74)$$

Typical profiles are set forth in Fig. 4-19 for the cited value of F and rate constants.   Overall selectivity and yield versus F are shown in Table 4-1.

4-9   TUBULAR-REACTOR ANALYSIS AND DESIGN

A nonisothermal gas-phase reaction of a complex nature will be assumed, which of course requires numerical solution by digital or analog computer; or in the case of simple adiabatic reaction, graphical solution may suffice.   Steady state is also assumed since except for start-up or imposed upsets, transient operation of tubular reactors is largely limited to heterogeneous catalytic tubular reactors wherein catalyst deactivation demands a time-on-stream dependency of concentration and temperature profiles.

Abandoning the 26 elements of the alphabetic periodic table, consider the

[1] M. M. Lund and R. C. Seagrave, *Ind. Eng. Chem. Fundam.*, **10**: 494 (1971).
[2] B. Gillespie and J. J. Carberry, *Chem. Eng. Sci.*, **21**: 475 (1966).

gas-phase chlorination of propylene with Cl_2 to produce allyl chloride. The reaction network is of the

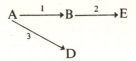

variety; specifically

$$Cl_2 + C_3H_6 \xrightarrow[A \to B]{1} CH_2{=}CHCH_2Cl + HCl$$
$$\text{Allyl chloride}$$

$$Cl_2 + B \xrightarrow[B \to E]{2} CHCl{=}CHCH_2Cl + HCl$$
$$\text{1,3-Dichloropropane}$$

$$Cl_2 + C_3H_6 \xrightarrow[A \to D]{3} CH_2Cl{=}CHClCH_3$$
$$\text{1,2-Dichloropropane}$$

Smith[1] treats this problem in terms of reactions 1 and 3, assuming therefore that consecutive overchlorination does not occur. Here we consider all three reactions. While process details have been made public[2], kinetic details are not available except for a few qualitative observations: (1) Reaction 3 occurs at temperatures as low as 100°C, while step 1 and therefore step 2 are of negligible velocity below 200°C. (2) At 500°C step 1 is faster than step 3.

We can conclude that $E_3 < E_1$ and E_2. While the paucity of specific rate data denies us the freedom to fashion a reaction model, some relative insights into reactor-model behavior can be gained by simulation based upon assumed kinetic formulation which reflect the trends noted above.

Supposing that the reactions are each first order in Cl_2 and organic, the following rate equations are postulated, where C is concentration of Cl_2 and A, B, D, and E are concentrations as identified above and where $r = $ g mol/(h)(liter)

$$r_1 = 5.4 \times 10^9 \exp\left(-\frac{15{,}840}{RT}\right) CA$$

$$r_2 = 1.6 \times 10^{12} \exp\left(-\frac{23{,}760}{RT}\right) CB$$

$$r_3 = 3.6 \times 10^5 \exp\left(-\frac{7920}{RT}\right) CA$$

The heats of reactions, in calories per mole, are

$$-\Delta H_1 = 28{,}000 \qquad -\Delta H_2 = 36{,}000 \qquad -\Delta H_3 = 20{,}000$$

Given the assumed kinetics, we shall explore isothermal, adiabatic, and non-isothermal behavior of a plug-flow tubular reactor. The isothermal case is one admittedly realized on a plant scale with difficulty unless an economic justification

[1] J. M. Smith, "Chemical Engineering Kinetics," 2d ed., McGraw-Hill, 1970.
[2] A. W. Fairbairn, H. A. Chenay, and A. J. Cherniavsky, *Chem. Eng. Prog.*, **43**: 280 (1947).

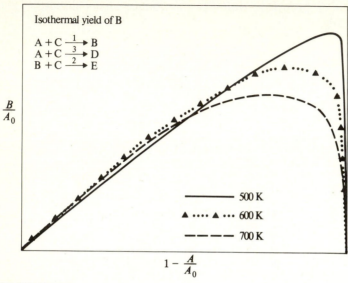

FIGURE 4-20
Isothermal yield of allyl chloride (B) versus conversion of A.

for a multitube, high-surface-to-volume reactor exists. However, the yield-conversion profiles at various levels of isothermality provide insights into precisely how temperature profile will affect yield. Under isothermal conditions overall yield of allyl chloride as a function of propylene conversion is, for a product-free feed ($B_0 = 0$),

$$\frac{B}{A_0} = \frac{K_1}{K_2 - 1}\left[\frac{A}{A_0} - \left(\frac{A}{A_0}\right)^{K_2}\right]$$

where

$$K_1 = \frac{k_1}{k_1 + k_3} \quad \text{and} \quad K_2 = \frac{k_2}{k_1 + k_3}$$

As an exercise, justify this linear yield-conversion relationship in view of the assertion of overall second-order kinetics for chlorination.

Allyl chloride isothermal yield B/A_0 is shown as a function of conversion, $1 - A/A_0$, in Fig. 4-20 at temperature levels of 500, 600, and 700 K for values of K_1 and K_2 computed from the given k-versus-T Arrhenius functionalities.

For the plug-flow, nonisothermal, nonadiabatic homogeneous tubular reactor, the continuity equations are, for negligible variation in u,

Mass:
$$-u\frac{dA}{dz} = k_1 AC + k_3 AC \tag{4-75}$$

$$-u\frac{dB}{dz} = -k_1 AC + k_2 BC \tag{4-76}$$

$$-u\frac{dC}{dz} = k_1 AC + k_2 BC + k_3 AC$$

Heat:

$$\rho u C_p \frac{dT}{dz} = -\Delta H_1 \, r_1 + (-\Delta H_2)r_2 + (-\Delta H_3)r_3 - Ua(T - T_c) \qquad (4\text{-}77)$$

We shall assume that momentum loss is of minor importance.

The coolant temperature may well be a function of z:

$$\pm (\rho C_p)_c \frac{dT_c}{dz} = Ua(T - T_c) \qquad (4\text{-}78)$$

The sign $\pm$ dictates whether cocurrent $(+)$ or countercurrent $(-)$ cooling is imposed. U is the overall wall heat-transfer coefficient. Implicit in the specification of heat transfer in terms of an overall wall coefficient and a bulk (T)-to-coolant-temperature (T_c) driving force is the asumption that a radial temperature gradient does not exist in turbulent flow in an unpacked tubular reactor; the conditions which justify such an assumption have been cited earlier in this chapter, namely, that except for the extraordinary instance of an extremely high wall flux and a very large exothermic homogeneous reaction, the turbulent-flow analogy between heat, mass, and momentum transport can be expected to be obeyed. A turbulent (flat) radial velocity profile will ordinarily be accompanied by a flat temperature profile; that is, $dT/dr = 0$.

It is of some convenience to cast the governing continuity equations into dimensionless form, designating $\bar{a} = A/A_0$, $\bar{b} = B/A_0$, $\bar{c} = C/A_0$, $t = T/T_0$, and $\theta = z/u$, $\theta_0 = L/u$, and $\tau = \theta/\theta_0$:

$$-\frac{d\bar{a}}{d\tau} = k_1 \theta_0 A_0 \, \bar{c}\bar{a} + k_3 \theta_0 A_0 \, \bar{c}\bar{a} \qquad (4\text{-}79)$$

$$\frac{d\bar{b}}{d\tau} = k_1 \theta_0 A_0 \, \bar{c}\bar{a} - k_2 \theta_0 A_0 \, \bar{b}\bar{c} \qquad (4\text{-}80)$$

$$-\frac{d\bar{c}}{\tau} = k_1 \theta_0 A_0 \, \bar{c}\bar{a} + k_2 \theta_0 A_0 \, \bar{b}\bar{c} + k_3 \theta_0 A_0 \, \bar{c}\bar{a}$$

and

$$\frac{dt}{d\tau} = \bar{\beta}_1 \frac{r_1 \theta_0}{A_0} + \bar{\beta}_2 \frac{r_2 \theta_0}{A_0} + \bar{\beta}_3 \frac{r_3 \theta_0}{A_0} - \bar{U}(t - t_c) \qquad (4\text{-}81)$$

$$\pm \frac{dt_c}{d\tau} = \bar{\bar{U}}(t - t_c) \qquad (4\text{-}82)$$

where

$$\bar{\beta}_i = \frac{-\Delta H_i}{\rho C_p T_0} A_0 \qquad \bar{U} = \frac{Ua\theta_0}{\rho C_p} \qquad \bar{\bar{U}} = \frac{Ua\theta_c}{(\rho C_p)_c}$$

Resolution of these dimensionless equations generates profiles of reduced concentrations as well as process and coolant reduced temperatures vs. dimensionless contact time $\tau = \theta/\theta_0$ or (since $\tau = z/L$) dimensionless reactor length. In terms of reduced temperature t, the rate coefficient is

$$k_i = k_{io} \exp\left[-\frac{E_i}{RT_0}\left(\frac{1}{t} - 1\right)\right] = k_{io} \exp\left[-\epsilon_i\left(\frac{1}{t} - 1\right)\right] \qquad (4\text{-}83)$$

Solution in the general case will be governed by the following dimensionless parameters for reactions $i = 1, 2, 3$:

$$\text{Da}_i = k_{i_0} \theta_0 A_0 \qquad \bar{\beta}_i = \frac{-\Delta H_i A_0}{\rho C_p T_0} \qquad \epsilon_i = \frac{E_i}{RT_0} \qquad (4\text{-}84)$$

$t_c = T_c/T_0$, $t = T/T_0$, $\tau = z/L$, and

$$\bar{U} = \frac{Ua\theta_0}{\rho C_p} \qquad \text{and} \qquad \bar{U} = \frac{Ua\theta_c}{(\rho C_p)_c} \qquad (4\text{-}85)$$

Solutions generated in terms of the key dimensionless groups are marked by a generality which will suggest to the analyst quite specific manipulations of plant parameters in order to optimize STY. How then is a specific reactor length and diameter determined for particular feed conditions of flow composition and temperature? We note that feed composition A_0 and/or temperature T_0 are contained in Da, $\bar{\beta}$, and ϵ. Tube surface-to-volume ratio a is found in $\bar{U}$ and $\bar{U}$, while the overall heat-transfer coefficient U is a function of fluid Reynolds number, itself a function of tube diameter and fluid velocity. Consequently dimensionless profiles of conversion $\bar{a}$ and yield $\bar{b}/\bar{a}$ versus τ for diverse values of Da, $\bar{\beta}$, ϵ, $\bar{U}$, and $\bar{U}$ are readily translated into dimensional parameters, as is the case in translating dimensionless heat, mass, and momentum correlations into particular dimensional design specifications.

Limiting cases follow from the general set of continuity equations:

Case 1: $t = 1$ at all τ isothermal PFR

Case 2: $t \neq 1$ at all τ and $\bar{U} = 0$ adiabatic PFR

Case 3: $t \neq 1$ at all τ and $\bar{U} \neq 0$ nonisothermal, nonadiabatic PFR

Case 1, the isothermal one, has been resolved and results set forth in Fig. 4-20. Case 2, adiabatic PFR operation, is resolved by numerical solution of the governing continuity equations (4-79) to (4-81). Typical results are displayed in Fig. 4-21. Case 3, the general instance of nonisothermal, nonadiabatic operation, is solved numerically, and typical results are also presented in Fig. 4-21 in terms of allyl chloride yield Y_B and fluid temperature.

What is set before us in Figs. 4-20 and 4-21 provides the analyst an insight with respect to the influences of key determining parameters upon reactor conversion and yield. Although a firm kinetic model of the complex reactions involved is not available, very definite relative trends can be validly detected in the light of even the most primitive of qualitative insights, as specified in the definition of this chlorination problem.

While we assumed a kinetic model for each of the three steps presumed to govern propylene chlorination by Cl_2 and then generated particular solutions, the determination or at least estimation of kinetic parameters from experimental, global, conversion, yield, and temperature data should not be underestimated. Given x, Y, and temperature data (terminal or intermediate values), the analyst, with the aid of either digital or analog computer, can attempt a fit of experimental,

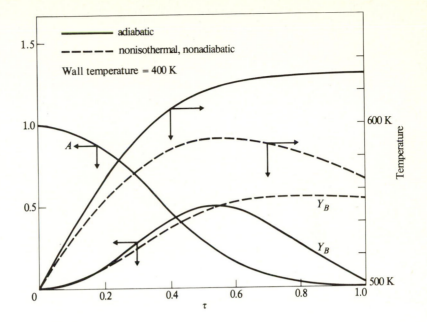

FIGURE 4-21
Adiabatic and nonadiabatic PFR profiles of yield and temperature for allyl chloride synthesis.

$$Da_1 = 0.1 \qquad \epsilon_1 = 16 \qquad \bar{\beta}_1 = 0.28$$
$$Da_2 = 0.01 \qquad \epsilon_2 = 24 \qquad \bar{\beta}_2 = 0.36$$
$$Da_3 = 0.02 \qquad \epsilon_3 = 8 \qquad \bar{\beta}_3 = 0.20$$

often integral, data with assumed kinetic models and their associated parameters. In a given instance of nonisothermal, nonadiabatic homogeneous reactor performance, estimates of $\bar{U}$, $\bar{\bar{U}}$, and $\bar{\beta}$'s can surely be made; thus manipulation of Da and ϵ values for each of the suspected steps in a complex network in principle makes the analyst capable of fitting overall performance data.[1]

4-10 DESIGN IN THE ABSENCE OF A KINETIC MODEL

In many instances, plant-design estimates are required where there exists a paucity of data from which a reaction-rate model can be fashioned. Typically, only conversion as a function of real or contact time may be on hand. Since for batch reaction, for example,

$$t = C_0 \int_0^x \frac{dx}{r} \qquad (4\text{-}86)$$

[1] See, for example, G. Emig and M. Köppner, *Chem. Eng. Sci.*, **29**: 2339 (1974).

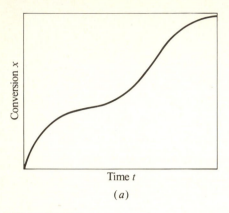
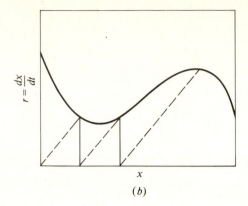

FIGURE 4-22
Graphical illustration of Jones' design method.

one can, in principle, differentiate the conversion-time data to secure rate-vs.-conversion data, from which graphical integration provides the holding or real time required to achieve required plant conversion.

Consider a case in which the design objective is that of specifying the number of series CSTRs required to achieve a specific conversion when only batch conversion-time data are available. As Jones[1] shows, the problem reduces to one of securing the derivatives of the x-versus-t curve for various values of x, from which, as illustrated in Fig. 4-22, the rate r is obtained versus x. Now for the first in a series of CSTRs

$$\frac{1}{\theta_1} = \frac{r}{C_0 x_1} \qquad (4\text{-}87)$$

Thus in Fig. 4-22b the conversion — rate from the first CSTR is given by a straight line of slope $1/\theta_1$ for an inlet conversion of zero. The holding time θ_2 for the second CSTR establishes its exit conversion x_2, and so on. The number of CSTRs and their respective volumes are thus specified to achieve the desired final conversion. Should each CSTR in the train be of equal volume, then, of course, the slopes are all equal. Note that the equal-volume train is not necessarily optimum but depends upon intrinsic kinetics. As kinetics is assumed to be unknown, for a given r-versus-x curve a simple trial-and-error procedure will quickly yield the optimum or near-optimum CSTR reactor-volume distribution with respect to production rate per total reactor-train volume.

When yield of a particular product in an ill-understood complex reaction is of concern, the analysis of Denbigh[2] proves to be most pertinent and, combined with the design method suggested by Jones, permits a designation of operation by (1) a number of CSTRs in series or (2) a batch PFR such as to maximize the yield. Following Denbigh, we designate the amount of reactant A which has reacted as

[1] R. W. Jones, *Chem. Eng. Prog.*, **47**: 46 (1951).
[2] K. G. Denbigh, *Chem. Eng. Sci.*, **13**: 25 (1961).

A_r and the total amount employed as A_t. Hence point or instantaneous yield of product P is

$$Y = \frac{a\ dP}{b\ dA} \qquad (4\text{-}88)$$

where a and b are stoichiometric coefficients. Overall, integral yield is then for a PFR or batch reactor

$$Y_o = \frac{1}{A_r} \int_0^{A_r} Y\ dA = \frac{P}{A_0 x} \qquad \text{or} \qquad Y_o = \frac{1}{A_t} \int_0^{A_r} Y\ dA = \frac{P}{A_0} \qquad (4\text{-}89)$$

It follows that the area beneath the curve described by a plot of point yield Y versus conversion of A must equal overall yield.

Again, following Denbigh, we can define an overall yield in terms of n CSTRs as

$$Y_o = \frac{1}{A_r} \sum_1^n Y_i\ \Delta A_i \qquad (4\text{-}90)$$

or

$$Y_o = \frac{1}{A_t} \sum_1^n Y_i\ \Delta A_i \qquad (4\text{-}91)$$

Referring to Fig. 4-23, we note that the total area beneath the Y-versus-x curve represents the batch or PFR yield. In contrast, should an n-CSTR network be envisioned, then in accord with Eq. (4-90) or (4-91), the shaded area described by the rectangles defines overall n-CSTR yield. Two limiting cases of isothermal point yield vs. conversion follow.

CASE 1   As schematized in Fig. 4-24a, the point yield Y decreases with conversion, in which case (save for the redundant instance of invoking a near-infinite train of CSTRs) the CSTR train always provides an overall yield (summation of rectangles) less than that granted by batch or PFR operation.

CASE 2   As schematized in Fig. 4-24b, if point yield increases and then decreases with conversion, a train of n CSTRs may generate a greater yield than batch or PFR operation, up to the point of maximum point yield.

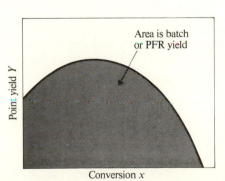

FIGURE 4-23
Point-yield-vs.-conversion data.

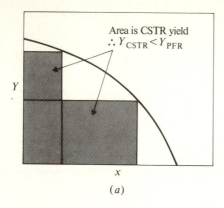

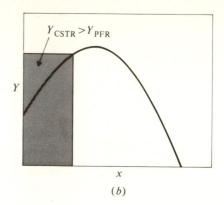

FIGURE 4-24
Denbigh's illustration of CSTR and PFR yield. [*K. G. Denbigh, Chem. Eng. Sci.*, **13**: 25 (*1961*).]

So it is that even for a complex reaction, in the absence of a kinetic model, sound design judgments are possible if data for conversion vs. time *and* point yield vs. conversion are at the disposal of the analyst. An illuminating illustration of such a design procedure is lucidly set forth by Denbigh in the pertinent instance in which detailed kinetic data were totally unknown.[1]

Should the Y-versus-x behavior indicate that an n-CSTR network will prove of benefit with respect to yield, temperature control, and general ease of operation, it must be borne in mind that the PFR-recycle network operated at a recycle ratio between the limits equivalent to CSTR and pure PFR behavior may be entertained as a feasible plant-reactor model (see Chap. 3). If a gas-phase reaction is involved, CSTR operation must yield to the PFR-recycle system.

Finally, insofar as not only conversion but yield can be expected to be temperature-sensitive, the graphical techniques of Jones and Denbigh can surely be utilized, in the common instances of unknown kinetics, to fashion an optimum n-CSTR train in which each of the CSTRs of that train is operated at a different temperature. Design then requires only x-versus-t and Y-versus-x data at diverse temperature levels.

A hypothetical situation of this nature is schematized in Fig. 4-25, where r-versus-x data (derived from x-versus-t data) and point yield Y-versus-x data are displayed at two temperature levels. Dwelling upon the r-versus-x data with a mind toward the Y-versus-x profile, one sees that the first of a three-stage CSTR train should be operated at a temperature T_2, corresponding to a yield commensurate with Fig. 4-25 while the second and third stages should be operated at a temperature T_1 to guarantee a maximum in STY.[2]

---

[1] Ibid.
[2] A CSTR-PFR train is suggested by Fig. 4-25.

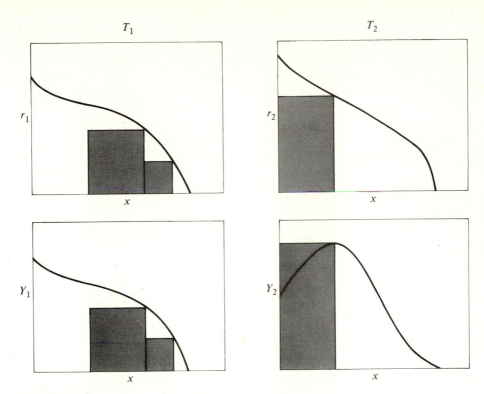

FIGURE 4-25
Schematic illustration of optimization of yield in a series of CSTRs of unequal temperature.

4-11 TRANSIENT BEHAVIOR OF CONTINUOUS REACTORS

As we remarked earlier, transient behavior of a continuous reactor in which homogeneous reactions occur is not ordinarily undertaken by design on a plant scale save, of course, during start-up and shutdown. In contrast, heterogeneous catalytic continuous reactors often operate in a transient fashion by reason of gradual deactivation of the catalyst, a topic which is treated in Chaps. 8 to 10.

Here we shall confine our attention to transient behavior of a chain of CSTRs set in series. For a large number of equal-volume CSTRs ($n > 20$) plug-flow behavior is simulated; hence, in principle, the n-CSTR transient analysis admits to description of the transient PFR.

Following Mason and Piret,[1] for the nth CSTR and linear kinetics

$$QA_{n-1} = QA_n + VkA_n + V\frac{dA_n}{dt} \qquad (4\text{-}92)$$

[1] D. R. Mason and E. Piret, *Ind. Eng. Chem.*, **42**: 817 (1950).

or

$$\frac{dA_n}{dt} + \frac{K_n A_n}{\theta_n} = \frac{A_{n-1}}{\theta_n} \tag{4-93}$$

where $K = 1 + k\theta$ and $\theta = V/Q$.

In terms of the Laplace transform

$$sf_n(s) - \mathcal{L}(t = 0) + \frac{K_n}{\theta_n} f_n(s) = \frac{f_{n-1}(s)}{\theta_n}$$

Solving for $f_n(s)$ gives

$$f_n(s) = \frac{A_n}{s + K_n/\theta_n} + \frac{f_{n-1}(s)/\theta_n}{s + K_n/\theta_n} \tag{4-94}$$

For the first vessel

$$f_1(s) = \frac{A_1}{s + (K/\theta)_1} + \frac{A_0/\theta_1}{s[s + (K/\theta)_1]} \tag{4-95}$$

for the second

$$f_2(s) = \frac{A_2}{s + (K/\theta)_2} + \frac{f_1(s)/\theta_2}{s + (K/\theta)_2} \tag{4-96}$$

and so on for succeeding CSTRs in the train. In general by successive substitution one obtains

$$f_n(s) = \frac{A_n}{(s + K/\theta)_n} + \frac{A_{n-1}/\theta_n}{(s + K/\theta)_n(s + K/\theta)_{n-1}} + \cdots$$

$$+ \frac{A_1/\theta_n \theta_{n-1} \cdots \theta_2}{\prod_i^n \left(s + \frac{K}{\theta}\right)_i} + \frac{A_0/\theta_n!}{s \prod_i^n \left(s + \frac{K}{\theta}\right)_i} \tag{4-97}$$

If isothermality prevails throughout the series network of equal-volume vessels, we have

$$f_n(s) = \frac{A_n}{s + K/\theta} + \frac{A_{n-1}/\theta}{(s + K/\theta)^2} + \cdots + \frac{A_1/\theta^{n-1}}{(s + K/\theta)^n} + \frac{A_0/\theta^n}{s(s + K/\theta)^n} \tag{4-98}$$

When we take the inverse transform, the concentration in, and issuing from, the nth CSTR, as a function of time, is

$$C_n = \frac{A_0}{K^n} + \exp\left(-\frac{Kt}{\theta}\right)\left\{\left[A_n + A_{n-1}\frac{t}{\theta} + \frac{A_{n-2}}{2!}\left(\frac{t}{\theta}\right)^2 + \cdots + \frac{A_1}{(n-1)!}\left(\frac{t}{\theta}\right)^{n-1}\right]\right.$$

$$\left. - \frac{A_0}{K^n}\left[1 + \frac{Kt}{\theta} + \left(\frac{Kt}{\theta}\right)^2\frac{1}{2!} + \cdots + \left(\frac{Kt}{\theta}\right)^{n-1}\frac{1}{(n-1)!}\right]\right\} \tag{4-99}$$

where A_n refers to the concentration in the nth vessel at $t = 0$.

Example: Start-Up of a CSTR Train

Consider a four-CSTR system, each vessel of equal volume and temperature. For $A_0 = 1$, $k = 1$, and $\theta = 2.16$, we wish to examine a start-up problem, i.e., let us contrast (1) batch start-up of each vessel to bring each to its steady-state four-

CSTR value and then starting flow with (2) simply filling each vessel at $A_0 = 1$ and starting flow to ultimately achieve steady state. (a) What are the steady-state values in each reactor? (b) For $A_0 = 1$ in each vessel at $t = 0$, how much time is required for the effluent from the fourth reactor to reach 10 percent of the steady-state value? (c) At the time found in the answer to part (b), how close is C_1 to its steady-state value? (d) What batch time is required to bring the fourth CSTR to its steady-state value? (e) What is the best start-up policy, batch or continuous?

SOLUTION (a) In steady state the effluent concentration from the nth CSTR is

$$C_n = \frac{1}{(1 + k\theta)^n}$$

Therefore, since $1 + k\theta = 3.16$,

$$C_1 = \frac{1}{3.16} = 0.316 \qquad C_2 = \frac{1}{(3.16)^2} = 0.1$$

$$C_3 = \frac{1}{(3.16)^3} = 0.0317 \qquad C_4 = \frac{1}{(3.16)^4} = 0.01$$

(b) At $t = 0$, $A_1 = A_2 = A_3 = A_4 = 1.0$. Substituting into Eq. (4-99) the governing numerical values, we obtain

$$C_4 = 0.01 + (0.99 + 0.4484t + 0.09575t^2 + 0.0115t^3) \exp(-1.46t)$$

C_4 is plotted as a function of t, from which we find that $C_4 = 0.011$ (10 percent of its steady-state value) at $t = 6.4$ h.

(c) At $t = 6.4$ h, the concentration in the first CSTR is, from Eq. (4-99), $C_1 = 0.316$, which is its steady-state value.

(d) Since $-dC_4/dt = k_1 C_4$, batch time required to bring the fourth CSTR to its steady-state value is

$$t = \frac{\ln(1/C_4)}{k} = 4.6 \text{ h}$$

(e) It is obvious (as might have been anticipated) that batch start-up is to be preferred since (1) it is accomplished in less time, 4.6 versus 6.4 h, and (2) unreacted feed is not wasted nor is its recovery for recycle required, as would be the case in continuous start-up.

Example: Transient Polymerization

An interesting study of transient polymerization in a CSTR has been published by Taylor.[1] As shown in Fig. 4-26, the system consists of a CSTR fed with catalyst, diluent, and monomer while product of molecular weight index $\bar{M}$ is withdrawn with diluent and catalyst. Taylor's task was that of identifying the

[1] J. H. Taylor, *Chem. Eng. Prog.*, **58**(8): 42 (1962).

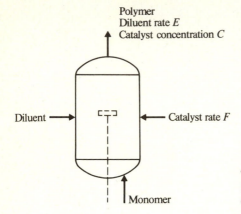

Polymer
Diluent rate E
Catalyst concentration C

Diluent ⟶

⟵ Catalyst rate F

Monomer

FIGURE 4-26
Taylor's polymerization reactor network.
[*J. H. Taylor, Chem. Eng. Prog.*, **58**(8):42
(*1962*).]

$\overline{M}$-versus-catalyst-concentration functionality. Instead of conducting a large number of lengthy steady-state CSTR experiments (holding time is of the order of hours in this system) one 35-h run was made, during which step changes in catalyst feed concentrations were made and the effluent values of $\overline{M}$ determined vs. time. A computer was employed to establish two empirical constants such that observed and computed values of $\overline{M}$ were in accord as a function of time on-stream.

Laboratory studies indicated that the product polymer molecular-weight index is related to catalyst concentration C by

$$M = a(C)^b f(I, T) \qquad (4\text{-}100)$$

where a and b are constants to be determined while $f(I, T)$ is some function of reaction temperature and feed-impurity level. For the unsteady-state CSTR, catalyst concentration is given by

$$\frac{dC}{dt} = \frac{F}{V} - \frac{C}{\theta} = C_0 - \frac{C}{\theta} \qquad (4\text{-}101)$$

where $F =$ catalyst feed rate
$V =$ reaction volume
$\theta =$ holding time
$C_0 =$ catalyst space velocity

We consider the CSTR at $t = 0$, when $\overline{M} = M_0$. A change in catalyst feed rate is then imposed (a step change). The value of $\overline{M}$ as a function of time for $t > 0$ is given by

$$\overline{M} = \left(M_0 + \frac{1}{\theta}\int M e^{t/\theta}\, dt\right) e^{-t/\theta} \qquad (4\text{-}102)$$

where M is given by Eq. (4-100).

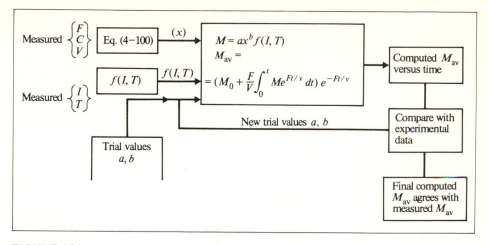

FIGURE 4-27
Taylor's flowsheet for the evaluation of rate parameters. [*J. H. Taylor, Chem. Eng. Prog.,*
58(8): 42 (1962).]

Armed with experimental $\overline{M}$-versus-t data, Taylor utilized a computer to
evaluate a and b as illustrated in the flowsheet reproduced in Fig. 4-27. There
resulted

$$M = 6.3(C)^{-1/2}f(I, T) = K\frac{1}{\sqrt{C}} \qquad (4\text{-}103)$$

for fixed impurity level and temperature.

Extension of Taylor's results to design Given Taylor's finding, suppose that we
are required to develop a relationship between $\overline{M}$ and t. First, we must express
C in Eq. (4-100) as a function of t. By Eq. (4-101) we find

$$C = \theta C_0 - \theta\left(C_0 - \frac{C_1}{\theta}\right)\exp\left(-\frac{t}{\theta}\right)$$

where C_1 is catalyst concentration at $t = 0$. By Eq. (4-103)

$$M = \frac{K}{\sqrt{C_0\theta - \theta(C_0 - C_1/\theta)e^{-t/\theta}}}$$

Hence
$$\overline{M} = \left(M_0 + \frac{K}{\theta}\int_0^t \frac{e^{t/\theta}\,dt}{\sqrt{A - Be^{-t/\theta}}}\right)e^{-t/\theta} \qquad (4\text{-}104)$$

where $A = C_0\theta$ and $B = \theta(C_0 - C_1/\theta)$. Let $y = e^{-t/\theta}$; then $dt = -\theta\,dy/y$, and the
integral in (4-104) is readily simplified to

$$\theta\int_0^y \frac{-(1/y^2)\,dy}{\sqrt{A - By}} \qquad (4\text{-}105)$$

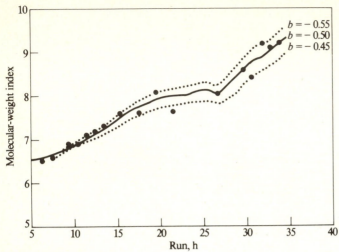

FIGURE 4-28
Comparison of computed and experimental molecular-weight indexes
secured by Taylor. [*J. H. Taylor, Chem. Eng. Prog., 58(8):42 (1962).*]

Hence

$$\overline{M} = \exp\left(-\frac{t}{\theta}\right)\left\{M_0 + K\left[\frac{\sqrt{A - By}}{Ay} - \frac{\sqrt{A - B}}{A}\right.\right.$$

$$\left.\left. + \frac{B}{A\sqrt{A}}\left(\tanh^{-1}\sqrt{\frac{A - By}{A}} - \tanh^{-1}\sqrt{\frac{A - B}{A}}\right)\right]\right\} \qquad (4\text{-}106)$$

Equation (4-106), due to Taylor, quite faithfully describes the $\overline{M}$-versus-t data, as set forth in Fig. 4-28.

As an exercise, derive Eqs. (4-101) and (4-102).

SUMMARY

General continuity equations for the cylindrical tubular reactor were developed and reduced to diverse limiting reactor types and modes of operation. The key dimensionless groups which dictate the spatial distributions of species and temperature were identified and their magnitudes indicated for both packed and unpacked tubular reactors; in the course of this discussion experimental and theoretical inquiries were briefly noted.

Command of the material provided in this chapter in concert with the principles set forth in Chap. 3 should give the analyst and designer a sound basis for responding prudently to the following questions:

1 For a given reaction network, what reactor type would best provide the design goal with respect to conversion, yield, STC, and/or STY?

2 For a specified reactor type, what manner of operation would best realize the design goal in terms of temperature environment (isothermal, adiabatic, nonisothermal, nonadiabatic) and species concentration distribution (semibatch, side-stream addition)?

   3   Given discussions with respect to questions 1 and 2, how does one faithfully model the system with a minimum of mathematical compromises?

   4   Faced with a paucity of detailed kinetic data from which a reaction model can be fashioned, how does one design a reactor from only conversion and point-yield data for the ill-understood reaction system?

   5   How does one treat a start-up and shutdown problem for an n-CSTR system or its equivalent?

ADDITIONAL REFERENCES

Analyses of batch-reactor performance for complex reactions in terms of optimum scheduling of temperature with time are presented by J. J. Evangelista and S. Katz, *Ind. Eng. Chem.*, **60** (3): 24 (1968). Optimal design principles are provided by J. M. Douglas and M. Denn, *Ind. Eng. Chem.*, **57** (11): 19 (1965).

     A treatment of reactor scale-up is set forth by F. A. Holland, *Chem. Eng.*, April **1963**: 145, and W. M. Small, *Chem. Eng. Prog.*, **65**(7): 81 (1969); while simulation and optimization issues are found in G. K. Boreskov and M. G. Slin'ko, *Br. Chem. Eng.*, **10**(3): 170 (1965), and W. J. Dassau and G. H. Wolfang, *Chem. Eng. Prog.*, **58**(4): 43 (1963). R. Shinnar and S. Katz, *Adv. Chem.*, **109**: 56 (1972), review the matter of polymerization kinetics and reactor design.

     The use of a noncatalytic fluid bed to achieve near isothermality in the synthesis of chloromethanes is detailed by P. R. Johnson, J. L. Parsons, and J. B. Roberts, *Ind. Eng. Chem.*, **51**(4): 499 (1959). Simulation of a nonisothermal multipass shell-and-tube reactor is realized by J. T. Banchero and T. G. Smith, *J. Heat Transfer, Trans. ASME*, **C95**: 206 (1973).

     Mixing and liquid-phase agitation are reviewed by D. Hyman, *Adv. Chem. Eng.*, **3**: 120 (1962). See also F. Manning and R. H. Wilhelm, *AIChE J.*, **9**: 12 (1963). P. J. Trambouze and E. L. Piret, *AIChE J.*, **5**: 384 (1959), develop graphical and analytical techniques for resolution of homogeneous-reactor design problems in batch, CSTR, and PFR circumstances. Analytical solutions of some problems in homogeneous adiabatic reactors are set forth by J. M. Douglas and L. C. Eagleton, *Ind. Eng. Chem. Fundam.*, **1**: 116 (1962), and a comprehensive approach to the design and optimization of homogeneous-flow reactors is to be found in G. T. Westbrook and R. Aris, *Ind. Eng. Chem.*, **53**(3): 181 (1961).

PROBLEMS

   4-1   Cite in order of importance the assumptions which underlie the continuity equations for mass and heat [Eqs. (4-11) and (4-12)].

   4-2   Develop, in dimensionless form, the mass continuity equations for laminar flow with first-order reaction *and* axial and radial diffusion. State the boundary conditions for (*a*) homogeneous reaction and (*b*) homogeneous and wall-catalyzed heterogeneous reaction.

   4-3   Reduce the general equation developed in response to Prob. 4-2 to describe the case in which:

       (*a*)   Radial diffusion is quite large while axial diffusion is negligible (comment on the consistency of this postulate if molecular diffusion prevails).

       (*b*)   Both radial and axial diffusion are of significant, telling magnitude.

4-4 Consider a reacting (a) liquid stream then (b) a gas stream flowing through a tubular reactor at Reynolds number of 100. What is the relative importance of radial and axial molecular diffusion in each case?

4-5 Visualize a tubular reactor upon the inner wall of which a catalytic agent is deposited. This tube is packed with inert spheres. (For what purpose?) Assuming nonisothermal nonadiabatic operation:
   (a) Write the governing continuity equations for the general circumstance.
   (b) Repeat part (a) ignoring axial dispersion.
   (c) Write the boundary conditions for parts (a) and (b) if the reaction catalyzed by the wall is

   Case 1: A $\longrightarrow$ B

   Case 2: A $\longrightarrow$ B $\longrightarrow$ C

4-6 NO is oxidized homogeneously *and* (catalytically) by silica gel. In the general nonisothermal nonadiabatic case, set down the governing continuity equations and boundary conditions for NO and temperature in a fixed bed of silica gel. Simplify for the condition of negligible axial dispersion. Repeat for steady-state plug flow (zero radial gradient).

4-7 Suppose a single CSTR is host to an nth-order reaction in steady state. If, at $t = 0$, the CSTR is subject to a step input of inert material (flow rate is unchanged), derive and solve the equation to yield effluent-reactant concentration-time behavior.

4-8 You must use an available CSTR for the conduct of an autocatalytic reaction $A + C \rightarrow P + 2C$. A conversion of 60 percent is desired, and laboratory data indicate that a contact time of 150 s suffices to give this conversion in a PFR. However, for the desired production rate the available CSTR exhibits a holding time of only 75 s. What value of $m = (A_0 + C_0)/A_0$ should be employed to achieve the required STC?

4-9 Consider the reaction

$$A + B \xrightarrow{\;k_1\;} 2B + P \qquad \text{second order}$$
$$A \xrightarrow{\;k_2\;} D \qquad \text{first order}$$

Establish a design chart of yield P/A_0 for a CSTR relative to that in a PFR. For a given value of k_1/k_2, what variable can be manipulated to increase the yield in each reactor type?

4-10 Hovorka and Kendall [*Chem. Eng. Prog.*, **56**(8): 58 (1960)] report the following conversion-contact time data for the reaction NaOH + ethyl acetate → Na acetate + ethanol in a tubular reactor of 3.2 cm ID at 29.8°C in which the aqueous solution flow rate was varied between 440 and 2072 ml/min.

Time, min	Conversion to Na acetate
0	0
0.5	0.1675
1	0.271
1.5	0.351
2	0.412
2.5	0.462
2.75	0.4835

The feed consisted of 0.1 N solutions of NaOH and ethyl acetate. Analyze these data and then design (a) a batch reactor and (b) a CSTR for a production feed rate of 100 liters/min and a conversion of 65 percent.

4-11 The alert stockholders of your firm suggest that an open-loop recycle reactor be used for the second-order saponifaction reaction discussed in Prob. 4-10. As we saw in Chap. 3, the issue of micromixing and macromixing is of importance for nonlinear kinetics. Prove that the plug-flow recycle reactor is a nonsegregated, micromixed one at large values of the recycle ratio.

4-12 Methanol is to be produced by the high-temperature homogeneous reversible reactions

Step 1: $CO + H_2 \rightleftharpoons CH_2O$

Step 2: $CH_2O + H_2 \rightleftharpoons CH_3OH$

where step 1 is the rate-determining step. Step 2 is rapidly equilibrated, the equilibrium constant being so large that the CH_2O concentration is quite low. Develop the PFR continuity equation in terms of moles of CO reacted as a function of contact time for equal molar feed of H_2 and CO.

4-13 For the reactions

$$CH_2Cl_2 + HF \longrightarrow CH_2ClF + HCl$$

$$CH_2ClF + HF \longrightarrow CH_2F_2 + HCl$$

it is desired to produce a maximum of the difluoro compound with a minimum consumption of HF. Based upon the following single-CSTR laboratory data, compare the CH_2F_2/HF consumption ratio in a PFR and CSTR.

	Moles per mole of CH_2Cl		
Contact time, min	CH_2Cl	CH_2ClF	CH_2F_2
0	1	0	0
0.5	0.67	0.167	0.167
1	0.5	0.167	0.33
2	0.33	0.133	0.534
∞	0	0	1

Assume HF to be in vast excess, so that linear kinetics can be assumed.

4-14 In the manufacture of an alkyl resin the following batch procedure is employed: 104 kg of walnut oil (MW = 104), 67 kg of phthalic anhydride, and 50 kg of glycerol are charged in 10 min to a batch autoclave of 220 liters capacity and then heated for 40 min to the reaction temperature of 260°C, at which point the reactants are "cooked" for 30 min (final oil conversion, 95 percent). After a 30 min cooling period the contents are discharged in a 10-min period. Utilizing the same reactor as a CSTR, estimate the productivity of the CSTR relative to the present batch operation, assuming (a) first order and (b) second order (first in oil and anhydride). Assume, also, that batch reaction occurs to a negligible degree during heat-up and cooling. Finally estimate how many CSTRs, operated in series, are necessary to match batch productivity.

4-15   In excess glycerin, glyceride reacts by a first-order reaction to produce unsymmetrical glycerides (a stage in alkyl-resin manufacture). Conversion-time data secured in a 185-liter batch autoclave are:

At 240°C		At 280°C	
Time, h	$x, \%$	**Time, h**	$x, \%$
0	0	0	0
1	29	0.5	47
2	53	1	65
3	67	1.5	71
4	76	2	73
5	85		

At each temperature, design a train of CSTRs, each of 1850 liters volume, for a total feed rate of 3700 liters/h to achieve 64 percent conversion at 240°C and 73 percent conversion at 280°C.

4-16   A complex resinification reaction exhibits the following batch behavior at 250°C:

Time, h	$x, \%$	**Time, h**	$x, \%$
0	0	4	82
1	27	5	90
2	50	6	95
3	68	7	97

Predict the conversion in (*a*) a series of three CSTRs, each of 2 h holding time, (*b*) a series of six CSTRs, each of 1 h holding time.

4-17   It is proposed that an irreversible first-order reaction be conducted in a series of six available equal-volume CSTRs. For comparative purposes assume a rate constant of unity and determine which arrangement gives maximum productivity at a conversion of 99 percent:
   (*a*)   Six in parallel
   (*b*)   Three parallel lines of two in series
   (*c*)   Two parallel lines of three in series
   (*d*)   Six in series
   (*e*)   Two parallel lines of two in series, followed by one line of two in series

4-18   Three CSTRs, each of 1 h holding time, when operated in series provide a conversion of 96.3 percent. It is proposed that production rate be doubled in this system by doubling the feed rate. (How else might it be doubled?) Determine the new conversion and calculate how much transient time is required for the exit conversion to be within 0.005 of the final steady-state value.

4-19   For what values of n and k is

$$\frac{1}{(1 + k\theta/n)^n} = \exp(-k\theta)$$

within 1 percent?

4-20   The kinetics of the thermal dealkylation of alkylnaphthalenes is reported by Bixel et al. [*Ind. Eng. Chem. Process Des. Dev.*, **3**(1): 78 (1964)]. Utilizing their data, design a tubular reactor to maximize the production (moles per time × volume) of naphthalene under conditions of
   (*a*)   Isothermality

    (*b*)   Adiabaticity

    (*c*)   Operation which is nonisothermal and nonadiabatic

    Radial gradients may be presumed negligible.

4-21 Fashion a problem pertaining to matters treated in this chapter and present it to your colleagues and instructor for solution.

5

HETEROGENEOUS REACTIONS

"Ay, now the plot thickens very much upon us."
George Villiers, Second Duke of Buckingham "The Rehearsal"

Introduction

Reactions involving more than a single phase are termed heterogeneous. The actual site of reaction may be in one phase, e.g., liquid, yet if one of the reactants is supplied to that phase from another distinct phase, e.g., gas, the *system* is considered heterogeneous. The reaction between aqueous solution of NaOH and dissolved CO_2 occurs in the homogeneous liquid phase; however, the dissolved reactant CO_2 may be supplied from a gas phase in contact with the liquid, and this system is consequently heterogeneous. On the other hand, the reactant CO_2 could be dissolved in pure water and then later mixed with the alkaline solution to cause reaction, in which case the system is homogeneous, as two miscible liquids containing dissolved reactants are caused to mix and react.

Heterogeneous reactions may occur in neither bulk phase but at the interface, or boundary, separating the phases. The combustion of nonporous carbon by oxygen-bearing gas is an example of a heterogeneous reaction which occurs at the gas-solid interface.

5-1 CLASSIFICATION OF HETEROGENEOUS REACTIONS

In general, one distinguishes between catalytic and noncatalytic heterogeneous reactions. The catalytic system involves at least one phase which is not a net reactant;
yet its presence alters the velocity and possibly the path of reaction. This agent,
the catalyst, must exist as a distinct phase if the term heterogeneous catalysis is to
be applied. In catalytic polymerization of gaseous olefins in the presence of an
acid, the gas is contacted with the acid-containing liquid; it dissolves therein and
reacts in the liquid phase. The system, while heterogeneous (gas-liquid), involves homogeneous catalysis as the catalyst does not form a distinct phase in the
reaction medium. Consider, however, the hydrogenation of liquid benzene in the
presence of Raney nickel catalyst particles suspended in the liquid phase. Hydrogen is supplied in gaseous form, is contacted within the liquid phase and dissolves
in it, and then reacts with benzene at the solid-catalyst surface to produce cyclohexane. The system is heterogeneous (gas-solid-liquid), involving heterogeneous
catalysis (solid catalyst–liquid).

  If, on the other hand, the benzene liquid is caused to flow or trickle through a
bed of supported Pt catalyst in the presence of hydrogen at suitable pressure, temperature, and low liquid flow rates, a truly heterogeneous system exists (gas-solid
catalyst-liquid) involving two classes of catalysis (gas-solid catalysis and liquid-solid
catalysis). Because wetting of the catalyst by the liquid benzene is incomplete,
part of the catalyst is exposed to the gaseous reactants (benzene vapor and hydrogen)
and catalyzes them, while that part of the catalyst surface covered with liquid
benzene catalyzes the liquid-phase transformation.

  The unique difference between catalyzed and noncatalyzed reactions is the
fact that the catalyst, say X, while participating in one or more of the elementary
reaction steps, is effectively regenerated at the end of each reaction cycle. For the
reaction $A + B \xrightarrow{X} P$, catalyzed by X, the detailed mechanism might be

$$
\begin{array}{ll}
A + X \longrightarrow AX \\
\underline{B + AX \longrightarrow P + X} \\
A + B \longrightarrow P
\end{array}
\qquad \text{or possibly} \qquad
\begin{array}{ll}
A + X \longrightarrow AX \\
B + X \longrightarrow BX \\
AX + BX \longrightarrow PX + X \\
\underline{PX \longrightarrow P + X} \\
A + B \longrightarrow P
\end{array}
$$

While X clearly participates in the process, it is regenerated, though not necessarily
regenerated intact. Its activity usually declines with use over a more or less long
period of time (large number of cycles involving repeated participation and
regeneration).

  If the heterogeneous system is noncatalytic, the participant is a true reactant,
being consumed to an extent dictated by reaction kinetics and thermodynamic
limitations. For example, iron in the presence of oxygen reacts to form iron oxide
$2Fe + \frac{3}{2}O_2 \rightarrow Fe_2O_3$. Fe is a reactant. Consider Fe in the presence of N_2 and
H_2. At suitable temperature and pressure, the following overall reaction occurs:
$Fe + N_2 + 3H_2 \rightleftharpoons 2NH_3 + Fe$. Fe is a catalyst in this case. Both systems are

heterogeneous (gas-solid), yet the first is noncatalytic while the second is a classic case of a heterogeneous catalytic reaction.

Examples of Heterogeneous Reactions

Heterogeneous catalytic reactions generally involve a solid catalyst and a fluid phase (gas and/or liquid) which supplies reactants to the site of catalysis, the fluid-solid interface. A more detailed treatment of heterogeneous catalysis is presented in Chap. 8.

Noncatalytic heterogeneous reactions may involve

1 Gas and liquid
2 Gas and solid
3 Liquid and liquid
4 Liquid and solid
5 Solid and solid

Gas-liquid reaction A classic example of industrial importance is the reaction of nitrogen tetroxide with liquid water to produce nitric acid. Gas bearing the reactant N_2O_4 is contacted with liquid water or dilute HNO_3; the N_2O_4 diffuses to the interface, is absorbed, and then reacts in the liquid phase.[1] Actually while the chemical reaction step occurs in the homogeneous aqueous phase, the system is a heterogeneous one from a phenomenological point of view, as we are dealing with a gaseous source of reactant and a liquid-phase sink containing the coreactant. Further treatment of gas-liquid reactions is given in Chap. 6.

Gas-solid reaction A number of examples of gas-solid reaction exist including the burning of coal particles, ore reduction, the coking of catalyst particles, and regeneration of fouled catalysts by oxidation of deposited carbonaceous matter. The solid coreactant may ultimately be consumed, as in the case of coke or carbon burning; thus particle size changes during reaction. In catalyst regeneration via deposited coke burning, the solid reactant is actually a deposited layer existing largely within the porous catalyst, so that catalyst particle size does not change during the deposited solid-gas reaction. In modern fluid catalytic cracking units, in addition to heterogeneous catalysis (hydrocarbon cracking over silica-alumina-zeolite catalysts), two types of noncatalytic gas-solid reactions take place: (1) certain hydrocarbons overreact in the cracking unit to form solid carbonaceous deposits upon the catalyst surface, and (2) in the regenerator, oxygen contained in a carrier gas reacts with this solid coke deposit to form CO, CO_2, and H_2O, thus renewing (regenerating) the catalyst activity. Detailed treatment of gas-solid reactions involving fixed and changing particle size will be set forth in Chap. 7.

Liquid-liquid reactions Obviously if such a system is to be heterogeneous, a degree of immiscibility must exist. Benzene nitration typifies liquid-liquid reaction.

---

[1] J. J. Carberry, *Chem. Eng. Sci.*, **9:** 189 (1958); M. M. Wendel and R. L. Pigford, *AIChE J.*, **4:** 249 (1958).

Benzene, immiscible with the aqueous phase containing a mixture of nitric and sulfuric acids, is intimately contacted with this acid phase.   As benzene diffuses and dissolves in the acid phase, it reacts with a nitronium ion generated from the nitric acid by a complex mechanism involving sulfuric acid.   The product, nitrobenzene, then diffuses back into the organic phase.   Again the locale of reaction is within one (the acid) phase; however, the system is heterogeneous with respect to the source of reactant supply.   Liquid-liquid alkylation is discussed in Chap. 6.

Liquid-solid reactions   The common water softener is a fine example of a unit involving liquid-solid reaction.   Water softening is ion exchange.   Ions in the liquid phase are transported to the surface of a solid containing other exchangeable ions. Rapid exchange (reaction) takes place at the liquid-solid surface (internal and external) boundary.   The product then diffuses from the reaction site into the liquid phase, leaving a solid (the ion-exchange resin) less rich in reactant ions.   The dissolution of metals by aqueous acid is another example of a solid-liquid reaction. Ion-exchange kinetic analysis is illustrated in Chap. 7.

Solid-solid reactions   The reduction of powdered oxides using a reducing solid existing in a distinct solid phase, e.g., a mixture of a metal oxide and carbon, constitutes a heterogeneous noncatalytic solid-solid reaction.   Typical is iron-ore reduction by coke.

General Characteristics

As at least two phases are found in such systems, the reaction rate will depend upon the intimacy of contact between the phases as well as upon the factors (temperature, pressure, concentration) which govern homogeneous reaction.   Hence area of contact between the phases is an important reaction variable.   Diffusional steps are implicit components in heterogeneous systems since one or more reactants and/or products must be transported from their phase of supply to another phase, where, for reactants, reaction actually occurs.   Factors which govern interphase heat and mass transport therefore become important reaction parameters.

In noncatalytic systems, if one phase is fixed, e.g., packed bed of carbon particles, while the other is passed continuously over the fixed phase, e.g., oxygen-containing gas flowing through, and reacting with, a bed of carbon granules, a transient condition exists since one reactant (the carbon granule) is consumed and is not replaced. The introduction of the transient vastly complicates kinetic analysis of such systems. The operating cycle of the domestic water softener, while familiar to many laymen, can be described in precise detail only by solving rather complex partial differential equations specifying key ion concentration as a function of time, bed length, radius, and the spatial distribution of ions within individual particles.   When nonisothermal conditions prevail in a non-steady-state system, complete solution of the problem requires a digital computer.   Such conditions exist, for example, in regeneration of a fixed bed of coked catalyst pellets.   The transient systems cited are essentially semicontinuous operations and the time-on-stream dependency creates the need for analysis far more sophisticated than those required in steady-state situations.

In this chapter, the effects of transport phenomena upon heterogeneous reaction rates will be discussed in a general fashion. Basic definitions are first introduced to specify the regions of transport. There follows a treatment of simple diffusion in the external field and then diffusion within the zone of reaction. Rate alteration due to diffusional intrusions is then quantified in terms of external and internal effectiveness factors. The influence of diffusion upon yield and/or selectivity is also revealed in a simple fashion. As the purpose of this chapter is that of emphasizing principles, the most simple reaction networks are employed to dramatize effects which will occur in more complex reacting systems to be treated in subsequent chapters.

5-2 DEFINITIONS

Focus upon the situation where two immiscible phases are in contact within a reactor. As noted earlier (Chap. 4), gradients which persist throughout the gross confines of the reactor per se are termed *interparticulate*. Examples come speedily to mind:

 1 In the fixed-bed solid-catalytic reactor, as noted in Chap. 4, there may exist axial and radial gradients of both species concentrations and gas-solid temperatures.
 2 In a liquid-liquid extraction-reaction unit, say the nitration of benzene by mixed acid in a bubble-column reactor, axial and radial gradients may persist.

In these distributed-parameter systems, the appropriate continuity equations, e.g., Eq. (4-11), show that for a given generation function and boundary conditions, the axial and radial Peclet numbers dictate the magnitude and character of these long-range gradients. Should intensive mixing be imposed by design, these long-range, interparticulate gradients are damped, thus promoting uniformity of concentrations and temperature throughout the confines of the reactor volume. Hence:

 3 Due to the local particle motion and gross recirculation of solids in a fluidized-bed gas-solid catalyst system, axial and radial gradients are considerably smaller in magnitude than those known to prevail in the fixed-bed counterpart.
 4 In the conduct of, e.g., benzene nitration with mixed acid within an agitated vessel, the high intensity of induced mixing will eliminate long-range spatial gradients.

In contrast to long-range (interparticulate) gradients, gradients which persist *about* and *within* the locale of reaction are termed *short range* (Fig. 5-1).

Interphase gradients can exist between phases; i.e., in the region adjacent to the interface of two distinct phases, gradients of concentration and temperature may prevail; these are appropriately termed *interphase*. For example,

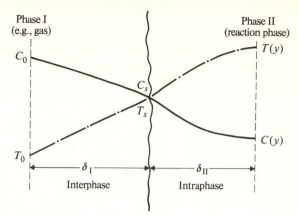

FIGURE 5-1
Steady-state concentration and temperature profiles in a two-phase reaction-diffusion network.

1   In dynamic contact of one fluid (i.e., a gas) with another (a liquid) within which absorption occurs, we are aware that in the so-called gas film and liquid film, gradients often exist due to transport limitations.

2   In the contact of a fluid and a solid (reactant or catalyst) the gradients about the solid, i.e., within the fluid film bathing the solid, are interphase in nature.

Interphase transport refers, then, to the diffusive-convection processes which link the source of reactants (bulk phase) to the sink of reaction (absorbing coreactant liquid, solid catalyst, or solid coreactant).   Diffusion and/or convection therefore occurs in *series* with reaction.

Intraphase gradients are, by definition, confined to the local reaction zone. In contrast with interphase transport, intraphase diffusion occurs simultaneously *with* reaction. (See Fig. 5-1.)

In this chapter, we shall confine our attention to the local transport processes, interphase and intraphase.   The nature of the transport coefficients will be cited, and the consequences of local gradients upon conversion and yield under realistic circumstances will be set forth in a simple fashion.   No specific system (gas-liquid, gas-solid, etc.) will be explicitly invoked at this juncture; instead diffusion-affected reaction of a general nature is emphasized.   Specific systems will be discussed in the chapters devoted to fluid-fluid reaction (Chap. 6), fluid-solid reaction (Chap. 7), and diffusion and heterogeneous catalysis (Chap. 9).

5-3   INTERPHASE AND INTRAPHASE TRANSPORT COEFFICIENTS

Suppose that a flowing gas (or liquid) is exposed to a fixed solid with which a component of the flowing fluid reacts, at a chemical rate $\mathscr{R}$.   In steady state, since the rate of reacting-species transport through the fluid film must equal the surface reaction rate $\mathscr{R}$,

$$k_g a(C_0 - C_s) = \mathscr{R}(C_s, T_s) \tag{5-1}$$

where k_g is a gas-phase mass-transport coefficient. If we consider a liquid-solid system, the coefficient is designated k_{L_0}. We recognize the interphase gradient in both concentration and temperature, as schematized in Fig. 5-1. For thermal transport equated to generation (say, exothermal of a simple nature), we have

$$ha(T_s - T_0) = -\Delta H\, \mathscr{R}(C_s, T_s) \tag{5-2}$$

where $\mathscr{R} = kC_s^n$ and, of course,

$$k = k_0 \exp\left[-\frac{E}{RT_0}\left(\frac{1}{t} - 1\right) \right] \tag{5-3}$$

where $t = T_s/T_0$.

In Eqs. (5-1) and (5-2) a is the external interface area per unit volume of the reaction phase. Note that the implication of the form of Eqs. (5-1) and (5-2) is that reaction rate is expressed as moles per time × volume, or if we deal with a solid reaction phase of known density, $\mathscr{R}$ is often expressed as gram moles per time × grams. In either case, k, the heterogeneous chemical-reaction-rate coefficient, retains the units of its homogeneous counterpart.

The precise nature of k_g, k_{L_0}, and h, the interphase transport coefficients, will depend upon the fluid mechanics of the system in the vicinity of the two-phase boundary. We illustrate this point in the following exposition, which, despite inherent oversimplifications, conveys some key insights into the interaction of mass and momentum transport.

Interphase-Transport-Coefficient Functionality

A phenomenological transport coefficient (k_g, k_{L_0}, h) is analytically defined by equating the coefficient to a flux divided by a convenient driving force. So

$$k_{L_0} \text{ or } k_g = \frac{\text{diffusive flux}}{C_0 - C_s} \quad \text{or} \quad h = \frac{\text{diffusive flux of heat}}{T_0 - T_s} \tag{5-4}$$

The diffusive flux is defined as (ignoring convective flux)

$$\text{Flux} = \begin{cases} -\mathscr{D}\left.\dfrac{dC}{dy}\right|_{y=\delta} & \text{mass diffusion} \\[2mm] -\lambda\left.\dfrac{dT}{dy}\right|_{y=\delta} & \text{heat diffusion} \end{cases} \tag{5-5}$$

where $\mathscr{D}$ is a molecular diffusivity and λ the thermal conductivity of the medium through which mass and thermal transport are occurring. Given a value of $\mathscr{D}$ and λ, the problem of analytically specifying k_g and h reduces to that of securing

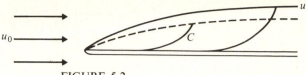

FIGURE 5-2
Boundary-layer-concentration and velocity profiles.

an analytical expression of the gradients dC/dy and dT/dy at the boundary $y = \delta$. Such an analytical specification requires (1) a model of transport, or mechanism, and (2) a sound definition of boundary conditions.   We might illustrate this point in the instance of simple mass transport *without reaction*, e.g., transport of mass through a stagnant gas or liquid film.   The film model invoked assumes a stagnant layer of thickness δ, within which all interphase transport resistances are confined. For gas mass transport, Eq. (5-4) defines the general situation (a definition), which is particularized once the model is invoked to permit analytical characterization of the flux dC/dy.

As Fig. 5-2 illustrates, when a fluid is passed over an immobile plate (sink or source of diffusing species), a laminar boundary layer develops in both concentra-tion (or temperature) and momentum.   Two continuity equations are demanded. For momentum

$$u\frac{\partial u}{\partial x} + v\frac{\partial u}{\partial y} = v\left(\frac{\partial^2 u}{\partial y^2} + \frac{\partial^2 u}{\partial x^2}\right) \tag{5-6}$$

and for mass

$$u\frac{\partial C}{\partial x} + v\frac{\partial C}{\partial y} = \mathscr{D}\left(\frac{\partial^2 C}{\partial y^2} + \frac{\partial^2 C}{\partial x^2}\right) \tag{5-7}$$

Equations (5-6) and (5-7) describe the spatial dependence of velocity and concen-tration in terms of diffusion and convection in the axial and perpendicular direction relative to a flat plate.   Solution of these equations provides the gradient at $y = \delta$, which (when multiplied by the molecular diffusivity) permits specification of the phenomenological coefficient by Eq. (5-4).

Equations (5-6) and (5-7), representing as they do a rather general description of local interphase continuity, are obviously susceptible to prudent simplifications, which generate, in order of convenience, a series of models of increasing primitive-ness yet, it is hoped, increasing usefulness.

For example, one can dismiss Eq. (5-6) as irrelevant, which is to assert that fluid motion is absent within the concentration boundary layer.   Abandoning transients, axial diffusion, and radial and axial convection of mass in Eq. (5-7) in-vites the ultimate simplification

$$\frac{d^2C}{dy^2} = 0 \tag{5-8}$$

*E*quation (5-8) then defines the universally admitted unreality known as film theory. Mass diffusion, in the spirit of this model, occurs through a stagnant layer of thickness δ.   After imposing the boundary conditions

At $y = \delta$: $C = C_0$

At $y = 0$: $C = C_s$

solution of Eq. (5-8) yields the linear concentration-vs.-distance relationship

$$C = (C_0 - C_s)\frac{y}{\delta} + C_s \qquad (5\text{-}9)$$

Differentiating to secure dC/dy at $y = \delta$, we obtain

$$\frac{dC}{dy}\bigg|_{y=\delta} = \frac{C_0 - C_s}{\delta} \qquad (5\text{-}10)$$

The flux, when equated to the phenomenological flux and expressed in terms of k_g, yields an analytical definition of k_g, in accord with Eqs. (5-4) and (5-10)

$$\mathscr{D}\frac{dC}{dy}\bigg|_{y=\delta} = \mathscr{D}\frac{C_0 - C_s}{\delta}$$

but

$$\mathscr{D}\frac{dC}{dy}\bigg|_{y=\delta} = k_g(C_0 - C_s)$$

and so

$$k_g = \frac{\mathscr{D}}{\delta} \qquad (5\text{-}11)$$

Equation (5-11) is an analytically secured definition of the gas-phase film mass-transfer coefficient.   Less severe compromises with Eqs. (5-6) and (5-7), lead to more complex analytical definitions of the phenomenological transport coefficient.

Having indicated the procedure whereby k_g (and, by inference, k_{L_0} and h) is specified, we consider solution of Eqs. (5-6) and (5-7), not rigorously, but under a modest atmosphere of restraint.   Specifically, we envision diffusion of matter to occur through a boundary layer within which fluid velocity is assumed to be an arbitrary function of axial and perpendicular distance, i.e.,

$$\frac{u}{u_0} = \frac{Ky^a}{x^b} \qquad (5\text{-}12)$$

The continuity equation for mass transport is

$$v\frac{\partial C}{\partial y} + u\frac{\partial C}{\partial x} = \mathscr{D}\left(\frac{\partial^2 C}{\partial y^2} + \frac{\partial^2 C}{\partial x^2}\right) \qquad (5\text{-}13)$$

Assuming constancy of physical properties and neglecting axial diffusion, $\partial^2 C/\partial x^2$, and the perpendicular component of velocity, $v\,\partial C/\partial y$, permits a simplification which renders the resulting mathematical model susceptible to analytical solution.[1]

[1] F. O. Mixon and J. J. Carberry, *Chem. Eng. Sci.*, **13**: 30 (1960).

The result provides the phenomenological average mass-transport coefficient k_g (or k_{L_o}) as a function of fluid velocity within the boundary layers. The velocity profile is defined by the arbitrarily assumed values of a and b in Eq. (5-12). The solution is

$$k_g = \frac{[K(b+1)/(a+2)^2]^{1/(a+2)}(a+2)^2 u_0\, P^{(-a-1)/(a+2)}}{\Gamma[1/(a+2)](a-b+1)\, Re^{(b+1)/(a+2)}} \qquad (5\text{-}14)$$

where K = coefficient determined by chosen values of a and b
$\quad P$ = physical-property group, a Schmidt or Prandtl number = $v/\mathscr{D}$ or v/α
$\quad Re$ = Reynolds number = Lu/v
$\quad \Gamma$ = gamma function

CASE 1 The velocity is invariant in x and y; that is, $a = b = 0$. Equation (5-14) reduces to

$$k_g \text{ or } k_{L_o} = 2\sqrt{\frac{\mathscr{D}}{\pi\theta}} \qquad (5\text{-}15)$$

where $\theta = x/u$, the time of transient diffusion. In terms of Reynolds and Schmidt numbers

$$j = \frac{k_g}{u}\, Sc^{1/2} = \frac{1.13}{\sqrt{Re}} \qquad (5\text{-}16)$$

which is merely Eq. (5-15) cast in convenient j-factor form. Note that in contrast with film theory [Eq. (5-11)] case 1 predicts that k_{L_o} or k_g is a square-root function of diffusivity $\mathscr{D}$. Case 1 actually describes the penetration theory of diffusion or conduction, as invoked for gas absorption into a liquid.[1]

CASE 2 Velocity within the boundary layer is a simple linear function of y and independent of x. So $a = 1$, $b = 0$, and Eq. (5-14) becomes

$$k_g \text{ or } k_{L_o} = 1.615\left(\frac{\mathscr{D}^2 \bar{u}}{2y_0 x}\right)^{1/3} \qquad (5\text{-}17)$$

which is the classic Leveque solution for heat transfer between the wall of a conduit of diamter $2y_0$ and a fluid in laminar flow of average velocity $\bar{u}$. Note that in contrast with film and penetration theory, k_g or k_{L_o} is a function of diffusivity raised to the two-thirds power. In j-factor form

$$j = \frac{k_g}{u}\, Sc^{2/3} = 1.615\, Re^{-2/3}\left(\frac{x}{2y_0}\right)^{-1/3}$$

CASE 3 Velocity is a function of both x and y. As in laminar boundary-layer theory, $a = 1$, $b = \frac{1}{2}$, and Eq. (5-14) reduces to

$$k_g \text{ or } k_{L_o} = 0.817\left(\frac{\mathscr{D}}{v}\right)^{2/3}\left(\frac{xu}{v}\right)^{-1/2} u \qquad (5\text{-}18)$$

[1] R. Higbie, *Trans. AIChE*, **31**: 365 (1935).

or in *j*-factor form

$$j = \frac{k_g}{u} Sc^{2/3} = \frac{0.817}{\sqrt{Re}} \qquad (5\text{-}19)$$

This brief discussion of the nature of interphase transport coefficients is offered simply to illustrate the complex dependency of the coefficient upon the fluid mechanics of the system as well as the variation in property-group (Schmidt, Prandtl number) exponents with the assumed nature of the velocity profile within the zone of interphase diffusion.

In general, the desired value of the mass and/or heat interphase transport coefficient is provided by the *j*-factor or Nusselt-number expression for the system of interest (packed bed, open tube, etc.).

The Intraphase Transport Coefficient

If the reaction phase is fixed, e.g., a porous solid, the governing continuity equation, in the instance of a flat-plate geometry *within* which reaction occurs (phase II in Fig. 5-1), is

$$\mathscr{D} \frac{d^2 C}{dy^2} = \mathscr{R}$$

By definition, of course, convection terms are absent in this fixed-reaction-phase case. If we are confronted, not with an immobile reaction phase, but, say, a re-acting-absorbing liquid phase, the continuity equation assumes the more complex form as convective terms intervene. These realities and their consequences will be dealt with subsequently. It suffices at this point to focus attention upon $\mathscr{D}$, the intraphase diffusivity and, of course, its analog λ for thermal intraphase conduction.

The precise specification of $\mathscr{D}$ waits upon greater insights into the diffusional events within presently ill-defined porous solids. Even greater caution must be urged with regard to λ, the thermal conductivity of porous media.

If the locale of reaction is a liquid phase wherein the simplicities of liquid-film theory may prudently be invoked, the intraphase liquid diffusivity $\mathscr{D}$ is readily de-termined in the light of a vast and comprehensive literature which reveals both useful data and provocative theory. It should be noted that whether the focal point of reaction-diffusion, i.e., intraphase events, is a porous solid (immobile) or a liquid phase characterized by complex fluid mechanics, one ultimately is confronted with the task of at least estimating the value of intraphase diffusivity $\mathscr{D}$. When dealing with the porous-solid intraphase diffusivity, due account of porosity and tortuosity must be taken, about which more will be said in Chap. 9. Such geometric considerations are, mercifully, irrelevant in the diffusion of dissolved gases within liquids. Data in such cases abound, and in the gas-liquid absorption-reaction system, the locale of difficulty is shifted to the problem of constructing a reasonable reaction-diffusion model, about which more will be said in Chap. 6.

5-4   INTERPHASE DIFFUSION AND REACTION[1]

It being our purpose to dramatize the qualitative influences of interphase mass and thermal diffusional intrusions upon heterogeneous reaction rates and yield, we choose to devote attention to those simple reaction networks which while perhaps not existentially akin to most real systems nevertheless retain the virtue of illustrating effects certain to be manifested in all real networks.   It is supposed that a flowing fluid containing a reactant is exposed to a fixed *nonporous* surface, which is capable of converting a single component of the fluid to product(s) at a rate $\mathscr{R}$. Under isothermal steady-state conditions, upon equating interphase mass transport to surface reaction of intrinsic order n, we have

$$k_g a (C_0 - C_s) = \mathscr{R} = kC_s^n \qquad (5\text{-}20)$$

It is intuitively obvious that to the extent that interphase mass-transport limitations conspire to deprive the reactive surface of the reactive species, a reduction in *surface* rate results.   Clearly the maximum local heterogeneous *surface* rate, for a given value of k, prevails when $C_s = C_0$, that is, when $k_g a$ becomes great enough for $C_0 - C_s$ to approach zero, and $\mathscr{R}$ retains its finite magnitude.   Given that rather intuitive generalization, it nevertheless seems worthwhile to explore the relative influence of reaction order n upon interphase diffusion-affected reaction, which is termed the *effective* or *global reaction rate* $\mathscr{R}$.

Isothermal Interphase Effectiveness

We consider first-order surface reaction, $n = 1$.   Solving Eq. (5-20) for the surface concentration (the unobservable, C_s), we readily obtain

$$C_s = \frac{C_0}{1 + k/k_g a} = \frac{C_0}{1 + \text{Da}_0}$$

and so the effective or global rate expressed in terms of *bulk* concentration (the observable C_0) is

$$\mathscr{R} = kC_s = \frac{kC_0}{1 + \text{Da}_0} \qquad (5\text{-}21)$$

Da, a Damköhler number, is the ratio of the specific surface chemical-reaction-rate coefficient to that of bulk mass transport.   Alternatively, since in general

$$k = \frac{1}{\tau_R} \quad \text{and} \quad k_g a = \frac{1}{\tau_D}$$

where τ_R and τ_D are time constants for the chemical and diffusive events, respectively, $\text{Da} = \tau_D/\tau_R$.   It naturally follows that when diffusion is rapid relative to reaction $\text{Da} \to 0$ and $\mathscr{R} = kC_0$, the system behaves as though it were homogeneous. At the other extreme, when surface reaction is rapid relative to the rate of reactant supply by diffusive mass transport, Da is large and in the limit $\mathscr{R} = k_g aC_0$.   In

---

[1] Hereafter, reaction orders are designated n and m, insofar as α and β now assume more important meanings.

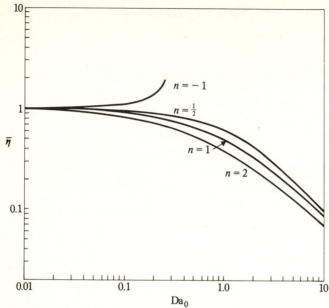

FIGURE 5-3
Isothermal external catalytic effectiveness for reaction order n.
(*G. Cassiere and J. J. Carberry, Chem. Eng. Educ., Winter, 1973: 22.*)

this latter instance, global rate is truly heterogeneous, depending as it does upon external surface area per unit of solid-phase volume a.   Thus the moles of reactant consumed per unit of time vary linearly with reaction-phase *volume* in chemical-reaction control in contrast with a linear variation with *area* when bulk mass transport governs the heterogeneous reaction.   Further, the variation of the experimental rate coefficient with temperature will be of the expected Arrhenius character when $Da \to 0$, and, at the extreme of mass-transport control, a near-zero activation energy (characteristic of $k_g a$) will become evident.

It will prove instructive to compare the global rate $\mathscr{R}$ with that of pure chemical-reaction control $\mathscr{R}_0$.   For linear kinetics

$$\bar{\eta} = \frac{\mathscr{R}}{\mathscr{R}_0} = \frac{kC_0}{1 + Da_0} \frac{1}{kC_0} = \frac{1}{1 + Da_0} \qquad (5\text{-}22)$$

This interphase or external effectiveness $\bar{\eta} = \mathscr{R}/\mathscr{R}_0$ is plotted in Fig. 5-3 as a function of Da_0.   For reactions of one-half and second-order, interphase effectiveness factors are[1]

$$n = \tfrac{1}{2}: \qquad \bar{\eta} = \frac{\mathscr{R}}{\mathscr{R}_0} = \sqrt{\frac{2 + Da^2}{2} \left[1 - \sqrt{1 - \frac{4}{(2 + Da^2)^2}}\right]}$$

$$n = 2: \qquad \bar{\eta} = \frac{\mathscr{R}}{\mathscr{R}_0} = \left[\frac{1}{2Da} (\sqrt{1 + 4Da} - 1)\right]$$

$$(5\text{-}23)$$

[1] G. Cassiere and J. J. Carberry, *Chem. Eng. Educ.*, Winter 1973: 22.

where

$$Da = \frac{kC_0^{n-1}}{k_g a} = \text{a Damköhler number} = \frac{\text{chemical-reaction velocity}}{\text{mass-transport velocity}}$$

Figure 5-3 reveals the effects of reaction order upon interphase effectiveness under isothermal conditions. Note that the higher the reaction order the more telling the effect of isothermal interphase mass-transport limitation. Recall that the same generalization was fashioned in our discussions (Chap. 3) of the influence of back-mixing upon normal reaction rates. The analogy is clear: for *normal* reactions, in which rate decreases with conversion, the desired gradient is provided by plug flow (under isothermal conditions). As backmixing alters the desired gradient, reactor efficiency decreases, the decrease being more severe the higher the reaction order. In the instances of concern in this chapter, the normal interphase condition is one of zero gradient, that is, $C_s \to C_0$, and an interphase mass-transport limitation prompts a deviation from the desired gradient to the detriment of interphase effectiveness. This taxation increases with reaction order. Note that no matter what the intrinsic surface reaction order may be, in mass-transport-control limit (Da large), the global rate becomes first order.[1]

Effectiveness in Abnormal Reactions

It will be recalled (Chaps. 2 and 3) that an abnormal reaction is one in which the rate increases with conversion over a range of conversion, e.g., in certain cases of autocatalysis and in reactions such as noble-metal catalysis of CO oxidation,[2] where in excess air, rate $= k/CO$, a rate model found to be valid at CO concentrations as low as 1 percent.[3]

As in normal kinetics, we equate bulk mass transport to surface reaction

$$k_g a(C_0 - C_s) = \frac{k}{C_s}$$

or

$$\frac{C_s}{C_0} = \frac{1}{2}\left(1 + \sqrt{1 - 4Da}\right)$$

Then

$$\bar{\eta} = \frac{k/C_s}{k/C_0} = \frac{C_0}{C_s} = \frac{2}{1 + \sqrt{1 - 4Da}}$$

where

$$Da = \frac{k}{C_0^2 k_g a}$$

As shown in Fig. 5-3, effectiveness for abnormal reaction actually increases, a result which one expects since a negative order implies inhibition of reaction rate. (No solution exists, for $Da > \frac{1}{4}$, as the assumed rate model is not valid at $C_s = 0$; see Sec. 5-15.)

---

[1] Da refers to bulk conditions, hence $Da \equiv Da_0$.
[2] D. G. Tajbl, J. Simons, and J. J. Carberry, *Ind. Eng. Chem. Fundam.*, **5**: 171 (1966).
[3] T. Mitani, M.S. thesis, University of Notre Dame, Notre Dame, Ind., 1969.

Effectiveness in Terms of Observables

Our interphase or external effectivenesses [Eqs. (5-22) and (5-23)] are expressed in terms of the Damköhler number Da, which, of course, contains the surface-reaction-rate coefficient k.   If k is known, $\bar{\eta}$ is evaluated and the global rate is calculated in terms of bulk concentration C_0 by

$$\mathscr{R} = \bar{\eta} k C_0 = \bar{k} f(C_0)$$

More often the global rate is measured in terms of C_0, establishing the experimental rate coefficient $\bar{k}$.   In such circumstances, Da cannot be calculated since the intrinsic value of k is not known, unless of course $\bar{k} = k$, a trivial case as $\bar{\eta}$ is then unity.   Given data for global rate vs. concentration, $\bar{\eta}$ can indeed be calculated by the following simple development.   Since $\bar{k} = \bar{\eta} k$, we have

$$\frac{\bar{k}}{k_g a} = \bar{\eta} \frac{k}{k_g a} = \bar{\eta} \, \text{Da} = \frac{\mathscr{R}}{k_g a C_0} \qquad (5\text{-}24)$$

$\bar{\eta}$Da is then a dimensionless observable as opposed to Da, usually an unobservable. Advantage is thus gained if $\bar{\eta}$ can be analytically related to the observable, $\bar{\eta}$Da, for diverse reaction orders.   Since a heat-transport limitation can be anticipated with the onset of a mass-diffusional intrusion, it would indeed prove useful if a general nonisothermal $\bar{\eta}$-versus-$\bar{\eta}$Da relationship could be established.   Such a development is presented in the following section.

Generalized Nonisothermal External Effectiveness[1]

In general $(T_0 \neq T_s,\ C_0 \neq C_s)$, the external effectiveness is

$$\bar{\eta} = \frac{k_s C_s^n}{k_0 C_0^n} = \frac{k_s}{k_0} \left(\frac{C_s}{C_0}\right)^n \qquad (5\text{-}25)$$

From Eq. (5-20) we find

$$\frac{C_s}{C_0} = 1 - \frac{\mathscr{R}}{k_g a C_0} \qquad (5\text{-}26)$$

Recall that the observed rate, in terms of Eq. (5-25), is

$$\mathscr{R} = k_s C_s^n = \bar{\eta} k_0 C_0^n$$

Equation (5-26) becomes

$$\frac{C_s}{C_0} = 1 - \bar{\eta}\text{Da}$$

and so Eq. (5-25) is

$$\bar{\eta} = \frac{k_s}{k_0} (1 - \bar{\eta}\text{Da})^n \qquad (5\text{-}27)$$

Since

$$\frac{k_s}{k_0} = \exp\left[-\epsilon_0\left(\frac{1}{t} - 1\right)\right]$$

[1] J. J. Carberry and A. A. Kulkarni, J. Catal., 3: 141 (1973).

we need only relate the reduced surface temperature T_s/T_0 to the observable in order to provide the general $\bar{\eta}$-versus-$\bar{\eta}$Da functionality.

Surface temperature is secured by a local heat balance, i.e., heat generated ($\pm$) due to surface reaction is equated, in steady state, to that transported by convection

$$ha(T_s - T_0) = -\Delta H \,\mathscr{R} \qquad (5\text{-}28)$$

Dividing by $k_g a C_0 T_0$ and invoking the heat and mass transport analogy:

$$j_D = \frac{k_g}{u}\,\text{Sc}^{2/3} = \frac{h}{\rho u C_p}\,\text{Pr}^{2/3} = j_H$$

we obtain

$$t = 1 + \frac{-\Delta H\, C_0}{\rho C_p T_0}\left(\frac{\text{Sc}}{\text{Pr}}\right)^{-2/3}\frac{\mathscr{R}}{k_g a C_0}$$

or

$$t = 1 + \bar{\beta}\cdot\bar{\eta}\text{Da} \qquad (5\text{-}29)$$

where

$$\bar{\beta} = \frac{-\Delta H\, C_0}{C_p T_0 \rho}\frac{1}{\text{Le}^{2/3}}$$

where

$$\text{Le} = \text{Lewis number for fluid} = \frac{\text{thermal diffusivity}}{\text{molecular diffusivity}}$$

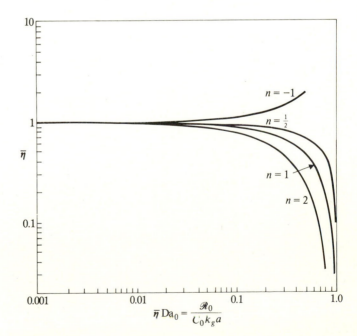

FIGURE 5-4
Isothermal external catalytic effectiveness in terms of observable for order n. (*G. Cassiere and J. J. Carberry, Chem. Eng. Educ., Winter,* 1973: 22.)

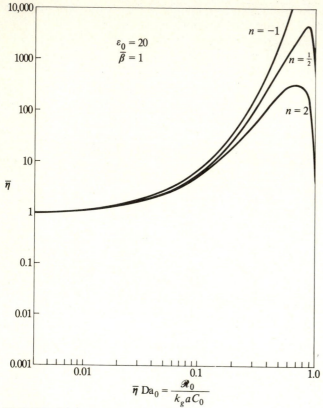

FIGURE 5-5
External nonisothermal effectiveness for nonlinear kinetics vs. observables.   [*J. J. Carberry and A. A. Kulkarni, J. Catal.*, **31**: 41 (1973).]

Substituting for t in the Arrhenius function, we then secure the nth-order nonisothermal external-effectiveness factor as a function of the observable, $\bar{\eta}\mathrm{Da}$:

$$\bar{\eta} = (1 - \bar{\eta}\mathrm{Da})^n \exp\left[-\epsilon_0\left(\frac{1}{1 + \bar{\beta}\bar{\eta}\mathrm{Da}} - 1\right)\right] \qquad (5\text{-}30)$$

This relationship is displayed for various orders and values of $\bar{\beta}$ and ϵ_0 in Figs. 5-4 to 5-7.

Certain characteristics of the nonisothermal external-effectiveness behavior should be noted:

1   The higher the intrinsic reaction order, the more telling the influence of a mass-diffusional limitation (Fig. 5-4).

2   For orders less than zero, $\bar{\eta}$ will always be greater than unity (Fig. 5-4).

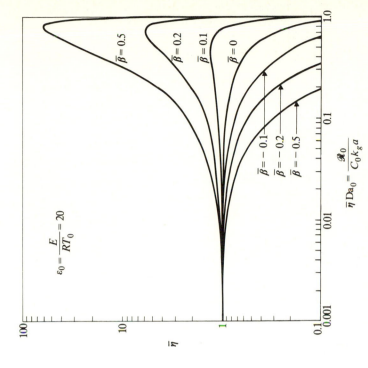

$\epsilon_0 = \dfrac{E}{RT_0} = 20$

FIGURE 5-7

External nonisothermal $\bar{\eta} - \bar{\eta}$Da relation for first order ($\epsilon_0 = 20$).
[J. J. Carberry and A. A. Kulkarni, J. Catal., **31**:41 (1973).]

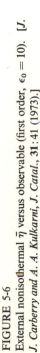
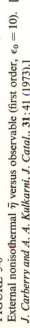

$\epsilon_0 = \dfrac{E}{RT_0} = 10$

FIGURE 5-6

External nonisothermal $\bar{\eta}$ versus observable (first order, $\epsilon_0 = 10$). [J.
J. Carberry and A. A. Kulkarni, J. Catal., **31**:41 (1973).]

3 $\bar{\eta}$ values well in excess of unity can be found for any kinetics when a heat-transport limitation is manifest for an exothermic reaction (Figs. 5-5 to 5-7).

4 In the nonisothermal case, $\bar{\eta}$ is far more sensitive to ϵ_0 than to β (Figs. 5-6 and 5-7).

5-5 ISOTHERMAL YIELD OR SELECTIVITY IN INTERPHASE DIFFUSION-REACTION

As noted earlier (Chaps. 2 and 3), the most complex of chemical reaction networks is profitably viewed as an assembly of consecutive, simultaneous, and parallel steps. For example, the catalytic oxidation of naphthalene may be described by the network.

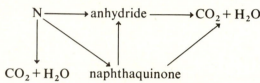

a composite of consecutive and simultaneous events.   In exploring the influence of interphase transport upon point yield and selectivity, it will be recalled that

$$Y = \text{yield} = \frac{\text{rate of desired-product generation}}{\text{rate of key-reactant consumption}}$$

$$S = \text{selectivity} = \frac{\text{yield of desired product}}{\text{yield of undesired product}}$$

We shall focus attention upon diffusional intrusions in each of these component networks of a complex network.

Consecutive Reactions

Let us first dwell upon the consecutive network of linear kinetics $A \overset{k_1}{\to} B \overset{k_2}{\to} C$, in which system, product B is assumed to be the desired species.   For catalysis over a nonporous solid, we equate the steady-state mass transport to surface reaction, assuming equality of gas-phase mass-transport coefficients for each of the transported species, ie.,

$$(k_g a)_A = (k_g a)_B = (k_g a)_C$$

For A: $k_g a(A_0 - A_s) = k_1 A_s$ (5-31)

For B $k_g a(B_s - B_0) = k_1 A_s - k_2 B_s$ (5-32)

For C: $k_g a(C_s - C_0) = k_2 B_s$ (5-33)

Designating the Damköhler numbers Da as Da_1, Da_2, ... for each respective step, we find, as before,

$$A_s = \frac{A_0}{1 + Da_1}$$

Substituting the above into Eq. (5-32) and solving for B_s in Eq. (5-32) results in

$$B_s = \frac{Da_1 A_0}{(1 + Da_1)(1 + Da_2)} + \frac{B_0}{1 + Da_2} \qquad (5\text{-}34)$$

where
$$Da_1 = \frac{k_1}{k_g a} \qquad \text{and} \qquad Da_2 = \frac{k_2}{k_g a}$$

The point yield of B is therefore

$$Y_B = -\frac{dB}{dA} = 1 - \frac{Da_1}{K_0(1 + Da_2)} - \frac{1}{K_0}\frac{B_0}{A_0}\frac{1 + Da_1}{1 + Da_2} \qquad (5\text{-}35)$$

where $K_0 = k_1/k_2$. Note that $Da_1/K_0 = Da_2$. The subscript zero indicates K at T_0.

Equation (5-35) reduces to

$$Y_B = \frac{1}{1 + Da_2} - \frac{1}{K_0}\frac{B_0}{A_0}\frac{1 + Da_1}{1 + Da_2} \qquad (5\text{-}36)$$

When Da_1 and $Da_2 = 0$, Eq. (5-36) assumes the form identical to that for point yield of B in a homogeneous reaction, i.e.,

$$Y_B = 1 - \frac{1}{K_0}\frac{B_0}{A_0}$$

When the desired product is not present in the gas phase, $B_0 = 0$, the initial point yield is solely a function of the Damköhler number for the second step [Eq. 5-36]

$$-\frac{dB}{dA} = \frac{1}{1 + Da_2} = Y_B \qquad \text{for} \quad B_0 = 0 \qquad (5\text{-}37)$$

Recall that

$$Da_2 = \frac{\tau_D}{\tau_{R_2}} \qquad \text{and so} \qquad Y_B \equiv \frac{\tau_{R_2}}{\tau_D}$$

The yield of B then depends upon the surface-reaction time for its annihilation to C, relative to the diffusion time for its escape from the catalytic surface.

Analogous to the situation in which isothermal backmixing taxes the yield of intermediate in the consecutive network by reason of intermediate overexposure, in the diffusion-affected heterogeneous reaction, intermediate overexposure upon the surface due to an escape limitation invites overreaction to the ultimate product. If total reaction is wanted, e.g., total catalytic combustion of autoexhaust species, mass-transport intrusions are, of course, desirable. On the other hand, in oxidizing naphthalene to valuable anhydride, transport intrusions are to be avoided.

Simultaneous Reflections

When B is the desired product in

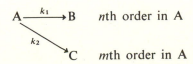

we have

$$\mathscr{R}_B = k_g a(A_0 - A_s) = k_1 A_s{}^n \qquad \mathscr{R}_C = k_g a(A_0 - A_s) = k_2 A_s{}^m$$

Selectivity is solely a function of surface kinetics

$$S_s = \frac{dB}{dC} = \frac{k_1 A_s{}^n}{k_2 A_s{}^m} = K_0(A_s)^{n-m} \qquad (5\text{-}38)$$

If the surface concentration of reactant A is that of the bulk A_0 (no diffusion limitation), then

$$S_0 = \frac{dB}{dC} = K_0(A_0)^{n-m} \qquad (5\text{-}39)$$

hence diffusion-affected selectivity relative to the diffusion-unaffected case is

$$\frac{S_s}{S_0} = \left(\frac{A_s}{A_0}\right)^{n-m} \qquad (5\text{-}40)$$

In consequence of Eq. (5-40) we observe that (1) when diffusional intrusions are negligible, selectivity assumes the value of its homogeneous counterpart. Equation (5-39) dictates selectivity, a function of k_1/k_2, bulk concentration, and relative orders. (2) With a diffusional modification, the effect upon selectivity depends solely upon the relative orders. In sum, for a mass-transfer limitation, $A_s < A_0$:

When $n > m$, selectivity decreases.
When $n < m$, selectivity increases.
When $n = m$, selectivity is unaffected.

These generalizations are to be expected in view of our earlier conclusions concerning diffusional-gradient effects as a function of reaction order (Fig. 5-4): the higher the order the more telling the reduction in interphase effectiveness. So without formal analysis, we can say of the simultaneous network that that reaction step of highest order will be taxed most severely by a mass-transport intrusion. Again the analogy with the influence of backmixing upon selectivity in a simultaneous reaction network of unequal orders is clear.

Parallel Reactions

$$A \xrightarrow{k_1} B \qquad n\text{th order}$$

$$C \xrightarrow{k_2} D \qquad m\text{th order}$$

Again B is the desired species. Since we have here two independent reactions, the yield and selectivity will depend upon the relative interphase effectiveness factors. As reaction order influences effectiveness, again (without recourse to formal proof) one can assert that the reaction step of the highest order will be most adversely affected by a diffusional gradient. Formally

$$S_s = \frac{k_1 A_s{}^n}{k_2 C_s{}^m} \qquad (5\text{-}41)$$

Unlike the simultaneous case, where a common reactant exists, the ratio of A to C in the parallel network can be exploited to engineer selectivity for given reaction orders. In principle, one might exploit differences in the mass-transport coefficient $k_g a$ for this network. In point of fact, little difference is usually found in such coefficients for various species unless there are significant molecular-weight differences between reactants A and C.

In general, for the parallel network selectivity is governed by the ratio of interphase effectiveness factors multiplied by the homogeneous rate for each independent reaction

$$S_s = \frac{\bar{\eta}_1 k_1 A_0^n}{\bar{\eta}_2 k_2 C_0^m} \qquad (5\text{-}42)$$

where $\bar{\eta}$ values are given by Eq. (5-30).

When each reaction is first order,

$$S_s = \frac{1 + \mathrm{Da}_2}{1 + \mathrm{Da}_1} \frac{k_1}{k_2} \frac{A_0}{C_0} \qquad (5\text{-}43)$$

For large values of Da (Da $\gg$ 1) selectivity in this mass-diffusion-controlled limit is, as expected,

$$S_{\mathrm{Da}\to\mathrm{large}} = \frac{(k_g a)_1}{(k_g a)_2} \frac{A_0}{C_0} \qquad (5\text{-}44)$$

As noted earlier, little difference between $(k_g a)_1$ and $(k_g a)_2$ can be anticipated in practice.

5-6 NONISOTHERMAL YIELD OR SELECTIVITY IN INTERPHASE DIFFUSION-REACTION

Noting that for consecutive, simultaneous, and parallel reaction networks, the rate-coefficient ratio $K = k_1/k_2$ appears in the yield (or selectivity) functionality [Eqs. (5-36), (5-38), and (5-42)], under conditions in which a temperature difference exists between fluid phase and the solid surface due to an interphase heat-transfer limitation, yield or selectivity will be dictated by the K value commensurate with the usually unobervable surface temperature. In the light of the general Arrhenius function

$$k = k_0 \exp\left[-\epsilon_0 \left(\frac{1}{t} - 1\right)\right]$$

for $K = k_1/k_2$, we have

$$K = K_0 \exp\left[-(\epsilon_1 - \epsilon_2)\left(\frac{1}{t} - 1\right)\right] \qquad (5\text{-}45)$$

where $K_0 = (k_1/k_2)_0$, that is, the rate-coefficient ratio at bulk-fluid temperature T_0, and, as before, $t = T_s/T_0$. Equation (5-45) is conveniently reexpressed as

$$K = K_0 \exp\left[(\epsilon_1 - \epsilon_2)\frac{t-1}{t}\right] \qquad (5\text{-}46)$$

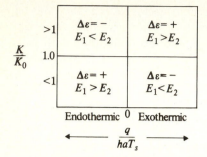

FIGURE 5-8
Quadrant of K/K_0 versus endothermal and exothermal reaction.

Heat generated at the surface q is, in steady state, equal to that transported in the interphase zone

$$q = ha(T_s - T_0) \quad \text{or} \quad q = haT_0(t - 1) \qquad (5\text{-}47)$$

and so

$$t - 1 = \frac{q}{haT_0}$$

and Eq. (5-46) becomes, in terms of generation (+ for exothermic, − for endothermal reaction),

$$K = K_0 \exp\left[(\epsilon_1 - \epsilon_2)\frac{q}{haT_s}\right] \qquad (5\text{-}48)$$

Yield and selectivity for nonisothermal interphase transport-influenced surface reaction is then governed by Eqs. (5-36) (consecutive reaction), (5-38) (simultaneous reaction), and (5-42) (parallel reaction), in which the local, surface value of K is, according to Eq. (5-48), a function of K_0, q/haT_s, and $\Delta\epsilon = \epsilon_1 - \epsilon_2$.

An interphase temperature gradient can be either a benefit or deficit to yield or selectivity. The normalized rate-coefficient ratio K/K_0 will vary as illustrated in Fig. 5-8.

Again the kinship between interphase heat and mass diffusional effects upon reaction yield or selectivity and the influence of backmixing of heat and mass upon reactor yield or selectivity is apparent (see Fig. 3-25).

Summary

An interphase effectiveness factor $\bar{\eta}$ is readily developed, which for a first-order surface reaction is

$$\bar{\eta}_i = \frac{1}{1 + \text{Da}_i} \qquad (5\text{-}49)$$

where Da_i is the Damköhler number

$$\text{Da}_i = \frac{k_i}{k_g a} \qquad (5\text{-}50)$$

Under nonisothermal conditions

$$k_i = k_{i_0} \exp\left[-\epsilon_0 \left(\frac{1}{t} - 1\right)\right] \qquad \begin{matrix} \epsilon_0 = E/RT_0 \\ t = T_s/T_0 \end{matrix} \qquad (5\text{-}51)$$

In terms of the observable parameter

$$\bar{\eta} \, \mathrm{Da} = \frac{\mathscr{R}}{k_g a A_0} = \frac{\text{observed rate}}{\text{bulk mass-transport rate}} \qquad (5\text{-}52)$$

the external effectiveness for any order and nonisothermal conditions is given by Eq. (5-30). Yield or selectivity in a complex chemical reaction network will be determined by relative activational energies $\Delta\epsilon$, the thermal parameters, $\bar{\beta}_1$ and $\bar{\beta}_2$, which affect relative rate-coefficient values (Fig. 5-8), and, of course, the mass-diffusional parameters.

The treatment set forth here is rooted in simple film theory or the Frank-Kamenetskii quasi-stationary model.[1] Rosner[2] has treated the isothermal external-effectiveness problem for nth-order kinetics in laminar-boundary-layer flows. When we compare $\bar{\eta}$ values for the film and boundary-layer postulates, the error in using the simple film model is at most about 15 percent. Errors in parameters derived from $\bar{\eta}$, such as activation energy, can, of course, be greater with the film model. However, the fact that a film model can be negotiated with ease to embrace nonisothermal situations for any reaction order gives it a distinct advantage.

Nor should the practical merits of Eq. (5-30) be overlooked. Several gas–nonporous-solid catalytic reactions of considerable importance exist:

1 NH_3 is oxidized over Pt-alloy wires to produce NO and thence HNO_3.
2 HCN is produced catalytically over Pt-alloy wires from NH_3, O_2, and a hydrocarbon.
3 Methanol is oxidized to formaldehyde over a nonporous silver catalyst.
4 In the automotive-exhaust catalytic muffler, the catalytic oxidation of CO and hydrocarbons at high temperature occurs on the external, or superficial, surface of the catalyst.

Significantly, in each of the above examples, external diffusional gradients in temperature and reactant concentrations are known to exist.

5-7 INTRAPHASE ISOTHERMAL DIFFUSION-REACTION

We now consider reaction occurring simultaneously with diffusion (intraphase diffusion-reaction), in contrast to the series reaction-diffusion event considered above (interphase diffusion-reaction). In intraphase diffusion-reaction, the mathematical models assume greater complexity. It was noted earlier in this chapter that in mass or heat transport in the absence of reaction, several solutions

---

[1] D. A. Frank-Kamenetskii, *Zh. Fiz. Khim.*, **13**: 756 (1939); see also his "Diffusion and Heat Exchange in Chemical Kinetics," Princeton University Press, Princeton, N.J., 1955.
[2] D. E. Rosner, *AIChE J.*, **9**: 321 (1963).

of increasing complexity can be generated, the level of complexity being commensurate with the sophistication of the model invoked to describe reality. Greater complexities result when chemical reaction is assumed to occur simultaneously with convection diffusion. Happily, the most simple of mass- or heat-transfer models is the surprisingly effective film theory. When chemical reaction occurs within the film, the resulting mathematical model is easily solved to yield results of great practical value.

It may be supposed that a reactant-bearing fluid flows past another immiscible phase, e.g., gas-solid, gas-liquid, liquid-liquid. The issue of interphase transport has been dealt with in the previous section. We dwell now upon the reacting species as it diffuses and reacts in the second phase, as schematized in Fig. 5-1. Should that second phase be solid or another fluid wherein transport resistance can be adequately described by film theory, then in steady state, equating diffusion to reaction, we have, for isothermal conditions and flat-plate geometry,

$$\mathcal{D}\frac{d^2C}{dy^2} = \mathcal{R}(C, T) \qquad (5\text{-}53)$$

The solution of Eq. (5-53) is, of course, a function of boundary conditions assumed to govern the system. Three rather general systems suggest themselves:

1 Diffusion and simultaneous reaction within a porous catalyst pellet
2 Diffusion and simultaneous reaction within a droplet of a dispersed coreactant fluid
3 Diffusion and simultaneous reaction within a bulk continuous phase, in which, in contrast with case 2, the source of diffusing reactant is dispersed, e.g., a gas being bubbled through a continuous coreacting phase

The physical situations are schematized in Fig. 5-9.

It is evident that cases 1 and 2 are similar insofar as the boundary conditions are the same; i.e., we have symmetry within the catalyst or droplet, and at the interface a concentration governed by interphase transport is specified. In steady state, a constant surface concentration prevails. These boundary conditions are for, say, a symmetrical flat-plate geometry of half thickness L:

$$y = 0, \text{ centerline:} \qquad \frac{dC}{dy} = 0 \qquad\qquad (5\text{-}54)$$

$$y = L, \text{ surface:} \qquad\quad C = C_s$$

What is sought in solving Eq. (5-53) for cases 1 and 2 is a measure of intraphase utilization or effectiveness. As before, we seek the diffusion-affected rate relative to that rate which would prevail within the reaction phase if intraphase diffusion were rapid enough to ensure a zero concentration gradient within the reaction phase; that is, $C = C_s$ throughout the droplet or catalyst pellet. So

$$\eta = \frac{1}{L}\int_0^L \frac{kf(C)\,dy}{k_0 f(C_s)} \qquad (5\text{-}55)$$

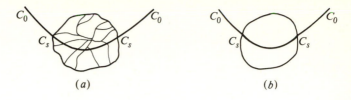

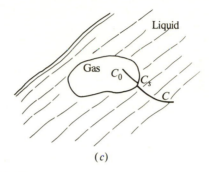

FIGURE 5-9
Diverse diffusion-reaction situations: (*a*) porous catalyst; (*b*) liquid droplet; (*c*) dispersed bubble in bulk liquid.

Under isothermal conditions, $k = k_0$

$$\eta = \frac{1}{L} \int_0^L \frac{f(C)\,dy}{f(C_s)} \qquad (5\text{-}56)$$

Given this definition [Eq. (5-56)] of isothermal intraphase effectiveness, Eq. (5-53) is readily solved for the boundary conditions set forth [Eq. (5-54)] to secure reactant concentration as a function of y. Integration according to Eq. (5-56) is then carried out to obtain η. Analytical resolution is obviously assured by the assumption that isothermal first-order kinetics prevails, and so

$$\mathcal{D}\frac{d^2C}{dy^2} = kC$$

or in dimensionless form, $f = C/C_s$, $z = y/L$,

$$\frac{d^2f}{dz^2} = \left(L^2\frac{k}{\mathcal{D}}\right)f \qquad (5\text{-}57)$$

Let $L^2 k/\mathcal{D} = \phi^2$, where ϕ is the Thiele modulus. Physically

$$\phi^2 = \frac{\text{chemical-reaction rate}}{\text{intraphase diffusion rate}}$$

Once η is analytically specified, the global rate of intraphase reaction is given by

$$\mathcal{R} = \eta k C_s$$

Example

Derive an analytical expression for isothermal intraphase effectiveness (phase-utilization factor) for a sphere (catalyst or droplet).

At a point within the sphere, the derivative of the product of flux and area in steady state must equal generation due to reaction. For a sphere of radius R, at any radial position, the area perpendicular to diffusive flux is $4\pi r^2$, and the differential spherical volume is $4\pi r^2\,dr$, so that

$$\frac{d(\text{flux} \times \text{area})}{dV} = \text{generation} \qquad (a)$$

Assuming a diffusive flux $= -\mathscr{D}\,dC/dr$, we have

$$\frac{d[-\mathscr{D}(dC/dr)4\pi r^2]}{4\pi r^2\,dr} = -kC^n$$

or

$$\mathscr{D}\left(\frac{d^2C}{dr^2} + \frac{2}{r}\frac{dC}{dr}\right) = kC^n \qquad (b)$$

Rendering (b) dimensionless by setting $f = C/C_s$ and $\rho = r/R$, we obtain

$$\frac{d^2f}{d\rho^2} + \frac{2}{\rho}\frac{df}{d\rho} = \frac{R^2 kC_s^{n-1}}{\mathscr{D}}f^n \qquad (c)$$

Designate $R^2 kC_s^{n-1}/\mathscr{D} = \phi^2$, where ϕ is the Thiele modulus. Let $f = z/\rho$; this substitution reduces Eq. (c) to

$$\frac{d^2z}{d\rho^2} = \phi^2 z^n \rho^{1-n} \qquad (d)$$

For linear kinetics $(n = 1)$, the result is

$$\frac{d^2z}{d\rho^2} = \phi^2 z \qquad (e)$$

for which the solution is

$$z = f\rho = C_1 \exp \phi\rho + C_2 \exp(-\phi\rho)$$

The boundary conditions are

At $\rho = 0$: $\dfrac{df}{d\rho} = 0$

At $\rho = 1$: $z = 1$

By the first condition $C_1 = -C_2$, and so

$$z = C_1[\exp \phi\rho - \exp(-\phi\rho)] = 2C_1 \sinh \phi\rho$$

Utilizing the second boundary condition gives

$$1 = 2C_1 \sinh \phi$$

and thus

$$z = f\rho = \frac{\sinh \phi\rho}{\sinh \phi}$$

or
$$f = \frac{C}{C_s} = \frac{\sinh \phi \rho}{\rho \sinh \phi} \tag{f}$$

Equation (f) provides the concentration profile within the spherical reaction phase. The utilization of that phase, as noted earlier, is expressed under isothermal conditions by

$$\eta = \frac{\dfrac{1}{R} \displaystyle\int_0^R C \, dr}{C_s} \tag{g}$$

Hence we have merely to integrate (f) in accord with (g) to obtain

$$\eta = \frac{3}{\phi} \left(\frac{1}{\tanh \phi} - \frac{1}{\phi} \right) \tag{h}$$

An alternative development involves expressing effectiveness η as the ratio of actual flux through the external surface of the sphere to the rate which would prevail in the absence of diffusional retardation

$$\eta = \frac{4\pi R^2 \mathscr{D}(-dC/dr)_{r=R}}{\frac{4}{3}\pi R^3 (k_0 C_s)} = \frac{3}{R} \frac{\mathscr{D}}{k_0 C_s} \frac{dC/dr_{r=R}}{}$$

Differentiating (f) to secure dC/dr evaluated at $r = R$ $(\rho = 1)$ yields equation (h).

In Fig. 5-10, the internal or intraphase effectiveness is displayed for a spherical reaction phase, first order, a flat-plate geometry, first order, and a second-order reaction in a flat plate. As Aris has shown,[1] geometric differences are minimized in Fig. 5-10 by defining the characteristic dimension of the reacting phase (catalyst or droplet) in terms of

$$\frac{\text{Phase volume}}{\text{Interfacial area}} = \frac{R}{3} \text{ (sphere)} = L \text{ (flat plate)}$$

Recall that $L = 1/a$, where a is the surface-to-volume ratio encountered in heat- and mass-transfer computations.

The behavior of the effectiveness or phase-utilization factor as a function of order deserves comment. The trends revealed in Fig. 5-10 for the internal effectiveness follow in spirit those set forth for external effectiveness (Fig. 5-3), which is to say (1) the higher the reaction order the more telling the influence of diffusional intrusions. We can assert, therefore, without recourse to formal mathematics, that an effectiveness greater than unity will be realized in the case of intraphase diffusion-reaction of negative order; formal proof is provided by Aris.[2] (2) For small values of the Thiele modulus ϕ (for intraphase effectiveness) and the Damköhler number Da (for interphase effectiveness), phase utilization or efficiency

---

[1] R. Aris, *Chem. Eng. Sci.*, **6**: 262 (1957).
[2] R. Aris, "The Mathematical Theory of Diffusion and Reaction in Permeable Catalysts," vol. 1, sec. 38, Clarendon, Oxford, 1975.

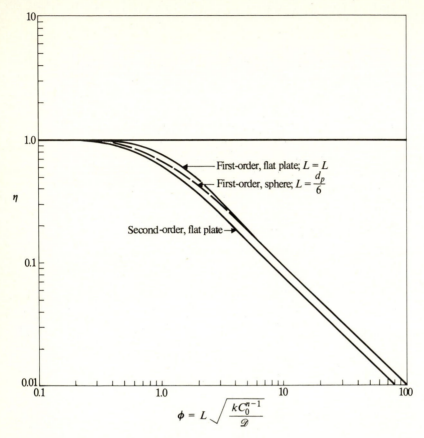

$$\phi = L \sqrt{\frac{k C_0^{n-1}}{\mathscr{D}}}$$

FIGURE 5-10
Intraphase η versus Thiele modulus for indicated geometries and reaction orders.

η approaches unity. (3) At large values of these moduli, both effectiveness factors approach

$$\bar{\eta} = \frac{1}{\text{Da}} \quad \text{(external or interphase)}$$

$$\eta = \frac{1}{\phi} \quad \text{(internal or intraphase)}$$

Since $\mathscr{R}$-global rate $= \eta k C_0^n$, for $\phi > 3$,

$$\eta = 1/\phi \qquad \text{and} \qquad \mathscr{R} = \frac{k C_0^n}{L \sqrt{\frac{k}{\mathscr{D}} C_0^{n-1}}} = \frac{\sqrt{k \mathscr{D}}}{L} C_0^{(n+1)/2} \qquad (5\text{-}58)$$

As a consequence of Eq. (5-58) in the domain of strong intraphase diffusion influence (1) the apparent reaction order n' is related to the intrinsic value n by

$$n' = \frac{n+1}{2} \qquad (5.59)$$

and (2) the experimentally observed rate coefficient $\bar{k}$ is

$$\bar{k} = \frac{\sqrt{k\mathscr{D}}}{L} \qquad (5-60)$$

and so the observed activation energy will be

$$\exp\left(-\frac{E_x}{RT}\right) = \exp\left(-\frac{E_t + E_D}{2RT}\right) \qquad (5-61)$$

i.e., the experimental activation energy E_x will be one-half the true value E_t (the diffusional activation energy E_D is usually very small). (3) The observed rate will be a linear function of $1/L$, the external-surface area-to-volume ratio.

5-8 INTERPHASE AND INTRAPHASE ISOTHERMAL EFFECTIVENESS

Assuming a flat-plate geometry, an overall isothermal effectiveness for linear kinetics is easily derived by simply utilizing the surface boundary condition

$$k_g(C_0 - C_s) = -\mathscr{D}\left.\frac{dC}{dy}\right|_{\text{surface}} \qquad (5-62)$$

instead of $C_s = C_0$ at the surface. Solution for intraphase diffusion-reaction in a flat plate for an interphase resistance [boundary condition (5-62)] yields the overall effectiveness η_0:

$$\eta_0 = \frac{\tanh \phi}{\phi[1 + (\phi \tanh \phi)/\text{Bi}_m]} \qquad (5-63)$$

where $\qquad \text{Bi}_m = \text{mass Biot number} = \dfrac{k_g L}{D} = \dfrac{\text{internal gradient}}{\text{external gradient}}$

While Eq. (5-63) gives overall effectiveness in terms of internal and external concentration gradients, its utilization requires an a priori knowledge of the intrinsic rate coefficient k in order that ϕ can be computed and η_0 evaluated for a given value of the mass Biot number Bi_m. We seek, then, as we did in the treatment of the external effectiveness, an observable modulus. Following Wheeler's[1] treatment of the internal observable, we can readily handle the internal-external problem. Let us assume linear isothermal kinetics. The observed global rate is

$$\mathscr{R} = \eta_0 k_0 C_0 = \frac{\tanh \phi(k_0 C_0)}{\phi[1 + (\phi \tanh \phi)/\text{Bi}_m]} \qquad (5-64)$$

---

[1] A. Wheeler, in P. H. Emmett (ed.), "Catalysis," vol. 2, Reinhold, New York, 1955.

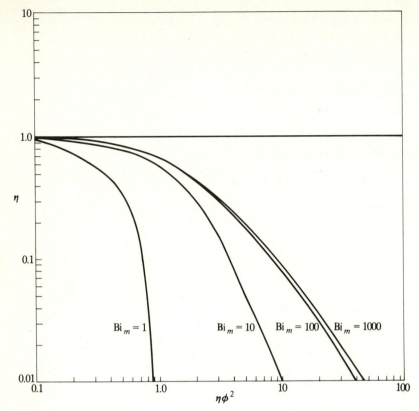

FIGURE 5-11
Interphase-intraphase catalytic effectiveness vs. the observable $\eta\phi^2 = \mathcal{R}L^2/\mathcal{D}C_0$ for various values of Bi_m for a first-order reaction.

If we multiply by $L^2/\mathcal{D}C_0$,

$$L^2 \frac{\mathcal{R}}{\mathcal{D}C_0} = \frac{\phi \tanh \phi}{[1 + (\phi \tanh \phi)/Bi_m]} = \eta_0\phi^2 \text{ an observable}$$

We note that Fig. 5-10 can be refashioned in terms of the observable $\eta_0 \phi^2$. Hence η_0 can be secured by a value of the observed, global, rate $\mathcal{R}$, bulk concentration C_0, particle or droplet size, and an estimate of $\mathcal{D}$. In Fig. 5-11 the η_0-observable relationship is displayed for various values of the mass Biot number for a first-order reaction.

Behavior of Global Rate

For first-order reaction, as ϕ becomes large ($\phi > 3$), $\tanh \phi \to 1.0$ and

$$\eta_0 = \frac{1}{\phi(1 + \phi/Bi_m)}$$

Under this condition ($\phi > 3$ and finite Bi_m)

$$\mathcal{R} = \frac{\sqrt{k\mathcal{D}}\,C_0}{L(1 + \phi/Bi_m)}$$

[Compare with Eq. (5-58).]   When $\phi/Bi_m > 1$, we find in the limit

$$\mathcal{R} = k_g a C_0$$

Table 5-1 summarizes key characteristics of interphase-intraphase diffusion-reaction for the three regimes of rate control.

5-9 NONISOTHERMAL INTRAPHASE EFFECTIVENESS[1]

In general

$$\eta = \frac{1}{L} \int_0^L \frac{kf(C)\,dy}{k_0 f(C_s)} \qquad (5\text{-}65)$$

We noted that under isothermal circumstances for normal kinetics, $\eta \leq 1.0$, since the numerator of Eq. (5-56) is always equal to or less than the denominator for normal reactions (rate decreases with conversion).

   If the reaction phase is nonisothermal, Eq. (5-65) applies and some interesting consequences become evident.   For an exothermic reaction, the intraphase temperature rise might be such that given a finite Arrhenius number, $\epsilon = E/RT_0$, k in the numerator of Eq. (5-65) could assume such a magnitude that, in spite of a reactant concentration drop well below C_s, the numerator exceeds the denominator in Eq. (5-65).   Therefore nonisothermal intraphase effectiveness may be equal to, less than, or greater than unity: η (nonisothermal) $\lessgtr 1.0$.

Table 5-1   CHARACTERISTICS OF INTERPHASE AND INTRAPHASE DIFFUSION-REACTION (ISOTHERMAL)

Regime of control	Observed		Effect of†	
	Order	Activation E	Particle size	Velocity
Chemical reaction	n	E	None	None
Intraphase diffusion-reaction	$\dfrac{n+1}{2}$	$\dfrac{E}{2}$	$\dfrac{1}{L}$	None
Interphase diffusion	1	≈ 0	$\left(\dfrac{1}{L}\right)^a$	u^b

† Exponents a and b are functions of nature and geometry of reactor system, e.g., fixed bed.

[1] The problem was first resolved iteratively by R. E. Schilson and N. R. Amundson, *Chem. Eng. Sci.*, **13:** 226 (1961).

Unlike interphase effectiveness, which pertains to diffusion-reaction events in series, the intraphase phenomenon involves reaction within the field of concentration *and* temperature gradients; in consequence the rate coefficient, sensitive as it is to temperature, may assume values which more than compensate in terms of rate for the deprivation of reactant due to intraphase mass-diffusion limitation.   Recall again the situation in the totally backmixed CSTR (Chap. 3), in which it was noted that backmixing of heat may indeed predominate relative to reactant dilution effects sponsored by mass backmixing.   Specifically, the benefit of thermal backmixing depends on two dimensionless groups.

$$\bar{\beta} = \frac{-\Delta H\, C_0}{\rho C_p T_0} \qquad \text{dimensionless adiabatic temperature rise}$$

and
$$\epsilon = \frac{E}{RT_0} = \text{Arrhenius number}$$

The nonisothermal intraphase diffusion-reaction situation is described by two continuity equations.   In dimensionless form, for linear kinetics

$$\frac{d^2 f}{dz^2} = \phi_0{}^2 f \exp\left[-\epsilon\left(\frac{1}{t} - 1\right)\right] \tag{5-66}$$

$$\frac{d^2 t}{dz^2} = -\frac{-\Delta H}{\lambda T_s}\, C_s \phi_0{}^2 f \exp\left[-\epsilon\left(\frac{1}{t} - 1\right)\right] \tag{5-67}$$

where $-\Delta H$ = reaction enthalpy change
$\quad\quad \lambda$ = intraphase thermal conductivity
$\quad\quad \mathscr{D}$ = intraphase mass diffusivity
The reduced intraphase temperature t is readily found by eliminating the rate of reaction between Eqs. (5-66) and (5-67), with the result, *valid for any kinetics*[1] *and geometry,*

$$-\frac{d^2 t}{dz^2} = \frac{-\Delta H\, \mathscr{D} C_s}{\lambda T_s}\, \frac{d^2 f}{dz^2} \tag{5-68}$$

Two integrations yield the Prater temperature

$$t = \frac{T}{T_s} = 1 + \frac{-\Delta H\, \mathscr{D}}{\lambda T_s}\, (C_s - C) \tag{5-69}$$

Equation (5-69) defines t at a point within the reaction phase at which the reactant concentration is C.

Nonisothermal intraphase effectiveness, secured by simultaneous numerical solution of Eqs. (5-66) and (5-67), will clearly be a function of three dimensionless

[1] C. D. Prater, *Chem. Eng. Sci.*, 8: 284 (1958).

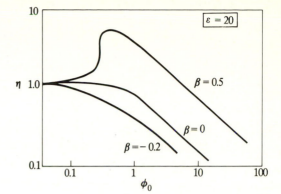

FIGURE 5-12
Qualitative behavior of η versus ϕ under nonisothermal conditions; $\beta = +$: exothermic; $\beta - 0$: isothermic; $\beta = -$: endothermic.

parameters, specifically, η (nonisothermal) $= f(\phi_0, \epsilon, \beta)$, where β is the intraphase dimensionless adiabatic temperature rise; i.e., by Eq. (5-69)

$$\beta = \frac{-\Delta H \, \mathscr{D} C_s}{\lambda T_s} \qquad (5\text{-}70)$$

Solutions of Eqs. (5-66) and (5-67) have been generated for the boundary conditions

$z = 1$ (surface): $\qquad f = 1 \qquad t = 1$

$z = 0$ (centerline): $\qquad \dfrac{df}{dz} = 0 \qquad$ and $\qquad \dfrac{dt}{dz} = 0 \qquad\qquad (5\text{-}71)$

Results are shown schematically in Fig. 5-12 for both endothermic and exothermic reaction and for isothermal first-order reaction. Effectiveness or reaction-phase-efficiency factors greater than unity are found as expected for large positive values of ϵ and β. (Compare with Figs. 5-6 and 5-7.)

5-10 INTERPHASE AND INTRAPHASE NONISOTHERMAL EFFECTIVENESS

An overall effectiveness factor under nonisothermal conditions can be fashioned in terms of interphase and intraphase heat and mass diffusional resistances. We noted above that intraphase nonisothermal effectiveness will be a function of ϕ, ϵ, and β. When interphase mass diffusion is anticipated, another governing dimensionless group becomes relevant, the mass Biot number

$$\text{Bi}_m = \frac{k_g L}{\mathscr{D}}$$

which is generated by equating bulk mass transport to mass diffusive flux across the interface

$$k_g(C_0 - C_s) = -\mathscr{D} \left.\frac{dC_s}{dy}\right|_{y=L}$$

In dimensionless form, $f = C/C_0$, $z = y/L$,

$$k_g(1 - f_s) = -\frac{\mathscr{D}}{L}\frac{df}{dz}\bigg|_{z=1} \qquad (5\text{-}72)$$

or

$$\frac{1 - f_s}{(df/dz)_{z=1}} = \frac{1}{k_g L/\mathscr{D}} = \frac{1}{\text{Bi}_m} \qquad (5\text{-}73)$$

If we now anticipate an interphase temperature gradient, the appropriate boundary condition at the two-phase boundary will be

$$h(T_s - T_0) = -\lambda\frac{dT}{dy}\bigg|_{y=L}$$

In dimensionless form, where $t = T/T_0$,

$$\frac{t - 1}{(dt/dz)_{z=1}} = \frac{1}{hL/\lambda} = \frac{1}{\text{Bi}_h} \qquad (5\text{-}74)$$

The thermal Biot number is

$$\text{Bi}_h = \frac{hL}{\lambda} \qquad (5\text{-}75)$$

Consequently in the general instance of overall effectiveness for surface reaction involving interphase and intraphase diffusional limitations of both heat and mass (and so interphase and intraphase nonisothermality), we conclude that the overall or global effectiveness is dependent upon several dimensionless groups:

$$\eta_0 = f(\phi, \epsilon, \beta, \text{Bi}_m, \text{Bi}_h) \qquad (5\text{-}76)$$

It is worth noting [as suggested by Eqs. (5-73) and (5-74)] that the Biot number expresses the ratio of intraphase to interphase gradients. Thus its magnitude suggests whether the major seat of thermal or mass resistance resides in the fluid film (interphase) or within the reaction phase (intraphase); or possibly the resistances may be distributed between both phases.

Solutions have been generated to yield the nonisothermal interphase and intraphase global effectiveness factor. A schematic representation is set forth in Fig. 5-13. The following points are worthy of note:

1 For exothermic reaction, an interphase heat-transport limitation (small value of Bi_h) increases η_0, while an interphase mass-transport resistance (small Bi_m) decreases η_0.

2 For endothermic reaction, effectiveness is decreased by both heat and mass interphase resistance (small values of Bi_h and Bi_m).

3 Large values of both Biot numbers reduce the η_0-versus-ϕ_0 relationship to that set forth earlier (Fig. 5-12) for nonisothermal intraphase effectiveness.

4 For large values of Biot numbers and small values of ϵ and β, the simple isothermal η-versus-ϕ relationship prevails (Figs. 5-10 and 5-11).

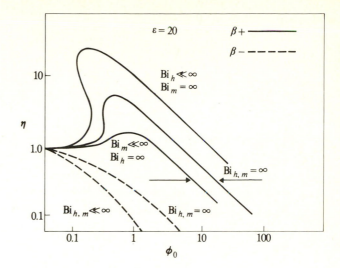

FIGURE 5-13
Schematic of nonisothermal intraphase and interphase effectiveness vs. Thiele-modulus behavior for diverse magnitudes of the mass Bi_m and heat Bi_h Biot numbers ($\pm\beta$).

5-11 PHYSICAL IMPLICATIONS

Since it has been established that global effectiveness is a function of ϕ, ϵ, β, Bi_m, and Bi_h, it seems appropriate here to set forth some estimates of the magnitudes of these parameters for gas-solid, gas-liquid, and liquid-solid systems.

When the focal point of reaction is a solid phase, e.g., porous catalyst, the Thiele modulus for $n = 1$, $\phi = L\sqrt{k/\mathcal{D}}$, is defined such that L is the dimension of catalyst formulation per se; hence if the catalytic agent, say Pt, is uniformly deposited throughout a porous pellet, L is the characteristic dimension of the pellet.

If a fluid phase is the intraphase region of reaction, L is more difficult to specify; generally it is recognized to be equivalent to fluid-film thickness δ. Recall that for liquid-film mass transfer without reaction

$$k_{L_0} = \frac{\mathcal{D}}{\delta} \qquad (5\text{-}77)$$

Letting $L = \delta$ in the Thiele modulus, we have

$$\phi = \delta\sqrt{\frac{k}{\mathcal{D}}} = \frac{\sqrt{k\mathcal{D}}}{k_{L_0}} \qquad (5\text{-}78)$$

The various governing dimensionless groups under discussion contain transport coefficients (λ, $\mathcal{D}$) and physical properties; their magnitudes for gases, liquids, and porous solids are set forth in Table 5-2.

Table 5-2 RANGE OF PHYSICAL PARAMETERS $\mathscr{D}$, λ, AND ρC_p FOR GAS, LIQUID, AND SOLID SYSTEMS†

	Phase		
Variable	**Gas**	**Liquid**	**Solid (porous)**
$\mathscr{D}$	1–0.1	10^{-5}–10^{-6}	10^{-1}–10^{-3}
λ	10^{-4}–10^{-5}	10^{-2}–10^{-4}	10^{-3}–10^{-4}
ρC_p, cal/(cm³)(°C) or cal/(ml)(°C)	10^{-2}–10^{-4}	10^{-1}–2	1–0.4

† Transport properties are pressure-dependent. $\mathscr{D}$ = diffusivity of a reactant within gas, liquid, and porous-solid phase (cm/s²), λ = thermal conductivity of each phase [(cal)/(cm)(s)(°C)], C_p = heat capacity for each phase [cal/(mol)(°C)].

In the light of the range of values set forth in Table 5-2, the spectra of anticipated values of the dimensionless groups which dictate global effectiveness are indicated in Table 5-3. As far as the Biot numbers are concerned, the ratio of the mass to thermal Biot numbers is given, as that ratio, in view of Eqs. (5-73) and (5-74), is indicative of the locale (interphase or intraphase) of the major gradient. Assuming that the analogy between heat and mass transfer applies to the fluid phase, we have

$$\frac{\text{Bi}_m}{\text{Bi}_h} = \frac{\lambda}{\mathscr{D}\rho C_p \, \text{Le}^{2/3}} = r = \bar{\beta}/\beta \qquad (5\text{ }79)$$

Inspection of the values of r likely to be encountered suggests that the internal dimensionless adiabatic temperature change β_i is significantly less than the external value and that in a liquid-solid catalyst system temperature gradients will be less pronounced than in gas-solid systems. The value of r [Eq. (5-79)] suggests that the major concentration gradient will be an intraphase one while the seat of the major temperature gradient may be interphase (external) in gas-solid systems.

Table 5-3 SPECTRUM OF ANTICIPATED VALUES OF ϵ, β, $\bar{\beta}$, AND r FOR GAS-SOLID AND LIQUID-SOLID CATALYZED SYSTEMS

	Gas-solid	Liquid-solid
$\epsilon = E/RT_0$	5–30	5–30
β_i, internal	0.001–0.3	0.001–0.1
β_x, external	0.01–2.0	0.001–0.05
$r = \text{Bi}_m/\text{Bi}_h$	10–10^4	10^{-4}–10^{-1}

5-12   INTERPHASE AND INTRAPHASE TEMPERATURE GRADIENTS[1]

Given the expected range of parameters in Table 5-3 for gas-solid and liquid-solid systems, coupled with prior analyses which are directed toward evaluating the effectiveness or phase-utilization factor under nonisothermal circumstances, an analysis seems justified to ascertain the relative importance of interphase and intraphase nonisothermality; i.e., for a given fluid-fluid, e.g., gas-liquid, or fluid-solid, system can we predict the comparative magnitude of interphase and intraphase temperature gradients?

We seek first an expression for the overall ΔT_0 between the centerline of the reaction phase (liquid droplet or catalyst pellet; see Fig. 5-9) and the surrounding bulk phase T_0, which is readily established in terms of observables. One then develops the ratio of external (interphase) ΔT_x relative to the total as follows.

Internal (Intraphase) ΔT

As developed earlier (Sec. 5-9), the intraphase ΔT_i is given in terms of the Prater number

$$\frac{\Delta T_i}{T_0} = \beta = \frac{-\Delta H \, \mathscr{D}(C_s - C)}{\lambda T_0} \qquad (5\text{-}80)$$

or when $C \to 0$ and $C_s = C_0(1 - \bar{\eta}\mathrm{Da})$,

$$\frac{\Delta T_i}{T_0} = \frac{-\Delta H \, \mathscr{D} C_0}{\lambda T_0}(1 - \bar{\eta}\mathrm{Da}) = \beta(1 - \bar{\eta}\mathrm{Da}) \qquad (5\text{-}81)$$

Recalling that the external (interphase) ΔT is, in terms of the observable, $\bar{\eta}\mathrm{Da}$,

$$\frac{\Delta T_x}{T_0} = \bar{\eta}\mathrm{Da}\bar{\beta}$$

where

$$\bar{\beta} = \frac{-\Delta H \, C_0}{\rho C_p T_0 \, \mathrm{Le}^{2/3}} \qquad (5\text{-}82)$$

then the overall temperature difference ΔT_0 is

$$\frac{\Delta T_0}{T_0} = \bar{\eta}\mathrm{Da}\bar{\beta} + \beta(1 - \bar{\eta}\mathrm{Da}) \qquad (5\text{-}83)$$

Now β is related to $\bar{\beta}$ by

$$\frac{\bar{\beta}}{\beta} = \frac{\lambda}{\mathscr{D}\rho C_p \, \mathrm{Le}^{2/3}} = \frac{\mathrm{Bi}_m}{\mathrm{Bi}_h} = r \qquad (5\text{-}84)$$

or

$$\frac{\Delta T_x}{\Delta T_0} = \frac{r(\bar{\eta}\mathrm{Da})}{1 + \bar{\eta}\mathrm{Da}(r - 1)} \qquad (5\text{-}85)$$

[1] J. J. Carberry and A. Kulkarni, loc. cit.; J. J. Carberry, *Ind. Eng. Chem. Fund.*, **14**: 129 (1975).

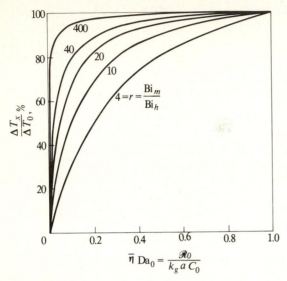

FIGURE 5-14
Ratio of external to total ΔT_0 versus the observable for
fluid-solid catalyzed reactions where r is the parameter.
[*J. J. Carberry, Ind. Eng. Chem. (Fund.)*, **14** (2): 129 (1975).]

This relationship is displayed in Fig. 5-14 for various values of r.

In the light of values of r in Table 5-3, we conclude that:

1   In a gas-solid system the major seat of thermal resistance is likely to be
external (interphase).
2   In a liquid-solid system, the converse situation will prevail.
3   Further, with respect to the overall ΔT_0, values of $\bar{\beta}$ given in Table 5-3 suggest
a small deviation from isothermality for liquid-solid systems, in contrast to
gas-solid systems, in which, by reason of the low ρC_p value for gases, nonisothermality is to be anticipated.

With respect to gas-liquid systems the spectrum of r values is similar to the
liquid-solid case, and the low values of β for the liquid phase prompt the expectation
of isothermality.

In fact, for high rates of intraphase reaction within a liquid droplet, phase
utilization η will become $\eta = 1/\phi$.   Under nonisothermal conditions, it has been
shown that the nonisothermal phase-utilization factor is adequately described for
$\phi > 3$ by[1]

$$\eta_{\text{non}} = \eta_{\text{iso}} \exp \frac{\alpha}{5} = \frac{\exp(\alpha/5)}{\phi_0} \qquad (5\text{-}86)$$

[1] J. J. Carberry, *AIChE J.*, **7**: 350 (1961); *Chem. Eng. Sci.*, **21**: 951 (1966).

where
$$\alpha = \beta\epsilon$$

or
$$\alpha = \frac{-\Delta H \, \mathscr{D} C_0}{\lambda T_0} \frac{E}{R T_0} \qquad (5\text{-}87)$$

As will be illustrated in Chap. 6, α for gas-liquid reactions is negligible, and so nonisothermality can usually be ignored in the gas-liquid reaction zone of diffusion-reaction.

5-13 YIELD IN INTRAPHASE DIFFUSION-REACTION

Signal features of intraphase diffusion influence upon yield or selectivity can be very simply set forth by bearing in mind the fact that the apparent rate coefficient for any reaction step in the interphase-intraphase diffusion-reaction system is $\bar{k} = \eta_{0_i} k_i$. For example, an apparent rate-coefficient ratio is

$$\bar{K} = \frac{\bar{k}_1}{\bar{k}_2} = \frac{\eta_{0_i} k_1}{\eta_{0_2} k_2} = \frac{\eta_{0_i}}{\eta_{0_2}} K \qquad (5\text{-}88)$$

Consecutive Reactions: $A \xrightarrow{\ k_1\ } B \xrightarrow{\ k_2\ } C$

In the total absence of diffusional intrusions, the point yield of B, as we found for homogeneous reaction in Chap. 2, is

$$-\frac{dB}{dA} = 1 - \frac{k_2}{k_1} \frac{B^m}{A^n} = 1 - \frac{1}{K} \frac{B^m}{A^n}$$

Point yield decreases, of course, as K decreases. In the heterogeneous reaction, assuming (reasonably) that diffusivities are equal and that $k_1 > k_2$, then with the manifestation of intraphase diffusion intrusion, $\eta_{0_1} < \eta_{0_2}$ since $\phi_1 > \phi_2$. The apparent rate-coefficient ratio [Eq. (5-88)] will then be less than the intrinsic, diffusion-uninfluenced value. Hence yield taxation of desired species B. If $k_1 \gg k_2$, in the diffusional regime as η_1 assumes values less than unity, η_2 is likely to remain unity and

$$\bar{K} = \eta_{0_1} \frac{k_1}{k_2} \qquad (5\text{-}89)$$

or
$$\bar{K} = \frac{k_1 \tanh \phi}{\phi[1 + (\phi \tanh \phi)/\text{Bi}_m] k_2}$$

For $\phi > 3$, we find for large Bi_m

$$\bar{K} - \frac{k_1}{\phi k_i} = \frac{\sqrt{k_1 \mathscr{D}}}{L k_2} \qquad (5\text{-}90)$$

While for a large Damköhler number (ϕ large, Bi_m small)

$$\bar{K} = \frac{k_g a}{k_2} \qquad (5\text{-}91)$$

When ϕ_1, ϕ_2, and Bi_m are large,

$$\overline{K} = \sqrt{\frac{k_1}{k_2}} \qquad (5\text{-}92)$$

Parallel Reactions

$$A \xrightarrow{k_1} B$$

$$X \xrightarrow{k_2} Y$$

Point selectivity is

$$\frac{dB}{dY} = \frac{\eta_1 k_1 A^n}{\eta_2 k_2 x^m}$$

or

$$S_B = \frac{\eta_1}{\eta_2} \text{ (homogeneous selectivity)}$$

For large Bi_m, in the limit of strong intraphase diffusional limitation and $\mathscr{D}_A = \mathscr{D}_B$

$$S_B = \sqrt{\frac{k_1}{k_2}} \frac{A^{(n+1)/2}}{x^{(m+1)/2}} \qquad (5\text{-}93)$$

Simultaneous Reactions

$$A \xrightarrow{k_1} B \qquad n\text{th order}$$
$$\searrow^{k_2}$$
$$C \qquad m\text{th order}$$

$$S = \text{point selectivity} = \frac{\eta_1 k_1 A_s^n}{\eta_1 k_2 A_s^m} = \frac{k_1}{k_2} A_s^{n-m}$$

For the diffusion-uninfluenced case

$$S_0 = \frac{k_1}{k_2} A_0^{n-m} \qquad \text{or} \qquad \frac{S}{S_0} = \left(\frac{A_s}{A_0}\right)^{n-m}$$

As in interphase diffusion, when $A_s < A_0$ under an intraphase diffusional gradient,

$$\frac{S}{S_0} \begin{cases} = 1 & n = m \\ < 1 & n > m \\ > 1 & n < m \end{cases}$$

Nonisothermal Intraphase Yield

Intraphase temperature gradients affect the rate-coefficient ratio k_1/k_2, and the quadrant diagram (Fig. 5-8) remains a valid guide as to whether the nonisothermal-ities within the reaction phase are to be a benefit or taxation relative to isothermal yield or selectivity. However, unlike interphase transport limitation, intraphase diffusional limitations lead, in the limit, to square-root dependency upon intrinsic coefficients. For example, in Eq. (5-45) the activation-energy difference $\Delta\epsilon$ in the expression (for $\phi > 3$)

$$K = K_0 \exp\left[-\Delta\varepsilon\left(\frac{1}{t} - 1\right)\right]$$

is not $\Delta\epsilon = E_1 - E_2$ but $\Delta\epsilon = E_1/2 - E_2$, and when both reactions become rapid (ϕ_1 and ϕ_2 large), $\Delta\epsilon = (E_1 - E_2)/2$.

Intraphase diffusion-influenced yield or selectivity under nonisothermal conditions may then be governed by activation-energy differences which are themselves diffusionally modified.

We conclude this chapter with a discussion of steady-state multiplicity and stability. These analyses for a heterogeneous reacting system are totally analogous to treatments set forth in Chap. 3 for the CSTR, a consequence readily perceived by comparing the CSTR material balance $(C_0 - C)/\theta = \mathscr{R} = kC^n$ with that for a single particle, where surface reaction occurs in series with bulk mass transport, that is, $k_g a(C_0 - C) = \mathscr{R} = kC^n$. The heat balances bear the same resemblances. Consequently a review of multiplicity and stability in the CSTR analyses (Chap. 3) will enhance comprehension of the following section.

5-14 STEADY-STATE MULTIPLICITY

As noted in Chap. 3 in our consideration of the CSTR, steady state exists when heat generation equals heat removal, in the exothermic example cited there. In a heterogeneous reaction at a point within any reactor, we realize again that local steady state is achieved by a balance of generation and removal (positive or negative, depending upon reaction exothermicity or endothermicity).

Let us consider a single nonporous catalytic solid pellet bathed by a flowing reactant-bearing stream of reactant bulk concentration A_0 and temperature T_0. We presume the reaction to be of a complex, yet linear, nature

$$A \xrightarrow{\ k_1\ } B \xrightarrow{\ k_2\ } C$$
$$\quad \searrow {\scriptstyle k_3}$$
$$\qquad\qquad D$$

Heat generation is

$$q = \left\{-\Delta H_1\, A k_1{}^0 \exp\left[-\epsilon_1\left(\frac{1}{t} - 1\right)\right] + (-\Delta H_3) A k_3{}^0 \exp\left[-\epsilon_3\left(\frac{1}{t} - 1\right)\right]\right.$$
$$\left. + (-\Delta H_2) B k_2{}^0 \exp\left[-\epsilon_2\left(\frac{1}{t} - 1\right)\right]\right\} \qquad \text{cal/(cm}^3\text{)(time)} \qquad (5\text{-}94)$$

Dividing by $\rho C_p T_0 u a$, where $a = 1/L$, the external surface per unit volume, and u is fluid velocity, we get a dimensionless heat-generation function; surface concentrations A and B are found in terms of bulk values by equating film mass transport to surface reaction, as illustrated in Sec. 5-4.

$$A = \frac{A_0}{1 + \text{Da}_1 + \text{Da}_3} \tag{5-95}$$

and

$$B = \frac{\text{Da}_1 A_0}{(1 + \text{Da}_1 + \text{Da}_3)(1 + \text{Da}_2)} + \frac{B_0}{1 + \text{Da}_2} \tag{5-96}$$

where

$$\text{Da} = \frac{k}{k_g a}$$

and

$$k = k^0 \exp\left[-\epsilon\left(\frac{1}{t} - 1\right)\right]$$

The dimensionless heat-generation function becomes for $B_0 = 0$ and $W = 1/t - 1$

$$\frac{q}{\rho u C_p T_0 a} = \frac{\bar{\beta}_1 K_1 \exp(-\epsilon_1 W) + \bar{\beta}_3 K_3 \exp(-\epsilon_3 W)}{1 + \text{Da}_1{}^0 \exp(-\epsilon_1 W) + \text{Da}_3{}^0 \exp(-\epsilon_3 W)}$$

$$+ \frac{K_2 \bar{\beta}_2 \text{Da}_1{}^0 \exp[-(\epsilon_1 + \epsilon_2)W]}{[1 + \text{Da}_1{}^0 \exp(-\epsilon_1 W) + \text{Da}_3{}^0 \exp(-\epsilon_3 W)][1 + \text{Da}_2{}^0 \exp(-\epsilon_2 W)]} \tag{5-97}$$

where

$$K_i = \frac{k_i{}^0 L}{u} \qquad \text{Da}^0 = \frac{k^0}{k_g a} \qquad \bar{\beta} = \frac{-\Delta H A_0}{\rho C_p T_0} \qquad \epsilon = \frac{E}{R T_0}$$

Volumetric heat removal in dimensionless form is

$$\frac{q}{\rho u C_p T_0 a} = \frac{h}{\rho u C_p}(t - 1) = \text{St}(t - 1) \tag{5-98}$$

where St is the Stanton number.

Figures 5-15 to 5-17 set forth typical heat-generation-versus-t profiles for a simple irreversible reaction (Fig. 5-15), a consecutive event (Fig. 5-16), and consecutive endothermic and exothermic reactions (Fig. 5-17).   As in the CSTR analysis

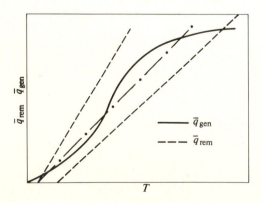

FIGURE 5-15
Heat generation $\bar{q}_{\text{gen}}$ and heat removal $\bar{q}_{\text{rem}}$ versus temperature T for a simple nonporous-surface-catalyzed exothermic reaction A → B.

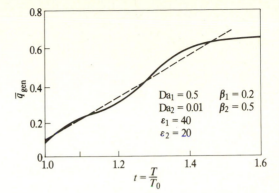

FIGURE 5-16
$\bar{q}_{gen}$ and $\bar{q}_{rem}$ versus t for consecutive
exothermic reaction $A \to B \to C$.

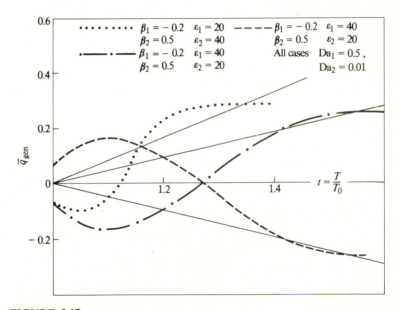

FIGURE 5-17
$\bar{q}_{gen}$ and $\bar{q}_{rem}$ versus t for consecutive exothermic and endothermic catalyzed
reactions $A \overset{1}{\to} B \overset{2}{\to} C$.

of Chap. 3, ignition, extinction, and unstable intersections of the heat removal-
generation functionalities are evident.

5-15   ISOTHERMAL MULTIPLICITY

In Chap. 3 it was noted that in the case of abnormal kinetic behavior, as found in
Pt-catalyzed oxidation of CO, two steady and one unstable state may prevail in an
isothermal CSTR.   Utilizing the general kinetic model for Pt-catalyzed oxidation

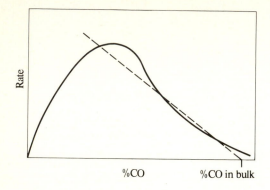

FIGURE 5-18
Isothermal multiplicity in Pt-catalyzed
oxidation of CO to CO_2.

%CO   %CO in bulk

of CO, and equating this rate to external mass transport, we find

$$k_g a(C_0 - C) = \frac{kC}{(1 + kC)^2} \qquad (5\text{-}99)$$

The right-hand side of Eq. (5-99) is plotted in Fig. 5-18, while the left-hand side is simply a straight line of slope $k_g a$, anchored at the bulk concentration C_0 and intersecting the right-hand-side function at two points (Fig. 5-18) corresponding to steady-state solutions and a third midpoint which is unstable.

5-16 STABILITY OF THE LOCALLY CATALYZED REACTION

We have cited the fact that steady-state multiplicity may prevail, as revealed by a plot of heat generation versus removal for a single catalyst pellet (Figs. 5-15 to 5-17). It has been noted that such a generation-removal functionality for the catalyst pellet resembles in qualitative form the behavior of the CSTR with respect to generation and removal ($\pm$). The stability problem for the CSTR was analyzed in Chap. 3, and criteria for stability were set forth for small deviations from the steady state. Therefore it should not be surprising that a stability analysis for the single catalyst particle bathed by a reactant-bearing fluid should resemble, in essence, that formulated for the CSTR.

We envision a single catalyst pellet within which a first-order reaction occurs, the pellet being bathed by a fluid at reactant concentration A_0 and at a bulk temperature T_0. In the steady state we have for mass

$$k_g a(A_0 - A_s) = k_s A_s \qquad (5\text{-}100)$$

and for heat
$$ha(T_0 - T_s) = -\Delta H k_s A_s \qquad (5\text{-}101)$$

where the subscript s refers to the steady-state condition of A and T and therefore k.

The transient relationships for heat and mass transport *in series* with surface first-order reaction are for mass

$$k_g a(A_0 - A) = kA + \frac{dA}{d\theta} \qquad (5\text{-}102)$$

and for heat
$$ha(T_0 - T) = -(-\Delta H)kA + \rho C_p \frac{dT}{d\theta} \quad \text{(5-103)}$$

where T, A, and k without subscript refer to transient values.

As in the analysis of CSTR stability, the term kA will be expressed by a Taylor expansion, then substituted into the transient equations (5-102) and (5-103), from which the steady-state equations (5-100) and (5-101) will be subtracted. The resulting relationships will then be analyzed with respect to their stability. Remember that k and k_s represent effective rate coefficients, and thus, in principle, they reflect whatever intraphase diffusion-affected behavior may prevail in the instance of porous-solid catalysis.

Expanding kA in terms of the Taylor series gives

$$kA = k_s A_s + (A - A_s)k_s + \frac{k_s A_s}{T_s} \frac{E}{RT_s} (T - T_s) \quad \text{(5-104)}$$

which is substituted into the transient equations (5-102) and (5-103), from which the steady-state equations (5-100) and (5-101) are subtracted. The result, expressed in dimensionless form, is

$$-\frac{dt}{d\phi} = -\bar{\beta}y + (H - \bar{\beta}\epsilon)t \quad \text{(5-105)}$$

and
$$-\frac{dy}{d\phi} = \left(1 + \frac{1}{\text{Da}}\right)y - \epsilon t \quad \text{(5-106)}$$

where
$$t = \frac{T - T_s}{T_s} \qquad y = \frac{A - A_s}{A_s} \qquad \phi = k_s \theta$$

and
$$\bar{\beta} = \frac{-\Delta H A_s}{\rho C_p T_s} = \text{adiabatic } \frac{\Delta T}{T_s}$$

$$H = \frac{ha}{\rho C_p k_s} = \frac{\text{heat transferred}}{\text{heat generated}}$$

$$\epsilon = \frac{E}{RT_s} = \text{Arrhenius number}$$

$$\text{Da} = \frac{k_s}{k_g a} = \text{Damköhler number}$$

Equations (5-105) and 5-106) are to be compared with those developed in Chap. 3 for CSTR stability analysis.

Invoking precisely the same method of analysis as set forth for the CSTR (Chap. 3), we readily conclude that stability of the single reacting catalyst pellet is assured when

$$1 + \frac{1}{\text{Da}} + H - \bar{\beta}\epsilon > 0 \quad \text{(5-107)}$$

and
$$\left(1 + \frac{1}{\text{Da}}\right)(H - \bar{\beta}\epsilon) + \epsilon\bar{\beta} > 0 \quad \text{(5-108)}$$

It is not fortuitous that CSTR analyses with respect to multiplicity of the steady state and stability per se bear an amazing resemblance to comparable problems associated with the single catalyst pellet bathed by a reactant-bearing fluid. Aris[1] has revealed this equivalence in its most universal scope.

Inspection of the criteria for stability as defined by Eqs. (5-107) and (5-108) leads us to conclude that the key dimensionless parameter is $\epsilon\bar{\beta}$, which, while of potential significance in a gas-phase–solid catalytic system, is surely of small magnitude in a liquid-phase–solid catalyst network. We must thus be ever alert to instabilities in a gas-solid system and, in general, anticipate stability in liquid-solid reacting systems. Our generalization, like all generalizations, is subject to exception.

SUMMARY

The extent to which a mass- and/or heat-transport resistance affects conversion and yield in a heterogeneous reaction system has been demonstrated by utilizing, first, the simple example of external diffusion in series with nonporous-surface reactions. Arithmetically negotiable functions emerge, which, while rooted in the primitive film theory of diffusional transport, nevertheless reveal behavior in qualitative accord with the more complex internal diffusion-reaction events. We learn that:

1 Under isothermal conditions, external mass-transport intrusions tax the surface reaction in proportion to its intrinsic order.
2 This taxation can be amplified or nullified under nonisothermal conditions.
3 The general external-effectiveness functionality is expressed neatly in terms of the observed global rate, local bulk concentration, and interphase mass-transport coefficient by

$$\bar{\eta} = (1 - \bar{\eta}\mathrm{Da})^n \exp\left[-\epsilon_0 \left(\frac{1}{1 + \bar{\beta}\bar{\eta}\mathrm{Da}} - 1\right)\right]$$

where n is reaction order, $\leqq 0$ and

$$\bar{\eta}\mathrm{Da} = \frac{\mathcal{R}_0}{k_g a C_0} \qquad \bar{\beta} = \frac{-\Delta H\, C_0}{\rho C_p T_0\, \mathrm{Le}^{2/3}} \qquad \epsilon_0 = \frac{E}{RT_0}$$

4 Yield in a complex surface-reaction network can be affected positively or negatively by external mass- and/or heat-diffusional intrusions, the effects depending upon the orders of various steps, activation-energy differences, and thermal parameters.
5 The internal (intraphase) diffusion-reaction behavior, although resolved with more mathematical difficulty, mirrors that of the interphase situation with respect to conversion and yield. Internal effectiveness, phase-utilization effectiveness, or phase-utilization efficiency can be phrased in terms of an observable

$$\Phi = \frac{\mathcal{R}_0 L^2}{\mathcal{D} C_0} = \eta\phi^2$$

[1] R. Aris, "The Mathematical Theory of Diffusion and Reaction in Permeable Catalysts," vol. 2, Clarendon, Oxford, 1975; see also *Chem. Eng. Sci.*, **24**: 149 (1969).

which is related to the external observable $\bar{\eta}$Da by the mass Biot number, $\eta\phi^2/\text{Bi}_m = \bar{\eta}\text{Da}$.

6 In the general instance of interphase and intraphase heat- and mass-transport–affected reaction, effectiveness and (in complex networks) yield are dictated by the Thiele moduli, $\phi_1, \phi_2, \ldots$, activational energies, $\epsilon_1, \epsilon_2, \ldots$, the Prater numbers, $\beta_1, \beta_2, \ldots$, the Biot numbers for heat and mass transport, and point bulk composition and temperature C_0 and T_0.

7 The magnitudes of the Biot numbers give an indication of the relative magnitudes of internal to external concentration and temperature gradients.

8 The intervention of interphase and/or intraphase gradients alters apparent or global kinetic parameters so that for an intrinsic nth-order reaction the apparent, measured order becomes, for internal diffusion, $n' = (n + 1)/2$, while the apparent activation energy becomes $E' = E/2$. External mass gradients, when governing, lead to linear kinetics and a near-zero activation energy. Unsuspected temperature gradients will of course render measured activational energies meaningless.

9 An examination of the range of thermal-transport kinetic parameters suggests that nonisothermality can be anticipated in gas-solid systems while such a threat is less severe in liquid-solid and gas-liquid reactions.

10 As indicated in Chap. 3 for the CSTR, a multiplicity of steady states and unstable states can prevail in a heterogeneous reaction, as illustrated in this chapter for a gas-solid reaction.

ADDITIONAL REFERENCES

An analysis of external diffusion with surface reaction in terms of boundary-layer flow is provided by D. E. Rosner, *AIChE J.*, **9**: 321 (1963); whereas selectivity alteration due to diffusional limitations in boundary-layer flow is analyzed by P. L. Chambre and A. Acrivos, *J. Appl. Phys.*, **27**: 1322 (1956); *Ind. Eng. Chem.*, **49**: 1025 (1957). A fine discussion of surface reaction with external boundary layers is to be found in E. E. Petersen, "Chemical Reaction Analysis," Prentice-Hall, Englewood Cliffs, N.J., 1965, which also sets forth valuable asymptotic solutions and criteria for diverse reaction-diffusion situations. D. Loffler and L. Schmidt, *AIChE J.*, **21**: 786 (1975), analyze the interesting cases of mass-transport-affected activity and selectivity on surfaces characterized by active and inactive regions.

M. C. Mercer and R. Aris, *Lat. Am. J. Chem. Eng. Appl. Chem.*, **2**: 149 (1971), provide a survey of realistic parameter values encountered in interphase and intraphase diffusion and reaction. Finally, the influence of homogeneous chemical reaction upon heat transfer is elucidated by D. M. Mason et al., *Chem. Eng. Sci.*, **26**: 1689 (1971).

PROBLEMS

5-1 Fashion a list of industrially important heterogeneous reactions in each of the categories cited in Sec. 5-1.

5-2 Express the j-factor correlations derived in Sec. 5-3 in the form of Nusselt (heat) and Sherwood (mass) numbers. Recall that $\text{Nu} = hL/\lambda$ and $\text{Sh} = k_g L/\mathscr{D}$.

5-3 Derive Eq. (5-23) for $n = \frac{1}{2}$ and 2.

5-4   For negative first-order kinetics we note that no solution of the $\bar{\eta}$-versus-Da relation exists for $Da > 0.25$.   Discuss this situation.

5-5   For a gas–nonporous-solid-catalyzed reaction, e.g., NH_3 oxidation over Pt-wire gauze, how would you obtain the observable, $\bar{\eta}Da$?

5-6   What is the isothermal and nonisothermal effectiveness factor for zero-order surface reaction?   Comment upon the meaning of a diffusion-affected zero-order reaction.

5-7   A supported Ni catalyst (56% Ni on kieselguhr) is utilized to catalyze the following reactions:

(a)   Vapor-phase hydrogenation of benzene at 125°C in excess H_2.   The Reynolds numbers at various contact times in a fixed bed are 2, 20, and 200.   The pellets are 3 mm in diameter, and the intraphase diffusivity is 5×10^{-2} cm²/s.

(b)   Oxidation of CO in excess O_2 at 250°C at the same Reynolds numbers as above. Intraphase diffusivity is 8×10^{-2} cm²/s.

For each system compute the fraction of overall interphase-intraphase ΔT which is in the external gas film at $\bar{\eta}Da = 0.2, 0.5$, and 0.7. $\lambda = 10^{-3}$ cal/(°C)(cm)(s).

5-8   Smith ("Chemical Engineering Kinetics," 2d ed.) presents the following observed rate data for Pt-catalyzed oxidation of SO_2 at 490°C obtained in a differential fixed-bed reactor at atmospheric pressure and bulk density of 0.8 g/cm³.

Mass velocity, g/(h)(cm²)	Bulk partial pressure, atm			Observed rate, g mol SO_2/(h)(g cat)
	SO_2	SO_3	O_2	
251	0.06	.0067	0.2	0.1346
171	0.06	.0067	0.2	0.1278
119	0.06	.0067	0.2	0.1215
72	0.06	.0067	0.2	0.0956

The catalyst pellets were 3.2 by 3.2 mm cylinders, and the Pt was superficially deposited upon the external surface.   Compute both external mass and temperature gradients.   If the reaction activation energy is 30 kcal, what error in rate measurement attends neglect of an external ΔT?   What error prevails if, assuming linear kinetics in SO_2, external concentration gradients are ignored?

5-9   Those who labor in the reaction-diffusion area are fond not only of zero-order kinetics but also of isothermal linear kinetics in a flat-plate geometry.   Join the club by deriving the intraphase η relation for the flat-plate situation.

5-10   Repeat Prob. 5-9 for overall (internal-external) effectiveness; i.e., derive Eq. (5-63).

5-11   How important is the mass Biot number in Prob. 5-10 with respect to its influence upon Eq. (5-63) for values of ϕ of 0.1, 1, 5, and 10?

5-12   In the synthesis of HNO_3 from NH_3 and air, the 10% NH_3–air mixture is passed over a Pt-wire gauze (80 mesh) of 30 layers depth.   Assuming the reaction to be totally controlled by gas-phase mass transfer, compute the NH_3 conversion.   The j factor of Satterfield-Cortez for gauzes is

$$j = 0.94 \, Re^{-0.7} \qquad \frac{d_w u}{\nu \epsilon} = Re$$

and contact time is for n layers

$$\theta \approx \frac{d_w \epsilon}{u} n$$

where $\epsilon = 0.5$ is the void fraction and d_w is the wire diameter. Assume (a) plug flow and (b) CSTR behavior. Finally for a bulk gas temperature of 800°C, calculate the surface temperature at the point of 50 percent conversion. The Reynolds number $d_w u/\nu$ is 30. Total pressure is 7 atm. The reaction is $2NH_3 + \frac{5}{2}O_2 \rightarrow 2NO + 3H_2O$.

5-13 What assumptions are implicit in Eqs. (5-6) and (5-7)?

5-14 Referring to Sec. 5-4, what is the maximum surface rate? The maximum possible global rate?

5-15 In Sec. 5-4, isothermal effectiveness $\bar{\eta}$ is derived for a negative first-order reaction. Derive the relationship for conversion in a CSTR in which a negative-order reaction occurs. Comment on the two systems and the range of real solution.

5-16 In Sec. 5-13 intraphase diffusion-reaction is discussed for consecutive reaction. Derive an expression for the yield of B, that is, B/A_0 as a function of A/A_0 for (a) linear reactions with intraphase diffusion, (b) the same but for finite external mass transfer. (c) Plot B/A_0 versus conversion of A for the following parameters, where the intrinsic coefficient ratio $k_1/k_2 = 16$.

Case	ϕ_1	Bi_m
1	0.01	1000
2	5	1000
3	5	10

5-17 Sherwood and Maak [*Ind. Eng. Chem. Fundam.*, **1**(2): 111 (1962)] report rate data on the reaction of ammonia with carbon at elevated temperatures. From their data, compute the external-effectiveness factor under nonisothermal conditions for both HCN formation and ammonia dissociation.

6

GAS-LIQUID AND LIQUID-LIQUID REACTION SYSTEMS

"Grown men, worried about bubbles!"
Waldo Bushnell Hoffman

Introduction

In this chapter an important class of reaction will be considered, namely, that involving a gaseous or immiscible liquid reactant which must be exposed to a coreactant contained in a liquid-phase medium. As the coreactant must enter the other phase in order to meet and react with the dissolved reactant, the problem before us is one of analyzing diffusion with chemical reaction. Gas-liquid reaction will be treated first.

Simple film theory has been reviewed in Chap. 5. The design equation for physical mass transfer will now be derived in terms of local and overall continuity equations. The influence of liquid-phase chemical reaction upon physical mass transfer in the liquid phase will then be developed. A consequence of this development is the specification of key regimes of mass-diffusion influence in terms of observable quantities. Models more sophisticated than the simple film postulate will also be cited to demonstrate that these more realistic models yield results which differ only negligibly from predictions rooted in the admittedly over-simplified film model. In consequence the prediction of reaction-influenced

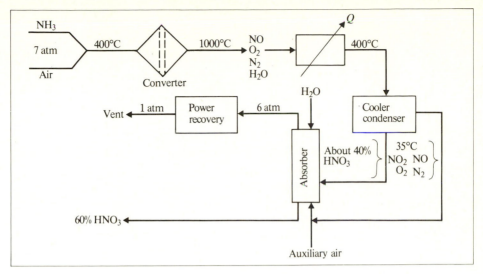

FIGURE 6-1
Simplified flowsheet of ammonia-oxidation process for HNO_3 manufacture.

absorption coefficients relative to pure physical-absorption values is readily realized in terms of simple film theory.

Following the development of continuity equations which govern diverse multiphase reactor types, some examples involving application of theory and extrapolation of laboratory rate data are presented to illustrate both the powers and limitations of available models.   Reaction-diffusion in immiscible liquid-liquid systems is treated with a brevity necessitated by the paucity of theory and data in this area.   Finally some interesting aspects of selectivity in immiscible fluid-fluid reaction systems are noted.

6-1   GAS-LIQUID REACTIONS

Consideration of this topic is justified in view of the vast number of industrially prominent systems which involve gas absorption with simultaneous reaction in the liquid phase.   Nitric acid, for example, is produced in the following manner (see Fig. 6-1):

1   Catalytic oxidation of NH_3 in air to produce nitric oxide, NO; water; N_2; and unreacted O_2 at about 1000°C.
2   Cooling of the gases with subsequent condensation of water.
3   With step 2, NO is oxidized to NO_2 (N_2O_4), which is then partially absorbed and reacts with condensed water to produce a weak nitric acid.
4   The major portion of NO, NO_2, and N_2O_4 which is not absorbed and reacted in the cooler-condenser (steps 2 and 3) is enriched with air and then contacted countercurrently in a bubble-cap absorption tower, where these

oxides of nitrogen are absorbed and reacted to yield a 60 percent by weight HNO_3 product.   A simplified process flowsheet is shown in Fig. 6-1.

The ammonia-oxidation process illustrates a case of absorption with reaction

$$1.5N_2O_4 + H_2O \rightleftharpoons 2HNO_3 + NO(g)$$

in which the objective is creation of a desired product, HNO_3.   Absorption with reaction is often undertaken to remove an undesirable species from the gas stream. Here the objective is purification.   For example, NH_3, a reactant in the ammonia-oxidation process, is synthesized over a promoted iron catalyst from a N_2-H_2 stream.   Since CO_2 is a catalyst poison, it must be removed from the ammonia-synthesis feed stream.   Absorption with chemical reaction is usually employed to purify the N_2-H_2 stream.   Specifically, the CO_2 content in the N_2-H_2 stream is reduced by contacting the gas with various amine solutions or with alkali or bicarbonate solutions.   Reaction occurs in the liquid phase between the absorbed CO_2 and the liquid-phase coreactant.

6-2   PHYSICAL ABSORPTION

Logically a treatment of absorption with reaction presupposes a familiarity with physical absorption per se.   Since the design procedure for absorption with reaction is identical to that employed in sizing physical-absorption apparatus, these principles will be outlined.   A countercurrent gas-liquid packed absorber will be considered.   This is but one type of gas-liquid contactor, as shown in Fig. 6-2.

Physical-Absorber Relations

We consider a large column packed with ceramic rings designed to provide intimate contact between an absorbing liquid and a gas containing a component being absorbed by the liquid (see Fig. 6-3).

Now we apply the continuity equation to gas and then liquid film (B).

Steady state   None

Gas phase: $\dfrac{dC_A}{dt} + \nabla N_A = \text{generation}$

Therefore $N_A = \text{const}$ (6-1)

By definition

$$N_A = \text{const} = Y_A(N_A + N_B) + J_A \qquad (6\text{-}2)$$

$$N_B = 0$$

then $$N_A = \frac{J_A}{1 - Y_A} = \frac{-C_T \mathscr{D}}{1 - Y_A} \frac{dY_A}{dx} = \frac{-C_T \mathscr{D}}{\delta} \ln \frac{Y_B}{Y_{B_i}}$$

Evaluate Y_B average and obtain

$$N_A = \frac{C_T \mathscr{D}}{(Y_B)_{lm} \delta} (Y_A - Y_{Ai}) = k_g(Y_A - Y_{Ai}) \qquad (6\text{-}3)$$

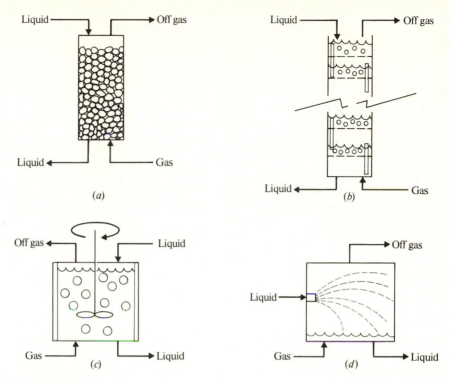

FIGURE 6-2
Diverse gas-liquid contactors: (a) packed column, (b) multistaged tray contactor, (c) stirred tank, and (d) spray chamber.

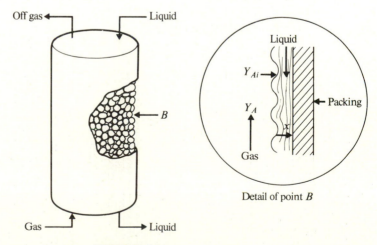

FIGURE 6-3
Overall and detailed view of a packed-column absorber.

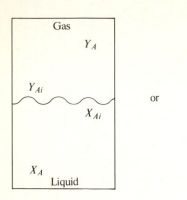
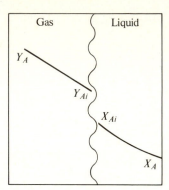

FIGURE 6-4
Physical-absorption situation.

We treat the liquid film in same fashion and obtain (X_{Ai} = mole fraction of A in liquid at interface)

$$N_A = -k_{L_0}(X_A - X_{Ai}) \qquad (6\text{-}4)$$

An overall transfer coefficient, K_0 and a driving force expressed in terms of distance from equilibrium must be fashioned:

$$N_A = K_0(Y_A^* - Y_A) \qquad (6\text{-}5)$$

For example, consider a gas A exposed to liquid containing some dissolved A at the moment of exposure, that is, $t \to 0$ (see Fig. 6-4). The usual relationship between Y_{Ai} and X_{Ai} is a simple linear one (Henry's law); i.e.,

$$Y_{Ai} = MX_i \qquad (6\text{-}6)$$

As time $t \to \infty$, bulk gas and liquid come to complete equilibrium

$$Y_A^* = MX_A \qquad (6\text{-}7)$$

At any point in our system, during transfer, the rate of transfer from gas phase must equal that into liquid phase. From Eqs. (6-3) and (6-4)

$$k_g(Y_A - Y_{Ai}) = -k_{L_0}(X_A - X_{Ai})$$

or

$$\frac{Y_A - Y_{Ai}}{X_A - X_{Ai}} = \frac{-k_{L_0}}{k_g} \qquad (6\text{-}8)$$

We now evaluate a driving force, $Y_A^* - Y_A$:

$$Y_A^* - Y_A \equiv (Y_A^* - Y_{Ai}) - (Y_A - Y_{Ai}) \qquad \text{an equality}$$

By Eqs. (6-6) and (6-7)

$$Y_A^* - Y_A \equiv MX_A - MX_{Ai} - (Y_A - Y_{Ai}) = M(X_A - X_{Ai}) - (Y_A - Y_{Ai})$$

By Eq. (6-8)

$$Y_A^* - Y_A = -M \frac{k_g}{k_{L_0}}(Y_A - Y_{Ai}) - (Y_A - Y_{Ai})$$

or

$$-(Y_A^* - Y) = (Y_A - Y_{Ai})\left(M \frac{k_g}{k_{L_0}} + 1\right)$$

Since

$$N_A = k_g(Y_A - Y_{Ai}) = \frac{k_g(Y - Y_A^*)}{1 + Mk_g/k_{L_0}} = K_{OG}(Y_A - Y_A^*) \qquad (6\text{-}9)$$

the overall mass-transfer coefficient based on gas-phase driving force is

$$K_{OG} = \frac{1}{1/k_g + M/k_{L_0}} \qquad (6\text{-}10)$$

or a mass-transfer expression for *a point* in our exchange tower. To determine tower dimensions we must solve the continuity equation for the tower. For a cylindrical tower we have, for the gas phase,

$$\frac{\partial C}{\partial t} + \frac{\partial N_{Ah}}{\partial h} + \frac{1}{r} \frac{\partial r\, N_{Ar}}{\partial r} = G \qquad (6\text{-}11)$$

G is generation into or out of the gas phase. As A is leaving the gas to enter the liquid, $G = -k_g a(Y_A - Y_{Ai})$, where a is surface area per volume, i.e., interfacial exchange area per volume of tower:

$$N_{Ah} = u_h C_A + J_{Ah} \qquad N_{Ar} = u_r C_A + J_{Ar} \qquad \text{and} \qquad C_T Y_A = C_A$$

Therefore

$$C_T \frac{\partial Y_A u}{\partial h} - \mathscr{D}_z \frac{\partial^2 C_A}{\partial h^2} + \cancelto{0}{u_r \frac{C_T \partial Y_A}{\partial r}} - \frac{\mathscr{D}_r \, \partial r \, \partial C_A/\partial r}{r\, dr} = G$$

$$C_T \frac{\partial Y_A u}{\partial h} = C_T \mathscr{D}_r\left(\cancelto{0}{\frac{\partial^2 Y_A}{\partial r^2} + \frac{1}{r}\frac{\partial Y_A}{\partial r}}\right) + \cancelto{0}{C_T \mathscr{D}_z \frac{\partial^2 Y_A}{\partial h^2}} + G \qquad (6\text{-}12)$$

Radial gradient　　　Axial diffusion

Assume that u is uniform across the tower, that is, u is not $f(r)$; but since u can change as material is absorbed from gas, $u = u_0/1 - Y_A$, where u_0 is velocity of carrier gas, free of A. Thus

$$C_T u_0 \frac{dY/(1-Y)}{dh} = -k_{gA}(Y_A - Y_{Ai}) \qquad (6\text{-}13)$$

We integrate (6-13) to get h, the height of the tower (cross-sectional area is fixed by u for given volumetric flow rate):

$$h = -\int_{Y_{in}}^{Y_{out}} \frac{C_T u_0 \, dY}{k_g a(Y - Y_i)(1-Y)^2} \qquad (6\text{-}14)$$

Now $G_M = C_T u_0$ is in moles per area $\times$ time and $k_g a(Y - Y_i) = K_{OG} a(Y - Y^*)$, so that

$$h = \int_{Y_{in}}^{Y_{out}} \left(\frac{G_M}{K_{OG} a}\right) \frac{dY}{(Y^* - Y)(1 - Y)^2} \qquad (6\text{-}15)$$

to be evaluated graphically. But if $G_M/K_{OG} a = \text{constant} = \text{height of transfer unit (HTU)}$ and dilute solution exists, $1 - Y \to 1.0$,

$$h = \text{HTU} \underbrace{\int_{Y_1}^{Y_2} \frac{dy}{Y^* - Y}}_{\text{NTU}} = \text{HTU} \ln \frac{Y^* - Y_{out}}{Y^* - Y_{in}}$$

$$= (\text{HTU})(\text{NTU}) \qquad \text{if } Y^* - Y \text{ is linear in } Y$$

where NTU is number of transfer units, or

$$h = \frac{\text{HTU}(Y_{in} - Y_{out})}{(Y^* - Y)_{lm}} \qquad (6\text{-}16)$$

Note that in the above derivation we have assumed a total absence of radial gradients of mass and momentum, negligible axial mixing of gas and liquid phases, and isothermal conditions. In the discussion of absorption with reaction which follows, isothermality will be assumed, and in all design illustrations (unless otherwise stated) the intrusions of radial gradients and finite axial mixing will be ignored. In packed columns the axial-mixing effects are small for large length-to-packing-diameter ratios, and uniformity of flow across the bed is generally assumed, though not proved. On the other hand, plate columns require careful attention: a liquid-phase concentration gradient often exists across large-diameter plates. Stirred-tank gas-liquid and liquid-liquid contactors may not be perfectly mixed and if the reactions involved are nonlinear, one must anticipate some segregated flow, particularly in reactions between immiscible liquid phases in stirred vessels. The influence of reaction in the liquid phase upon k_{L_0} will now be examined.

6-3 GAS-LIQUID REACTION MODELS[1]

When physical absorption of a gas into a liquid is accompanied by simultaneous reaction in the liquid phase, an enhancement of the absorption rate may occur.[2] The chemical reaction can cause the species gradient in the liquid phase to steepen. Since the mass flux is

$$N_A = -\mathscr{D} \frac{dC_A}{dx}\bigg|_{x=0}$$

[1] J. Bridgwater and J. J. Carberry, *Br. Chem. Eng.*, **12** (1): 58; (2): 217 (1967).
[2] If, as we shall see, chemical reaction is extremely slow relative to physical mass transfer, the overall conversion process can be slower than simple physical absorption; i.e., the reaction phase is saturated with the absorbing gas.

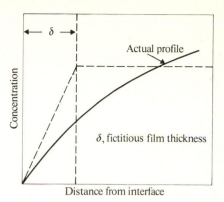

FIGURE 6-5
Actual and equivalent (fictional) transport gradients.

the increase in dC/dx leads to an increase in the mass-transfer flux N_A. What we seek in this section is a theoretical foundation which will permit prediction of the mass-transfer enhancement due to chemical reaction. Since several models of physical mass transfer exist, it follows that an equal number of models of physical absorption with reaction can be devised.

Film Theory

This model essentially attributes all resistance to heat and mass transfer to a thin film at the transfer boundary, as in Fig. 6-5. The concept is hardly novel, having been previously known as the diffusion layer, attributed to Noyes and Whitney and Nernst. Two such films in series, e.g., gas-liquid interface, constitute a two-film situation permitting the specification of an overall resistance in terms of each resistance.

As the film is considered a stagnant layer within which heat and/or mass gradients exist, a simple description is possible [see Eq. (6-1)]

$$\mathcal{D}\frac{d^2C}{dx^2} = 0 \qquad (6\text{-}17)$$

That is, simple diffusion in steady state is assumed to apply. Hydrodynamic gradients are ignored, and solution of the above equation yields

$$N_A = \frac{\mathcal{D}}{\delta}(C_0 - C_1) = k_{L_0}(C_0 - C_1) \qquad (6\text{-}18)$$

Hence the physical mass-transfer coefficient is, by definition,

$$k_{L_0} = \frac{\mathcal{D}}{\delta} \qquad (6\text{-}19)$$

Film theory therefore predicts that k_{L_0} varies linearly with species diffusivity, a result not verified experimentally. The thickness δ is beyond measurement, as it is essentially fictional. Qualitatively δ should decrease with increasing Reynolds number. We now seek k_L, the coefficient affected by chemical reaction; i.e., what is k_L/k_{L_0}?

Diffusion and Reaction in a Film

Assume the reaction between dissolved substances in the liquid phase: $A + zB \rightarrow P$. The film "thickness" is δ and a is interfacial area per volume of apparatus; let V be the total liquid volume per volume of apparatus. The bulk-to-film volume is then $V/a\delta = \alpha$. For diffusion with simultaneous second-order reaction we have

$$\mathscr{D}_A \frac{d^2 A}{dx^2} = kAB \qquad \mathscr{D}_B \frac{d^2 B}{dx^2} = kABz \qquad (6\text{-}20)$$

with the boundary conditions

$$x = 0 \qquad A = A_0$$

$$\frac{dB}{dx} = 0 \qquad \text{B nonvolatile}$$

$$x = \delta \qquad B = \bar{B} \qquad \text{bulk concentration of B} \qquad (6\text{-}21)$$

$$-a\mathscr{D}_A \frac{dA}{dx} = k\bar{B}A(V - a\delta)$$

This last boundary condition simply states that whatever A passes from the film into the bulk phase, it reacts there with B; that is, the diffusive flux equals the rate of reaction in the bulk phase. This bulk phase is assumed to be well mixed. From the boundary condition it is evident that two extreme conditions will prevail: (1) if the reaction-rate constant is large relative to diffusion, A will be consumed largely in the film and the bulk phase will never be utilized or (2) a slow rate of chemical reaction relative to diffusion will cause the locus of reaction to be shifted from the film to within the body (bulk) of the liquid.

Pseudo-First-Order Reaction and Diffusion

Let $f = A/A_0$, $y = x/\delta$, and $B = \bar{B}$, a constant. Then (6-20) and (6-21) become

$$\frac{d^2 f}{dy^2} = \delta^2 \frac{k\bar{B}}{\mathscr{D}} f \qquad \left| \begin{array}{l} y = 0, f = 1 \\[2mm] y = 1, \ -\dfrac{df}{dy} = \delta \dfrac{k\bar{B}f}{aD_A}(V - a\delta) \end{array} \right. \qquad (6\text{-}22)$$

Define

$$\phi = \delta \sqrt{\frac{k\bar{B}}{\mathscr{D}}} = \frac{\sqrt{k\bar{B}\mathscr{D}}}{k_{L_0}} \qquad \text{since } k_{L_0} = \frac{\mathscr{D}}{\delta} \text{ by Eq. (6-19)}$$

Thus

$$\phi = \frac{\text{rate of reaction with simultaneous diffusion}}{\text{rate of physical mass transfer}}$$

Then $d^2 f/dy^2 = \phi^2 f$ and

At $y = 0$: $\quad f = 1$

At $y = 1$: $\quad -\dfrac{df}{dy} = \phi^2 f(\alpha - 1) \qquad \alpha = \dfrac{V}{a\delta}$ $\hspace{2cm}$ (6-23)

k_{L_0} is the liquid-phase mass-transfer coefficient in the absence of reaction. Note that α can be expressed in terms of this transfer coefficient as

$$\alpha = \frac{V}{a\delta} = \frac{V}{a\mathscr{D}} k_{L_0} = \frac{V}{aL} \text{Sh}$$

where Sh = Sherwood number = $k_{L_0} L/\mathscr{D}$ and L is some characteristic dimension of the apparatus, e.g., column diameter or drop size. This specification of α is due to Kramers and Westerterp.[1]

The general solution of the basic differential equation is

$$f = C_1 e^{\phi y} + C_2 e^{-\phi y}$$

Since at $y = 0$, $f = 1$, we have $C_1 = 1 - C_2$. Applying the second boundary condition gives

$$C_2 = \frac{[\phi(\alpha - 1) + 1]e^{\phi}}{(e^{\phi} + e^{-\phi}) + \phi(\alpha - 1)(e^{\phi} - e^{-\phi})}$$

Since $2 \sinh \phi = e^{\phi} - e^{-\phi}$ and $2 \cosh \phi = e^{\phi} + e^{-\phi}$, after some minor manipulation we obtain

$$f = \frac{\cosh \phi(1 - y) + \phi(\alpha - 1) \sinh \phi(1 - y)}{\cosh \phi + \phi(\alpha - 1) \sinh \phi} \hspace{2cm} (6-24)$$

We want the

$$\text{Rate per unit area (flux)} = -\mathscr{D} \left. \frac{dA}{dx} \right|_{x=0} = -\frac{\mathscr{D} A_0}{\delta} \left(\frac{df}{dy} \right)_{y=0}$$

$$-\frac{\mathscr{D} A_0}{\delta} \left. \frac{df}{dy} \right|_{y=0} = \frac{\mathscr{D} A_0}{\delta} \frac{\phi[\phi(\alpha - 1) + \tanh \phi]}{(\alpha - 1)\phi \tanh \phi + 1} = J = k_L A_0$$

where k_L is the liquid-phase mass-transfer coefficient in the presence of chemical reaction. As $\mathscr{D}/\delta = k_{L_0}$, the enhancement factor E' is

$$E' = \frac{J}{\mathscr{D} A_0/\delta} = \frac{k_L}{k_{L_0}} = \phi \frac{\phi(\alpha - 1) + \tanh \phi}{(\alpha - 1)\phi \tanh \phi + 1} \hspace{2cm} (6-25)$$

The diffusive flux J can also be expressed relative to the bulk-phase homogeneous-reaction rate $\mathscr{R} = k\bar{B}A_0 V/a$

$$\frac{J}{\mathscr{R}} = \frac{\mathscr{D}\phi}{\delta k \bar{B} V/a} \frac{\phi(\alpha - 1) + \tanh \phi}{(\alpha - 1)\phi \tanh \phi + 1}$$

or $\qquad \dfrac{J}{\mathscr{R}} = \dfrac{1[\phi(\alpha - 1) + \tanh \phi]}{\alpha\phi[(\alpha - 1)\phi \tanh \phi + 1]} = \eta = \text{phase utilization}$ $\hspace{1cm}$ (6-26)

---

[1] H. Kramers and K. R. Westerterp, "Elements of Chemical Reactor Design and Operation," Academic, New York, 1963.

Comparing (6-25) and (6-26), we note that

$$\frac{J}{\mathscr{R}}\alpha = \frac{k_L/k_{L_0}}{\phi^2}$$

Kramers and Westerterp[1] interpret $J/\mathscr{R}$ as a measure of degree of utilization of the entire liquid phase. When $J = \mathscr{R}$, reaction occurs throughout the entire liquid volume, while under severe diffusion limitations, $J \ll \mathscr{R}$ and utilization is low, the reaction being confined to within the film[2] ($\eta < 1$).

Equations (6-25) and (6-26) simply indicate that the effect of chemical reaction upon liquid-phase absorption coefficient is governed by the diffusion-reaction modulus ϕ, expressing the ratio of diffusion-influenced reaction in the film to physical mass transfer, and α, the ratio of bulk to film volume. A number of important limiting situations can be deduced from (6-25) and/or (6-26). (Recall that for small values of the argument ϕ, $\tanh \phi \to \phi$, while for large values of ϕ, $\tanh \phi \to 1$.)

CASE 1 When reaction-velocity constant $k\bar{B}$ is large, $\tanh \phi \to 1$ and, by (6-25), $k_L/k_{L_0} = \phi$ and, by (6-26), $J/\mathscr{R} = 1/\phi$; thus $k_L = \sqrt{k\bar{B}\mathscr{D}}$ and $J = \sqrt{k\bar{B}\mathscr{D}} A_0$, and the fraction of total liquid phase which is utilized is small (large ϕ), reaction being confined to the film; $\eta < 1.0$

CASE 2 When the reaction-velocity constant is nonzero but small, $\tanh \phi \to \phi$ and

$$\frac{k_L}{k_{L_0}} = \frac{\alpha\phi^2}{\alpha\phi^2 - \phi^2 + 1} \qquad (6\text{-}27)$$

$$\frac{J}{\mathscr{R}} = \frac{1}{\alpha\phi^2 - \phi^2 + 1} \qquad (6\text{-}28)$$

The behavior of (6-27) and (6-28) depends not only on ϕ but also on α. We note that

$$\phi^2 = \frac{k\bar{B}\mathscr{D}}{k_{L_0}{}^2} = \left(\frac{\text{reaction-diffusion rate in film}}{\text{physical mass-transfer rate}}\right)^2 \qquad (6\text{-}29)$$

$$\alpha\phi^2 = \frac{k\bar{B}}{k_{L_0}}\frac{V}{a} = \frac{\text{bulk chemical-reaction rate}}{\text{physical mass-transfer rate}} \qquad (6\text{-}30)$$

While the product $k\bar{B}\mathscr{D}$ may be much smaller than k_{L_0}, that is, ϕ small, two possibilities exist with respect to $\alpha\phi^2$: (a) k_{L_0} may be smaller than $k\bar{B}V/a$, or (b) k_{L_0} may be larger than $k\bar{B}V/a$.

CASE 2a When ϕ is small and $\alpha\phi^2$ large,

$$\frac{k_L}{k_{L_0}} = \frac{\alpha\phi^2}{\alpha\phi^2 - \phi^2 + 1} \to 1 \qquad \text{and} \qquad k_L \to k_{L_0}$$

---

[1] Ibid.
[2] Compare with Thiele catalyst effectiveness (Chap. 5).

while

$$\frac{J}{\mathscr{R}} = \frac{1}{\alpha\phi^2 - \phi^2 + 1} \to \frac{1}{\alpha\phi^2}$$

Reaction occurs to some extent in the bulk phase ($J/\mathscr{R} = 1/\alpha\phi^2$), and the rate is governed by physical mass transfer. For example, if $\phi = 0.1$, $\alpha = 1000$, then

$$\frac{k_L}{k_{L_0}} = \frac{10^3 \times 10^{-2}}{10 - 10^{-2} + 1} = 1$$

$\eta = J/\mathscr{R} = 0.1$; 10 percent phase utilization.

CASE 2b   When ϕ is small and $\alpha\phi^2$ is small,

$$\frac{k_L}{k_{L_0}} < 1 \qquad \frac{J}{\mathscr{R}} \to 1$$

so that $k_L < k_{L_0}$ and $J = \mathscr{R}$, the rate of chemical reaction in the entire bulk phase. For example, let $\phi = 0.1$, as in case 2a, but $\alpha = 10$; then

$$\frac{k_L}{k_{L_0}} = \frac{10 \times 10^{-2}}{0.1 - 10^{-2} + 1} \approx 0.1$$

$$\eta = \frac{J}{\mathscr{R}} = \frac{1}{0.1 - 10^{-2} + 1} \approx 1$$

Here reaction occurs throughout the entire bulk phase, and the absorption rate is governed exclusively by the homogeneous-reaction rate.

Note that in all cases except 2b the amount absorbed per unit time is proportional to the interfacial area between phases, while in case 2b bulk volume rather than interfacial area governs the amount absorbed per unit of time.

The assumption of pseudo-first-order reaction was made to secure analytical solution of the absorption-reaction differential equation. This is tantamount to supposing that B, the coreactant, is uniformly accessible in film and bulk phases. When, however, the reaction-velocity constant k becomes very large, not only is A consumed completely within the film but its coreactant B will surely be depleted unless its diffusivity is so great that it can be sustained in excess in the film. For infinitely rapid reaction in the film, B must suffer depletion. This situation and the other cases treated above are depicted in Fig. 6-6.

Case 3 (Fig. 6-6) cannot be solved analytically as both differential equations are nonlinear; however, case 4 can be treated as the reaction is infinitely rapid. We are then dealing with diffusion of each component toward a reaction zone of zero thickness. For the reaction $A + zB \to P$, considering the gas-film as well as the liquid-phase process,

$$N_A = k_g(p - p_0) = \frac{\mathscr{D}_A}{x_1} A_0 = \frac{\mathscr{D}_B}{zx_2} \bar{B} \qquad (6\text{-}31)$$

Case 1

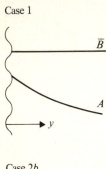

Case 2a

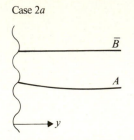

Case 2b

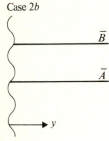

Case 3

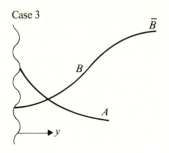

Case 4

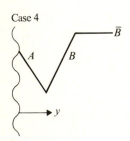

FIGURE 6-6
Diverse concentration gradients in absorption-reaction within one phase.
Case 1: Pseudo-first-order reaction in film (regime B). Case 2a: Slow reaction
(regime C). Case 2b: Extremely slow reaction (regime D). Case 3: Fast
second-order reaction (regime E). Case 4: Infinitely rapid nth-order reaction
(regime A).

where **B** is nonvolatile.   The situation is shown in Fig. 6-7.   Then

$$\frac{\mathcal{D}_A}{x_1} = \frac{x_L}{x_1}\frac{\mathcal{D}_A}{x_L} = \frac{x_L}{x_1}k_{L_0} = \frac{k_{L_0}}{W} \qquad \text{where } W = \frac{x_1}{x_L} \qquad (a)$$

and $\gamma = \mathcal{D}_B/\mathcal{D}_A$

$$\frac{\mathcal{D}_B}{x_2} = \frac{x_L}{x_2}\frac{\mathcal{D}_A\gamma}{x_L} = \frac{x_L\gamma}{x_2}k_{L_0} = \frac{\gamma k_{L_0}}{1-W} \qquad (b)$$

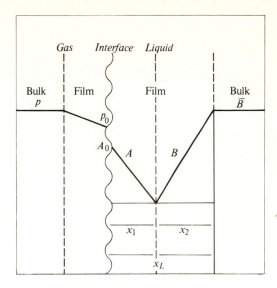

FIGURE 6-7
Gradients in the limiting case of nth-order infinitely rapid reaction diffusion within a film.

so that

$$N_A = k_g(p - p_0) \qquad (6\text{-}32a)$$

$$= \frac{k_{L_0}}{W} M p_0 \qquad (6\text{-}32b)$$

$$= \frac{\gamma k_{L_0} \bar{B}}{z(1 - W)} \qquad (6\text{-}32c)$$

Now we solve (6-32b) and (6-32c) for $1/W$, substitute the result in Eq. (6-32b) and then solve (6-32a) and (6-32b) for p_0; the result is substituted in Eq. (6-32a) to yield

$$\frac{1}{W} = 1 + \frac{\gamma \bar{B}}{zM p_0}$$

Then

$$p_0 = \frac{k_g p - (k_{L_0}/z)\gamma \bar{B}}{k_g + k_{L_0} M}$$

and finally

$$N_A = \frac{p}{1/k_g + 1/k_{L_0} M}\left(1 + \frac{\gamma \bar{B}}{zMp}\right) \qquad (6\text{-}33)$$

In terms of liquid-phase resistance only

$$N_A = k_{L_0}\left(1 + \frac{\gamma B}{zM p_0}\right)p_0 M = k_L p_0 M = \text{definition of } k_L$$

so that

$$E' = \frac{k_L}{k_{L_0}} = \left(1 + \frac{\mathscr{D}_B \bar{B}}{z\mathscr{D}_A A_0}\right) \qquad (6\text{-}34)$$

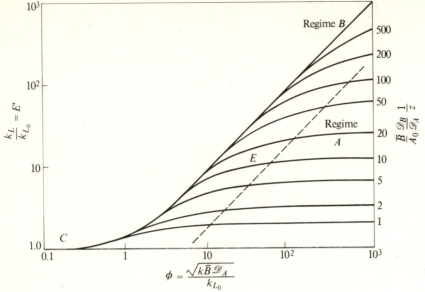

FIGURE 6-8
Van Krevelen–Hoftyzer plot of k_L/k_0 versus ϕ. (Regime D at $E' < 1.0$.)

$$A_0 = Mp_0 \qquad x_1 + x_2 = x_L \qquad \gamma = \frac{\mathscr{D}_B}{\mathscr{D}1}$$

Note that in this infinitely rapid nth-order-reaction case k_L increases linearly with $\bar{B}$ and k_L varies inversely with interfacial concentration of solute gas A_0. Both these effects follow from the fact that diffusive path lengths x_1 and x_2 are altered as A_0 and $\bar{B}$ are varied.

The case intermediate between pseudo first order and case 4, that is, case 3, was analyzed numerically and resolved approximately by van Krevelen and Hoftyzer,[1] thus establishing the k_L/k_{L_0} behavior in the intermediate region of finite second-order reaction. In Fig. 6-8 k_L/k_{L_0} is plotted against ϕ for pseudo-first-order reaction where $\alpha\phi^2 > 1$. Equation (6-34) is also plotted.

This display, in a form originally suggested by van Krevelen and Hoftyzer, reveals their intermediate second-order solutions. It is important to bear in mind that Fig. 6-8 applies only where the chemical-reaction rate $k\bar{B}V/a > k_{L_0}$, that is, $\alpha\phi^2 > 1$. As illustrated earlier, k_L can be less than k_{L_0} when the liquid-phase reaction rate is slower than the physical mass-transfer rate commensurate with the hydrodynamics and geometry of the system. In Fig. 6-8, it is observed that the infinitely rapid nth-order situation leads to k_L/k_{L_0} independent of ϕ (as would be expected), while with increase in

$$\frac{\bar{B}}{A_0} \frac{\mathscr{D}_B}{\mathscr{D}_A} \frac{1}{z}$$

[1] D. W. van Krevelen and P. J. Hoftyzer, *Rev. Trav. Chim. Pays-Bas*, **67**: 563 (1948).

pseudo-linear kinetics in A becomes manifest. When $\alpha\phi^2 < 1$ the case is treated simply as a homogeneous liquid-phase reaction, and for all practical purposes the system is no longer heterogeneous from the point of view of design if it is assumed that little can be done to alter $\alpha\phi^2$. Actually for a given value of ϕ, some flexibility in manipulating α is possible. As $\alpha = V/a\delta$, δ is fixed by hydrodynamics and geometry, but V/a, the bulk volume per interfacial area, can be altered by a wise choice of absorption equipment. A packed column, for example, would be characterized by a smaller V/a value than a stirred-tank contactor. Consequently if ϕ is quite large, volume utilization is low [Eq. (6-26)] and area per volume is more important than unemployed bulk volume. When, on the other hand, ϕ is quite small, the designer should contrive to establish $k_L \geq k_{L_0}$ rather than $k_L < k_{L_0}$. Thus V/a differences between various absorber types may profitably be exploited to achieve the condition $\alpha\phi^2 > 1$, with a consequent reduction in equipment size.

Another rather obvious point concerns the futility of enhancing the liquid-phase absorption coefficient if in fact the system is controlled by gas-phase mass transport even in the absence of reaction. We refer here, of course, to those circumstances in which reaction is caused to occur simply to enhance absorption rate. If $k_g \ll Mk_{L_0}$, there is no point in making k_{L_0} larger. On the other hand, if a significant equilibrium backpressure retards absorption in the absence of reaction, driving force as well as the transfer coefficient can be increased by consuming the reactant with reaction. Before discussing experimental data and comparing them with film theory, other models of physical adsorption, with and without reaction, deserve treatment.

6-4 PENETRATION THEORY OF HIGBIE

Instead of viewing absorption as a steady-state diffusion of a species through a fictitious film [Eqs. (6-1) to (6-3)], Higbie[1] viewed the process in terms of transient "conduction," or diffusion of the species into a stagnant liquid. As a function of time of exposure, the concentration gradient in the liquid would be as shown in Fig. 6-9. The governing differential equation describing this physical situation is

$$\mathcal{D}\frac{\partial^2 C}{\partial x^2} = \frac{\partial C}{\partial t} \qquad (6\text{-}35)$$

with the boundary conditions

At $x = 0$: $t = t$ $C = C_1$

At $x = \infty$: $t = t$ $C = 0$

At $x = x$: $t = 0$ $C = 0$

Using these boundary conditions, one solves for $C(x, t)$ and then determines the flux, $-\mathcal{D}(dC/dx)_{x=0}$, which is

$$N = -\mathcal{D}\left(\frac{dC}{dx}\right)_{x=0} = C_1\sqrt{\frac{\mathcal{D}}{\pi t}} = k'_{L_0}C_1 \qquad (6\text{-}36)$$

[1] R. Higbie, *Trans. AIChE*, **31**: 365 (1935).

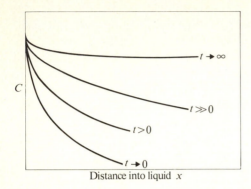

FIGURE 6-9
Schematic of C-versus-x gradients as a function of time according to the penetration model.

Therefore the instantaneous value of the liquid-phase mass-transfer coefficient is

$$k'_{L_0} = \sqrt{\frac{\mathscr{D}}{\pi t}} \qquad (6\text{-}37)$$

Note that a consequence of penetration theory is that $k'_{L_0} \propto \sqrt{\mathscr{D}}$, while film theory predicts $k_{L_0} \propto \mathscr{D}$. Integrating Eq. (6-37) over the range of t from zero to θ, the time of exposure, yields the average value k_{L_0}

$$k_{L_0} = 2\sqrt{\frac{\mathscr{D}}{\pi\theta}} \qquad (6\text{-}38)$$

or k_{L_0} is twice the value of k'_{L_0} at $t = 0$. It should be borne in mind that Higbie's development assumes that a body of liquid is exposed for a time θ. This is a batch exposure time, if you will, so that by analogy with our treatment of residence-time distributions in reaction vessels, the Higbie batch case of holding time θ may be viewed as the equivalent of a PFR residence-time distribution. The lifetime is identical for all elements. It is important to understand this point, for we can now visualize a large number of liquid elements each being exposed to absorbing gas for a *different* period of time. In the limit, then, an exponential distribution might exist comparable to the CSTR. As noted by Perlmutter,[1] this view of physical-absorption exposure-time distribution invites numerous analyses for cases where elements of absorbing liquid enjoy exposure-time distributions intermediate between PFR and CSTR extremes, analyses not unlike those undertaken in Chap. 3.

6-5 SURFACE-RENEWAL MODEL

Danckwerts[2] considered the more realistic situation involving exposure of a gas to a turbulent liquid interface. Due to turbulence, eddies, or liquid elements, are brought from the bulk to the interface, where they are exposed to the absorbing

[1] D. D. Perlmutter, *Chem. Eng. Sci.*, **16**: 287 (1961).
[2] P. V. Danckwerts, *Ind. Eng. Chem.*, **43**: 1460 (1951).

gas. If each eddy were exposed for exactly the same time θ, the Higbie model would apply, as transient diffusion would occur into each eddy for a time θ. Danckwerts assumed, however, that an exponential (CSTR) distribution of eddy exposure times prevails.

Danckwerts solved Eq. (6-35) and then evaluated the average absorption rate by

$$\text{Rate} = \int_0^\infty N\phi(t)\, dt \qquad (6\text{-}39)$$

where N is the instantaneous flux and $\phi(t)$ the eddy-age frequency function, which for the case at hand, for τ equal to the mean exposure time, is

$$\phi(t) = \frac{1}{\tau} \exp\left(\frac{-t}{\tau}\right) \qquad (6\text{-}40)$$

This is to be compared with the Higbie implication, namely, $\phi(t) = 1/\tau$. The average mass-transfer coefficient derived by Danckwerts is

$$k_{L_0} = \sqrt{\frac{\mathscr{D}}{\tau}} = \sqrt{\mathscr{D}S} \qquad (6\text{-}41)$$

where S is the surface-renewal rate.

As Perlmutter demonstrates, between the Higbie plug-flow eddy-age distribution and the CSTR distribution assumed by Danckwerts there exists little difference (the ratio being $2/\sqrt{\pi}$). Clearly other age distributions between these limits will yield k_{L_0} values lying between the Higbie and Danckwerts prediction. The significant fact is the predicted relation $k_{L_0} \propto \sqrt{\mathscr{D}}$, as found by Higbie and Danckwerts, in contrast with film theory, which asserts that $k_{L_0} \propto \mathscr{D}$.

6-6 TRANSIENT ABSORPTION WITH FIRST-ORDER REACTIONS

For first-order reaction and absorption

$$\mathscr{D}\frac{\partial^2 C}{\partial x^2} = \frac{\partial C}{\partial t} + kC \qquad \text{where } k = k_1 \bar{B} \qquad (6\text{-}42)$$

Based upon penetration theory, the solution of (6-42) is

$$\frac{k_L}{k_{L_0}} = \left(\frac{\mathscr{D}k}{k_{L_0}^2}\right)^{1/2} \left\{ \left(1 + \frac{k_{L_0}^2}{\mathscr{D}k}\right) \operatorname{erf}\left[\frac{2}{\sqrt{\pi}} \left(\frac{\mathscr{D}k}{k_{L_0}^2}\right)\right]^{1/2} + \frac{1}{2}\left(\frac{\mathscr{D}k}{k_{L_0}^2}\right)^{1/2} \exp\left(\frac{-4\mathscr{D}k}{k_{L_0}^2}\right) \right\}$$

$$(6\text{-}43)$$

while the Danckwerts surface-renewal model with first-order reaction yields

$$\frac{k_L}{k_{L_0}} = \left(1 + \frac{\mathscr{D}k}{k_{L_0}^2}\right)^{1/2} = (1 + \phi^2)^{1/2} \qquad (6\text{-}44)$$

Compare this with film theory [Eq. (6-25)], where ϕ is large

$$\frac{k_L}{k_{L_o}} = \left(\frac{k\mathscr{D}}{k_{L_o}^2}\right)^{1/2} = \phi \qquad (6\text{-}45)$$

In each model the absorption coefficient with reaction k_L relative to that for pure physical absorption k_{L_o} is a function of one dimensionless group $\phi = (k\mathscr{D}/k_{L_o}^2)^{1/2}$. Note that for large values of ϕ ($\phi > 3$) Eqs. (6-43) to (6-45) become identical, i.e.,

$$\frac{k_L}{k_{L_o}} = \phi \qquad (6\text{-}46)$$

Absorption with Second-Order Reaction

For a rapid second-order reaction with transient absorption, we have the situation described earlier except that the narrow zone of reaction is not fixed, as we assumed in Eqs. (6-31) to (6-34), but moves inward with time. Solution of the differential equations describing this transient situation yields,[1] for equality of species diffusivities,

$$\frac{k_L}{k_{L_o}} = 1 + \frac{\bar{B}}{A_0} \qquad (6\text{-}47)$$

which is identical to the film-model result [Eq. (6-34)] when $\mathscr{D}_B/\mathscr{D}_A = 1$. When $\mathscr{D}_B/\mathscr{D}_A \neq 1$, Eq. (6-34) is still an excellent approximation to the transient case.

For the finite second-order reaction with absorption, computer solution[2] of the appropriate equations reveals, once again, little significant difference between the transient- and film-theory result. Further, for a variety of reactions *at* equilibrium Olander[3] demonstrates the virtual equivalance of film-model and surface-renewal- (or penetration-) model results.

In sum it appears that when one is comparing the absorption coefficient with reaction with that for pure physical absorption (k_L/k_{L_o}), the results are virtually independent of the particular absorption model being invoked.

6-7   TEMPERATURE EFFECTS IN ABSORPTION-REACTION

The developments presented here rest implicitly upon the assumption that isothermal conditions prevail within the zone of diffusion-reaction. In principle, finite enthalpy of reaction invites the possibility of nonisothermality, and thus the reaction-influenced mass-transport coefficient k_L can be temperature-dependent. Now it is clear that an extreme test of nonisothermal effects would be a consideration of the case of rapid reaction (and therefore rapid heat generation or abstraction) within a

[1] T. K. Sherwood and R. L. Pigford, "Absorption and Extraction," 2d ed., McGraw-Hill, New York, 1952.
[2] P. L. T. Brian et al., *AIChE J.*, **7**: 226 (1961); R. M. Secor and J. A. Beutler, *AIChE J.*, **13**: 365 (1967).
[3] D. R. Olander, *AIChE J.*, **6**: 233 (1960).

minimum volume, i.e., in the film. We consider then the case of rapid pseudo-first-order reaction within the liquid film (regime B). For this circumstance the phase-*utilization* factor $J/\mathcal{R}$ assumes its limiting value, $J/\mathcal{R} = 1/\phi$. As $J/\mathcal{R}$ is identical to the Thiele effectiveness factor η for strong diffusion-influenced catalytic reaction (Chap. 5), whatever may be said of nonisothermal influences upon η clearly applies to $J/\mathcal{R}$.

Carberry[1] has shown that the asymptotic value of the nonisothermal η relative to its isothermal value is

$$\frac{\eta}{\eta_0} = \exp \frac{\epsilon\beta}{5}$$

or

$$\eta = \frac{\exp(\epsilon\beta/5)}{\phi_0} \qquad (6\text{-}48)$$

where

$$\epsilon\beta = \frac{-\Delta H \, A_0 \mathcal{D}}{\lambda T_0} \frac{E}{R T_0}$$

In the absorption-reaction case

$$\phi_0 = \frac{\sqrt{k_0 \, \overline{B} \mathcal{D}}}{k_{Lo}}$$

As k_L^0 (isothermal reaction-influenced coefficient) is equal to ϕ_0 in regime B (Fig. 6-8), the nonisothermal coefficient k_L relative to the isothermal value is

$$\frac{k_L}{k_L^0} = \exp \frac{\epsilon\beta}{5}$$

We consider a case treated by Danckwerts in terms of penetration theory, specifically the absorption of CO_2 into buffer solution.[2] Using appropriate parameters as set forth by Danckwerts, we compute $\beta = 14 \times 10^{-6}$. Though E is not known, it is clear that when we use an improbably large value of E/RT_0 of, say, 10^2, the resulting $\epsilon\beta$ is small enough to assure that $k_L/k_L^0 = 1.0$; a result in accord with Danckwerts' analysis. Inspection of the range of thermal-kinetic parameters which determine $\epsilon\beta$ for reaction-diffusion in the liquid phase shows that the assumption of isothermality within the liquid film is indeed a safe postulate.

6-8 COMPARISON WITH EXPERIMENT (PHYSICAL ABSORPTION)

The work of Sherwood and Hollaway,[3] which involved an exhaustive study of liquid-film-controlled mass transfer in a packed column, demonstrated that $k_{Lo} \propto \sqrt{\mathcal{D}}$, thus lending support to the penetration or surface-renewal model of physical adsorption of a gas into a liquid *under the hydrodynamic conditions pre-*

[1] J. J. Carberry, *AIChE J.*, **7**: 350 (1961).
[2] P. V. Danckwerts, *Appl. Sci. Res.*, **A3**: 385 (1953).
[3] T. K. Sherwood and F. A. L. Hollaway, *Trans. AIChE*, **36**: 39 (1940).

vailing in a packed absorber. Emphasis upon the hydrodynamic condition is required as the k_{L_0}-versus-$\mathcal{D}$ relation is a function of the velocity gradient in the absorbing phase. In the Higbie-Danckwerts model, diffusion is independent of velocity gradients since either a stagnant liquid or eddies are the absorption sinks. In general the k_{L_0}-versus-$\mathcal{D}$ functionality varies between $k_{L_0} \propto \sqrt{\mathcal{D}}$ and $k_{L_0} \propto \mathcal{D}^{2/3}$ as one proceeds from the stagnant or eddy model to diffusion into a flowing stream in which a velocity gradient exists at the interface. As Mixon[1] shows, a general relationship for k_{L_0} or k_g and fluid velocity as a function of two directions can be fashioned and reduced to various limiting cases, e.g., the cases treated by Higbie and Danckwerts as well as situations conforming to boundary-layer flow (see Chap. 5).

6-9 VERIFICATION OF ABSORPTION-REACTION MODELS

Consider the Danckwerts model for first-order reaction accompanying absorption

$$\frac{k_L}{k_{L_0}} = \sqrt{1 + \frac{k\mathcal{D}}{k_{L_0}^2}} = \frac{\sqrt{k_{L_0}^2 + k\mathcal{D}}}{k_{L_0}} \qquad (6\text{-}49)$$

Recall that in terms of surface renewal S, $k_{L_0} = \sqrt{\mathcal{D}S}$. Then

$$k_L = \sqrt{\mathcal{D}(S + k)} \qquad (6\text{-}50)$$

and the rate of absorption per unit volume is

$$\mathcal{R}a = aC\sqrt{\mathcal{D}(S + k)} \qquad \text{where } a = \frac{\text{interfacial area}}{\text{vol of apparatus}} \qquad (6\text{-}51)$$

If one determines $\mathcal{R}a$ experimentally as a function of k, a plot of $(\mathcal{R}a)^2$ versus k should yield a linear relationship of slope $a^2(C\sqrt{\mathcal{D}})^2$ and intercept $a^2(C\sqrt{\mathcal{D}})^2 S$. With an independent source of $C\sqrt{\mathcal{D}}$, where C is the concentration of dissolved reactant and $\mathcal{D}$ its diffusivity, a and S can be determined for the apparatus. In principle, then, for another system one need only determine $C\sqrt{\mathcal{D}}$ and k to predict the reaction-influenced absorption rate for the new system. This means measuring the homogeneous rate constant k as well as C. Some simplification results if for the test system $k \gg S$; in this case a is determined by a single measurement of $\mathcal{R}a$. Danckwerts et al.[2] have provided experimental verification of Eq. (6-50) for a packed-column absorber, absorbing CO_2 into carbonate-bicarbonate buffer solutions containing arsenite catalyst. Values of S and a were inferred from $\mathcal{R}a$-versus-k data (Fig. 6-10).

[1] F. O. Mixon and J. J. Carberry, *Chem. Eng. Sci.*, **13**: 30 (1960).
[2] D. Roberts and P. V. Danckwerts, *Chem. Eng. Sci.*, **17**: 961 (1962); P. V. Danckwerts, A. M. Kennedy, and D. Roberts, *Chem. Eng. Sci.*, **18**: 63 (1963); G. M. Richards, G. A. Ratcliff, and P. V. Danckwerts, *Chem. Eng. Sci.*, **19**: 325 (1964).

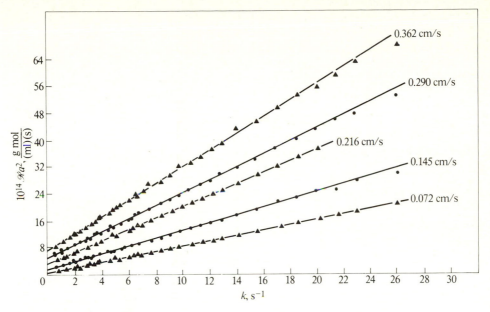

FIGURE 6-10
Verification of Danckwerts' model of surface renewal for the determination of interfacial area per unit volume. Parameter is liquid velocity. [*D. Roberts and P. V. Danckwerts, Chem. Eng. Sci.*, **17**: *961 (1962).*]

6-10 PRACTICAL UTILITY OF FILM THEORY[1]

Consider Fig. 6-8 once more. Four extremes of rate control are designated *A*, *B*, *C*, and *E*. Andrew[2] designed a clever gas-liquid contactor consisting of a disk column fully wetted by liquid which is exposed to gas circulated at a high rate. From the known wetted area, measured absorption rate, and driving force the mass-transfer coefficient for the liquid phase in the presence of reaction is determined. The dependence of k_L on liquid flow rate L and liquid-phase reactant and absorbing-gas concentrations are clues as to which mode of rate control (*A*, *B*, *C*, *E*, or an intermediate one) prevails in a given system. Since liquid holdup (volume of liquid) is known to vary with $L^{1/3}$ while the liquid-phase physical-absorption coefficient k_{L_0} varies as about $L^{3/4}$, a table of experimental criteria can be established. Such criteria, as set forth by Andrew, are given in Table 6-1.

Example: CO_2 Absorption into NH_3 Solution

Andrew demonstrated the utility of his disk column with the CO_2-NH_3 solution system. His k_L values were independent of L at two temperature levels and proportional to $\sqrt{NH_3}$, thus establishing mechanism *B* as the predominant one in the

[1] S. P. S. Andrew, *Accad. Naz. Lincei 5th Corso Estivo Chim. Varese, Sept. 26–Oct. 8, 1960.*
[2] S. P. S. Andrew, *Chem. Eng. Sci.*, **3**: 279 (1954); S. P. S. Andrew and D. Hanson, *2d Eur. Symp. Chem. React. Eng. 1960*, pap. D1.

disk column. A value of k_g can be found, of course, by studying NH_3 absorption into pure water. From the data Andrew determined the chemical rate constant and known (or computed) diffusivities and Henry's-law constants (Table 6-2).

With this information, the diffusion-reaction modulus of Fig. 6-8 is known for a given k_{L_0}; that is,

$$\phi = \frac{\sqrt{k\mathscr{D}C_L}}{k_{L_0}} = \frac{\sqrt{k\mathscr{D}(NH_3)}}{k_{L_0}}$$

and for a given C_L and partial pressure of absorbing gas the term (Fig. 6-8)

$$\frac{\bar{B}1}{A_0 z}$$

is known, assuming $\mathscr{D}_B = \mathscr{D}_A$. For this CO_2–dilute-NH_3-liquor system, the disk-column study provides all key information whereby k_L/k_{L_0} can be determined from Fig. 6-8 for another type of absorption apparatus. One need only know or measure k_{L_0}, k_g, and interfacial area a.

Suppose that a bubble-cap absorber is to be designed for the absorption of CO_2 into weak NH_3 liquor. What is required, of course, is the plate efficiency for each stage. Assuming a uniform, well-mixed liquid on the bubble-cap plate, the

Table 6-1 EFFECT UPON K_L (OBSERVED) OF CHANGES IN NEGOTIABLE VARIABLES†

			Changes in:	
Regime	Mechanism	L	Gas concentration	Liquid concentration
A	Infinitely fast second order	$L^{3/4}$	$\sim (A_0)^{-1}$	$\sim C_L^{1}$
B	Rapid first-order reaction	None	None	$\sim C_L^{1/2}$
C	Slow reaction controlled by physical absorption	$L^{3/4}$	None	Small
D	Bulk-chemical run control	$L^{1/3}$	None	$\sim C_L$

These criteria follow from

A
$$k_L = k_{L_0}\left(1 + \frac{C_L}{A_0}\frac{\mathscr{D}_L}{\mathscr{D}_g}\frac{1}{z}\right) = g\left(L^{3/4}, \frac{C_L}{A_0}\right)$$

B $k_L = \sqrt{k\mathscr{D}C_L} = f'\sqrt{C_L}$

C $k_L = k_{L_0} = f''L^{3/4}$

D‡ $k_L = kC_L V = f'''(C_L, L^{1/3})$ where $V = g(L^{1/3})$

† S. P. S. Andrew, *Accad. Naz. Lincei 5th Corso Estivo Chim. Varese, Sept. 26–Oct. 8, 1960*
‡ Where $E' < 1.0$ in Fig. 6-8.

Murphree efficiency $\mathscr{E}$ is related to the overall transfer coefficient K_0, interfacial area a, and a superficial gas velocity u by

$$\ln (1 - \mathscr{E}) = -\frac{K_0\, a}{u} \qquad \text{where } \frac{1}{K_0} = \frac{1}{k_g} + \frac{1}{Mk_L} \qquad (6\text{-}52)$$

and
$$k_L = k_{L_0} E' \qquad \text{where } E' = k_L/k_{L_0}$$

as determined from Fig. 6-8.   Andrew has reexpresssed the AIChE k_{L_0} correlations for bubble caps of submergence $\mathscr{S}$ in cgs units as

$$k_g = \frac{7u^{1/4}}{\sqrt{\mathscr{S}}}\,\mathscr{D}_g^{\,1/2} \qquad k_{L_0} = \frac{11u^{1/4}}{\sqrt{\mathscr{S}}}\,\mathscr{D}_e^{\,1/2} \qquad (6\text{-}53)$$

The interfacial area a, *per area of plate* is isolated from the usual correlations of $k_g a$ and $k_{L_0} a$ and was determined by Andrew for bubble-cap plates in the following manner:

1 CO_2 is absorbed into dilute NaOH on the plate.
2   Under hydrodynamic conditions comparable to those of the proposed design, E' is measured.
3   For CO_2 absorption into dilute caustic, the chemical rate constant (and thus ϕ) is known from independent sources.
4   When k_L and E' are known, a is determined.   Andrew recommends the relation

$$a = 0.7\sqrt{u}\,\mathscr{S}^{5/6} \qquad (6\text{-}54)$$

Continuing, for $p_{CO_2} = 0.1$ atm, $NH_3 = 10^{-3}$ g mol/cm³, $\mathscr{S} = 10$ cm, and $u = 20$ cm/s.   At 20.7°C, Andrew's disk-column study gives for

$$\mathscr{D} = 1.82 \times 10^{-5} \text{ cm}^2/\text{s} \qquad k = 3.56 \times 10^5 \text{ cm}^3/(\text{mol})(\text{s})$$
$$M = 3.85 \times 10^{-5} \text{ mol}/(\text{cm}^3)(\text{atm}) \qquad \text{and} \qquad MR_g T = 0.95$$

Assume equality of diffusivities in liquid phase. $\mathscr{D}_{CO_2}$ in gas phase $= 0.14$ cm²/s. Then

$$\frac{B}{Az} = \frac{B}{pMz} = \frac{NH_3}{p_{CO_2} M \times 1} = \frac{10^{-5}}{(3.85 \times 10^{-5})(\frac{1}{10})} = 260$$
$$\sqrt{k\mathscr{D}_e(NH_3)} = \sqrt{(3.56 \times 10^5)(1.82 \times 10^{-5})(10^{-3})} = 0.08$$

Table 6-2

	Temperature, °C	
	20.7	60.5
$\mathscr{D}$, 10^{-5} cm²/s	1.82	4.53
M, 10^{-5} mol/(ml)(atm)	3.85	1.55
k, 10^5 ml/(mol)(s)	3.56	38.6

From Eqs. (6-53)

$$k_g = 7 \frac{20^{1/4}}{\sqrt{10}} (0.14)^{1/2} = 1.7 \text{ cm/s}$$

and

$$k_{L_0} = \frac{11(20)^{1/4}}{\sqrt{10}} (1.82 \times 10^{-5})^{1/2} = 0.032$$

and so

$$\phi = \frac{\sqrt{k \mathscr{D}_e(NH_3)}}{k_{L_0}} = \frac{0.08}{0.032} = 2.5$$

From Fig. 6-8, for $\phi = 2.5$ and $B/Az = 260$, we see

$$E' = \frac{k_L}{k_{L_0}} = \phi = 2.5$$

Since

$$\frac{1}{K_0} = \frac{1}{k_g} + \frac{1}{MRTk_{L_0}E'}$$

we have

$$\frac{1}{K_0} = \frac{1}{1.7} + \frac{1}{0.95(0.08)} = 13.8 \quad \text{s/cm}$$

Now by Eq. (6-54)

$$a = 0.7\sqrt{20} \times 10^{5/6} = 21.3 \text{ cm}^2/\text{cm}^2 \text{ of plate area}$$

$$\ln(1 - \mathscr{E}) = -\frac{21.3}{20(13.8)}$$

and so $\mathscr{E} = 7$ percent, a poor efficiency indeed. Examination of the factors involved reveals that there is much to be gained by increasing NH_3, as the condition of rapid pseudo-first-order reaction already prevails. Clearly ϕ must be increased to increase E' and thus significantly decrease $1/K_0$. Note that an increase in p_{CO_2} actually will reduce E' since the rapid second-order reaction is approached with increasing p_{CO_2}. Operating then at 60°C, where

$$\phi = \frac{\sqrt{(38.6 \times 10^5)(4.53 \times 10^{-5})(10^{-3})}}{0.05} = \frac{0.42}{0.05} = 8.4$$

so that $k_{L_0} E' = 0.42$ at 60°C vs. 0.08 at 20°C and MRT $= 0.44$, we have

$$\frac{1}{K_0} = \frac{1}{1.7} + \frac{1}{0.44(0.42)} = 6 \quad \text{and} \quad \mathscr{E} = 16\%$$

An increase in temperature enhances k significantly, ϕ by a factor of 8.4/2.5 and E' accordingly. However, CO_2 solubility is decreased and the net result is an efficiency increase of about twofold. An increase in NH_3 concentration by one order of magnitude will be beneficial if an excess of liquid-phase coreactant can be tolerated economically. It should be observed that the enhancement in absorption rate is not linear in the value of the reaction velocity constant or coreactant (NH_3) concentration, but by reason of simultaneous rapid reaction-diffusion a square-root relationship exists. Had the condition of infinitely rapid second-order reaction existed (regime A), obviously nothing would be gained by altering temperature, though of course M would be somewhat different.

Figure 6-8, based upon film theory and irreversible reaction, is only slightly altered in the light of penetration theory and reversible reaction; corrections can be made as outlined by McNiel.[1] Secor and Beutler[2] present an exhaustive study comparing film and penetration models and conclude that little difference can be found providing $\mathscr{D}_B/\mathscr{D}_A$ does not depart markedly from unity.

As the criteria cited in Table 6-1 indicate, each regime of absorption-reaction is characterized by interfacial area, bulk volume, gas- and liquid-phase compositions, and their respective diffusional resistances. An ideal laboratory gas-liquid reactor, as suggested in Chap. 2, would be one in which bulk composition is uniform with respect to time and spatial coordinates, and an ideal two-phase laboratory reactor is one in which each phase conforms to CSTR behavior yet with variable interfacial area. Such an ideal has been designed and elevated to a reality by Levenspiel and Godfrey,[3] and a schematic of this two-phase CSTR is to be found in Fig. 6-11. Since steady-state CSTR behavior is easily achieved in both phases, both phase compositions can be varied from one experiment to another. Perforated plates of diverse interfacial areas can be fashioned to vary a', and, in principle, liquid-phase volume can be varied, as can temperature. In consequence, a series of carefully designed experiments, all in the steady state, permit the determination of k_g, k_{L_0}, k_L, and liquid-phase-reaction kinetic parameters.[4]

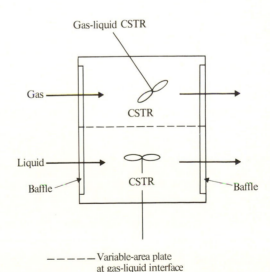

FIGURE 6-11
Levenspiel's gas-liquid twin CSTR contactor. [*O. Levenspiel and J. H. Godfrey, Chem. Eng. Sci.*, **29**: *1723 (1974).*]

———— Variable-area plate
at gas-liquid interface

[1] K. McNiel, Ph. D. thesis, Chemical Engineering, University of Cambridge, 1966.
[2] R. M. Secor and J. A. Beutler, *AIChE J.*, **13**: 365 (1967).
[3] O. Levenspiel and J. H. Godfrey, *Chem. Eng. Sci.*, **29**: 1723 (1974).
[4] J. H. Godfrey, Ph.D. thesis, Oregon State University, Corvallis, 1973.

6-11 REGIME IDENTIFICATION IN TERMS OF OBSERVABLES

For pseudo-first-order reaction the phase-utilization factor η is

$$\eta = \frac{J}{\mathcal{R}_0} = \frac{1}{\alpha\phi} \frac{(\alpha - 1)\phi + \tanh \phi}{(\alpha - 1)\phi \tanh \phi + 1} \tag{6-55}$$

$$\eta = \frac{\text{actual, observed, absorption-reaction rate}}{\text{homogeneous rate at } k_0, A_0}$$

The observed rate of gas-liquid reaction is

$$aJ = \eta\mathcal{R}_0 = \eta k_0 A_0 \bar{B} \tag{6-56}$$

Recall that

$$\alpha = \frac{V}{a\delta} = \frac{\text{total reactor-phase liquid volume}}{\text{film volume}}$$

Now, in the spirit of Chap. 5, we multiply both sides of Eq. (6-56) by $\delta^2/\mathcal{D}A_0 = \mathcal{D}/k_{L_0}^2 A_0$ since $\delta = \mathcal{D}/k_{L_0}$; we obtain

$$\frac{\mathcal{D}Ja}{k_{L_0}^2 A_0} = \eta \frac{k_0 \bar{B}\mathcal{D}}{k_{L_0}^2} = \eta\phi^2 \tag{6-57}$$

Figure 6-12 reveals η plotted vs. the observable $\eta\phi^2$ for diverse values of α. This figure provides the analyst with the means whereby the locale of fluid-fluid immiscible-phase linear reaction-diffusion can be determined solely in terms of:

1 Measured rate Ja

2 The physical mass-transfer coefficient k_{L_0}

3 Reactant concentration in the parent phase A_0

4 The diffusion coefficient for A in the reaction phase

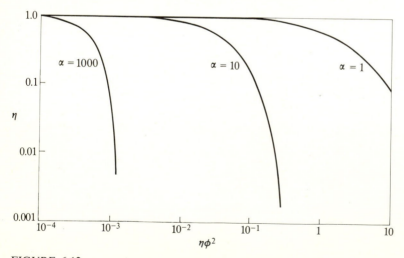

FIGURE 6-12
Intraphase effectiveness (efficiency) vs. the observable $\eta\phi^2$.

We now turn attention to the issue of multiphase reactor models, in which discussion the two-phase continuity equations will be developed. The problem of multiplicity of the steady state will then be addressed, followed by discussion and illustration of gas-liquid reactor design rooted in laboratory observations.

6-12 MULTIPHASE-REACTOR MODELS

Figure 6-2 shows diverse gas-liquid reactor types. In earlier sections of this chapter, emphasis was focused on gas-liquid reaction per se, i.e., details to elucidate the influence of chemical reaction upon gas absorption, that local process being termed generation G in the continuity equation developed in Sec. 6-2 for the packed-column absorber. It is now appropriate that the global issue of the reactor per se be examined. For it must be emphasized that the scale-up process need not require a commitment, on the large (plant) scale, to the reactor type used to secure rate data on the small (laboratory) scale. Convenience may well dictate the semibatch procurement of laboratory gas-liquid reaction-rate data, e.g., continuous flow of a gas through a batch of coreactant-bearing liquid. However, a radically different reactor network may well prove economically more advantageous at production levels.

Let us assume, then, that we have at our disposal a phenomenological reaction-rate expression and we have now to engineer matters; i.e., we must now specify the optimum reactor type to achieve an economically attractive productivity. The theory which must guide us is, in essence, identical to that set forth in Chap. 4; i.e. we must develop continuity equations for diverse multiphase reactor systems.

Two-Phase-Reactor Models

In our consideration of two-phase-reactor environments we can logically envision the possibilities shown in Table 6-3 with respect to spatial distribution of each phase in the axial coordinate. In batch and semibatch operation (cases 2 and 3) conversion is a function of real time, while in the continuous-reactor networks, holding or residence time dictates performance for a given temperature and concentration and their distributions. Of course, the continuous network may operate under transient circumstances, and in all cases the dispersed-phase droplets or bubbles may change dimension as they journey through the coreactant phase.

Table 6-3

Operation		Phase I		Phase II
1 Continuous	*a*	PFR	*A*	PFR
	b	CSTR	*B*	CSTR
2 Semibatch	*a*	PFR	*B*	Well-mixed
	b	CSTR		
3 Batch	*b*	Well-mixed	*B*	Well-mixed

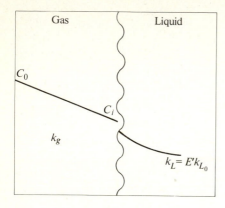

FIGURE 6-13
Local view of two-phase gradients.

In Fig. 6-13 a local view of a two-phase system is presented in which phase I supplies reactant to its locale of diffusion-affected reaction in phase II. Recalling the definition of phase utilization η and the enhancement factor E', and anticipating a phase I mass-transfer resistance, the interfacial value of C_i, for example, a gas, is

$$C_i = \frac{C_0}{1 + \eta kM/k_g a} = \frac{C_0}{1 + E'k_{L_0}M/k_g} \qquad (6\text{-}58)$$

so that global rate $\mathcal{R}$ at a point is

$$\mathcal{R} = \frac{\eta kMC_0}{1 + \eta kM/k_g a} = \frac{E'k_{L_0}MC_0}{1 + E'k_{L_0}/k_g} \qquad (6\text{-}59)$$

where M is Henry's-law coefficient

$$M = \frac{A_i}{C_i} \qquad (6\text{-}60)$$

and η and E' are as given earlier [Eqs. (6-26) and 6-25)] for pseudo-first-order reaction-absorption. For infinitely rapid nth-order reaction, E' is defined by Eq. (6-34).

The matter of two-phase reactor design now becomes that of specifying the spatial distribution of each phase.

Gas phase (I):
$$\mathcal{D}\,\frac{\partial^2 C}{\partial z^2} - u\,\frac{\partial C}{\partial z} = k_g a(C - C_i) + \frac{\partial C}{\partial t} \qquad (6\text{-}61)$$
$$\quad\ (a)\qquad\ (b)\qquad\qquad (c)\qquad\quad (d)$$

Liquid phase (II):
$$\mathcal{D}\,\frac{\partial^2 A}{\partial z^2} - \bar{u}\,\frac{\partial A}{\partial z} = \mathcal{R} + \frac{\partial A}{\partial t} \qquad (6\text{-}62)$$

Plug flow: $\qquad$ Term $(a) = 0 \qquad$ since $\mathcal{D} = 0$

Perfectly mixed: Term $(a) = 0$ since $\dfrac{\partial^2 A}{\partial z^2} = 0$

$$\text{Term } (b) = \frac{A_0 - A}{\theta}$$

Steady state: Term $(d) = 0$

Batch operation: Term $(a) = 0$ term $(b) = 0$

Assuming steady state, we have

	Phase I	*Phase II*	
Case 1aA:	$-\dfrac{dC}{d\theta} = k_g a(C - C_i)$	$-\dfrac{dA}{d\theta} = \mathscr{R}$	(6-63)
Case 1aB:	$-\dfrac{dC}{d\theta} = k_g a(C - C_i)$	$\dfrac{A_0 - A}{\bar{\theta}} = \mathscr{R}$	(6-64)
Case 1bB:	$\dfrac{C_0 - C}{\theta} = k_g a(C - C_i)$	$\dfrac{A_0 - A}{\bar{\theta}} = \mathscr{R}$	(6-65)
Case 2aB:	$-\dfrac{dC}{d\theta} = k_g a(C - C_i)$	$-\dfrac{dA}{dt} = \mathscr{R}$	(6-66)
Case 2bB:	$\dfrac{C_0 - C}{\theta} = k_g a(C - C_i)$	$-\dfrac{dA}{dt} = \mathscr{R}$	(6-67)
Case 3bB:	$-\dfrac{dC}{dt} = k_g a(C - C_i)$	$-\dfrac{dA}{dt} = \mathscr{R}$	(6-68)

The equations derived above are so general that they are useless in gas-liquid reactor design since, unlike the homogeneous counterparts, such physical factors as bubble volume, interfacial area, spray phenomena in the gas phase, etc., are involved.

To illustrate, consider case 1aB in further detail. Assuming that the ideal-gas law governs, we have for a mole fraction y

$$-\frac{d(PV_b/RT)y}{d\theta} = k_g a P V_b \left(y - \frac{AM}{P} \right) \qquad (6\text{-}96)$$

where θ is the bubble residence time. Clearly y, a, and V_b, the bubble volume, are point values, and except when y is small or equal to unity for small extent of absorption, the average values must be employed, e.g.,

$$\bar{y} = \frac{1}{\theta_r} \int_0^\theta y(\theta) \, d\theta \qquad (6\text{-}70)$$

Olson's Generalized Multiphase-Reactor Model

A design approach which admirably embraces not only direct-contact mass-transport processes but reaction and mass transport in multiphase systems is set forth by Pavlica and Olson.[1] Since two-phase reactors (gas-liquid, liquid-liquid, and gas-fluidized solid) are generally characterized by backmixing in one or both phases yet well mixed radially, the Olson model consists of two continuity equations, one for the dispersed phase

$$\frac{1}{Pe_D}\frac{d^2a}{dz^2} = \frac{da}{dz} + \frac{k_D\epsilon f L}{u_D}a + \frac{K_D a'L}{u_D}(a-A)$$

or

$$\bar{A}\frac{d^2a}{dz^2} = \frac{da}{dz} + Ba + \bar{E}(a-A) \qquad (6\text{-}71)$$

where a = reduced dispersed-phase concentration
$\quad f$ = its fractional holdup
$\quad \epsilon$ = packed-bed void fraction ($\epsilon = 1$ for unpacked contactors)
$\quad L$ = reactor length
$\quad u_D$ = dispersed-phase velocity
$\quad k_D$ = reaction rate constant
$\quad K_D$ = mass-transfer coefficient
$\quad a'$ = interfacial area

For the continuous phase at reduced concentration A

$$F\frac{d^2A}{dz^2} = \pm\frac{dA}{dz} + GA + H(A-a) \qquad (6\text{-}72)$$

where

$$F = \frac{1}{Pe_c} \qquad Pe = \frac{uL}{\mathscr{D}} = \text{Peclet number}$$

$$G = \frac{k_c\epsilon(1-f)L}{u_c} \qquad H = \frac{K_D M a'L}{u_c}$$

where M = Henry's-law constant
$\quad u_c$ = continuous-phase velocity

In Eq. (6-72), the first term on the right-hand side is negative for countercurrent flow and positive for the cocurrent situation.

Note that these two equations anticipate mass transfer *in series* with bulk-phase reaction, a limiting situation which prevails for small values of B and G, respectively. As shown earlier, the general two-phase diffusion-reaction case, particularly for gas absorption *with* reaction, is most realistically described in terms of a reaction-affected mass-transport model; hence

$$K_D = E'K_{D_0} \qquad (6\text{-}73)$$

[1] R. T. Pavlica and J. H. Olson, *Ind. Eng. Chem.*, **62**(12): 45 (1970).

where E' is the enhancement coefficient or, alternatively,

$$k_D = \eta k_{D_0}$$

where η is the phase-utilization factor.[1]   In general, three cases may occur:

1   *Rapid reaction* B and/or $G \to \infty$ since $\eta \to 0$ and reaction is incorporated into the transport coefficient [Eq. (6-73)].
2   *Slow reaction* $\bar{E}$ and H terms exist *in series* with bulk reaction, B and G.
3   *Negligible reaction* $\bar{E}$ and H then govern a purely mass-transfer process.

We have six dimensionless groups which characterize a wide class of two-phase reaction systems, namely $\bar{A}$, B, C, D, $\bar{E}$, and F.   As Olson points out, further simplification is realized by noting that

$$\Lambda = \frac{H}{\bar{E}} = M \frac{u_D}{u_c} = \text{extraction factor}$$

$$\text{Da}_D = \frac{B}{\bar{E}} = \text{Damköhler number, dispersed phase} \qquad (6\text{-}74)$$

$$\text{Da}_c = \frac{G}{H} = \text{Damköhler number for continuous phase}$$

Numerical solutions and fruitful generalizations are presented by Olson for a diversity of two-phase contacting devices (spray columns, packed beds, etc.), for which the governing physical dimensionless groups have been well correlated. Reaction parameters are, of course, unique; yet, as will be shown in Chap. 10, the Olson model proves to be of promise in the estimation of fluidized-catalytic-reactor performance.

The reader should consult the papers of Russell and his students for a detailed exposition of tank and tubular gas-liquid reactor models as well as parameter-estimate guidelines.[2]

The problem of possible multiplicity of the steady state will now be briefly treated for gas-liquid reaction in an adiabatic CSTR in which pseudo-first-order reaction occurs.   A more exhaustive analysis for the second-order case is provided by Luss and his students.[3]

6-13 MULTIPLICITY OF THE STEADY STATE

While the essential features of steady-state multiplicity were discussed and illustrated in Chaps. 3 and 5, it is of interest to explore the situation involving two phases with interphase and intraphase mass-diffusional intrusions.

We consider a gas-liquid absorption-reaction system in a CSTR in which both phases are assumed to be perfectly mixed.   Assume that at the gas-liquid interface,

---

[1] Note that E' and η can be functions of extent of reaction.
[2] T. W. F. Russell et al., *Ind. Eng. Chem.*, **60**: 12 (1968); **61**: 6, 15 (1969).
[3] L. A. Hoffman, S. Sharma, and D. Luss, *AIChE J.*, **21**: 318 (1975).

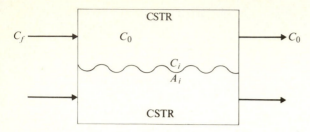

FIGURE 6-14
CSTR model of gas-liquid reaction-absorption.

a temperature-dependent Henry's-law distribution coefficient dictates interfacial equilibrium as in Fig. 6-14, a schematic of the system

$$\frac{A_i}{C_i} = M_0 \exp{(\bar{h})}\left(\frac{1}{t} - 1\right) \qquad (6\text{-}75)$$

where $\bar{h}$ is the dimensionless enthalpy of solution h/RT_0. Locally the heat generation is

$$q = -\Delta H \, \eta k_0 \, M_0 \, C_i \exp\left[-(\epsilon - \bar{h})\left(\frac{1}{t} - 1\right)\right] \qquad (6\text{-}76)$$

where $\epsilon = E/RT_0$ and E is the activation energy. Should a gas-phase mass-transfer gradient exist,

$$k_g a(C_0 - C_i) = \eta k M C_i$$

or

$$C_i = \frac{C_0}{1 + \eta \overline{\text{Da}} \exp{[-(\epsilon - \bar{h})(1/t - 1)]}} \qquad (6\text{-}77)$$

The perfectly mixed gas-phase concentration C_0 is related to its feed value C_f by a simple CSTR balance:

$$C_0 = \frac{C_f}{1 + \dfrac{\text{Da} \, \eta \exp{[-(\epsilon - \bar{h})(1/t - 1)]}}{1 + \overline{\text{Da}}\eta \exp{[-(\epsilon - \bar{h})(1/t - 1)]}}} \qquad (6\text{-}78)$$

where

$$\overline{\text{Da}} = \frac{k_0 \, M_0}{k_g a} \qquad \text{and} \qquad \text{Da} = k_0 \, M_0 \, \theta_0$$

When we substitute (6-78) into (6-77) and that into (6-76) and normalize, the dimensionless heat-generation function emerges as

$$\frac{q\theta_0}{\rho C_p T_f} = \frac{\beta \text{Da}\eta \exp{[-\Delta(1/t - 1)]}}{1 + (\overline{\text{Da}} + \text{Da})\eta \exp{[-\Delta(1/t - 1)]}} \qquad (6\text{-}79)$$

where

$$\beta = \frac{-\Delta H \, C_f}{\rho C_p T_f} \qquad \text{and} \qquad \Delta = \epsilon - \bar{h}$$

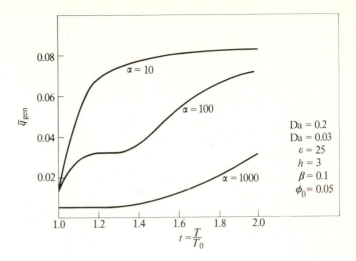

FIGURE 6-15

$\bar{q}_{gen}$ versus reduced temperature for a gas-liquid reaction.

and η is the phase-utilization efficiency factor, given by Eq. (6-26).   Of course η is a function of temperature since

$$\phi = \phi_0 \exp\left[-\frac{\epsilon}{2}\left(\frac{1}{t} - 1\right)\right]$$

which assumes that the activation energy for diffusion is negligible.

Note that the form of Eq. (6-79) is the familiar one encountered earlier; however, the behavior will not be analogous except for $\eta = 1$ throughout the temperature range explored.   For in the absence of intraphase diffusion-reaction, the usual sigmoidal q_{gen}-versus-t relation would be apparent.   In the present instance, the behavior of the temperature-dependent product $\eta \exp\left[-\Delta(1/t - 1)\right]$ proves to be interesting since η decreases with temperature in a complex manner governed by ϕ_0 and α, while the exponential Arrhenius function increases as governed by the difference between reaction activation energy and the enthalpy of solution.   For a range of diffusion-reaction parameters the dimensionless heat-generation function is displayed as a function of $t = T/T_f$ in Fig. 6-15.   As Luss' work shows, more than three steady states can exist in a gas-liquid CSTR.[1,2]

6-14 SCALE-UP OF BENCH-SCALE DATA

To demonstrate the virtues of an intelligent bench-scale study, we shall detail the kinetic study of the reaction of gaseous HCl with liquid lauryl alcohol as reported by Kingsley and Bliss.[3]   Not only does their study provide a comprehensive rate

---

[1] D. Luss, *AIChE J.*, **21**: 318 (1975).

[2] Our treatment, while accounting for the influence of temperature on M, does not include heat of solution in the heat balance. Luss does so (see Prob. 6-15).

[3] H. A. Kingsley and H. Bliss, *Ind. Eng. Chem.*, **44**: 2479 (1952).

expression which allows the designer to evaluate diverse reactor networks but some insight into the reaction mechanism is provided in the light of our earlier discussions.

The hydrochlorination of lauryl alcohol to yield lauryl chloride is typical of a gas-liquid reaction requiring intimate contact of gas with liquid. The product is of industrial importance as it is a raw material for manufacture of dodecyl mercaptan, a modifier for synthetic rubber. Scale-up to commercial level requires a comprehensive rate expression and a basis for optimum design to assure efficient utilization of gas.

Kingsley and Bliss studied the reaction in a semicontinuous bench-scale reactor. HCl was continuously fed to a batch charge of alcohol in a well-stirred vessel. By appropriate sampling of the liquid phase as a function of time, the data for alcohol concentration vs. time were fitted to the integrated expression for nth-order reaction with respect to the alcohol. The rate data are best assembled for $n = \frac{1}{2}$, that is,

$$-\frac{dA}{dt} = k\sqrt{A}$$

Since the reaction is catalyzed by dissolved zinc chloride, k will be a function of catalyst and HCl concentration as well as temperature and gas-liquid interfacial area. Interfacial area was not determined, but gas rate G was varied, as were other key variables. k was found to be a unique function of G/V, V being the liquid-phase volume; the activation energy between 110 and 150°C is 16 kcal/g mol. At 150°C

$$k = y_0(0.645)(0.43 + 64f) \left(\frac{G}{V}\right)^{0.55} \tag{6-80}$$

where f = weight fraction of catalyst
y_0 = inlet HCl mole fraction
G = moles HCl/time

Equation (6-80) is valid at catalyst concentrations above 0.5 percent. We note that the hydrochlorination rate is proportional to the square root of liquid-phase coreactant and G/V raised to the 0.55 power. In the light of our earlier discussions it is apparent that the Kingsley-Bliss findings are in excellent accord with the model of rapid pseudo-first-order reaction with diffusion, the zone of reaction being confined to the film (regime B). The G/V function reflects enhancement of interfacial area and possibly the gas-film contribution.

Example: Semibatch Operation

Application of the Kingsley-Bliss results will now be illustrated. We consider the problem of designing[1] a semicontinuous unit capable of handling a 22.65-kg mol charge of alcohol. The time required to achieve 96 percent conversion is to be determined, as well as the efficiency of HCl utilization at 150°C for three conditions:

---

[1] Problem taken from H. Bliss, chemical engineering kinetics course, Yale, 1955.

(1) $G/V = 3$, (2) $G/V = 1.5$, and (3) $G/V = 3$ with two vessels in series. Conditions 1 and 2 involve one vessel. One knows the average molecular weight of lauryl alcohol is 204 and its density at $150°C = 0.75$ g/ml. Let $x = $ conversion of alcohol. At $150°C$ and 1 percent by weight catalyst

$$\frac{dA}{dt} = -0.68 \left(\frac{G}{V}\right)^{0.55} \sqrt{A_0} \sqrt{1-x} \qquad (a)$$

Since A_0 is the initial alcohol concentration, $A = A_0(1-x)$ and

$$\frac{dx}{dt} = \frac{0.68}{\sqrt{A_0}} \left(\frac{G}{V}\right)^{0.55} \sqrt{1-x} \qquad (b)$$

Integrating gives

$$1 - \sqrt{1-x} = \frac{0.68}{2\sqrt{A_0}} \left(\frac{G}{V}\right)^{0.55} t \qquad (c)$$

$$A_0 = \frac{0.75(1000)}{204} = 3.66 \text{ g mol/liter}$$

The charge is 22,650 g mol of alcohol. The volume of charge is

$$\frac{22,650(204)}{0.75(1000)} = 6160 \text{ liters} = V$$

CASE 1   For $G/V = 3$, by Eq. (c), and for $x = 0.96$

$$1 - \sqrt{0.04} = \frac{0.68}{2\sqrt{3.66}} 3^{0.55} t \qquad (d)$$

$$t = 2.44 \text{ h}$$

For 100 percent yield, the moles HCl reacted equal the moles of alcohol reacted:

$$0.96(22,650) = 21,700 \text{ g mol HCl reacted}$$

$$\text{Moles HCl fed} = \frac{G}{V}(V)t = 3(6160)(2.44) = 45,400 \text{ g mol HCl}$$

$$\text{HCl utilization efficiency} = \frac{21,700}{45,400} \times 100 = 48\% \qquad (e)$$

CASE 2   This is the same as case 1, except that $G/V = 1.5$. Again, by Eq. (c), $t = 3.64$ h. Then

$$\text{HCl fed} = 1.5(6160)(3.64) = 33,600 \text{ g mol HCl}$$

$$\text{Efficiency} = \frac{21,700}{33,600} \times 100 = 64.6\% \qquad (f)$$

This result could have been anticipated since a slower addition rate, while reducing reaction rate, leads to the escape of less unreacted HCl.

CASE 3   We can split the charge so that 96 percent conversion is realized in each at the same time (thus we would have a series of two vessels of unequal volumes) (case 3), or we might split the charge equally and determine the conversion required in the second vessel, the contents of which when finally combined with that of the first vessel gives a product of 96 percent conversion (case 4).   Now as $G/V = 3$ for the first reactor, G/V will be less and a function of time in the second reactor.   If we assume product water vapor is condensed between reactors, $y_0 = 1$ in each reactor.

Let $\bar{\alpha} =$ fraction total volume in the first reactor.

$$\text{Total volume of system} = V_T = 6160 \text{ liters} \qquad (g)$$

HCl balance:

$$\text{Input} = \text{output} + \text{accumulation}$$

$$G_1 = G_2 + \bar{\alpha} A_0 V_T \frac{dx}{dt} \qquad (h)$$

$$A_0 V_T = \text{initial total moles alcohol} = 22{,}650$$
$$G_1 = 3 V_T \bar{\alpha} = 18{,}570\bar{\alpha}$$

Then

$$G_2 = 18{,}570\bar{\alpha} - 22{,}650\bar{\alpha} \frac{dx}{dt}$$

Now for $G/V = 3$ and $A_0 = 3.66$ g/liter,

$$\frac{dx}{dt} = 0.655\sqrt{1 - x} \qquad \text{as in case 1}$$

and $\sqrt{1 - x} = 1 - 0.328t$, so that

$$\bar{\alpha}22{,}650 \frac{dx}{dt} = \bar{\alpha}22{,}650(0.655) - \bar{\alpha}22{,}650(0.655)(0.328)t \qquad (i)$$

Therefore $G_2 = \bar{\alpha}(3770)(1 + 1.3t)$ and $V_2 = 6160(1 - \bar{\alpha})$

The rate equation for reactor 2 is then

$$\left(\frac{dx}{dt}\right)_2 = \frac{0.68}{\sqrt{3.66}}\left(\frac{\bar{\alpha}}{1 - \bar{\alpha}}\right)^{0.55}\left(\frac{3770}{6160}\right)^{0.55}(1 + 1.3t)^{0.55}\sqrt{1 - x} \qquad (j)$$

Combining constants and integrating and then solving for $\bar{\alpha}/(1 - \bar{\alpha})$ gives

$$\frac{\bar{\alpha}}{1 - \bar{\alpha}} = \left[\frac{(1 - \sqrt{1 - x})(14.8)}{(1 + 1.3t)^{1.55} - 1}\right]^{1/0.55} \qquad (k)$$

For $x = 0.96$, we know that $t = 2.44$ h; then

$$\frac{\bar{\alpha}}{1 - \bar{\alpha}} = 1.945 \qquad \text{or} \qquad \bar{\alpha} = 0.66$$

Therefore for equal conversion in a time of 2.44 h, the reactor volumes should be split so that

$$V_1 = 4100 \text{ liters} \qquad V_2 = 2090 \text{ liters} \qquad (l)$$

The HCl consumed is 21,700 g mol, and the HCl fed is 3(4100)(2.44) = 30,000 g mol, so that

$$\text{HCl efficiency} = \frac{21,700}{30,000} \times 100 = 72.3\% \qquad (m)$$

an increase over previously considered modes of operation. If greater time of reaction can be tolerated, further reduction in the volume of reactor 1 increases HCl efficiency. Thus if $\bar{\alpha} = 0.25$, $t = 5.6$ h and HCl efficiency becomes 84 percent. At some point t becomes great enough to offset the gain in HCl utilization. The limit is, since $(G/V)_1 V_T \bar{\alpha} t = 21,700$ for 100 percent utilization, for

$$\left(\frac{G}{V}\right)_1 = 3 \qquad \text{and} \qquad V_T = 6160 \qquad (n)$$

$$\bar{\alpha} t = 1.17 \text{ for } 100\% \text{ HCl efficiency}$$

In case 4, involving two equal-volume reactors, if

$$3\frac{V_T}{2}t_4 < 3(0.66V_T)t_3$$

HCl efficiency will be greater than that found in case 3.

CASE 4   Assuming that conversion in the first reactor is complete ($x_1 \approx 1$), conversion x_2 in the second reactor must be found by simple material balance:

$$A_0(0.96) = \frac{A_0}{2}(x_1 + x_2) \qquad (o)$$

If $x_1 = 1$ (to be checked later), then $x_2 = 0.92$. The time required for a 92 percent conversion in the second reactor must now be determined, using the integrated form of Eq. (h), where $\bar{\alpha}$ is now 0.5. After finding t, a conversion calculation for the first reactor will be made to check our assumption of $x_1 = 1$. For $\bar{\alpha} = 0.5$, the integrated form is

$$1 + (1 - \sqrt{1 - x_2})(14.8) = (1 + 1.3t)^{1.55}$$

For $x_2 = 0.92$ we find

$$t = 3 \text{ h} \qquad (p)$$

For reactor 1, at $t = 3$

$$x_1 = 1 - 0.000225 \approx 1$$

so that our assumption was sound, and 3 h of reaction at $G_1/V = 3$ in a series of two equal-volume reactors provides, after mixing the final contents of each, the desired 96 percent conversion.   HCl utilization for case 4 is

$$\text{HCl fed} = 3 \, \frac{6160}{2} \, 3 = 27,800 \text{ g mol} \tag{q}$$

$$\% \text{ HCl efficiency} = \frac{21,700}{27,800} \times 100 = 78\%$$

An alternative mode of semicontinuous operation, suggested by Kingsley and Bliss, involves programming the HCl feed rate so that throughout the reaction the off gas is always in excess by a constant value.   This means reducing the gas rate as reaction progresses, which of course further reduces the rate through the G/V functionality.   This manner of operation was actually demonstrated experimentally by the authors, and computed conversions based upon the rate equation agreed very well with experimental data secured during programmed gas-rate operation. We illustrate programmed addition of HCl with an example due to Bliss.

Let it be assumed that throughout reaction in a single vessel the HCl addition rate will be scheduled so that it is always in 15 percent excess of its rate of consumption

$$\frac{G}{V} = 1.15 \, \frac{dA}{dt} = 1.15 \, A_0 \, \frac{dx}{dt}$$

Since

$$\frac{dx}{dt} = \frac{0.68}{\sqrt{A_0}} \left(\frac{G}{V}\right)^{0.55} \sqrt{1-x} \qquad \text{at } 150°\text{C and } 1\% \text{ catalyst}$$

then for $A_0 = 3.66$, as in the previous problem, on substituting for G/V in terms of dx/dt,

$$\frac{dx}{dt} = \frac{0.68}{1.9} (1.15)(3.66) \left(\frac{dx}{dt}\right)^{0.55} \sqrt{1-x} \tag{r}$$

or

$$\frac{dx}{dt} = 0.58(1-x)^{1.11}$$

The time required for a 96 percent conversion by this mode of operation, on integration, is

$$0.58t = \frac{9.1}{(1-x)^{0.11}} - \frac{9.1}{1} \qquad \text{or} \qquad t = 6.73 \text{ h}$$

and HCl utilization is 85 percent.

Example: Continuous Operation

The above example provides some insight into factors of importance in design, i.e., reactor volume, time, and reagent-consumption efficiency.   In the semicontinuous method of operation, however, charging and discharge times must be added to

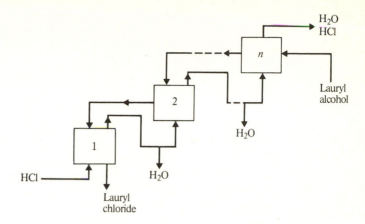

FIGURE 6-16
n-stage countercurrent network for a gas-liquid reaction.

reaction time and associated labor costs.   Continuous operation eliminates these factors in addition to providing product qualities associated with steady-state operation.   We shall now investigate possible means of continuous production of lauryl chloride.

Assume that a countercurrent system is to be designed, as schematized in Fig. 6-16.   Each stage may be a stirred vessel with appropriate gas-dispersion device or a sieve tray at the bottom of the vessel.   With H_2O removal between stages, G/V will decrease progressively, but because of countercurrent liquid flow, reactant concentration is highest in the last stage, where G/V is lowest.   Further, the possibility of realizing very high, virtually 100 percent HCl utilization exists with countercurrent operation under the assumption that in each stage the liquid phase is perfectly mixed while the gas bubbles pass up through the liquid in plug flow.[1]   Thus in the last stage G/V will be small, consisting of pure HCl feed, while vapor leaving the last stage will contain virtually pure water vapor.

The design goal is to produce 48 mol[†] of lauryl chloride per hour at 96 percent conversion level; hence 50 mol of lauryl alcohol must be fed per hour.   We wish to determine the number of contacting stages required under continuous, steady-state operation.   A 100 percent HCl utilization is desired.   A holdup time per stage of 1 h will be assumed, $T = 150°C$, and the solution contains 1 percent by weight of dissolved catalyst.   Each stage is visualized as shown in Fig. 6-17. In view of the assumption of perfect mixing in the liquid phase, the Kingsley-Bliss rate equation for liquid-phase reaction applies in terms of the effluent liquid-phase concentration and the G/V value entering the stage, so that

$$-\frac{dA}{dt} = \frac{A_{n+1} - A_n}{\theta} = k\left(\frac{G}{V}\right)_n^{0.55}\sqrt{A_n} \qquad (a)$$

---

[1] Compare with two-phase fluidized-bed model in Chap. 10.
† The choice of pound, gram, or kilogram moles is left to the instinct of the reader.

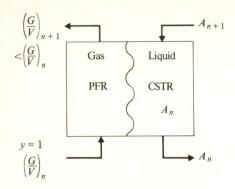

FIGURE 6-17
Stage model of gas-liquid contacting.

where $\theta = V/Q$ is the stage reaction volume per volume feed rate.   Conversion x is $x = (A_{n+1} - A_n)/A_{n+1}$ or, solving for A_{n+1},

$$A_{n+1} = \frac{A_n}{1 - x}$$

Substituting and dividing through by A_n gives

$$\frac{1}{1 - x} - 1 = k\left(\frac{G}{V}\right)_n^{0.55} \left(\frac{\theta}{\sqrt{A_n}}\right) = \beta \qquad (b)$$

or, solving for x,

$$x = \frac{\beta}{\beta + 1}$$

Specifying θ, G/V, and A_n for the first stage allows calculation of x for given temperature k, and hence A_2 and G/V for second stage are then computed.   One simply proceeds in this fashion until inlet lauryl alcohol concentration is reached and an HCl balance is realized.   The inlet value of G/V is dictated by the HCl requirement and utilization efficiency.   For virtually complete utilization

$$\frac{G}{V} 6190 = 21,700 \qquad (c)$$

Thus $(G/V)_{n=1} = 3.5$. G/V entering the next, second, stage is (moles/$t \cdot V$)

$$\left(\frac{G}{V}\right)_2 = \left(\frac{G}{V}\right)_1 - \frac{(A_2 - A_1)Q}{V}$$

or in general

$$\left(\frac{G}{V}\right)_{n-1} - \frac{(A_n - A_{n-1})Q}{V} = \left(\frac{G}{V}\right)_n \qquad (d)$$

Now $V = 6190$ liters, $Q = 6190$ liters/h, and $\theta = 1$ h.

Stage 1:

$$A_1 = A_0(1 - x_0) = 3.66(0.04) = 0.146 \text{ mol/liter}$$

where x_0 is overall conversion and A_0 fresh alcohol-feed concentration, computed in the earlier example. At 150°C

$$\beta_1 = k\left(\frac{G}{V}\right)^{0.55}\frac{\theta}{\sqrt{A_1}} = 0.68(3.5)^{0.55}\frac{1}{\sqrt{0.146}} = 3.54$$

$$x_1 = \frac{\beta}{1+\beta} = 0.78 = 1 - \frac{A_1}{A_2} \qquad \text{so } A_2 = 0.664 \qquad (e)$$

Therefore in the first stage, 78 percent of incoming alcohol from stage 2 at a concentration of 0.664 mol/liter is converted to final product of 0.146 mol/liter of unconverted alcohol. Therefore A_2, average composition in and leaving the second stage, is 0.664 mol/liter and G/V entering stage 2 is

$$\left(\frac{G}{V}\right)_2 = 3.5 - (0.664 - 0.146) = 2.98$$

Stage 2:
$$\beta = \frac{0.68(2.98)^{0.55}(1)}{\sqrt{0.664}} = 1.52$$

$$x_2 = \frac{1.52}{2.52} = 0.602 = 1 - \frac{A_2}{A_3} \qquad \text{so } A_3 = 1.67 \qquad (f)$$

and
$$\left(\frac{G}{V}\right)_3 = 2.98 - (1.67 - 0.664)(1) = 1.874$$

Stage 3:
$$\beta = \frac{0.68(1.874)^{0.55}}{1.67} = 0.743$$

$$x_3 = \frac{0.743}{1.743} = 0.426 \qquad A_4 = 2.9 \qquad (g)$$

$$\left(\frac{G}{V}\right)_4 = 1.874 - (2.9 - 1.67) = 0.644$$

Stage 4:
$$\beta = \frac{0.68(0.644)^{0.55}}{\sqrt{2.9}} = 0.31$$

$$x_4 = \frac{0.31}{1.31} = 0.24 \qquad (h)$$

Thus
$$A_5 = \frac{2.9}{1 - x_4} = 3.8 \text{ mol/liter}$$

This value of A_5 is greater than the fresh feed value of 3.66 mol/liter, and a recalculation would be required using an adjusted G/V value entering stage 1 until the material balance is closed. Actually greater precision than provided by slide rule is required to obtain an exact balance. There is little point in going beyond this computation for the present purposes. A four-stage countercurrent unit seems necessary, and actual design recommendations would probably specify five stages.

Deviations from CSTR behavior in any stage would be an advantage, of course. Note that in continuous countercurrent operation the production rate is

$$48 \frac{\text{mol}}{\text{(h)(volume)}} = 1.93 \times 10^{-3} \frac{\text{mol product}}{\text{(h)(reaction volume)}}$$

which is to be compared to the semicontinuous operation rate of

$$\frac{48 \text{ mol}}{2.44(6190)} = 3.17 \times 10^{-3} \frac{\text{mol product}}{\text{(h)(reaction volume)}}$$

However, if due allowance is made for charging, heat-up, cooling, and discharging, the semicontinuous production rate will be far below the continuous value. In addition, HCl utilization is far higher in continuous operation. It might be noted that the staged continuous system allows one to distribute HCl feed between stages, thus enhancing rate and reducing the number of stages (reaction volume) with, of course, a corresponding reduction in HCl utilization efficiency.

Nitrogen Oxides Absorption-Reaction

A classic and highly important absorption-reaction system is that involving the absorption of nitrogen oxides (NO_2, N_2O_4, NO, N_2O_3, HNO_2) into water and nitric acid to yield a 60 percent by weight HNO_3. At finite acid strength an equilibrium reaction is involved:

$$N_2O_4 + H_2O \rightleftharpoons HNO_3 + HNO_2$$

$$HNO_2 \rightleftharpoons \tfrac{1}{3}H_2O + \tfrac{1}{3}HNO_3 + \tfrac{2}{3}NO$$

In a commercial absorber, HNO_3 strength varies from about 60 percent by weight in the bottom tray to zero percent at the top tray, the gases passing upward countercurrent to the descending liquid. Andrew[1] notes that a number of liquid-phase reactions are possible involving N_2O_4, NO_2, N_2O_3 (in equilibrium with NO, NO_2), and HNO_2. Gas-phase reactions between water vapor and NO_2 and/or N_2O_4 are also cited by some workers; however, some controversy exists on this point.[2]

Andrew[3] reports the results of a detailed analysis of single sieve-plate efficiency carried out by procedures similar to that detailed above for the CO_2–NH_3 solution system. All candidate reactions were analyzed by Andrew; plate efficiency as a function of gas strength for various irreversible reactions involved were computed, and an excellent comparison with experimental data was found. In accord with experience in plant absorbers, plate efficiency drops with gas strength from a value of over 50 percent at column bottom to only several percent at the top of the absorber. Actual absorbers involve reversible reactions, as mentioned

[1] S. P. S. Andrew and D. Hanson, *2d Eur. Symp. Chem. React. Eng. 1960*, pap. O1.
[2] J. J. Carberry, *Chem. Eng. Sci.*, **9:** 189 (1959); see also M. M. Wendel and R. L. Pigford, *AIChE J.*, **4:** 249 (1958).
[3] Loc. cit.

above, and a complete analysis of the nitrogen oxides absorption-reaction system over the entire range of acid strength encountered commercially is yet to be made, although an admirable contribution is provided by Denbigh and Prince.[1]

In this study of the rate of absorption over a wide range of acid strength, Denbigh made what appears to be pioneering use of the short wetted-wall column. In the absence of high partial pressures of NO, the absorption-reaction rate in acid between 0 to 60 percent HNO_3 by weight is found by Denbigh to be

$$-\frac{d(N_2O_4)}{dt} = k[N_2O_4 - C(N_2O_4)^{1/4}(NO)^{1/2}]$$

so C is a function of the equilibrium constant; its explicit dependence is set forth by Carberry,[2] while Gray and Yoffe[3] discuss the chemical, ionic mechanism of the reaction and derive an a priori rate expression identical to Denbigh's. The role of NO in the absorption was studied by Caudle and Denbigh.[4]

It can be supposed that if interfacial area for a particular contactor is known or susceptible to measurement, Denbigh's rate expression can be used to compute absorption rates. In a shell-and-tube cooler condenser turbannular flow can be anticipated, so that the interfacial area between the core of gas and liquid film at the wall can be taken as roughly equal to the tube-wall area. Absorption-reaction rates may then be computed, while gas-phase conversion of NO to NO_2 can be determined from Bodenstein's kinetic data. As the cooler condenser is non-isothermal, involving condensation of water vapor as well as reaction in both gas and liquid phases, calculations are tedious. In the absorber, however, some simplifications are permissible; the design procedure is outlined below.

Commercial Absorption of Nitrogen Oxides

As conducted commercially, the plate-absorber-column volume is determined largely by a homogeneous-gas-phase reaction occurring between the absorption plates. Since on the plate we have

$$\tfrac{3}{2}N_2O_4 + H_2O \;\rightleftharpoons\; 2HNO_3 + NO \qquad (6\text{-}81)$$

the desorbed NO must be oxidized to NO_2 (N_2O_4) to realize maximum fixed-nitrogen absorption. In the presence of O_2, the reaction

$$NO + \tfrac{1}{2}O_2 \;\longrightarrow\; NO_2 \qquad (6\text{-}82)$$

occurs in the gas phase between the plates. The NO oxidation is rather slow, exhibiting negative temperature dependence for reasons cited by Carberry.[5]

An effective absorber design procedure involves:

1 Assuming chemical equilibrium to be established on each plate for a specific acid product. For, say, a 60 percent by weight HNO_3 product issuing from

[1] K. G. Denbigh and A. J. Prince, *J. Chem. Soc.*, **1947**: 790.
[2] J. J. Carberry, *Chem. Eng. Sci.*, **9**: 189 (1959).
[3] P. Gray and A. D. Yoffe, *Chem. Rev.*, **55**: 1069 (1955).
[4] P. G. Caudle and K. G. Denbigh, *Trans. Faraday Soc.*, **49**: 39 (1953).
[5] J. J. Carberry, *Chem. Eng. Sci.*, **9**: 189 (1959).

the first, bottom, tray, the gas composition leaving that tray is computed for a fixed feed-gas composition using the appropriate equilibrium relation, which is expressed analytically in Chilton's treatise.[1]

2   Given the gas composition leaving the plate, the gas-phase oxidation of NO is computed for the contact time determined by plate spacing and gas velocity.

3   A material balance establishes the acid strength on the next plate, and step 1 is repeated for the gas composition as modified by interplate oxidation.

The nitrogen oxides–absorption system best exemplifies the demands of design and the limited supply of fundamental data required for rigorous a priori design.   Only Denbigh's study covers the entire range of commercial acid strength of interest.   Yet to apply his rate equation, more accurate interfacial area data are needed for bubble-cap absorbers.   Further, in commercial units the assumption of a uniform liquid composition across the plate is questionable, as is that of plug flow in the interplate gas phase, where a third-order reaction (NO oxidation) occurs. Nor is the system isothermal, the heat of absorption being considerable in the lower third of the column.   Finally, the product acid strength is crucially governed by the partial pressure of N_2O_4 in the gas entering the bottom plate, suggesting that the cooler-condenser preceding the absorber plays a key role in not only condensing water of reaction created via NH_3 oxidation and cooling the gases but also in oxidizing NO to NO_2 (N_2O_4) to establish the state of oxidation commensurate with production of the desired acid strength.   The cooler-condenser is then the heart of the nitrogen oxides–absorption system in the NH_3 oxidation route to HNO_3, and its design is far more challenging than that of the absorption column.   Referring to Fig. 6-1, we see that gases at perhaps 300 to 400°C enter the cooler-condenser, which must produce a cooled gas (about 25 to 30°C) of high NO_2 content relative to NO. The following events take place within this unit:

1   Hot gas is cooled to its dew point.

2   Water condenses.

3   With further cooling NO is oxidized to NO_2 (N_2O_4).

4   The N_2O_4 is transported to the gas-liquid interface and there absorbed with reaction to produce HNO_3 and NO.

5   Product NO on entering the gas phase is oxidized to NO_2, and so on.

The cooler-condenser thus involves heat transfer, condensation, gas-phase reaction, and diffusion and reaction in the condensed phase.

Equipment volume (cooler-condenser and absorber) in the ammonia-oxida-tion-process system is clearly governed by mixed events.   The fact that ammonia-oxidation-process units operate at well over 90 percent efficiency is testimony to the intelligence, judgment, and resourcefullness of those responsible for its development at a time when the term simultaneous absorption-reaction kinetics was not yet

[1] T. H. Chilton, Dupont Pressure Process for Manufacture of HNO_3, *AIChE Monogr.* 13, 1960.

coined, much less an object of research. Details on other absorption-reaction systems of commercial interest can be found in Norman's text.[1]

Example: N_2O_4-Absorber Design

As illustrated in Fig. 6-1 and described above, design of a NO_2-N_2O_4 absorber involves the assumption of chemical equilibrium on the tray and the computation of the extent of gas-phase oxidation of NO between trays. The equilibrium absorption-reaction is

$$3NO_2 + H_2O(l) \rightleftharpoons 2HNO_3(l) + NO \qquad (a)$$

also

$$2NO_2 \rightleftharpoons N_2O_4 \qquad (b)$$

Reaction (a) is equilibrated slowly; (b) instantaneously.

The irreversible rate of gas-phase oxidation of NO in terms of partial pressures is

$$-\frac{d(NO)}{d\theta} = k_1(NO)^2(O_2) \qquad (c)$$

where k_1, as determined by Bodenstein,[2] is

$$\log k_1 = \frac{635}{T} - 1.026 \qquad T \text{ in Kelvins} \qquad (d)$$

Let it be assumed that a 60 percent by weight HNO_3 is to be produced in a bubble-cap plate absorber of 1.6 m ID and 30-cm plate spacing (between liquid level and bottom of the plate above). Let $T = 30°C$, and total pressure be 7 atm. Gross weight of the acid desired is 4345 kg/h equivalent to 91.16 mol/h of HNO_3.[3] Feed rate and composition entering the first, bottom, plate is

Species	Feed rate, mol/h
NO	3.4
$NO_2 + N_2O_4$	68.2
O_2	60.0
N_2	868.4
Total	1000

Referring to Fig. 6-1, we note that some HNO_3 is produced in the cooler-condenser and is directed to the absorber at a plate of equal acid strength. Assuming that 20.9 mol/h of HNO_3 is created in the condenser, the balance, 70.2 mol/h, is to be produced in the absorber. Thus absorber recovery efficiency is $70.26/71.6 \times 100 = 98.2$ percent since 71.6 mol of NO-NO_2-N_2O_4 is fed. Let the partial pressures in atmospheres be $a = NO$, $b = NO_2$, $c = N_2O_4$, and $s = O_2$. If

[1] W. S. Norman, "Absorption, Distillation and Cooling Towers," Wiley, New York, 1961.
[2] M. Bodenstein, Z. Phys. Chem., **100**: 68 (1922).
[3] Pound-mol, of course.

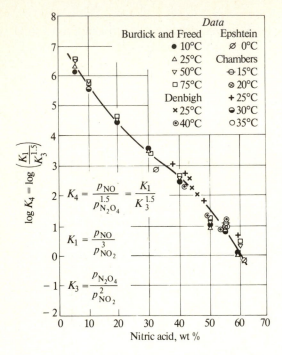

$$\log K_4 = \log\left(\frac{K_1}{K_3^{1.5}}\right)$$

$$K_4 = \frac{p_{NO}}{p_{N_2O_4}^{1.5}} = \frac{K_1}{K_3^{1.5}}$$

$$K_1 = \frac{p_{NO}}{p_{NO_2}^3}$$

$$K_3 = \frac{p_{N_2O_4}}{p_{NO_2}^2}$$

Nitric acid, wt %

FIGURE 6-18

Generalized equilibrium relationship for the HNO_3-NO-N_2O_4 system. Data from C. L. Burdick and E. S. Freed, *J. Am. Chem. Soc.*, **43**: 518 (1921); D. A. Epshtein, *J. Gen. Chem. USSR*, **9**: 792 (1939); F. S. Chambers and T. K. Sherwood, *Ind. Eng. Chem.*, **29**: 1415 (1931); K. G. Denbigh and A. J. Prince, *J. Chem. Soc.*, **1947**: 790. [*J. J. Carberry, Chem. Eng. Sci.*, 9:189 (1959).]

subscript 1 refers to entering and 2 to exiting species on the plate, then, in accord with Eq. (*a*),

$$3(a_2 - a_1) = (b_1 + 2c_1) - (b_2 + 2c_2) \qquad (e)$$

Composition leaving a plate is dictated by equilibria (*a*) and (*b*). For (*a*) the correlation of experimental data in terms of N_2O_4 (Fig. 6-18) can be expressed in terms of weight fraction w of HNO_3 by

$$\log K_4 = 7.412 - 20.29w + 32.47w^2 - 30.87w^3 \qquad (f)$$

while Bodenstein's relationship for (*b*) is

$$\log K_3 = \frac{2993}{T} - 9.226 \qquad T \text{ in Kelvins} \qquad (g)$$

Note that our K_4 is independent of T in the manner of the correlation in Fig. 6-18. Now

$$K_4 = \frac{a}{c^{1.5}} \qquad \text{and} \qquad K_3 = \frac{c}{b^2} \qquad (h)$$

Therefore

$$c = K_3 b^2 \qquad \text{and} \qquad a = K_4 K_3^{3/2} b^3$$

which are equilibrium values, so that Eq. (*e*) becomes

$$3(K_4 K_3^{3/2} b_2^3 - a_1) = (b_1 + 2c_1) - (b_2 + 2K_3 b_2^2) \qquad (i)$$

while a_1, b_1, and c_1 are computed from the entering composition and total pressure; for example, $a_1 = 3.4/1000 \times 7 = 0.0238$ atm. For $w_1 = 0.6$, by (*f*) we find $K_4 = 1.82$ and by (g) $K_3 = 4.5$, so that the value of NO_2 in equilibrium with 60% acid on the first plate is given by

$$52.5b_2{}^3 + 9b_2{}^2 + b_2 - 0.549 = 0 \qquad (j)$$

By trial and error we find

$$b_2 = NO_2 = 0.1524 \qquad N_2O_4 = c_2 = K_3 b_2{}^2 = 0.1045$$

and

$$NO = a_2 K_4 c_2{}^{3/2} = 0.061 \text{ atm}$$

Total nitrogen peroxide (NO_2, N_2O_4) absorbed is

$$(b_1 + 2c_1) - (b_2 + 2c_2) = 0.116 \text{ atm} \qquad (k)$$

or $\quad \dfrac{0.116}{7} \times 1000 \times (\frac{2}{3}) = 11.1 \text{ mol } HNO_3$ produced on plate 1

Therefore, the moles of HNO_3 descending from plate 2 is 80, from which one computes the acid strength on plate 2 to be $w_2 = 0.562$.

In its time of residence in the gas space between plates 1 and 2, NO is oxidized to NO_2 according to Eq. (c). The residence time computed from interplate volume and volumetric gas flow at 7 atm and 30°C is 1.25 s. At 30°C, k_1 is, by Eq. (d), 11.75. Since O_2 is in vast excess, Eq. (c) becomes

$$-\frac{d(NO)}{d\theta} = \bar{k}_1 (NO)^2$$

which is readily integrated to

$$\frac{1}{a_1} = \frac{1}{a_2} + (2s_2 - a_2)\frac{k\theta}{2} \qquad (l)$$

Substituting the known values, we find the partial pressure of NO entering plate 2 to be $a = 0.0224$ atm.

The procedure is now repeated for the next plate, and so on, taking into account the side-stream addition of cooler-condenser acid at that plate at which its composition approximates that of the side stream.

Detailed profiles, as computed by the author for T. H. Chilton, are set forth in Table 6-4, and overall results reflecting the influences of plate spacing, temperature, and pressure upon the number of theoretical plates are given in Table 6-5.

The question remains, however: How many actual plates are required to realize the design specifications? In other words, what is the plate efficiency? Not surprisingly, plate efficiency depends both upon gas- and liquid-phase composition, as the work of Andrew and Hanson indicates for the nitrogen oxides–nitric acid absorption process.[1] Their data would suggest that an overall efficiency of about 50 percent would suffice for prudent design.

Finally, this absorption-reaction system is exothermic, and cooling coils are required on the trays, particularly the lower plates, where the major fraction of the

---

[1] S. P. S. Andrew and D. Hanson, *Chem. Eng. Sci.*, **14**: 105 (1961).

acid is produced, as Table 6-4 indicates.   A temperature rise on one or more of the lower plates is detrimental to product acid strength since both the absorption equilibria (a) and (b) and the rate of NO oxidation are adversely affected by high temperature.

In this section we have attempted to set forth both theoretical and practical aspects of reaction between gases and liquids.   It appears that in spite of its inherent simplifications, the film theory nevertheless provides the designer with a powerful basis upon which meaningful performance can be predicted.   In essence the effect of chemical reaction upon physical absorption of a gas into a liquid can be beneficial or negligible, depending upon the values of key dimensionless parameters which provide ratios for reaction with diffusion and reaction per se relative to physical mass transfer (ϕ and α).   More sophisticated absorption models, which involve a transient term, reveal an influence of reaction upon absorption which differs insignificantly from that predicted by simple film theory, due originally to Noyes, Whitney, and Nernst and later propagated by Whitman.

By cleverly designed experiments which reveal interfacial area or equivalent information, scale-up is possible.   Great flexibility in design is possible in terms of reagent deployment, e.g., interstage addition, as well as reactor type (packed column, sieve-tray absorber, etc.).   The treatment presented in this chapter is not intended to be exhaustive: its purpose is to present in outline the key factors

Table 6-4 N_2O_4-ABSORBER-COMPUTATIONS †

Plate	HNO_3, wt %	Oxides entering, mol	State of oxidation, % $\dfrac{NO_2 + 2N_2O_4}{NO + NO_2 + 2N_2O_4}$	Total oxides, % recovered	
				Per plate	In column, cumulative
1	60.0	71.6	95.0	15.5	15.5
2	56.2	60.5	91.7	21.1	36.6
3	50.2	45.4	86.5	22.2	58.8
4§	42.3	29.6	77.6	15.4	74.2
5	36.8	18.6	66.7	8.17	83.6
6	27.3	12.8	58.3	5.46	87.7
7	19.4	8.9	47.5	3.22	91.0
8	14.2	6.5	39.0	1.99	92.9
9	10.7	5.1	32.9	1.31	94.2
10	8.2	4.2	28.4	0.92	95.1
11	6.4	3.5	24.9	0.69	95.8
12	5.0	3.0	22.3	0.52	96.3
13	3.9	2.7	20.2	0.41	96.5
14	3.1	2.3	18.5	0.32	97.0
15	2.4	2.1	17.1	0.27	97.2
16	1.8	1.88	16.0	0.22	97.3
17	1.3	1.73	15.0	0.20	97.5
18	0.8	1.63	14.1	0.17	97.7
19	0.5	1.49	13.4	0.14	98.1
20	0.2	1.41	12.8	0.12	98.2

† T. H. Chilton, Chem. Eng. Prog. Monogr. 13, p. 56, 1960.
§ Cooler-condenser acid added on this plate.

Table 6-5   INFLUENCE OF KEY PARAMETERS UPON
N_2O_4-ABSORBER EFFICIENCY †

Effect of total pressure, § $T = 30°C$, 30-cm plate spacing

Pressure, atm	No. of plates
5	57
7	20
9	12

Effect of temperature, pressure = 7 atm, 30-cm plate spacing

Temp, °C,	No. of plates
20	18
30	20
40	28

Effect of plate spacing, § $T = 30°C$, pressure = 7 atm

Plate spacing, cm	No. of plates
30	20
60	15
90	12

† T. H. Chilton, *Chem. Eng. Prog. Monogr.* 13, p. 56, 1960.
§ These effects point to the importance of gas-phase reaction between trays in dictating column size.

governing gas-liquid reaction and some aspects of the design problem. Further details should be sought in available texts and journal literature to which reference has been made.

6-15 HETEROGENEOUS LIQUID-LIQUID REACTIONS

Reaction between partially or totally immiscible liquids involves, in principle, simultaneous diffusion and thus in many key ways resembles gas-liquid reaction. However, while diffusion-reaction is generally confined to the liquid phase in a gas-liquid system, it is possible to anticipate the process of diffusion-reaction occurring in both phases in a heterogeneous liquid-liquid system.

In spite of the prevalence of two-phase liquid-liquid reaction in industry, remarkably little fundamental work has been performed to clarify these complex events. The nitration of benzene in a two-phase liquid-liquid system involving an aqueous mixed acid phase in contact with an organic phase is a classic example involving diffusion-reaction. So too is the olefin-alkylation system as carried out in the petroleum industry. The principles offered earlier in this chapter for gas-liquid absorption-reaction remain valid for liquid-liquid systems, and the criteria by which various extremes of rate control can be ascertained are certainly relevant. We noted earlier that useful conclusions can be reached by viewing reaction-diffusion and reaction per se relative to physical mass transfer. As it happens, many liquid-liquid reactions (two-phase of course) are conducted at relatively low temperatures. Liquid-phase olefin alkylation, for example, is carried out at temperatures well below 0°C. The comparatively low boiling points of many organics naturally limit their liquid-phase reaction environment to temperatures which are low compared with most reaction systems. In consequence actual chemical reaction rates in one or both phases tend to be rather low with the result that (1) chemical reaction per se governs the overall rate or (2) reaction can be considered to occur in series with physical mass transfer rather than simultaneously. These are regimes C and D of Fig. 6-8.

In this very brief discussion of reaction-diffusion in immiscible liquid-liquid systems, we intend to be even less exhaustive than we were in discussions of gas-liquid reaction. A few case histories will be treated to direct attention to some aspects of the problem and to focus upon the similarities between liquid-liquid and gas-liquid reaction systems.

Very Slow Reaction with Immiscible Liquids

When reaction in one of two phases is very slow relative to physical mass transport, we have simple homogeneous kinetics in *one* phase; however, one reactant, at least, is distributed between the two phases. A distribution coefficient thus governs the equilibrium relationship between phases. Klein, McKelvey, and Webre[1] treat the problem in a manner which permits simultaneous measurement of distribution coefficient and reaction-rate constant. Consider an organic (*o*) and aqueous (*a*) phase in intimate contact, of volumes V_o and V_a, respectively. For a first-order reaction in phase *a*,

$$\frac{-dA_a}{dt} = kA_a \qquad (6\text{-}83)$$

while the reactant concentration in equilibrium in the organic phase A_o is related by the distribution coefficient

$$K = \frac{A_o}{A_a} \qquad (6\text{-}84)$$

A differential material balance for A yields

$$V_o \frac{dA_o}{dt} + V_a \frac{dA_a}{dt} + V_a kA_o = 0 \qquad (6\text{-}85)$$

By Eq. (6-84)

$$\frac{dA_o}{dt}\left(V_o + \frac{V_a}{K}\right) = -V_a \frac{k}{K} A_o \qquad (6\text{-}86)$$

[1] E. Klein, J. B. McKelvey, and B. G. Webre, *J. Phys. Chem.*, **62**: 286 (1958).

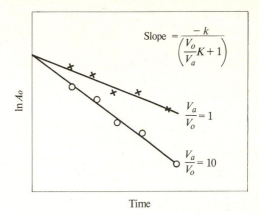

FIGURE 6-19
Graphical display of two-phase reaction
data permitting the extraction of key rate
parameters. [*E. Klein, J. B. McKelvey,
and B. G. Webre, J. Phys. Chem.*, **62**:286
(*1958*).]

which, upon integration, where $A_o{}^0$ is initial concentration of A, that is at $t = 0$,
after equilibrium is established, i.e. interfacial

$$\ln A_o{}^0 - \ln A_o = \frac{kt}{(V_o/V_a)\,K + 1} \qquad (6\text{-}87)$$

A plot of $\ln A_o$ versus t provides a slope of $-k/[(V_o/V_a)K + 1]$. Variation
in V_o/V_a will provide another slope in agreement with Eq. (6-87), so that both k and
K can then be determined (Fig. 6-19). The method is obviously applicable to gas-
liquid reaction, as the authors note. It would appear, in fact, that as long as
pseudo-first-order reaction in A prevails, the method is not restricted to the
assumption of slow chemical reaction, for the measured k can well be a complex
function of reaction-diffusion and coreactant concentration, as outlined earlier.

Klein et al. evaluated their model in a study of the reaction of butadiene-
diepoxide with cotton cellulose in the presence of aqueous sodium hydroxide and
successfully determined the aqueous-phase rate constant and distribution coefficient
for the diepoxide in this batch liquid-liquid reaction study.

Liquid-Phase Alkylation

We shall now devote attention to a very practical liquid-liquid reaction system
involving mass transfer and selectivity, namely, the H_2SO_4-catalyzed alkylation
of butylene with isobutane. A C_8 alkylate is produced as well as higher alkylate.
Since the quality of alkylate, such as octane rating in ultimate gasoline tests, is
essentially proportional to C_8, the ratio of C_8 to all other alkylate is a measure of
selectivity.

The reaction itself occurs in the continuous acid-catalyst phase, in which both
olefin molecules and carbonium ions are solvated, while the isobutane is transported from the organic to acid phase in which it reacts. Thus, in principle, mass
transfer of isobutane can affect overall reaction rate. In practice such alkylations
are often carried out continuously in well-stirred stages, the overflowing product

emulsion being ultimately settled and separated. The following analysis is that of Jernigen, Gwyn, and Claridge.[1] A simplified version of the various reactions involved begins with the creation of a carbonium ion by reaction of acid proton with olefin. Let

$$C_4{}^{2-} = \text{olefin, butylene}$$
$$H^+ = \text{acid proton}$$
$$C_4^*, C_8^* = \text{carbonium ions of } C_4{}^{2-} \text{ and } C_8{}^{2-}$$
$$B = \text{isobutane}$$
$$C_8, C_{12} = \text{saturated alkylate}$$

Thus

Step 1: $\qquad C_4{}^{2-} + H^+ \longrightarrow C_4^*$

Step 2: $\qquad C_4^* + C_4{}^{2-} \longrightarrow C_8^*$

Step 3: $\qquad C_8^* + B \longrightarrow C_8 + C_4^* \qquad$ primary product, C_8

Step 4: $\qquad C_8^* + C_4{}^{2-} \longrightarrow C_{12}^* \qquad$ undesired by-product alkylate and polymer

Step 5: $\qquad C_{12}^* + B \longrightarrow C_{12} + C_4^*$

First an olefin, $C_4{}^{2-}$, balance for a single CSTR of volume $\overline{V}$, flow rate Q, and difference between olefin feed and effluent concentration $\Delta C_4{}^{2-}$ is

$$\frac{Q \, \Delta C_4{}^{2-}}{\overline{V}} = S = k_1(C_4{}^{2-}) + k_2(C_4{}^{2-}C_4^*) + k_4(C_4{}^{2-}C_8^*) \qquad (6\text{-}88)$$

where S is olefin space velocity.

If the rate of production and consumption of carbonium ions is equal in steady state, $k_1(C_4{}^{2-}) = k_6(C_x^*)$, where k_6 is a composite constant representing all routes to carbonium-ion loss. In fact Eq. (6-88) can be simplified to

$$S = k_1(C_4{}^{2-}) + K_2(C_4{}^{2-}C_x^*)$$

By the relation $k_1(C_4{}^{2-}) = k_6(C_x^*)$, we find

$$S = k_1(C_4{}^{2-}) + \frac{K_2(C_4{}^{2-})^2 k_1}{k_6}$$

Assuming that

$$k_1(C_4{}^{2-}) < \frac{K_2 k_1}{k_6}(C_4{}^{2-})^2$$

then

$$C_4{}^{2-} = \sqrt{\frac{k_6 S}{K_2 k_1}} \qquad (6\text{-}89)$$

[1] E. C. Jernigen, J. E. Gwyn, and E. L. Claridge, Simultaneous Mass Transfer and Reaction Rate Effects in Butylene Alkylation, *San Francisco AIChE Meet.*, *May 1965*.

Since the product quality is known to depend upon the ratio of the reaction of C_8^* with isobutane to that of C_8^* with olefin, for a CSTR

$$\frac{C_8}{C_x} = \frac{k_3(C_8^*)B_a}{k_4(C_8^*)(C_4{}^{2-})} = \frac{k_3}{k_4}\frac{B_a}{C_4{}^{2-}} \qquad (6\text{-}90)$$

where B_a is isobutane concentration in the acid phase. If it is supposed that mass transport of isobutane from organic to acid phase occurs *in series* with its consumption in the acid phase, then

$$k_{L_0} a(K_b B_o - B_a) = k_3(C_8^*)B_a + k_5(C_{12}^*)B_a \qquad (6\text{-}91)$$

where $K_b = (B_a/B_o)_i$ Equation (6-91) can be written in terms of non-C_4 carbonium ion

$$k_{L_0} a(K_b B_o - B_a) = K_3 B_a(C_8^* + C_{12}^*) \qquad (6\text{-}92)$$

Solving (6-92) for B_a and noting that $C_8^* + C_{12}^* = C_x^* - C_4^*$, where C_x^* is total steady-state carbonium-ion concentration, we have

$$B_a = \frac{K_b B_o}{1 + (K_3/k_{L_0} a)(C_x^* - C_4^*)}$$

Defining K_1 as the fraction of non-C_4 carbonium ions gives $C_x^* - C_4^* = K_1(C_x^*)$. Recalling that

$$C_x^* = \frac{k_1}{k_6}C_4{}^{2-}$$

by (6-89) we find

$$B_a = \frac{K_b B_o}{1 + (K_3 K_1 k_1/k_{L_0} ak_6)\sqrt{k_6 S/K_2 k_1}} \qquad (6\text{-}93)$$

Substituting (6-93) and (6-89) into Eq. (6-90) gives

$$\frac{C_8}{C_x} = \frac{k_3 K_b B_o}{k_4\sqrt{k_6 S/K_2 k_1}[1 + (k_1 K_3 K_1/k_{L_0} ak_6)\sqrt{k_6 S/K_2 k_1}]} \qquad (6\text{-}94)$$

where $k_{L_0} a$ is very large, mass transport is rapid relative to reaction, and

$$\frac{C_8}{C_x} = \frac{k_3 K_b B_o}{k_4\sqrt{k_6 S/k_1 K_2}} = K_r J \qquad (6\text{-}95)$$

$$K_r = k_3 K_b/k_4 \sqrt{\frac{k_6}{k_1 K_2}} \qquad \text{and} \qquad J = B_o\sqrt{S}$$

In terms of K_r, Eq. (6-94) becomes

$$\frac{C_8}{C_x} = \frac{JK_r}{1 + K_1 K_r k_4\sqrt{S/K_2 K_b k_{L_0} a}} = \frac{J}{1/K_r + K_1 k_4\sqrt{S/K_2 K_b k_{L_0} a}} \qquad (6\text{-}96)$$

The various rate and equilibrium constants are, of course, temperature-dependent, while k_{L_0} and a are functions of fluid-mechanical parameters. C_8/C_x is measured; J

is governed by isobutane concentration in the organic phase and olefin space velocity. In an experiment of fixed power input, a variation in a and measurement of $J/(C_8/C_x)$ would permit determination of K_r and $K_1 k_4/K_2 K_b k_{L_0}$ according to the linear relation

$$\frac{J}{C_8/C_x} = \frac{1}{K_r} + \frac{K_1 k_4 \sqrt{S}}{K_2 K_b k_{L_0} a} \qquad (6\text{-}97)$$

Jernigan et al. relate a to macroscopic measurable variables in the following manner. Define

a = surface area of drops per volume of acid
s = surface area of drops per volume of organic
Φ = fractional volume of dispersed phase
H = volume fraction of organic
H_e = volume fraction of organic in emulsion
V = volume fraction acid; $H + V = 1$

When the organic phase is dispersed, s in terms of average drop diameter is $6/d$. At fixed power input $d(1 - \Phi) =$ constant, and so $s =$ (const)$(1 - \Phi)$ and

$$1 - \Phi = \frac{V}{H_e H + V}$$

where $H_e H$ and V are measurable. As organic exists in the emulsion and in the nonemulsed condition, $H_e \neq 1$. Since $a = s H_e H/V$, we have

$$a = \frac{C_1 H_e H}{H_e H + V} \qquad C_1 = \text{const}$$

Therefore

$$\frac{J}{C_8/C_x} = \frac{1}{K_r} + \frac{K_1 k_4}{K_2 K_b k_{L_0} C_1} G \qquad (6\text{-}98)$$

where the experimentally varied quantity G is

$$G = \sqrt{S} \, \frac{H_e H + V}{H_e H}$$

Experimental data plotted according to Eq. (6-98) are shown in Fig. 6-20. From slope and intercept the coefficients of Eq. (6-98) are determined for various acid strengths, temperatures, power input, etc., thus providing a range of K_r and $K_1 k_4/K_2 K_b k_{L_0}$ values which, according to Eq. (6-96), should correlate yield C_8/C_x as a function of rate, distribution parameters, and s and a, as shown in Fig. 6-21, where a is defined in terms of $H_1 H_e$ and V. Complex as this liquid-liquid reaction is, the correlation is gratifying and certainly provides a sound basis for scale-up and plant-performance analyses.[1]

---

[1] See J. B. Malloy and W. C. Taylor, Scale-up of Liquid-Liquid Reactors, *AIChE Meet., Boston, 1964.*

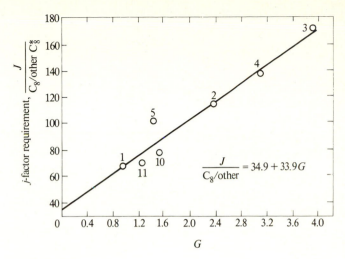

FIGURE 6-20
Correlation according to Eq. (6-98). [*E. C. Jernigen, J. E. Gwyn, and E. L. Claridge, San Francisco AIChE Meet., (1965).*]

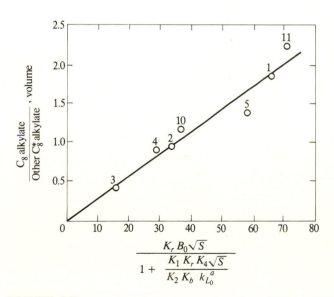

FIGURE 6-21
Overall yield correlation for liquid-liquid acid-catalyzed alkylation. [*E. C. Jernigen, J. E. Gwyn, and E. L. Claridge, San Francisco AIChE Meet., (1965).*]

6-16 SELECTIVITY IN FLUID-FLUID REACTION SYSTEMS

The previous example, although it concerned selectivity, was of a limited nature insofar as it involved mass transfer in *series* with reaction. At this point a more general approach to selectivity seems appropriate. Whereas the van Krevelen plot and the models supporting it provide a basis for predicting simple rates of absorption-reaction, a devotion to selectivity or yield would seem to be in order for the heterogeneous fluid-fluid reaction system. Bridgwater[1] and Szekely and Bridgwater[2] have concerned themselves with the matter of gas-liquid reaction selectivity in terms of both film[1] and penetration[2] models. Following Bridgwater[1] and then Szekely and Bridgwater[2] we shall consider the analytically negotiable cases of consecutive linear reaction confined to a liquid phase supplied by a coreactant-laden gas phase. The consecutive case $A \xrightarrow{k_1} B \xrightarrow{k_2} C$ is our concern, where B is the desired product.

Bridgwater[1] treated this system in terms of film theory, with boundary conditions conforming to Fig. 6-22. Both A and desired product B are volatile, and it can be supposed that the yield of B will depend upon the relative rates of transport of B into the bulk and gas phases, as well as upon the usual absorption-reaction parameters (ϕ, α) and intrinsic kinetics $(k_2/k_1 = \lambda)$. Relative rates of transport into the bulk and gas phases can be characterized by

$$\mu = \frac{k_{L_0} \mathscr{D}_B}{k_{gB} M_B \mathscr{D}_A} = \frac{\text{liquid-side coefficient of physical absorption}}{\text{gas-side coefficient of physical absorption}}$$

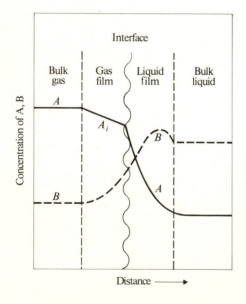

FIGURE 6-22
Bridgwater's film model of gas-liquid absorption-reaction for a consecutive reaction. [*J. Bridgwater, Chem. Eng. Sci.,* 22: 185 (1967).]

[1] J. Bridgwater, *Chem. Eng. Sci.,* **22**: 185 (1967).
[2] J. Szekely and J. Bridgwater, *Chem. Eng. Sci.,* **22**: 711 (1967).

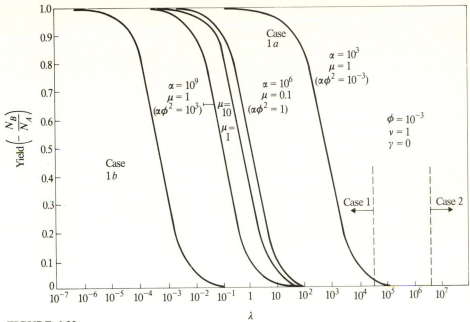

FIGURE 6-23

Yield profiles for diverse values of α and μ as defined in the text for irreversible consecutive linear kinetics (film model). [*J. Bridgwater, Chem. Eng. Sci.,* **22**: *185 (1967).*]

Bridgwater's solution provides point yield of B (N_B/N_A) as a function of

$$\phi = \frac{\sqrt{k_1 \mathscr{D}_A}}{k_{L_0}} \qquad \phi\sqrt{\frac{\lambda}{v}} \qquad v = \frac{\mathscr{D}_B}{\mathscr{D}_A}$$

where $\alpha = V/a\delta$ and μ is as defined above.

Figure 6-23 reveals the yield-versus-λ relationship for slow reaction ($\phi < 0.2$) for a range of α and μ values. $\gamma = 0$ implies negligible B in the gas phase. Similar relations are shown in Fig. 6-24 for rapid reaction ($\phi > 2$). Figure 6-24b corresponds to the yield-versus-λ relation for consecutive reaction yield of B in a homogeneous CSTR. Note that in the gas-liquid case yields of B greater than those realizable in a CSTR are possible for both slow and rapid reaction. Summarizing, we have:

Case 1 ϕ and $\phi\sqrt{\lambda/v}$ small (Fig. 6-23)
    Case 1a $\alpha\phi^2 \ll 1$
    Case 1b $\alpha\phi^2 \gg 1$
Yield of B is high if γ, $\alpha\phi^2$, and μ are small.

Case 2 ϕ small but $\phi\sqrt{\lambda/v}$ large (Fig. 6-23)
    Case 2a $\alpha\phi^2 \ll 1$
    Case 2b $\alpha\phi^2 \gg 1$
Yield of B is always low.

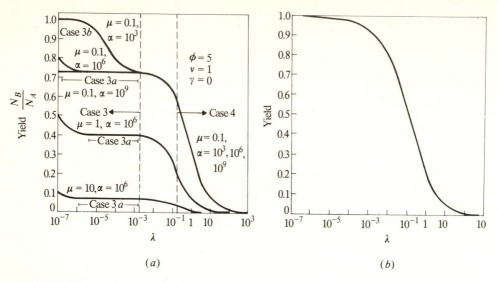

FIGURE 6-24
(a) The effect of α and μ upon consecutive linear reaction yield, $\phi > 2$ (film model); (b) maximum yield for a CSTR. J. [*Bridgwater, Chem. Eng. Sci.*, **22**:185 (*1967*).]

Case 3 ϕ large, $\phi\sqrt{\lambda/v}$ small (Fig. 6-24)
Case 3a $\alpha\phi^2\lambda(1 + \mu) \gg v$ and $\mu \leq 1$
Case 3b $\alpha\phi^2\lambda(1 + \mu) \ll v$
Yield in case 3a depends on μ, is independent of α, and is high for low values of γ and v.   In 3b, yield of B is high if α and γ are small.

Case 4 ϕ and $\phi\sqrt{\lambda/v}$ large (Fig. 6-24).   Yield of B is high if μ and γ are small.

Yield in the Light of Penetration Theory

Whereas it was shown that in the simpler instances of conversion $(A \rightarrow C)$ in absorption-reaction little difference exists in k_L/k_{L_0} values predicted by film and penetration (or surface-renewal) models, yield predictions in a consecutive system $(A \rightarrow B \rightarrow C)$ are apparently more sensitive to the model invoked, as shown by Szekely and Bridgwater.[1]   They compare predicted yield of B for a film model Y_f to that rooted in penetration theory Y_P for the case where bulk concentration in the liquid of A and product B are zero, $\mu = 0$, $\gamma = 0$, and $v = 1$.   Results are given in Fig. 6-25, where Y_P/Y_f is plotted versus λ for various values of ϕ.   Significantly, deviations between the predictions of the two theories can be as large as 21 percent. Deviations are evident over a wide range of ϕ and λ.

[1] J. Szekely and J. Bridgwater, *Chem. Eng. Sci.*, **22**: 711 (1967).

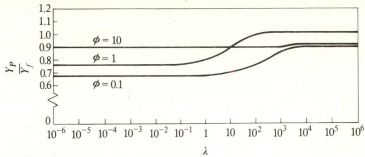

FIGURE 6-25
Yield contrast between film and penetration models. [*J. Szekely and J. Bridgwater, Chem. Eng. Sci.,* **22**: *711 (1967).*]

Yield in Simultaneous Reaction Systems

If gas A is absorbed into another phase, where it reacts to produce two products, i.e.,

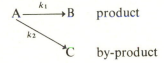

then if reaction orders are identical and transport coefficients essentially the same for volatile B and C, yield will be governed solely by intrinsic rate coefficients k_1 and k_2. If, on the other hand, B is produced via first-order kinetics in the liquid phase while C is the by-product of a second-order process, a significant concentration gradient in A in the liquid phase will retard the higher-order reaction. This conclusion is rooted in analyses presented in Chap. 3, where PFR and CSTR performance was discussed for simultaneous reactions as affected by backmixing (gradient alteration). As backmixing or axial diffusion taxes the higher-order reaction, so diffusion limitations in the liquid film affect the higher-order reaction to the benefit of the lower-order simultaneous reaction. (See also Chap. 5.)

Wendel[1] treats a number of interesting yield networks in terms of penetration theory. Both volatile and nonvolatile reaction products are anticipated in Wendel's analyses.

Simultaneous Absorption of Two Gases; Parallel Reaction in the Liquid Phase

Astarita and Gioia[2] studied both by experimental and theoretical means the industrially important problem of simultaneous absorption of CO_2 and H_2S into aqueous hydroxide solutions. The question to be answered is: What is the relative rate of absorption of CO_2 to H_2S? Treating the assumed reactions

[1] M. M. Wendel, Ph.D. thesis, University of Delaware, Newark, 1956.
[2] G. Astarita and F. Gioia, *Ind. Eng. Chem. Fundam.,* **4**: 317 (1965).

involved as approximately instantaneous and irreversible, an analysis in terms of diffusion of species to zones of reaction of near-zero thickness (regime A of van Krevelen plot), Astarita and Gioia developed film-concentration profiles for the reaction mechanism

$$CO_2 + 2OH \longrightarrow CO_2{}^{2-} + H_2O \qquad (6\text{-}99a)$$

$$H_2S + OH \longrightarrow HS^- + H_2O \qquad (6\text{-}99b)$$

$$H_2S + CO_2{}^{2-} \longrightarrow HS^- + HCO_3{}^- \qquad (6\text{-}99c)$$

$$HCO_3{}^- + OH \longrightarrow CO_2{}^{2-} + H_2O \qquad (6\text{-}99d)$$

The absorption-reaction model is assumed to involve the following steps:

1 CO_2 dissolves and diffuses to the primary reaction plane, where it reacts with OH^-.

2 The $CO_2{}^{2-}$ ions so formed diffuse in both directions.

3 H_2S dissolves and diffuses to the secondary reaction plane, where it reacts with $CO_2{}^{2-}$, forming HS^- and $HCO_3{}^-$.

4 HS^- diffuses into the bulk liquid; $HCO_3{}^-$ diffuses to the primary reaction plane and is converted [reaction (6-99d)] to $CO_2{}^{2-}$. We have then an internal diffusive recycling of $CO_2{}^{2-}$ and $HCO_3{}^-$ between reaction planes.

The situation is displayed in Fig. 6-26, from which it is apparent that the

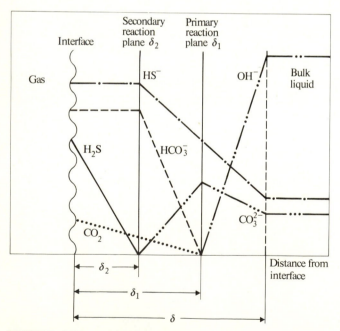

FIGURE 6-26
Film model of simultaneous absorption reaction of two species (H_2S and CO_2) into alkali. [*G. Astarita and F. Gioia, Ind. Eng. Chem. Fundam.,* **4**:*317 (1965).*]

relative absorption rate of CO_2 to H_2S is

$$\frac{(CO_2)_{interface}/\delta_1}{(H_2S)_{interface}/\delta_2}$$

not $(CO_2)_{interface}/(H_2S)_{interface}$, which would follow if CO_2 and H_2S both reacted at a common reaction plane. As $\delta_2 < \delta_1$, the observed selective absorption of H_2S from CO_2-H_2S mixtures is thus confirmed.

In conclusion, it should be emphasized that in our brief consideration of some aspects of gas-liquid and liquid-liquid reaction we have tacitly assumed interfacial equilibrium, probably justified in many cases; however, recent research has revealed instances in which equilibrium does not prevail at a two-phase boundary during absorption.

Finally, whereas shear at the gas-liquid interface is usually small, with the result that a velocity gradient in the liquid near the gas-liquid interface is likely to be zero, in the case of liquid-liquid reaction, shear may not be negligible, so that one must anticipate a velocity gradient in the diffusion-reaction zone which vastly complicates analysis. A simple square-root relationship between the coefficient and diffusivity no longer exists with a finite velocity gradient, and indeed the system actually involves boundary-layer flows with simultaneous reaction and diffusion, a topic far more complicated than those treated in this chapter.

SUMMARY

A study of the matter presented in this chapter, complemented by the literature citations and Additional References should enable one to:

1 Identify the regime of rate control which governs the absorption-reaction system
2 Utilize said identification for the manipulation of reaction conditions (temperature, coreactant concentrations, surface-to-volume ratio, fluid mechanics) to realize reasonable, if not optimum, plant-scale rates of absorption-reaction
3 Plan appropriate laboratory experiments designed to reveal the locale of rate control
4 Prudently scale up a two-phase system based upon either detailed understanding or, in the more usual case, phenomenological laboratory scale data
5 Appreciate those parameters and their interactions which govern selectivity in complex absorption-reaction networks
6 Exercise a prudent caution in treating these dispersed-phase systems marked by complexities with respect to fluid dynamics and the ill-understood but critical problems of coalescence and dispersion

ADDITIONAL REFERENCES

A comprehensive treatise on mass transfer is T. K. Sherwood, R. L. Pigford, and C. R. Wilke, "Mass Transfer," McGraw-Hill, New York, 1975, while G. Astarita, "Mass Transfer with Chemical Reaction," Elsevier, Amsterdam, 1967, treats absorption-reaction with telling lucidity, see also his paper in *Ind. Eng. Chem.*, **58**(8): 18 (1966). A particular emphasis upon the chemistry in absorption-reaction is found in P. V. Danckwerts, "Gas Liquid Reactions," McGraw-Hill, New York, 1970.

Critical reviews of mass transfer in gas-liquid contacting systems are provided by S. Sideman, P. Hortascu, and J. Fulton, *Ind. Eng. Chem.*, **58**(7): 32 (1966); F. Whitt, *Br. Chem. Eng.*, **12**(4): 554 (1967); and F. Valentin, *Br. Chem. Eng.*, **12**(8): 1213 (1967). P. V. Danckwerts and M. Sharma, *Br. Chem. Eng.*, **15**(4): 522 (1970), summarize methods of measuring interfacial area and mass-transfer coefficients in two-fluid systems. G. R. Hughmark, *Chem. Eng. Prog.*, **58**(4): 62 (1962), correlates holdup in gas-liquid flow, while B. Gal-Or and W. Resnick, *Ind. Eng. Chem. Process Des. Dev.*, **5**(1): 15 (1966), present correlations of gas residence time in an agitated gas-liquid contactor. Axial mixing in gas-spurged tubular vessels is analyzed by W. B. Argo and D. R. Cova, *Ind. Eng. Chem. Process Des. Dev.*, **4**(4): 322 (1965).

R. C. Miller, D. S. Noyce, and T. Vermeulen, *Ind. Eng. Chem.*, **56**(6): 43 (1964), provide a graphical correlation of kinetics of aromatic nitration. Ethylene glycol nitration kinetics in mixed acid is discussed by J. Roth, F. Stow, and D. Kouba, *Ind. Eng. Chem.*, **50**(9): 1283 (1958), while J. S. Naworski and P. Harriot, *Ind. Eng. Chem. Fundam.*, **8**(3): 397 (1969), report upon mechanisms and rates in the diffusion-affected oligomerization of 1-butene in sulfuric acid. Scale-up relations for a fluidized-droplet reactor are set forth by J. B. Malloy and W. C. Taylor, *Chem. Eng. Prog.*, **61**(7): 101 (1965). Mass-transfer effects upon selectivity data in hydrogenation of fatty oils are analyzed by K. Hashimoto, M. Teramoto, and S. Nagata, *J. Chem. Eng. Jpn.*, **4**(2): 150 (1971).

PROBLEMS

6-1 Cite some industrially important gas-liquid reactions in which the design goal is (*a*) removal of an undesirable gas-phase species and (*b*) creation of a desired liquid-phase species.

6-2 In a packed-tower unit employed to remove traces of ammonia from a process-gas stream by absorption into water the HTU for gas and liquid phases is $HTU_l = 0.1$ m and $HTU_g = 1$ m. It is proposed that addition of mineral acid to the water will enhance tower efficiency by increasing the rate of absorption. Comment upon this suggestion.

6-3 Danckwerts and Sharma [*Br. Chem. Eng.*, **15**(4): 522 (1970)] present data on surface-to-total-volume ratio for diverse gas-liquid and liquid-liquid contactors. In sum

a

Device	cm²/cm³ gas-liquid	cm²/ml liquid-liquid
Packed columns	0.2–3	1–5
Bubble columns	1–10	1–10
Plate columns	1–4	
Sieve trays	1–5	
Mechanically agitated tanks	2–30	50–800

Using Sherwood-number correlations for these devices, compute the range of anticipated values of α, defined in Sec. 6-3. Employ meaningful values of the characteristic length.

6-4 Nitric acid is produced by contacting N_2O_4 in an airstream with water, the reaction being $\frac{3}{2}N_2O_4 + H_2O \rightleftarrows 2HNO_3 + NO$. Estimate the influence of this exothermic solution-reaction process upon η for an N_2O_4 interfacial mole fraction of 0.05. Pertinent data are

$$\text{Total pressure} = 7 \text{ atm} \qquad T = 40°C$$

$$\text{Henry's law constant} = 158 \text{ atm/(mol } N_2O_4/\text{mol soln)}$$

$$\Delta H \text{ of solution-reaction} = -30 \text{ kcal}$$

$$\text{Activation energy} = 19 \text{ kcal}$$

Use the properties of water for the solution.

6-5 In a laboratory falling-film absorber, CO_2 is adsorbed and reacted with an ammonia solution. At high recycle of the gas stream, rates are measured at various gas-phase (A) and liquid-phase (B) concentrations.

T, °C	$\mathscr{R} \times 10^3$	$A \times 10^2$	$B \times 10^3$
10	3	1	1
	12	2	4
	8.4	2	2
50	4	2	1
	8.3	2	2
	16	2	4

Interpret these data in terms of the governing absorption-reaction regimes.

6-6 How might one increase the absorption-reaction rates in Prob. 6-5? Discuss particularly the efficacy of raising temperature.

6-7 In Sec. 6-10 a computation of plate efficiency for CO_2 absorption in ammonia solution was set forth. It has been suggested that the addition of a catalyst (homogeneous) or an increase in ammonia concentration should enhance efficiency. Compute plate efficiency for

(a)

T, °C	20.7	60
k_1	7×10^5	60×10^5

(b) At each temperature for $B = 5 \times 10^{-3}$ with and without the catalyst.

6-8 Repeat Prob. 6-7 for seal depths of 20 and 30 cm.

6-9 In Fig. 6-12 the phase-utilization factor η is displayed as a function of the observable $\eta\phi^2$ for a range of α values. Express the relationships between k_L/k_{L_0}, η, and $\eta\phi^2$. Reduce each to the case of $\alpha = 1$. To what diffusion-reaction situation does the $\alpha = 1$ limit apply?

6-10 Referring to Fig. 6-8, what is the significance of $\alpha\phi^2$ values less than unity?

6-11 Express the observable in terms of (a) the Sherwood number and (b) the j factor.

6-12 Naworski and Harriott [*Ind. Eng. Chem. Fundam.*, 8(3): 397 (1969)] made measurements of the rate of oligomerization of 1-butene in sulfuric acid. The observed rate at 25°C is $\mathscr{R} = J = 10$ mmol/(min)(ml). Assuming a value of the area of dispersed butene to be 10 cm²/ml (see Prob. 6-3 and its reference), a diffusivity of 4×10^{-6} cm²/s, A_i of 3×10^{-2} mol/ml, an average droplet size of 3×10^{-2} cm, and Sherwood numbers of 2 and 10, compute the range of phase-utilization efficiencies and the absolute rate coefficients for the reaction.

6-13   Repeat Prob. 6-12, assuming drop surface to total volume to be 100 and then 300 cm²/ml.

6-14   Rephrase the Pavlica-Olson continuity equations for (1) PFR, (2) CSTR, and (3) intermediate levels of backmixing for:
 (*a*)   Rapid pseudo-first-order reaction in the continuous phase and no reaction in the dispersed phase
 (*b*)   Rapid pseudo-first-order reaction in the dispersed phase and none in the continuous phase
 (*c*)   Rapid pseudo-first-order reaction in both phases

6-15   Multiplicity of the steady state was touched upon in Sec. 6-13.   Redevelop these relations to include heat of solution of the solute in the solvent.   In principle, how many stable and unstable states can be anticipated?   Refer to the work of Hoffman, Sharma, and Luss [*AIChE J.*, **21**: 318 (1975)].

6-16   Repeat the illustrations of hydrochlorination of lauryl alcohol production for $(G/V)^{0.25}$ and $(G/V)^{0.75}$.

6-17   Carry out the design computations for the continuous multistaged countercurrent lauryl alcohol–hydrochlorination reaction, assuming that water vapor is not removed from the HCl between stages.

6-18   For the conditions set forth in the example in Sec. 6-14 compute the number of theoretical trays required for N_2O_4 absorption-reaction assuming:
 (*a*)   Perfect backmixing between trays
 (*b*)   Pure O_2 rather than air as an oxidizing source
 (*c*)   Refrigeration of the top five trays under standard PFR conditions (assume 3°C as the refrigerated-tray temperature)
 (*d*)   Repeat part (*c*) for interstage CSTR conditions

6-19   For the gas-phase interstage reaction in nitrogen oxides absorption-reaction $(2NO + O_2 \rightarrow 2NO_2 \rightleftarrows N_2O_4)$ the gas-phase rate of oxidation, in terms of partial pressure in atm at 40°C, is

$$-\frac{d(NO)}{d\theta} = k(NO)^2 O_2$$

k at 40°C, according to Bodenstein, is $k = 12$ (atm² · s)$^{-1}$.   Assuming that excess O_2 is present, compute the contact time required to convert 10 percent of a 4 mole percent NO stream at 1, 5, 7, and 15 atm total pressure.   Compare your results with the influence of total pressure upon absorber volume as cited in Table 6-5.

6-20   Acrylonitrile is synthesized by reaction of acetylene gas with a solution of HCN in the presence of Nieuwland's-type homogeneous catalyst.   Utilizing the data and analyses reported by Spillane and Koons [*Chem. Eng. Prog.*, **62**(5): 92 (1966)], design a continuous reactor to produce 1000 kg/h of acrylonitrile at 7 atm total pressure of acetylene.   Comment on the temperature–rate-coefficient behavior exhibited in figure 11 of that article.   Compare a PFR with CSTR in terms of STY.

6-21   Wendel (Ph.D. thesis, University of Delaware, Newark, 1956) analyzed absorption-reaction in terms of penetration theory for (*a*) A $\underset{k_2}{\overset{k_1}{\rightleftarrows}}$ B, where A is the absorbing gas and product B is desorbed, and (*b*) A $\overset{k_1}{\longrightarrow}$ B $\rightleftarrows$ C, where A absorbs and C is volatile.   The second reaction is equilibrated.   Assuming the diffusivities to be equal in each case, in terms of penetration theory determine:
 (*i*)   The rate of desorption of B and adsorption of A in case *a*
 (*ii*)   The desorption rate of C in case *b*

6-22 Repeat Prob. 6-21 in terms of film theory and compare your results with penetration predictions.

6-23 Consider the absorption of A into a liquid where for reaction linearity

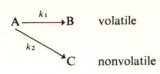

$$A \xrightarrow{\ k_1\ } B \qquad \text{volatile}$$

$$\searrow^{k_2} \ C \qquad \text{nonvolatile}$$

Derive an analytical expression for point selectivity dB/dC in terms of film theory.

7

FLUID-SOLID NONCATALYTIC REACTIONS

". . . se tu ben la tua Fisica note,
tu troverai . . .
che l'arte vostra quella, quanto pote,
segue . . ."

Dante, Inferno, xi, 101

Introduction

As indicated in Chap. 5, noncatalytic reaction of a heterogeneous nature may prevail
between a fluid (gas and/or liquid) and a solid coreactant or a solid containing a
coreactant. The reduction of Fe_2O_3 by H_2, the burning of coal, and ion exchange
such as water softening, are common examples of noncatalytic fluid-solid reaction.
A most important illustration of fluid-solid noncatalytic reaction involves
solid catalysts. Solid catalysts generally consist of porous support structures within
which catalytic agents are deposited. While catalysis and catalysts are discussed
in detail in the following chapter, it is not inappropriate to note at this point that
noncatalytic fluid-solid reactions play a signal role in the porous-solid-catalyst life
cycle. For example, in the catalytic cracking of large hydrocarbons by silica-
alumina-zeolite catalysts, overcracking occurs, i.e., some fraction of the feed is
cracked to solid coke, which remains deposited upon the catalytic sites with a consequent reduction in specific catalyst activity. From Fig. 7-1, a simplified schematic of a porous cracking catalyst, it is evident, in the light of principles set forth

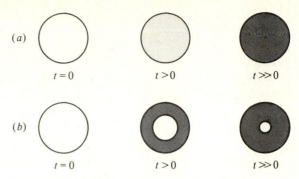

FIGURE 7-1
(a) Uniform and (b) shell-progressive coking.

in Chap. 5, that when coke precursers move (diffuse) rapidly into the catalyst pores relative to their conversion to coke on the pore walls, coke-deposition effectiveness will be unity, the net result being a uniform deposition of coke throughout the catalyst-pellet interior. If, on the other hand, coke-producing reactions are rapid relative to diffusive supply, the coke will preferentially deposit at the pore mouth and the zone of coking will move inward with time, a process known as pore-mouth poisoning (coking, in this instance). Both uniform and pore-mouth coking represent cases of noncatalytic gas-solid reaction.

To rectify catalyst deactivation by coking, regeneration via oxidation of the coke-laden catalyst is employed. Specifically the coked catalyst is exposed to an oxygen-bearing gas stream, in which process oxygen diffuses into the coked catalyst pore, where the surface combustion of coke occurs. Again the relative rates of diffusion and surface reaction dictate whether coke removal is uniform or of the pore-mouth (moving zone) type. Regeneration is therefore another example of noncatalytic gas-solid reaction.[1]

In this chapter fluid-solid reactions (gas-solid, liquid-solid) will be treated in terms of prudent engineering simplifications which focus attention upon key parameters known to govern such systems. The general case of fluid-solid noncatalytic reaction involves transient phenomena; however, the gas-solid system admits to the pseudo-steady-state treatment, which will receive particular emphasis. A brief discussion of ion exchange will be set forth by way of suggesting the immense complexity of fluid-solid noncatalytic reactions which do not yield to the pseudo-steady-state approximation.

An exhaustive treatment of noncatalytic fluid-solid reactors is not attempted in this text. Since the principal emphasis of our work relates to catalytic-reactor phenomena, the noncatalytic fluid-solid analyses outlined in this chapter are fashioned in terms of their relevance to heterogeneous catalysis and those reactors which entertain catalytic transformations.

---

[1] Regeneration of coked catalysts can be catalyzed by metals present on the support.

7-1   GENERAL CONSIDERATIONS

Coke burning, ore reduction by gaseous reducing agents, oxidation of solids, for example, ZnS, and metal dissolution by acids and ion exchange are typical fluid-solid noncatalytic reactions.   Unlike solid-catalyzed reactions (to be treated later), in the noncatalytic fluid-solid reaction a component of the solid reactant phase is consumed.   Thus a transient condition exists.   The complexity of the problem depends upon the relative magnitudes of the various physical and chemical rate processes involved.   If diffusion of reactant fluid through the fluid phase bathing the particle *and* through the solid phase is rapid relative to chemical reaction between fluid and solid, the process is characterized by the usual chemical-kinetic parameters.   For a porous sample of say, Fe_2O_3, exposed to H_2 gas, the rate will be proportional to concentrations of gas and unreduced solid since rapidity of gas diffusion may ensure total penetration of the H_2 throughout the porous iron oxide sample.

For batch reduction with chemical-reaction control

$$-\frac{d(H_2)}{dt} = kf(H_2)(Fe_2O_3) \qquad (7\text{-}1)$$

while in a flow system, e.g., fixed bed, PFR,

$$\frac{\partial(H_2)}{\partial t} + u\frac{\partial(H_2)}{\partial z} = -kf(H_2)(Fe_2O_3) \qquad (7\text{-}2)$$

If on the other hand, a diffusional step is rate-controlling, for a flow system we have, where a is surface-to-volume ratio,

$$\frac{\partial(H_2)}{\partial t} + u\frac{\partial(H_2)}{\partial z} = -a\mathscr{D}\frac{\partial(H_2)}{\partial r}\bigg|_{r=R} \qquad (7\text{-}3)$$

Should bulk mass transport control, the approximation

$$-\mathscr{D}\frac{\partial(H_2)}{\partial r}\bigg|_{r=R} = k_g(H_2) \qquad (7\text{-}4)$$

applies, while if the diffusional step is an intraparticle one, then in principle the diffusion equation, assuming chemical equilibrium,

$$\left[\frac{\partial(H_2)}{\partial t}\right]_{particle} = \mathscr{D}\,\nabla^2(H_2) \qquad (7\text{-}5)$$

must be solved for appropriate boundary conditions and the resulting flux incorporated into Eq. (7-3).

When, as is more generally true, no one step exclusively controls, the relation,

$$\left[\frac{\partial(H_2)}{\partial t}\right]_{particle} = \mathscr{D}\,\nabla^2(H_2) - kf(H_2)(Fe_2O_3) \qquad (7\text{-}6)$$

must be solved and incorporated into Eq. (7-3) with a surface boundary condition phrased in terms of the bulk mass-transport flux; i.e.,

$$k_g(H_{2_0} - H_{2_s}) = -\mathscr{D}\left.\frac{\partial(H_2)}{\partial r}\right|_{r=R} \qquad (7\text{-}7)$$

becomes a boundary condition, simply stating that fluid-film transport equals that of the intraparticle diffusive flux at the particle surface.

The chief difficulty in solving the general case [Eq. (7-6)] for boundary condition (7-7) involves the transient $\partial(H_2)/\partial t$ and the nonlinear intraparticle chemical reaction term, $kf(H_2)(Fe_2O_3)$. Often pseudo-first-order kinetics may be safely assumed, and if the problem can be phrased in terms of a product shell (porous) surrounding an impervious reactant core, the transient term can be dropped as shown momentarily. One then has the pseudo-steady-state, shell-progressive model (SPM). Hence for $t > 0$, reactant gas is transported through the fluid boundary layer and thence diffuses through a shell of product solid to the product-reactant interface. As Bischoff[1] shows, for a gas-solid system, the product shell migrates inward at a rate far slower than the transportation of gas to the reaction interface, and so $\partial(H_2)/\partial t \to 0$. Under these conditions, one can equate the series events, for batch operation,

$$-\frac{1}{a}\frac{d(Fe_2O_3)}{dt} = k_g(H_{2_0} - H_{2_s}) = -\mathscr{D}\left.\frac{d(H_2)}{dr}\right|_{r=R} = k(H_2)$$

If both product and reactant are porous solid, a distinct reactant-product interface may not necessarily exist. Then the zone of reaction may be a uniform one through the particle if chemical reaction controls or a smeared zone when mixed rate control exists. There must then be a critical rate-constant value below which a smeared or uniform zone of reaction exists; on the other hand, for this critical and larger value of the rate constant, a narrow zone exists clearly separating the product shell from the reactant core.

The three possibilities are schematized in Fig. 7-2 for a given value of time t, where R is the external surface. In general terms of gaseous reactant C and solid B, the SPM would reflect the case shown in Fig. 7-2b. If bulk mass transfer controls, $C_s \to 0$ and

$$-\frac{1}{a}\frac{dB}{dt} = k_g C_0 \qquad (7\text{-}8)$$

For shell-diffusion control, $C_1 \to 0$ (kC_1 is finite, since, by implication, $k \to \infty$) and $C_2 = C_1 = 0$

$$\frac{1}{a}\frac{dB}{dt} = \mathscr{D}\left.\frac{dC}{dr}\right|_{r=R} = \frac{\mathscr{D}C_s}{R - r} \qquad (7\text{-}9)$$

[1] K. Bischoff, *Chem. Eng. Science*, **18**: 711 (1963). A more general treatment is given by D. Luss, *Can. J. Chem. Eng.*, **46**: 154 (1968).

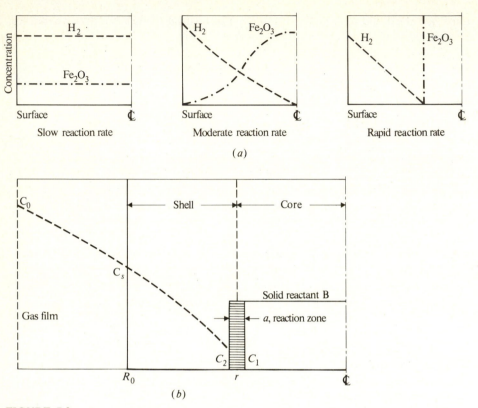

FIGURE 7-2
(a) General limits of gas-solid noncatalytic reaction; (b) general detail indicating the three regimes of rate controls.

While for chemical-reaction control (nonporous core)

$$-\frac{1}{a}\frac{dB}{dt} = kC_1 \qquad (7\text{-}10)$$

With a porous core, it must be shown that k must be great enough to ensure a sharp concentration drop ($C_2 = C_1$), thus establishing the shell-core interface.

The critical minimum k value is related to shell diffusion $k/\mathcal{D}$ or, in dimensionless form, $kR/\mathcal{D} =$ Damköhler number Da. In consequence, a critical Da value must be specified for application of the SPM. Gorring[1] states that if the SPM is to apply, the zone or band of reaction in a totally porous system must be quite narrow. That is, the concentration drop in the reaction zone C_1/C_2 should be of the order a/R of $\frac{1}{50}$, which means that catalytic effectiveness η *in the moving*

---

[1] J. J. Carberry and R. L. Gorring, *J. Catal.*, **5**: 529 (1966).

reaction zone must be at most $\frac{1}{200}$, implying that $\phi \approx 200$ since $\eta = 1/\phi$ for large ϕ values (>3). ϕ and the Damköhler number are related by $Da = \phi^2/LS$, where $1/L$ is the external surface-to-volume ratio and S is the total (BET)[1] area per unit volume; hence LS is the ratio of total (BET) area to geometric area of the pellet or particle. Since ϕ must be equal to or greater than 200, for the SPM to apply to a totally porous system we must have

$$Da \geq \frac{4 \times 10^4}{LS} \qquad (7\text{-}11)$$

When reactant solid is nonporous or virtually impervious to the diffusing reactant, reaction-zone thickness is, by definition, zero and SPM applies for any value of Da. Clearly the product shell must be permeable. For application of the SPM, then:

1 Da must be equal to or greater than the value give by Eq. (7-11).
2 The pseudo-steady-state approximation must be valid. Therefore only gas-solid systems can be treated by SPM for Bischoff's analysis indicates a substantial error involved in applying the pseudo steady state to liquid-solid systems.
3 For any value of LS, the SPM is applicable to a gas-solid system in which the core (unreacted solid) is virtually impervious, while the product shell is porous. Of necessity events must occur in series in this case, as the diffusion shell progresses at a rate governed by the chemical reaction between gas and nonporous reactant solid, thus producing the porous shell.

In summary, the general case of fluid-solid reaction (batch or flow system) would require solution of the complete diffusion equation with transient and (perhaps) nonlinear kinetics. That is, transient diffusion and simultaneous reaction within the solid must be considered [Eq. (7-6)]. For a flow system the solution of Eq. (7-6), yielding the surface flux, must then be substituted into Eq. (7-3) to yield the time-on-stream–reactor-length profile. More generally, radial and axial temperature and concentration profiles must be considered, as well as nonisothermality within the solid phase.

However, assuming isothermal batch operation, so long as the criteria specified above are obeyed, the problem can be vastly simplified in terms of simple linear series events occurring under pseudo-steady-state conditions. The SPM will now be employed to analyze a number of noncatalytic gas-solid reaction systems where particle shrinkage or expansion due to reaction will also be considered. The liquid-solid noncatalytic system (ion exchange) will be treated later with a prudence necessitated by its inherent transient complexity.

---

[1] The term BET is explained in Chap. 8. It includes internal plus external area.

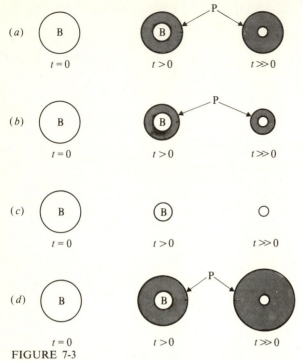

FIGURE 7-3
Schematic of diverse possibilities in a gas-solid noncatalyzed reaction network; B = solid reactant, P = product. (a) Constant particle size; (b) shrinking particle size; (c) totally volatile product; (d) expanding particle size.

7-2 KINETICS OF NONCATALYTIC GAS-SOLID REACTIONS[1,2]

A number of noncatalytic reactions between gas and solid exist which merit particular attention as both chemical and diffusional factors are involved. As shown in Fig. 7-3, at time zero (initial exposure) the reacting gas A in contact with the particle starts to react with the external surface. The reaction of A with solid B creates a product solid C and, say, a gas D. At $t > 0$ a layer of product C has been formed so that A must diffuse through this layer to react with unconverted B within the particle. In general the particle size may change with reaction if the molar density of B and C differ; or, in the limit, if reaction products are gases, the particle simply shrinks as reaction proceeds and no product layer exists.

For fixed and shrinking or expanding particle size we shall treat quantitatively the rate process involving fluid-film diffusion, product-layer diffusion, and gas-solid chemical reaction at the reactant interface.

These three resistances exist in series. Now in general, for a linear series

[1] Based upon SPM.
[2] D. White and J. J. Carberry, *Can. J. Chem. Eng.*, **43**: 334 (1965).

network, the total time for completion of the process is $t = t_1 + t_2 + t_3 + \cdots + t_n$ for n events in series. If one event or step is rate-controlling, say step 2, $t = t_2$. When no single step is slower than another, $t = \sum_1^n t_i$, and in consequence we need only specify the functionality for each step, add the times, and obtain the general relationship which assumes no one step to be rate-controlling.

Variation of Particle Size with Reaction

$$A(g) + bB(s) \rightarrow cC(s) + D(g)$$

The moles of B reacted at any time, for a sphere of radius R_0, are

$$N_B = \rho_B \frac{\frac{4}{3}\pi(R_0{}^3 - r_c{}^3)}{M_B} \qquad (7\text{-}12)$$

where r_c is the radius of the unreacted core. The moles of C created are

$$N_c = \rho_c \frac{4}{3}\pi \frac{R^3 - r_c{}^3}{M_c} \qquad (7\text{-}13)$$

where R is the radius of the sphere at any time. In the absence of shrinkage or expansion $R = R_0$. Reaction stoichiometry indicates

$$N_B = \frac{b}{c} N_c \qquad (7\text{-}13a)$$

Solving for R after equating (7-12) and (7-13) according to (7-13a), we obtain

$$R^3 = R_0{}^3 \left[\frac{\rho_B M_c c}{\rho_c M_B b} + \left(1 - \frac{\rho_B M_c c}{\rho_c M_B b}\right) \left(\frac{r_c}{R_0}\right)^3 \right] \qquad (7\text{-}14)$$

Let
$$z = \frac{\rho_B M_c c}{\rho_c M_B b} \qquad \text{and} \qquad \gamma = \frac{r_c}{R_0} \qquad (7\text{-}15)$$

Then
$$R = R_0[z + (1 - z)\gamma^3]^{1/3} \qquad (7\text{-}16)$$

$$z \begin{cases} >1 & \text{particle expansion} \\ <1 & \text{particle shrinkage} \\ =1 & \text{particle size constant} \end{cases}$$

Diffusion through product-layer control The continuity equation is

$$\mathscr{D} \nabla^2 A = \frac{\partial A}{\partial t} \qquad (7\text{-}17)$$

As noted, for *gas*-solid systems Bischoff[1] has shown that the pseudo-steady-state approximation ($\partial A/\partial t = 0$) is valid over virtually the entire range of γ; that is, the error attendant upon setting $\partial A/\partial t = 0$ in Eq. (7-17) is small in gas-solid systems. In liquid-solid systems a significant error is involved in such an approximation. In

[1] Loc. cit.

essence Bischoff's analysis shows that the boundary of the unreacted core moves inward at a rate much slower than the gas A. The rate of transport of A is about 1000 times the rate of movement of r_c, this factor corresponding roughly to the ratio of solid to gas densities. This ratio is very much smaller in a liquid-solid system; consequently $\partial A/\partial t = 0$ is not a justified assumption in the liquid-solid case, e.g., ion exchange. For gas–spherical-solid reactions, Eq. (7-17) becomes

$$\frac{d^2A}{dr^2} + \frac{2}{r}\frac{dA}{dr} = 0 \qquad (7\text{-}18)$$

We solve Eq. (7-18) for the boundary conditions

At $A = A_0$: $r = R$

At $A = 0$: $r = r_c$

and obtain the flux

$$-\mathscr{D}\frac{dA}{dr}\bigg|_{r=R} \qquad (7\text{-}19)$$

$$\frac{A}{A_0} = \frac{1 - r_c/r}{1 - r_c/R}$$

and

$$\frac{dA}{dr}\bigg|_{r=R} = \frac{A_0}{R^2(1/r_c - 1/R)} \qquad (7\text{-}20)$$

The flux multiplied by area $4\pi R^2$ must equal the rate of consumption of A; that is,

$$\frac{dN_A}{dt} = 4\pi R^2 \mathscr{D}\frac{dA}{dr}\bigg|_{r=R} = \frac{4\pi A_0 \mathscr{D}}{1/r_c - 1/R} \qquad (7\text{-}21)$$

Now

$$\frac{dN_A}{dt} = \frac{1}{b}\frac{dN_B}{dt}$$

and N_B is given by Eq. (7-12), so that

$$-\frac{\rho_B}{M_B}\frac{4\pi r_c^2}{b}\frac{dr_c}{dt} = \frac{dN_A}{dt} = \frac{4\pi A_0 \mathscr{D}}{1/r_c - 1/R} \qquad (7\text{-}22)$$

or

$$-\int_{R_0}^{r_c}\left(\frac{1}{r_c} - \frac{1}{R}\right)r_c^2\,dr_c = \frac{bA_0\mathscr{D}}{\rho_B'}\int_0^t dt \qquad (7\text{-}22a)$$

where $\rho' = \rho/M$, the molar density of B. Note that at $t = 0$, $r = R_0$, and that for the non-fixed-particle-radius case,

$$R = R_0[z + (1 - z)\gamma^3]^{1/3} = R_0\phi \qquad (7\text{-}23)$$

so that

$$-\int_{R_0}^{r_c} r_c\,dr_c\frac{+}{R_0}\int_{R_0}^{r_c}\frac{r_c^2}{\phi}\,dr_c = \frac{bA_0\mathscr{D}}{\rho_B'}t \qquad (7\text{-}24)$$

Since

$$r_c = \gamma R_0 \qquad dr_c = R_0 \, d\gamma \qquad \text{and} \qquad \phi = [z + (1 - z)\gamma^3]^{1/3} \qquad (7\text{-}25)$$

we have

$$\frac{R_0^2}{2}(1 - \gamma^2) + R_0^2 \int_{R_0}^{r_c} \frac{\gamma^2 \, d\gamma}{[z + (1 - z)\gamma^3]^{1/3}} = \frac{b\mathscr{D}A_0}{\rho'_B} t \qquad (7\text{-}26)$$

Let

$$\alpha = z + (1 - z)\gamma^3 \qquad \text{and} \qquad d\alpha = 3(1 - z)\gamma^2 \, d\gamma \qquad (7\text{-}27)$$

then

$$\frac{R_0^2}{2}(1 - \gamma^2) + \frac{R_0^2}{2} \frac{[z + (1 - z)\gamma^3]^{2/3} - 1}{1 - z} = \frac{b\mathscr{D}A_0}{\rho'_B} t \qquad (7\text{-}28)$$

Solving for t gives

$$t_D = \frac{\rho'_B R_0^2}{2b\mathscr{D}A_0} \left\{ \frac{[z + (1 - z)\gamma^3]^{2/3}}{1 - z} + (1 - \gamma^2) - \frac{1}{1 - z} \right\} \qquad (7\text{-}29)$$

For invariant particle size $z = 1$, and application of L'Hospital's rule yields

$$t_D = \frac{\rho'_B R_0^2}{b\mathscr{D}A_0} \left(\frac{1 - \gamma^2}{2} - \frac{1 - \gamma^3}{3} \right)_{R = \text{const}} \qquad (7\text{-}30)$$

Note that $1 - \gamma^3 = x$ is the ratio of particle volume reacted to total volume. In Fig. 7-4, x is plotted versus τ, defined as

$$\tau = \frac{1}{R_0} \sqrt{\frac{b\mathscr{D}A_0 t}{\rho'_B}} \qquad (7\text{-}31)$$

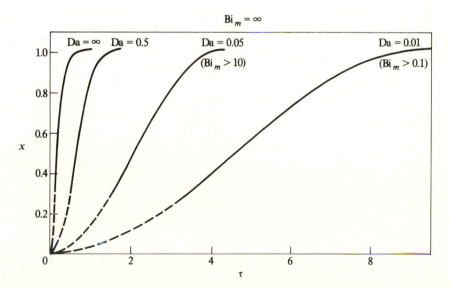

FIGURE 7-4
Conversion x-versus-τ relation for a gas-solid reaction. The limits of applicability of the flat-plate model are evident in dashed lines (fixed particle size). [*J. J. Carberry and R. L. Gorring, J. Catal., 5: 529 (1966).*]

In the case of flat-plate geometry and $z = 1$

$$x = \frac{1}{L}\sqrt{\frac{b\mathscr{D}A_0 2t}{\rho'_B}} \qquad (7\text{-}32)$$

Specifying L in terms of volume to external surface area for a sphere, we obtain

$$x = \frac{4.25}{R_0}\sqrt{\frac{b\mathscr{D}A_0 t}{\rho'_B}} \qquad (7\text{-}33)$$

Figure 7-4 reveals that the x-versus-τ relation for a sphere is linear in the range of x values of zero to about 0.3. In this range

$$x = \frac{4.25}{R_0}\sqrt{\frac{b\mathscr{D}A_0 t}{\rho'_B}}$$

Thus the flat-plate geometry provides a rather accurate description of the diffusion process in a sphere for a significant range of reaction x.

Chemical-reaction control If the reaction between gas and solid at the boundary r_c is rate-controlling, then

$$\frac{1}{4\pi r_c^2}\frac{dN_B}{dt} = bkA_0^m \qquad (7\text{-}34)$$

Again expressing N_B in terms of r_c [Eq. (7-12)], we have

$$-\rho'_B\frac{dr_c}{dt} = bkA_0^m \qquad (7\text{-}35)$$

Integrating from R_0 to r_c and solving for t gives

$$t_r = \frac{\rho'_B R_0}{bkA_0^m}(1 - \gamma) \qquad (7\text{-}36)$$

This expression is obviously independent of particle shrinkage or expansion.

Gas-film-diffusion control With variation of particle size it is evident that the gas-phase mass-transfer coefficient becomes a variable. Furthermore, if the reacting particle is falling through a gas environment, its velocity will change with shrinkage or expansion. As the mass-transfer coefficient k_g is a function of velocity and particle radius, k_g must be expressed in terms of ϕ. We shall consider k_g functionalities for two cases.

CASE 1: FIXED BED OF PARTICLES[1]

$$j_0 = \frac{k_g}{u}\left(\frac{v}{\mathscr{D}}\right)^{2/3} = 1.15\left(\frac{d_p u}{v}\right)^{-1/2} \qquad (7\text{-}37)$$

[1] J. J. Carberry, *AIChE J.*, 6: 460 (1960).

Assuming that velocity through the bed remains essentially constant, k_g is

$$k_g = k_{g0} \sqrt{\frac{R_0}{R_0 \phi}} = \frac{k_{g0}}{\phi^{1/2}} \qquad (7\text{-}38)$$

CASE 2: FREE-FALLING PARTICLE

$$\frac{k_g d_p}{\mathscr{D}} = 2 + 0.6 \left(\frac{\nu}{\mathscr{D}}\right)^{1/3} \left(\frac{d_p u}{\nu}\right)^{1/2} \qquad (7\text{-}39)$$

At very low Reynolds numbers

$$\frac{k_g d_p}{\mathscr{D}} = 2 \qquad \text{or} \qquad k_g = \frac{k_{g0}}{\phi} \qquad (7\text{-}40)$$

At high Reynolds numbers

$$k_g = k_{g0} \frac{1}{\phi^{1/2}} \left(\frac{u}{u_0}\right)^{1/2} \qquad (7\text{-}41)$$

The velocity of fall varies as $\sqrt{d_p/C_D}$, where C_D is the drag coefficient. At high Reynolds numbers $C_D = $ constant, and

$$k_g = \frac{k_{g0}}{\phi^{1/4}} \qquad (7\text{-}42)$$

In general we can write

$$k_g = \frac{k_{g0}}{\phi^n} \qquad (7\text{-}43)$$

We now equate the consumption rate of B with gas-film mass transfer

$$\frac{1}{4\pi R^2} \frac{dN_B}{dt} = -\frac{\rho_B' r_c^2}{R_0^2 \phi^2} \frac{dr_c}{dt} = \frac{bk_{g0} A_0}{\phi^n} \qquad (7\text{-}44)$$

or

$$-\frac{\rho_B' R_0 \gamma^2}{\phi^{2-n}} \frac{d\gamma}{dt} = bk_{g0} A_0 \qquad (7\text{-}45)$$

Integrating

$$-\int_1^{\gamma_c} \frac{\gamma^2 \, d\gamma}{\phi^{2-n}} = \frac{bk_{g0} A_0}{\rho_B' R_0} \int_0^t dt \qquad (7\text{-}46)$$

Let

$$\phi = [z + (1-z)\gamma^3]^{1/3} = \alpha^{1/3}$$

Recall that

$$\gamma^2 \, d\gamma = \frac{d\alpha}{3(1-z)} \qquad (7\text{-}47)$$

$$-\frac{1}{3(1-z)} \int_1^\gamma \frac{d\alpha}{\alpha^{(2-n)/3}} = \frac{bk_{g0} A_0 t}{\rho_B' R_0} \qquad (7\text{-}48)$$

For $n \neq -1$ we have

$$\frac{-1}{1-z}\left[\frac{\alpha^{(1+n)/3}}{1+n}\right]_1^\alpha = \frac{bk_{g0}A_0 t}{\rho_B' R_0}$$

$$\frac{1}{1-z}[1 - \alpha^{(1+n)/3}] = \frac{(1+n)bk_{g0}A_0 t}{\rho_B' R_0}$$

and so

$$t_{mt} = \frac{\rho_B' R_0}{(1+n)bk_{g0}A_0} \frac{1 - \alpha^{(1+n)/3}}{1-z} \qquad (7\text{-}49)$$

For the case of constant particle size, $z = 1$, $n = 0$, applying L'Hospital's rule we obtain

$$t_{mt} = \frac{\rho_B' R_0}{3bk_{g0}A_0}(1 - \gamma^3)_{R=\text{const}} \qquad (7\text{-}50)$$

The total time of reaction when mass transfer, diffusion through product layer, and first-order reaction contribute to rate control is given as a function of γ for variable particle size by $t_T = t_{mt} + t_r + t_D$ [Eqs. (7-49), (7-36), and (7-29)]

$$t_T = \frac{\rho_B' R_0^2}{b\mathscr{D}A_0}\left(\frac{1}{2(1-z)}\left\{\frac{1 - \alpha^{(1+n)/3}}{[(1+n)/2]\,\text{Bi}} - (1 - \alpha^{2/3})\right\} + \frac{1-\gamma^2}{2} + \frac{1-\gamma}{\text{Da}}\right) \qquad (7\text{-}51)$$

For constant particle size

$$t_T = \frac{\rho_B' R_0^2}{b\mathscr{D}A_0}\left[\frac{1-\gamma^3}{3}\left(\frac{1}{\text{Bi}}-1\right) + \frac{1-\gamma^2}{2} + \frac{1-\gamma}{\text{Da}}\right] \qquad (7\text{-}52)$$

where Bi is a Biot number $= k_{g0}R_0/\mathscr{D}$ and Da is a Damköhler number $= kR_0/\mathscr{D}$. Note that the Biot number expresses the ratio of *gas*-film coefficient to *solid*-phase diffusivity.

In Fig. 7-5, $x = 1 - \gamma^3$ is plotted versus $\tau = (1/R)\sqrt{b\mathscr{D}A_0 t/\rho_B'}$ for various values of Da and z, where Bi $= \infty$.

Case of Totally Volatile Product: $C + bB \rightarrow$ gas

In this instance no product layer exists, and reaction time is simply the sum of reaction and gas-film diffusion times, that is, $t_{mt} + t_r$. However, ϕ must be evaluated for the case where R at all times equals r_c. For $z = 0$, $\phi = \gamma$,

$$t = \frac{\rho_B' R_0}{bk_{g0}C_0}\left(\frac{1 - \gamma^{n+1}}{n+1} + \frac{1-\gamma}{\text{Da/Bi}}\right) \qquad \text{where} \qquad \frac{\text{Da}}{\text{Bi}} = \frac{k}{k_{g0}} \qquad (7\text{-}53)$$

Typical results are shown in Fig. 7-6, where C_0 refers to any bulk gas-phase reactant.

Flat-Plate Approximation $(z = 1)$

Equation (7-52) is implicit in $x = 1 - \gamma^3$ and is therefore of limited analytical value. For in general one seeks a relationship which predicts x as a function of time t rather than t as a function of x. As we have observed in the case of product-

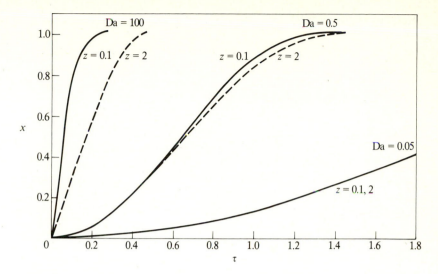

FIGURE 7-5
The influence of particle contraction ($z = 0.1$) and expansion ($z = 2$) upon particle reaction vs. time; $(Bi)_m = \infty$, $n = 0.25$. [*J. J. Carberry and D. White, Can. J. Chem. Eng.*,
43: *334 (1965)*.]

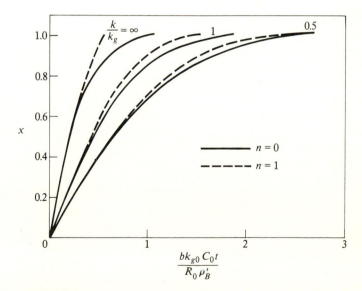

FIGURE 7-6
Graphical display of conversion according to Eq. (7-53) for a totally
volatile product. [*J. J. Carberry and D. White, Can. J. Chem. Eng.*,
43: *334 (1965)*.]

layer diffusion-rate control [Eqs. (7-30) to (7-32)], the simple flat-plate geometry nicely describes the sphere case for x values up to about 0.3. We shall now derive the general mixed-rate-control case for finite Bi and Da for a flat-plate geometry and then invoke the fact that flat-plate thickness L can be related to sphere dimensions by $L = R/3$. We then compare both solutions to determine the range of x over which the flat-plate expression adequately defines the rate process for spheres. Further, in the course of derivation, the validity of the expression $t_T = \sum_1^n t_i$ will be demonstrated.

There must, of course, be an engineering justification for the use of flat-plate geometry. As illustrated in the case of product-layer diffusion, only 30 percent of the total process is adequately defined in terms of Eq. (7-33), and this would seem to be a severe limitation. If total reaction of a particle is of concern, clearly the flat-plate model is inadequate. However, other processes of practical interest can be described by the diffusion-reaction transient relations, and the practical range of interest in these cases may be limited penetration of reactant. For example, consider the poisoning of catalysts and coking of catalysts processing hydrocarbons. As specified in the introduction to this chapter, pore-mouth or selective poisoning involves progressive penetration of the poisoning or coking agent into the particle. Wheeler treated the case where x is time-independent. However, in general a poisoning agent is imposed upon the reactor in fixed and continuous supply so that x is time-dependent. Equation (7-52) is general in nature and can realistically describe the pore-mouth poisoning-rate process. More importantly, however, in catalysis total poisoning of the particle is seldom realized, as one need only encounter pore-mouth poisoning of perhaps 10 to 40 percent ($x = 0.1$ to 0.4) before the rate of reactant transport becomes dictated by simple diffusion through the poisoned or coked pore. In this case the synthesis rate is low enough to be virtually uneconomical. Consequently, regeneration or disposal is called for when only a fraction of total particle volume is coked or poisoned. It follows, then, that if the more convenient flat-plate solution relating x to time on-stream t is known to apply rather accurately over the range of x of *practical* interest, the flat-plate development is justified and offers analytical convenience not found with the more rigorous sphere solution.

For the flat-plate geometry we seek a relationship between x and time on-stream. As particle size remains constant in the case of fouling, poisoning, or coking of a catalyst of fixed size, we shall confine outselves to the case where $z = 1.0$ (see Fig. 7-7). Let $\gamma = X/L$. Equate the rate of consumption of B to the rate of gas-film mass transfer

$$A_0 - A_s = -\frac{\rho'_B L \, d\gamma}{bk_g dt} = \frac{\rho'_B L(1 - \gamma)}{bk_g t} \qquad (7\text{-}54)$$

or

$$A_0 = A_s + \frac{\rho'_B L(1 - \gamma)}{bk_g t}$$

Now we equate the diffusion rate through the product layer to consumption of B.

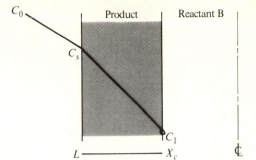

FIGURE 7-7
Flat-plate model of a gas-solid reaction:

$$\frac{1}{A}\frac{dN_B}{dt} = -\rho_B'\frac{dx}{dt}$$

$$= bk_g(c_0 - c_s) = b\mathscr{D}\frac{dc}{dx} = bkc_1$$

First we solve for the concentration distribution across the product layer

$$-\mathscr{D}\frac{dA}{dX} = \text{const} \qquad \text{At } X = X_c, A = A_c$$

$$\text{At } X = L, \quad A = A_s$$

$$A_s - A = \frac{(A_s - A_1)(1 - \bar{\gamma})}{1 - \gamma} \qquad \bar{\gamma} = \frac{X}{L} \qquad (7\text{-}55)$$

$$\mathscr{D}\left.\frac{dA}{dX}\right|_{X=L} = \frac{\mathscr{D}(A_s - A_c)}{L(1 - \gamma)} = -\frac{\rho_B' L}{b}\frac{d\gamma}{dt}$$

Integrating gives

$$\frac{b\mathscr{D}(A_s - A_c)t}{L^2\rho_B'} = 1 - \tfrac{1}{2} - \gamma + \frac{\gamma^2}{2}$$

$$\frac{b2\mathscr{D}(A_s - A_c)t}{L^2\rho_B'} = 1 - 2\gamma + \gamma^2 = (1 - \gamma)^2$$

$$A_s = A_c + \frac{(1 - \gamma)^2 L^2 \rho_B'}{b2\mathscr{D}t} \qquad (7\text{-}56)$$

Now we obtain A by equating the chemical-reaction rate to consumption of B

$$-\rho_B'L\frac{d\gamma}{dt} = bkA_c$$

Solving for A_c gives

$$A_c = \frac{(1 - \gamma)\rho_B'L}{bkt} \qquad (7\text{-}57)$$

Substituting (7-57) into (7-56) and that result into (7-54), we obtain

$$A_0 = \frac{(1 - \gamma)\rho_B'L}{bkt} + \frac{(1 - \gamma)^2 L^2 \rho_B'}{b2\mathscr{D}t} + \frac{(1 - \gamma)\rho_B'L}{bk_g t}$$

Solving for t, we have

$$t_T = \frac{\rho_B' L^2}{2b\mathscr{D}A_0}\left[2(1-\gamma)\left(\frac{1}{kL/\mathscr{D}} + \frac{1}{k_g L/\mathscr{D}}\right) + (1-\gamma)^2\right] \qquad (7\text{-}58)$$

Now

$$L = \frac{R}{3} \qquad \text{Bi} = \frac{k_g R}{\mathscr{D}} \qquad \text{Da} = \frac{kR}{\mathscr{D}}$$

$$t_T = \frac{\rho_B' R^2}{18b\mathscr{D}A_0}\left[2(1-\gamma)\left(\frac{3}{\text{Bi}} + \frac{3}{\text{Da}}\right) + (1-\gamma)^2\right] \qquad (7\text{-}59)$$

For flat plate $x = 1 - \gamma$; then

$$\frac{18tb\mathscr{D}A_0}{\rho_B' R^2} = \beta^2 = 2x\alpha + x^2 \qquad \text{where } \alpha = \frac{3}{\text{Bi}} + \frac{3}{\text{Da}} \qquad (7\text{-}60)$$

Equation (7-59) is equivalent to $t_T = \sum_1^n t_i$, where n is the number of series first-order steps. Equation (7-60) is readily solved for x

$$x = -\alpha + \sqrt{\alpha^2 + \beta^2} \qquad (7\text{-}61)$$

Equation (7-61) very accurately describes the x-versus-t relationship for spheres in the x range of 0 to about 0.30 over a wide range of Bi and Da values, as shown in Fig. 7-4.

As an exercise, for the constant-particle size case, create a generalized plot of reaction time t relative to the time for total conversion τ versus conversion for a sphere when (a) Bi $= \infty$ and Da $= \infty$, (b) Bi $= \infty$ and Da $= 100$, (c) Bi $= \infty$ and Da $= 10$, (d) Bi $= \infty$ and Da $= 1$, (e) Bi $= 100$ and Da $= \infty$, (f) Bi $= 10$ and Da $= \infty$, and (g) Bi $= 100$ and Da $= 100$.

Validation of SPM

While a number of experimental verifications of gas-solid noncatalytic reaction models might be cited, one example strikes us as both highly persuasive and relevant to issues which will be discussed subsequently, namely, regeneration of coke-laden porous-catalyst pellets by combustion in an oxygen-bearing gas stream.

Weisz and Goodwin[1] report upon experiments in which they measured the kinetics for combustion of coke deposited upon diverse porous cracking catalysts in both bead and powdered form. As has been noted, if diffusion of gaseous reactant is rapid relative to gas-solid reaction, the solid phase is uniformly consumed at all radial positions within the porous pellet. When reaction vigor is increased (say, by an increase in reaction temperature), gas-solid consumption becomes so great that ultimately the premises of the SPM should be obeyed.

As it happens, silica-alumina beads are rendered transparent when immersed in a high-refractive-index liquid such as carbon tetrachloride. With such immersion the distribution of deposited coke becomes apparent. Figure 7-8 shows the Weisz-Goodwin photographs of a partially regenerated coked Si-Al pellet when

[1] P. B. Weisz and R. D. Goodwin, *J. Catal.*, **2**: 397 (1963).

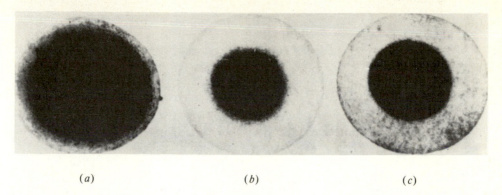

(*a*)   (*b*)   (*c*)

FIGURE 7-8
Photographs of partially regenerated coke catalyst beads at three temperature levels: (*a*) low, (*b*) intermediate, and (*c*) high temperature. [*P. B. Weisz and R. D. Goodwin, J. Catal.,* **2**: *397 (1963).*]

subjected to (*a*) intrinsic, or uniform, combustion; (*b*) combustion under near SPM conditions; and (*c*) SPM burning, a rather dramatic illustration of SPM gas-solid reaction.

    As for the criterion for the transition between the two regimes illustrated so graphically in Figs. 7-8 and 7-9, Weisz and Goodwin invoke the Wheeler-Weisz criterion, cited in Fig. 5-11; that is, for pellet radius R_p

$$\frac{\mathscr{R}L^2}{\mathscr{D}A_0} = \frac{(dA/dt)R_p^{\,2}}{\mathscr{D}A_0} > 1.0$$

where $\mathscr{R} = dA/dt$ is the observed rate. For an oxygen concentration A_0 of 3 μmol/cm^3 and a diffusivity $\mathscr{D}$ of 5×10^{-3} cm^2/s for a bead of 0.2 cm radius the

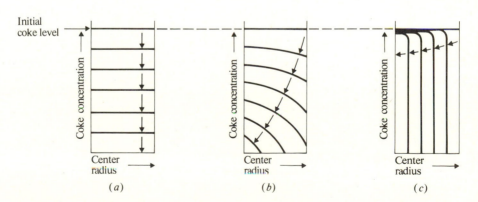

FIGURE 7-9
Concentration profiles in the regimes of (*a*) uniform, (*b*) intermediate, and (*c*) SPM burning of coke catalyst. [*P. B. Weisz and R. D. Goodwin, J. Catal.,* **2**: *397 (1963).*]

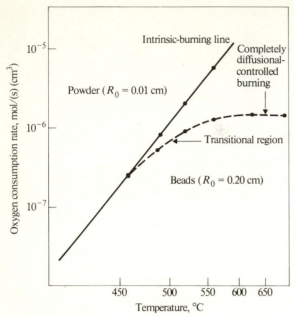

FIGURE 7-10
Intrinsic, transitional, and completely diffusion-controlled burning of coke in Si-Al catalysts. [*P. B. Weisz and R. D. Goodwin, J. Catal.*, **2**:397 (1963).]

criterion indicates that the observed rate below which diffusional effects will be absent is

$$\frac{dA}{dt} < \frac{A_0}{R_p^{\,2}} \mathscr{D} = 0.4 \; \mu mol/(cm^3)(s)$$

In Fig. 7-10 the observed burning rate is plotted versus reciprocal temperature for both the 0.2-cm-radius bead and its pulverized form. Magnificent agreement between the Wheeler-Weisz criterion and the data is evident for the beads, while, as expected, the powdered coked catalyst exhibits intrinsic-burning kinetic behavior.

Given evidence of the SPM mode of coke burning for beads at $T > 470°C$, which of the three possible SPM steps (itemized earlier) actually dictates the global rate? In terms of the fraction of coke remaining f, if intraphase diffusion (as opposed to bulk mass transport or coke-oxygen interfacial reaction) controls, then for spherical beads

$$\tfrac{1}{2}(1 - f^{2/3}) - \tfrac{1}{3}(1 - f) = \frac{b\mathscr{D}A_0}{R_p^{\,2}C} \theta = K\theta \qquad (7\text{-}62)$$

where C is coke concentration and θ reaction time. The experimental data secured by Weisz and Goodwin are displayed in accord with Eq. (7-62) for various

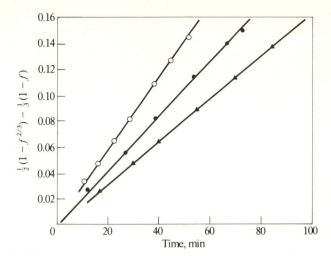

FIGURE 7-11
Burnoff function vs. time on three different beads. [*P. B. Weisz and R. D. Goodwin, J. Catal.*, **2**: *397 (1963)*.]

bead diameters and coke levels, thus verifying SPM intraphase diffusion-controlled kinetics (Fig. 7-11).

Weisz and Goodwin go on to verify the dependency of K in Eq. (7-62) upon C, R_p, and $\mathscr{D}$, the latter being varied by adjustments in the average pore radius of the beads.   It is to be recognized that coke-burning data for known values of R_p, b, A_0, and C permit the determination of intraphase diffusivity, $\mathscr{D}$, which, incidentally, reflects any dead-ended pore contribution.

Intrinsic coke-burning kinetics   In a companion paper, Weisz and Goodwin[1] present data and analyses for coke burning under diffusion-unaffected conditions, i.e., the intrinsic-burning kinetic regime.   They find that the rate is linear in coke concentration and oxygen, in contrast with zero-order coke dependency under SPM circumstances.   For coked beads, low-temperature ($<500°C$) burning is obviously necessary for diffusion-unaffected kinetics, while (as Fig. 7-10 suggests) the powdered coked catalyst, e.g., fluidizable catalyst, will exhibit intrinsic kinetic behavior at much higher temperatures.   These distinctions prove important in formulating a reactor-regeneration model, discussed in Chap. 10.

7-3 NONISOTHERMAL GAS-SOLID REACTION

In the previous sections of this chapter, we developed equations relating the time of reaction to SPM conversion as a function of Da and Bi under isothermal conditions.   Given the vigor of certain gas-solid reactions, the nonisothermal circum-

---

[1] P. B. Weisz and R. D. Goodwin, *J. Catal.*, **6**: 227 (1966).

stance deserves comment. Specifically, we address the problem of linear reaction in the regime in which the SPM applies, but the supposition of isothermality is now abandoned.

Let us consider a spherical solid containing a solid reactant exposed to a coreactant gaseous species. The diffusive fiux is then

$$\frac{1}{r^2} \frac{d}{dr} \left(r^2 \mathscr{D} \frac{dA}{dr} \right) = 0 \qquad \text{for } r_c < r < R_0 \qquad (7\text{-}63)$$

The boundary conditions are

At $\quad r = R_0$:

$$\mathscr{D} \frac{dA}{dr} \bigg|_{R_0} = k_g(A_0 - A_s) \qquad (7\text{-}64)$$

$$\text{Flux} \qquad \text{External mass transport}$$

At $\quad r = r_c$:

$$\mathscr{D} \frac{dA}{dr} \bigg|_{r_c} = kA_c \qquad (7\text{-}65)$$

$$\text{Flux} \qquad \text{Surface reaction}$$

where A_0, A_s, and A_c are concentrations in the bulk gas, external pellet surface, and reacting interface, respectively.

Assuming that the pseudo-steady-state approximation applies to heat transport (valid if C_p of the solid is small), we can fashion a heat balance as follows:[1]

$$\frac{1}{r^2} \frac{d(r^2 \lambda \, dT/dr)}{dr} = 0 \qquad r_c < r < R_0 \qquad (7\text{-}66)$$

with the boundary conditions

At $\quad r = R_0$:

$$-\lambda \frac{dT}{dr} \bigg|_{R_0} = h(T_s - T_0) \qquad (7\text{-}67)$$

At $\quad r = r_c$:

$$-\lambda \frac{dT}{dr} \bigg|_{r_c} = -\Delta H \, kA_c \qquad (7\text{-}68)$$

while

$$k = \mathscr{A} \exp \left(-\frac{E}{RT_c} \right) \qquad (7\text{-}69)$$

The mass continuity equation (7-63) is readily solved for the specified boundary conditions (7-64) and (7-65). We integrate Eq. (7-63) twice, utilizing the boundary conditions that $A = A_s$ at $r = R_0$ and $A = A_c$ at $r = r_c$:

$$A - A_c = (A_s - A_c) \frac{r_c/R_0 - r_c/r}{1 - r_c/R_0} \qquad (7\text{-}70)$$

which is the concentration profile throughout the solid product or ash layer. Equation (7-70) is differentiated with respect to r and evaluated at $r = r_c$ to yield the flux and thus dN_A/dt, the rate per particle in terms of A_s and A_c; so

$$\frac{dA}{dr} \bigg|_{r=r_c} = \frac{A_s - A_c}{r_c(1 - r_c/R_0)} \qquad (7\text{-}71)$$

---

[1] Ignoring heat-up of the pellet, an approximation; see D. Luss and N. R. Amundson, *AIChE J.*, **15**: 195 (1967).

Since in-series linear events in the pseudo steady state admit to

$$\frac{dN_A}{dt} = 4\pi R_0^2 k_g (A_0 - A_s) \quad (7\text{-}72a)$$

$$= 4\pi r_c^2 \mathscr{D} \left(\frac{dA}{dr} \right)_{r_c} \quad (7\text{-}72b)$$

$$= 4\pi r_c^2 k A_c \quad (7\text{-}72c)$$

we have

$$\frac{dN_A}{dt} = 4\pi r_c \mathscr{D} \frac{A_s - A_c}{1 - r_c/R_0} \quad (7\text{-}73)$$

We can eliminate A_s and dN_A/dt from Eqs. (7-72a) to (7-72c) to find A_c as a function of A_0 and r_c, with the result

$$\frac{A_c}{A_0} \equiv \frac{1}{1 + \gamma^2 \text{Da}/\text{Bi} + \text{Da}\gamma(1 - \gamma)} \quad (7\text{-}74)$$

where $\gamma = r_c/R_0$. Substituting A_c from above into (7-72c) gives the rate per pellet as

$$\frac{dN_A}{dt} = \frac{4\pi r_c^2 k A_0}{1 + \gamma^2 \text{Da}/\text{Bi} + \text{Da}\gamma(1 - \gamma)} \quad (7\text{-}75)$$

This rate in terms of moles of A, N_A is readily translated into γ by recognizing that by the stoichiometry of the assumed reaction

$$\frac{dN_A}{dt} = \frac{1}{b} \frac{dN_b}{dt} = -\frac{4\pi r_c^2 \rho}{bM} \frac{dr_c}{dt} \quad (7\text{-}76)$$

which is also equal to $4\pi r_c^2 k A_c$ [Eq. (7-72c)]. Therefore

$$-\frac{dr_c}{dt} = \frac{bMkA_0/\rho}{1 + \gamma^2 \text{Da}/\text{Bi} + \text{Da}\gamma(1 - \gamma)} \quad (7\text{-}77)$$

Recall that for the sphere, conversion is $x = 1 - \gamma^3$; hence Eq. (7-77) can be phrased in terms of conversion x as a function of time.

By similar reasoning the continuity equation for temperature [Eq. (7-66) with its boundary conditions, Eqs. (7-67) and (7-68)], can be resolved to yield

$$\frac{T_c}{T_0} = t_c = 1 + \text{Da}\beta \frac{A_c}{A_0} \left[\gamma(1 - \gamma) + \frac{\gamma^2}{\text{Bi}_h} \right] \quad (7\text{-}78)$$

(note that the units of k are centimeters per second), where

$$\beta = \frac{-\Delta H A_0 \mathscr{D}}{\lambda T_0} \qquad \text{adiabatic} \frac{\Delta T}{T_0}$$

$$\text{Bi}_h = \frac{h R_0}{\lambda} \qquad \text{thermal Biot number}$$

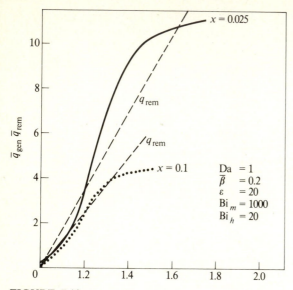

FIGURE 7-12
Heat generation and heat-removal functions vs. reduced
temperature t at various conversion levels.

Substituting Eq. (7-74) for A_c/A_0, we obtain

$$t_c = 1 + \left\{ \frac{\text{Da}\beta[\gamma(1 - \gamma) + \gamma^2/\text{Bi}_h]}{1 + \gamma^2\text{Da}/\text{Bi} + \text{Da}\gamma(1 - \gamma)} \right\} \qquad (7\text{-}79)$$

where $\text{Bi} = k_g R_0/\mathscr{D}$ is the mass Biot number.

Equation (7-79) is gainfully recast into a form which reveals a multiplicity of
steady-state solutions.   Heat generation, utilizing Eq. (7-74), is

$$q_{\text{gen}} = \frac{k(-\Delta H)A_0}{1 + \gamma^2\text{Da}/\text{Bi} + \text{Da}\gamma(1 - \gamma)}$$

In dimensionless form

$$\bar{q}_{\text{gen}} = \frac{R_0 k}{\mathscr{D}} \frac{-\Delta H \, A_0 \, \mathscr{D}}{\lambda T_0} \frac{1}{1 + \gamma^2\text{Da}/\text{Bi} + \text{Da}\gamma(1 - \gamma)} \qquad (7\text{-}80)$$

or

$$\bar{q}_{\text{gen}} = \text{Da}\beta \frac{A_c}{A_0} = \text{Da}_0\beta \frac{A_c}{A_0} \exp\left[-\epsilon_0\left(\frac{1}{t_c} - 1\right)\right]$$

Rearranging Eq. (7-79) gives

$$\frac{t_c - 1}{\gamma(1 - \gamma) + \gamma^2/\text{Bi}_h} = \frac{\text{Da}_0 \, \beta \, \exp\left[- \epsilon_0(1/t_c - 1)\right]}{1 + \gamma^2\text{Da}/\text{Bi} + \text{Da}\gamma(1 - \gamma)} \qquad (7\text{-}81)$$

The right-hand side of (7-81) is the dimensionless heat-generation function, while the left-hand side is the heat-removal function.   Note that the Damköhler number is

$$\text{Da} = \text{Da}_0 \exp\left[-\epsilon_0 \left(\frac{1}{t_c} - 1\right)\right] \qquad \epsilon_0 = \frac{E}{RT_0} \qquad (7\text{-}82)$$

At various levels of conversion, i.e., fixed values of γ, and diverse values of the governing dimensionless groups, β, Da_0, Bi, and Bi_h, the heat-generation and heat-removal functions are displayed in Fig. 7-12.   In contrast to gas-solid catalysis, both the heat-removal and heat-generation functions are themselves functions of γ and therefore of conversion.

Note also that unlike the CSTR multiplicity behavior, in gas-solid noncatalytic reaction both $\bar{q}_{\text{gen}}$ and $\bar{q}_{\text{rem}}$ functionalities are conversion-dependent.

7-4   GAS-SOLID-REACTION EFFECTIVENESS FACTORS

An effectiveness factor for a gas-solid noncatalytic reaction is defined precisely as in Chap. 5:

$$\eta = \frac{kA \text{ at reaction interface } r_c}{k_0 A_0, \text{ bulk fluid conditions}}$$

Under isothermal conditions, by Eq. (7-74),

$$\eta_{\text{iso}} = \frac{A_c}{A_0} = \frac{1}{1 + (\text{Da}/\text{Bi})\gamma^2 + \text{Da}\gamma(1 - \gamma)}$$

which is set forth in Fig. 7-13, versus conversion $x = 1 - \gamma^3$ for the indicated Da and Bi.

Nonisothermal effectiveness is

$$\eta = \frac{1}{\exp\left[\epsilon_0(1/t_c - 1)\right] + (\text{Da}/\text{Bi})_0\gamma^2 + \text{Da}_0\,\gamma(1 - \gamma)} \qquad (7\text{-}83a)$$

where the subscript zero refers to the dimensionless groups valuated at T_0, the bulk fluid temperature.   In terms of its isothermal value, the nonisothermal effectiveness is

$$\eta = \frac{1}{\exp\left[\epsilon_0(1/t_c - 1)\right] + 1/\eta_{\text{iso}} - 1} \qquad (7\text{-}83b)$$

The behavior of the nonisothermal gas-solid noncatalytic effectiveness factor is presented as a function of conversion in Fig. 7-13 for values of the dimensionless parameters cited therein.

In the light of Eqs. (7-81) and (7-83b)

$$\bar{q}_{\text{gen}} = \eta\,\text{Da}_0\,\beta$$

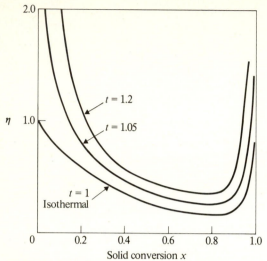

FIGURE 7-13
Isothermal and nonisothermal gas-solid noncatalytic
effectiveness factors; η versus conversion; $\text{Da}_0 = 10$,
$\text{Bi}_m = 100$, $\varepsilon = 20$.

7-5   GAS-SOLID REACTION IN TERMS OF OBSERVABLES

In view of the rather obvious fact that in instances of heterogeneous reaction, species concentrations and local temperature usually escape direct measurement, it is imperative that whenever feasible these variables be susceptible to computation or at least estimation in terms of observable, i.e., measurable, parameters.   This point has been illustrated in Chap. 5 for the case of nonporous interphase heat- and mass-transport–affected surface catalytic reaction, where effectiveness is gainfully and simply expressed as a function of the observable $\bar{\eta}\text{Da} = \mathcal{R}_0/k_g a A_0$, $\mathcal{R}_0$ being the *observed* rate of reaction.   For the porous solid catalyst, we learn that its effectiveness can be fruitfully displayed in terms of an observable, the Wheeler-Weisz number, an internal observable

$$\Phi = \frac{\mathcal{R}_0 L^2}{\mathcal{D} A_0} = \eta \phi^2 \qquad (7\text{-}84)$$

which is related to the external observable $\bar{\eta}\text{Da}$ by $\text{Bi}_m = k_g L/\mathcal{D}$

$$\bar{\eta}\text{Da} = \frac{\Phi}{\text{Bi}_m} \qquad (7\text{-}85)$$

In fashioning the SPM of gas-solid noncatalytic reaction under nonisothermal conditions, the temperature at the solid-reactant–gaseous-coreactant interface, $t_c = T_c/T_0$, is encountered in the analyses of both steady-state multiplicity [Eq. (7-81)] and effectiveness [Eq. (7-83b)].   This temperature is not ordinarily subject to experimental measurement save by inference in the isothermal case, where, of course, $T_c = T_0$.

Suppose that one has at one's disposal data for conversion versus time of reaction for a gas-solid noncatalytic reaction. Conversion defines γ for a sphere by the relation

$$x = 1 - \gamma^3 \qquad (7\text{-}86)$$

The derivative of the x-versus-t curve at a particular value of x defines the *observed* reaction rate $\mathscr{R}_0$ through Eq. (7-77) and

$$-\frac{dN_A}{dt} = \frac{4\pi \rho R_0^{\,3}}{bM} \gamma^2 \frac{d\gamma}{dt}$$

Assuming that $k_g a$ can be predicted by an appropriate governing j-factor correlation, $\bar{\eta}\mathrm{Da}$ is thus at hand as well as γ.

For the nonisothermal gas-solid noncatalytic reaction which progresses in conformity with the SPM we equate in the pseudo steady state the heat generated at the gas-solid shrinking-core interface with that transported by convection in the gas film[1]

$$-\Delta H\, k_c A_c = h(T_s - T_0) \qquad (7\text{-}87)$$

where subscript c refers to the core, s to the external pellet surface, and 0 to the bulk-fluid phase. Note that $k_c A_c = \mathscr{R}_0$, the observed rate.[2]

Equation (7-87) is normalized by $k_g a T_0$

$$\frac{A_0(-\Delta H)}{T_0} \frac{\mathscr{R}_0}{k_g a A_0} = \frac{h}{k_g}(t_s - 1) \qquad (7\text{-}88)$$

Since h and k_g are related by the j-factor analogy, that is, $j_H = j_D$

$$\frac{h}{k_g} = \rho C_p (\mathrm{Le})^{2/3}$$

where the Lewis number $\mathrm{Le} = \mathrm{Sc}/\mathrm{Pr}$. Therefore the reduced external surface temperature, in terms of the observable $\bar{\eta}\mathrm{Da}$ and bulk-fluid dimensionless adiabatic temperature change $\bar{\beta}$, is

$$t_s = \frac{T_s}{T_0} = 1 + \bar{\beta}(\bar{\eta}\mathrm{Da}) \qquad (7\text{-}89)$$

The pellet external surface temperature, usually unobservable, is thus predictable in terms of the thermal parameter $\bar{\beta}$ and the observable $\bar{\eta}\mathrm{Da}$.

But we seek the reduced temperature at the core interface $t_c = T_c/T_0$ since its value, not t_s, governs gas-solid noncatalytic reaction. The relationship between t_c

---

[1] As with Eq. (7-66), this interphase heat balance does not account for heat-up of the solid.

[2] Note that if observed rate is per unit area, $\bar{\eta}\,\mathrm{Da} = \dfrac{\mathscr{R}_0}{k_g A_0}$; per unit volume, $\dfrac{\mathscr{R}_0}{k_g a A_0}$.

and t_s is easily established by equating external fluid-film heat transfer to thermal conduction through the reacted shell of length $R_0 - r_c$

$$R_0^2 h(T_s - T_0) = \frac{\lambda(T_c - T_s)r_c^2}{R_0 - r_c} \qquad (7\text{-}90)$$

Normalizing in terms of $\gamma = r_c/R_0$, $t_c = T_c/T_0$, we obtain

$$\frac{hR_0}{\lambda}(t_s - 1) = (t_c - t_s)\frac{\gamma^2}{1 - \gamma} \qquad (7\text{-}91)$$

or

$$\mathrm{Bi}_h(t_s - 1) = (t_c - t_s)\frac{\gamma^2}{1 - \gamma}$$

Solving for t_s gives

$$t_s = \frac{t_c + \mathrm{Bi}_h(1 - \gamma)/\gamma^2}{1 + \mathrm{Bi}_h(1 - \gamma)/\gamma^2} \qquad (7\text{-}92)$$

Substitution of (7-92) into (7-89) and rearrangement give the unobservable core-interface reduced temperature t_c solely in terms of the observable $\bar{\eta}\mathrm{Da}$, conversion as a function of γ, and the thermal parameter $\bar{\beta}$:

$$t_c = [1 + \bar{\beta}(\bar{\eta}\mathrm{Da})]\left(1 + \mathrm{Bi}_h\frac{1 - \gamma}{\gamma^2}\right) - \frac{(1 - \gamma)\mathrm{Bi}_h}{\gamma^2} \qquad (7\text{-}93)$$

Equation (7-93) in concert with (7-82) permits estimation of the nonisothermal gas-solid noncatalytic effectiveness. Furthermore, the heat-generation–heat-removal relationship [Eq. (7-80)] is now readily expressed in terms of the observable and conversion.

7-6 REACTOR DESIGN

Bearing in mind the crucial distinction between a constant and variable bulk gas-phase concentration with respect to the manipulation of the basic gas-solid non-catalytic rate equation, we now direct attention to gas-solid noncatalytic reactors. Evidently when the solid reactant is bathed by a constant-gas-phase coreactant concentration, use of the simple t-versus-x, or γ, function specifies isothermal design. In contrast, should bulk-gas coreactant concentration be dependent upon time- and/or distance within the reactor, the resultant t-versus-x functionality will assume significantly greater complexity even under isothermal conditions. If nonisothermality persists, the reactor-design problem becomes, if not unwieldy, at least difficult. The forthcoming treatment of gas-solid reactors, which does not pretend to universality, is limited to instances which admit to instructive, negotiable, and meaningful illustration.

Gas-Solid Reactors

Diverse gas-solid reactor designs are sketched in Fig. 7-14. A typical fixed-bed gas-solid reaction is regeneration of deactivated catalyst pellets, e.g., the burning of coke deposits. Both gaseous- and solid-reactant concentrations are functions

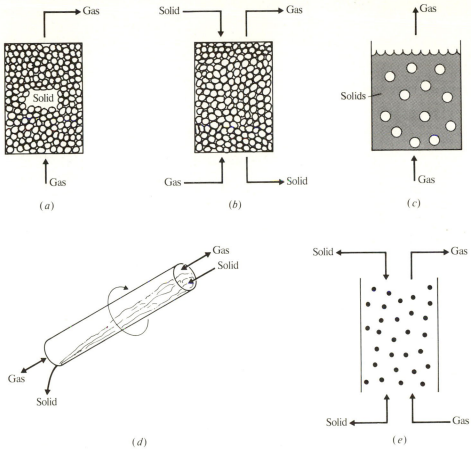

FIGURE 7-14
Gas-solid noncatalytic reactor types: (*a*) fixed bed, (*b*) moving bed, (*c*) fluid bed, (*d*) kiln with co-current or countercurrent gas flow, (*e*) dilute-phase transport line with cocurrent or countercurrent solids flow.

of time and position, and (particularly in coke burning) a temperature wave migrates through the bed; consequently analyses of this reactor are marked by some mathematical complexities.

The moving bed is specifically designed for catalytic reactions accompanied by catalyst deactivation since fresh catalyst is continuously added to replace spent species. Catalyst regeneration via gas-solid reaction can also be undertaken in a moving-bed unit. Such reactors are consequently steady-state units, and analyses are therefore somewhat simpler than for the fixed-bed mode of operation. The fluid bed, with continuous steady-state gas flow, may be operated in batch fashion with respect to solids or with continuous addition and removal of solids. The kiln is a special variation on the moving bed, while the transport-line reactor is a limiting case of fluidization.

Fixed-bed reactor For the fixed-bed gas-solid noncatalytic reactor the continuity equation for the gas phase, assuming plug flow, is

$$-u\frac{\partial A}{\partial z} = \mathscr{R}_p + \frac{\partial A}{\partial t} \qquad (7\text{-}94)$$

For gas temperature

$$\rho u C_p \frac{\partial T}{\partial z} = -\Delta H \mathscr{R}_p - \rho C_p \frac{\partial T}{\partial t} \qquad (7\text{-}95)$$

For spherical solids $\mathscr{R}_p$ is secured by Eq. (7-75). Should conversion limited to 30 percent be of interest, e.g., pore-mouth poisoning, regeneration of pore-mouth-coked catalyst, and coking itself, the simpler form applies. The core radius as a function of time is given by Eq. (7-77). Numerical solution of all governing equations is obviously required. A comprehensive analysis of adiabatic regeneration in a fixed bed is presented by Olson et al.,[1] while poisoning of fixed beds is analytically described by Olson.[2]

Moving-bed reactor In this reactor the solids are moved steadily and continuously from top to bottom, the withdrawal and addition being necessitated either by deactivation of catalyst pellets or by consumption (reaction) of solid-phase reactants. Thus if fresh or rejuvenated catalyst pellets suffer deactivation in the course of reaction, a steady state with respect to the catalyst can be achieved by continuously adding fresh catalyst to the reactor and removing spent catalyst from it. A moving packed bed represents such an implementation, and it can be assumed that the mobile bed flows in a plug-flow fashion. Generally, the reacting gas passes countercurrently with respect to the moving bed. A noncatalytic gas-solid system might involve regeneration of a coked catalyst as it is exposed in a moving bed countercurrent to an oxygen-bearing gas stream.

For the moving bed, the governing continuity equations assume the steady-state form. For the gas-phase composition

$$-u\frac{dA}{dz} = \mathscr{R}_p \qquad (7\text{-}96)$$

while for gas-phase temperature (adiabatic)

$$\rho u C_p \frac{dT}{dz} = -\Delta H \mathscr{R}_p \qquad (7\text{-}97)$$

where again, as in the fixed-bed case, $\mathscr{R}_p$ is given by Eq. (7-75) i.e. $\mathscr{R}_p$ is $\dfrac{1}{V}\dfrac{dN_A}{dt}$.

Two limiting cases may be considered: (1) the gas-phase composition is a function of z, that is, $dA/dz \neq 0$; (2) by reason of an excess of gaseous reactant, its concentration is virtually constant with respect to reactor length z; that is, $dA/dz = 0$.

[1] K. E. Olson, D. Luss, and N. R. Amundson, *Ind. Eng. Chem. Prod. Res. Dev.*, **7**: 96 (1968).
[2] J. H. Olson, *Ind. Eng. Chem. Fundam.*, **7**: 185 (1968).

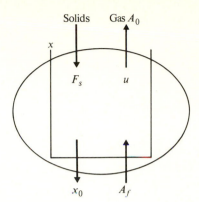

FIGURE 7-15
Material balance about a moving bed.

In the general case, case 1, we first express $\mathscr{R}_p$ in terms of solid-phase conversion, which for the SPM reduces to expressing core radius r_c as a function of time; so, by Eq. (7-77), for a sphere we have

$$-\frac{dr_c}{dt} = \frac{bMkA_0/\rho}{1 + \gamma^2 Da/Bi + Da\gamma(1 - \gamma)}$$

For the moving bed dt is easily expressed in terms of bed depth z by

$$t = \frac{z}{U_s} \qquad \text{so} \qquad dt = \frac{dz}{U_s}$$

where U_s is the solids velocity. Therefore, since conversion x for spheres is $1 - \gamma^3 = x$, Eq. (7-77) becomes

$$\frac{dx}{dz} = \frac{(3bMkA_0/\rho)(1 - x)^{2/3}}{U_s R_0\{1 + Da/Bi(1 - x)^{2/3} + Da(1 - x)^{1/3}[1 - (1 - x)^{1/3}]\}} \qquad (7\text{-}98)$$

CASE 1   If A_0 is a function of z, a material balance for the countercurrent moving-bed-reactor model establishes the A-versus-x relationship for a feed at A_f, as shown in Fig. 7-15.

$$u(A_f - A_0) = \frac{F_s}{b}(x_0 - x) \qquad (7\text{-}99)$$

where A_0 is the gaseous-reactant concentration at a point where solid-reactant conversion is x, x_0 being the exit solids conversion and F the molar velocity of the solids in moles per area time and b the stoichiometric coefficient for the reaction. When we substitute Eq. (7-99) into (7-98), solids conversion is expressed as a function of moving-bed depth. In Fig. 7-16 solids conversion is plotted versus dimensionless moving-bed depth for various values of Da, Bi, and the number of solid-phase transfer units (NTU), as defined by Olson;[1] i.e., where $Z = z/L$,

---

[1] Ibid.

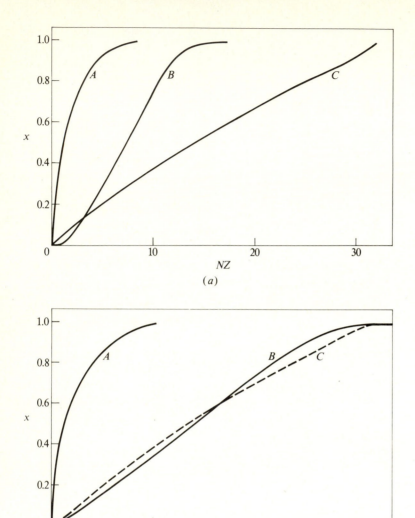

FIGURE 7-16
Conversion vs. moving-bed depth for diverse isothermal circumstances:
(a) Da = 10, $Bi_m = 100$; (b) Da = 100, $Bi_m = 100$. Curve A = constant gas
composition, curve B = countercurrent, $F = 0.5$, curve C = cocurrent, $F = 0.5$.

Eq. (7-98) becomes

$$\frac{dx}{dZ} = \frac{\text{NTU}_s(A_0/A_f)(1-x)^{2/3}}{1 + (\text{Da/Bi})(1-x)^{2/3} + \text{Da}(1-x)^{1/3}[1-(1-x)^{1/3}]} \tag{7-100}$$

where

$$\text{NTU}_s = \frac{bMkA_fL}{\rho U_s R_0}3$$

CASE 2   If A_0 is invariant with respect to z, solids conversion becomes a simple function of z or time of moving-solids residence. In this case, Eq. (7-100) is integrated at constant A_0 to yield a time-of-contact (or bed-length) relationship with conversion. In terms of dimensionless length Z and fraction of solid reactant remaining, $f = 1 - x$,

$$Z = \frac{1}{N}\left[3(1 - f^{1/3}) + \frac{\text{Da}}{\text{Bi}}(1 - f) + \text{Da}(f - \tfrac{3}{2}f^{2/3} + \tfrac{1}{2})\right]$$

Of course, the question of nonisothermality of the moving bed must be considered. Should either external mass transport or shell diffusion govern the local rate, gross temperature excursions will exert a minor influence upon these rate processes and in consequence both short- and long-range (interphase, intraphase, and interparticulate) gradients of temperature will affect local (and therefore overall) conversion only modestly. In contrast, chemically-rate-controlled gas-solid reaction will be markedly influenced by a nonisothermal reactor environment.

Fluidized-bed reactor   In general the fluid-bed reactor, consisting of solids suspended in a highly agitated condition by reason of fluid flow through the reactor, resembles a boiling liquid, two phases being apparent, a relatively solids-free bubble (gas) phase and a solids-rich emulsion phase. The solids phase may be retained within the reactor throughout the course of reaction which constitutes a batch fluidized bed. Gas-solid noncatalytic reaction in a batch fluid bed therefore involves a solids conversion which is a function of time on-stream. Should fluidized solid reactant be steadily and continuously added and withdrawn from the reactor, a steady state with respect to both fluid (gas) and solid phases prevails.

Insofar as the fluidized-bed reactor (batch or continuous) involves two distinct phases between which species transfer must occur, it is obvious that simple models which presuppose primitive CSTR behavior of the fluid bed represent severe compromises with reality. A detailed treatment of the fluidized-bed reactor will be presented in Chap. 10. For qualitative illustration, the simple CSTR model of the fluid-bed noncatalytic reactor will be tentatively entertained in this chapter. The seriousness of this compromise is schematized in Fig. 7-17, where the simple homogeneous CSTR model is contrasted with the more realistic two-phase fluid-bed model.

Qualitatively one can anticipate that insofar as the primitive CSTR model of the fluid bed implies an intimate, uniform contacting of perfectly mixed gas with perfectly mixed solids, the local and therefore integral rates of reaction will be greater than would be expected to prevail when reactant transport from a bubble to emulsion phase intrudes upon intrinsic kinetics to alter the global rate in a manner similar to transport modification of reaction, as described in Chap. 5.

In effect a rate coefficient invoked in the ideal CSTR model of a fluidized bed must be judged to be a maximum value rooted in the assumption that transport rates between the bubble-rich and solids-rich emulsion phase are rapid relative to local gas-solid reaction. Further, while near-perfect mixing of the solids phase may seem reasonable, the bubble phase is more likely to be in a near-plug-flow

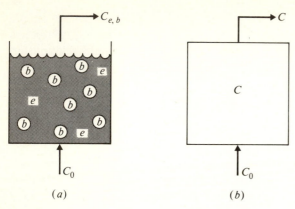

FIGURE 7-17
Schematic of a real fluid bed in contrast with a CSTR idealization. (*a*) Two-phase system, where e =solids-rich emulsion, b = solids-poor bubbles; (*b*) one-phase CSTR.

condition.   Therefore, the simple CSTR model of the fluidized bed is a severe compromise with reality with respect to both long-range and short-range concentration and (in the case of a large enthalpy of reaction) temperature gradients.

Bearing in mind the quite primitive nature of the CSTR model of both the batch and continuous fluid-bed reactor, we can set forth the governing continuity equations for the CSTR fluid-bed noncatalytic gas-solid reactor as follows.

Batch fluid bed   For the perfectly mixed gas phase, in steady state

$$\frac{A_f - A_0}{\theta_g} = \mathscr{R}_p \qquad (7\text{-}101)$$

For the batch solids phase, conversion x is a function of time on-stream t at a *fixed* gas-phase concentration A_0

$$\frac{dx}{dt} = \frac{(3bMkA_0/\rho R_0)f^{2/3}}{1 + (\text{Da}/\text{Bi})f^{2/3} + \text{Da}(f^{1/3} - f^{2/3})} \qquad (7\text{-}102)$$

These relations assume, of course, that an excess of gaseous reactant assures a constant value of A_0 throughout the course of reaction.   If this is not the case, A_0 is a function of time on-stream.   In view of the compromise inherent in the CSTR model of the fluidized-bed reactor, it does not seem worthwhile to pursue the case of A_0 as a function of time on-stream.

Continuous fluid bed   In this case, both gas and solid phases are continuously fed and withdrawn at a constant rate.   In the limiting compromise both phases are assumed to be perfectly mixed (CSTR behavior in both phases).   For the gas phase

$$\frac{A_f - A_0}{\theta_g} = \mathscr{R}_p$$

Since the continuously fed solids are surely completely segregated (as defined in Chap. 3), the simple material-balance formulation of the CSTR is inadmissible; for although the events in gas-solid noncatalytic reaction are linear, particle global conversion surely is not linear. For a segregated CSTR, we learned (Chap. 3) that exit average concentration is

$$\bar{C} = \frac{1}{\theta} \int_0^\infty C e^{-t/\theta} \, dt \qquad (7\text{-}103)$$

For perfectly mixed solids in a CSTR

$$1 - \bar{x} = \frac{1}{\theta} \int_0^t (1 - x) e^{-t/\theta} \, dt \qquad (7\text{-}104)$$

But $1 - x$ is an implicit function of t for mixed rate-controlled gas-solid reaction *unless* a conversion of 0.3 or less is of interest; in that case the flat-plate solution may be utilized since x is explicitly given as a function of t by Eq. (7-61)

$$x = 3\left(\frac{1}{\text{Bi}} + \frac{1}{\text{Da}}\right)\left[\sqrt{1 + \frac{2b\mathscr{D}A_0 t}{\rho'_B R_0^2 (1/\text{Bi} + 1/\text{Da})^2}} - 1\right] \qquad (7\text{-}105)$$

However, even in this limiting case the resulting integral proves to be beyond analytical resolution. One can, of course, solve Eq. (7-104) if only one of the three modes of rate control is assumed to govern the rate of solids conversion, as illustrated in Levenspiel's text,[1] however in view of the drastic simplifications inherent in assuming that a single-phase CSTR model can describe the two-phase fluid bed, the effort required to resolve Eq. (7-104) for the general or even specific case hardly seems justified.

Transport-line and raining-solids reactors As illustrated in Fig. 7-18, cocurrent and countercurrent gas-solids flow may be envisioned. The cocurrent gas-solids

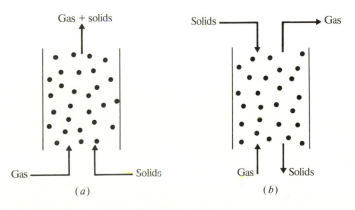

FIGURE 7-18
(*a*) Transport-line and (*b*) raining-solids reactors.

[1] O. Levenspiel, "Chemical Reaction Engineering," 2d ed., Wiley, New York, 1972.

system is truly a transport-line reactor insofar as reacting solids are carried through the reaction zone by the reactant-bearing gas. In contrast, should the solids flow or rain downward in countercurrent flow relative to the upward-flowing gas, we have a raining-solids reactor. In either case, steady state prevails with respect to both phases since both gas and solids conversion are functions only of reactor length. These reactors are therefore dilute-phase moving beds, and the relationships set forth for the moving bed apply, in principle, so long as modifications of solids holdup and velocity are entertained.

7-7 LIQUID-SOLID NONCATALYTIC REACTION

A Classic Example: Ion Exchange

It was emphasized early in this chapter that while gas-solid noncatalytic reaction admits (in general though not universally) to the pseudo-steady-state SPM, in liquid-solid noncatalytic reaction the transient terms must be respected for reasons cogently established by Bischoff.[1] Ion exchange between a flowing-liquid stream bearing ion A and a solid-particle batch phase containing an exchangeable ion B (fixed-bed ion exchange, e.g., the water softener) involves transients in both fluid and solid phases. The definition of terms commonly used in ion exchange seem so unusual that their specification deserves citation in terms of the general exchange situation (Fig. 7-19):

m = volume of voids per weight of bed
x = wt of exchange resin traversed
V = volumetric flow rate of liquid
t = real time on-stream
C = concentration of ion in liquid stream
q = resin concentration of reactant, meq/g, where meq is milliequivalents

As is indicated in Fig. 7-19 at $t > 0$ the front section of the exchanger is saturated, the end section is fresh, and an exchange zone exists within which a

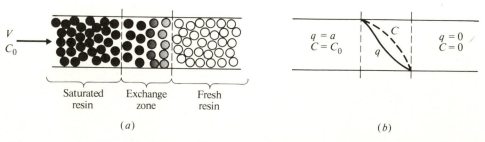

Saturated resin   Exchange zone   Fresh resin

(a)

$q = a$
$C = C_0$

C

q

$q = 0$
$C = 0$

(b)

FIGURE 7-19
(a) Ion-exchange bed; (b) liquid-solid concentration profiles at a given moment.

[1] Loc. cit.

solid and fluid concentration gradient exists. Assuming plug flow in the fixed-bed ion-exchange case, the isothermal continuity equation is

$$\left(\frac{\partial C}{\partial x}\right)_t + \frac{m}{V}\left(\frac{\partial C}{\partial t}\right)_x + \frac{1}{V}\left(\frac{\partial q}{\partial t}\right)_x = 0 \qquad (7\text{-}106)$$

There must be an equilibrium relationship between liquid and solid resin, which may assume diverse limiting forms. Consider divalent-monovalent exchange, such as $Cu^{2+} + 2HR \rightleftarrows CuR_2 + 2H$, where $R \equiv$ the resin. At equilibrium (terms in parentheses refer to concentration)

$$K_0 = \frac{(CuR_2)\gamma_{CuR_2}(H)^2\gamma_H{}^2}{(Cu)\gamma_{Cu}(HR)^2\gamma_{HR}{}^2}$$

where γ is the activity coefficient. Assuming that for dilute solution $\gamma_{Cu}/\gamma_H{}^2 = 1$, and defining a as total resin capacity at saturation, that is, $a = q$ at equilibrium, and ρ as grams of resin per cubic centimeter of wet resin (which swells upon wetting), we find

$$(CuR_2) = q\rho \qquad (HR) = (a-q)\rho$$

$$(H) = C_0 - (Cu) \qquad \text{since } (Cu) + (H) = C_0$$

$$(Cu) = C$$

Then

$$K = K_0 \frac{\gamma_{HR}{}^2}{\gamma_{CuR_2}} = \frac{q/a}{(1-q/a)^2}\frac{(1-C/C_0)^2}{C/C_0}\frac{C_0}{\rho a} \qquad (7\text{-}107)$$

or

$$\frac{q/a}{(1-q/a)^2} = \overline{K}\frac{C/C_0}{(1-C/C_0)^2} \qquad (7\text{-}108)$$

where

$$\overline{K} = K_0 \frac{\gamma_{HR}{}^2}{\gamma_{CuR_2}}\frac{\rho a}{C_0} \qquad (7\text{-}109)$$

An equilibrium diagram of fractional saturation of resin, q/a versus the fraction of fluid phase C/C_0 is schematized in Fig. 7-20 for large, small, and unit values of $\overline{K}$. By definition, we have for

$$\overline{K} \gg 1 \qquad \text{favorable, near-irreversible equilibrium}$$
$$\overline{K} = 1 \qquad \text{linear equilibrium}$$
$$\overline{K} < 1 \qquad \text{unfavorable equilibrium}$$

As an exercise for di-divalent equilibrium, derive the equilibrium expression in terms of q/a and C/C_0, for example, for the exchange $Cu + CaR_2 \rightleftarrows CuR_2 + Ca$.

In principle, numerical solution of the continuity equation (7-106) for the governing equilibrium function permits prediction of C and q as a function of bed depth x and time on-stream t. In fact a simplification of the continuity equation is realized by a change of variable

$$y = Vt - mx \qquad x = x \qquad (7\text{-}110)$$

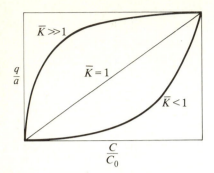

FIGURE 7-20
Ion-exchange equilibria.

y is the difference between liquid volume admitted and the amount retained by the resin. Now

$$\left(\frac{\partial C}{\partial x}\right)_t = \frac{\partial C}{\partial y}\frac{\partial y}{\partial x} + \frac{\partial C}{\partial x}\frac{\partial x}{\partial x} \qquad (7\text{-}111)$$

$$\left(\frac{\partial C}{\partial t}\right)_x = \frac{\partial C}{\partial y}\frac{\partial y}{\partial t} + \frac{\partial C}{\partial x}\frac{\partial x}{\partial t} \qquad (7\text{-}112)$$

$$\left(\frac{\partial q}{\partial t}\right)_x = \frac{\partial q}{\partial y}\frac{\partial y}{\partial t} + \frac{\partial q}{\partial x}\frac{\partial x}{\partial t} \qquad (7\text{-}113)$$

therefore by Eqs. (7-110) and (7-111) through (7-113)

$$\left(\frac{\partial C}{\partial x}\right)_t = -\frac{\partial C}{\partial y}m + \frac{\partial C}{\partial x} \qquad \left(\frac{\partial C}{\partial t}\right)_x = V\frac{\partial C}{\partial y} \qquad \left(\frac{\partial q}{\partial t}\right)_x = V\frac{\partial q}{\partial y}$$

which upon substitution in the continuity equation (7-106) yields

$$\frac{\partial C}{\partial x} + \frac{\partial q}{\partial y} = 0 \qquad (7\text{-}114)$$

the general ion-exchange equation which applies to a number of transient fluid-solid rate phenomena, such as fixed-bed catalyst poisoning, as Olson[1] has demonstrated (Chap. 10).

It is not possible within the limited scope of this chapter to treat the complex topic of ion exchange exhaustively. References to comprehensive analyses are cited at the end of this chapter. Here we merely provide a modest introduction to the subject by analyzing some rather primitive but nontrivial cases.

Insofar as ion exchange involves literally ionic reactions, it is reasonable to assume that such an ionic chemical reaction per se proceeds at a rather high specific intrinsic velocity, as is the wont of ionic reactions in general. Consequently, local rates of transport through the fluid film bathing the resin particle and diffusion within the exchange-site-bearing resin phase will surely prove to be of com-

[1] J. H. Olson, Ind. Eng. Chem. Fundam., 7: 185 (1968).

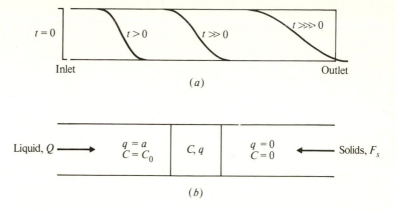

FIGURE 7-21
(*a*) Liquid-phase concentration profile at various times; (*b*) balance about exchange band.

manding importance in dictating local global (and therefore overall) reaction rate. Hence ion exchange can generally be viewed as a process governed by fluid-film (interphase) and/or solid-film (intraphase) rate which (by virtue of the comparatively small enthalpy of reaction and high-liquid-phase heat capacity involved) can safely be treated as an isothermal event.

In the ion-exchange case we shall treat in some detail the favorable equilibrium case where fluid-film mass transport governs the local global rate. This treatment is rooted in the analyses of Klotz[1] and Selke and Bliss.[2]

Referring to Fig. 7-21, the observer at the end of the bed will be witness to a fluid exchange-ion concentration of zero from the time $t = 0$ to a certain time at which unexchanged ions emerge. When, arbitrarily, $C/C_0 =$, say, 0.05, breakthrough is said to occur. Ion-exchange design involves the specification of exchanger size (length and diameter required to accommodate a specific weight of resin of equilibrium capacity a) and the time on-stream required to reduce a volumetric feed to a particular breakthrough value of C/C_0. As in any reaction-engineering problem, one is obliged to analyze small-scale laboratory-reactor data to fashion a model which is then cautiously extrapolated to predict values for a large-scale reactor.

The bases of our analyses of fluid-film-controlled exchange characterized by a very favorable equilibrium are depicted in Fig. 7-21*b*, where the exchange zone, or band, gradients and the input-output quantities are schematized. Upstream of the band, $q = a$ (saturation) and $C = C_0$, while downstream of the exchange band $q = 0$, $C = 0$. For a constant bandwidth, one visualizes the problem as

[1] I. M. Klotz, *Chem. Rev.*, **39**: 241 (1946).
[2] W. A. Selke and R. H. Bliss, *Chem. Eng. Prog.*, **46**: 509 (1950).

consisting of fluid flowing at a rate Q counter to solid resin moving at a rate F_s, so that an overall balance yields

$$\underset{\text{Into band}}{QC_0 + F_s(0)} = \underset{\text{Out of band}}{Q(0) + F_s a} \qquad (7\text{-}115)$$

while within the band

$$QC = F_s q \qquad (7\text{-}116)$$

Dividing (7-116) into (7-115), the fluid-solid nonequilibrium relation in the exchange band is

$$\frac{C}{C_0} = \frac{q}{a} \qquad (7\text{-}117)$$

Assuming (1) liquid-film mass-transport control and (2) a favorable equilibrium so that the equilibrium liquid-ion concentration at the resin surface is virtually zero, the local rate is

$$\frac{dq}{dt} = k_L s C \qquad (7\text{-}118)$$

where k_L = liquid-film coefficient, cm/min

$\qquad s = \text{cm}^2/\text{g resin}$

$\qquad C = \text{meq/ml}$

Now $\qquad\qquad\qquad dt = \dfrac{dy}{V} \qquad$ where y is in liters

By Eq. (7-118)

$$\frac{dq}{dt} = V\frac{dq}{dy} = \frac{aV}{C_0}\frac{dC}{dy} = k_L s C$$

or

$$\int_{C_0}^{C} \frac{dC}{C} = \frac{k_L s C_0}{aV} \int_{y \text{ at } C_0}^{y} dy \qquad (7\text{-}119)$$

An approximate material balance in the band gives (neglecting $q\,dx$)

$$y_{C_0} C_0 = ax + \int_{C=0}^{C=C_0} C\,dy \qquad (7\text{-}120)$$

However,

$$dy = \frac{dC}{C}\frac{aV}{k_L s C_0} \qquad (7\text{-}121)$$

hence by (7-120)

$$y_{c_0} = y_0 = \frac{ax}{C_0} + \frac{aV}{k_L s C_0} \qquad \text{lower limit of (7-119)}$$

so that integration of Eq. (7-119) results in

$$\ln\frac{C}{C_0} = \frac{k_L s C_0}{aV} y - \frac{k_L s x}{V} - 1 \qquad (7\text{-}122)$$

Therefore a semilog plot of breakthrough liquid-phase-reduced unexchanged ion concentration versus y, the volume of liquid issued from the bed, would be linear; the slope would be equal to $k_L s C_0/aV$ and the intercept equal to $k_L s x/V + 1$.

If laboratory ion-exchange data $C/C_0 - y$ conform to Eq. (7-122) the transport coefficient is secured and provides a basis for scale-up to plant ion-exchange levels involving a larger resin size and differing flow rate *assuming* that the mode of rate control does not change with scale-up. Such a procedure will shortly be illustrated, but first a question may be anticipated: If exchange may be solely controlled by fluid-film mass transfer, why is it not possible simply to employ a fixed-bed mass-transfer j-factor correlation to secure k_L and then design a priori?

We noted earlier that ion-exchange rates can safely be envisioned as being controlled by fluid-film and/or solid-phase diffusional steps. As an approximation, an overall external-internal mass-transport coefficient can be postulated in the usual fashion in terms of the fluid-film coefficient k_L and a solid-resin-phase "effective" mass-transfer coefficient k_s

$$\frac{1}{K_0} = \frac{1}{k_L} + \frac{1}{k_s} \qquad (7\text{-}123)$$

The Klotz-Selke-Bliss model, developed above, assumes that $K_0 = k_L$, while in fact since the resin is exposed to a liquid-phase concentration spanning the range of from zero to C_0, it cannot be expected that one mode of rate control will necessarily be manifest throughout the exchange band. One should not then expect a unique mechanism of rate control; thus k_L values extracted via the Klotz-Selke-Bliss model may not always exhibit the behavior predicted by j-factor correlations.

Example[1]: Ion Exchange

Given two sets of laboratory data for Cr-H exchange with Amberlite IR-100, we are required to estimate the dimensions of a plant-scale ion exchanger. The laboratory data are given in Table 7-1 and the breakthrough data in Table 7-2.

Table 7-1

	Run	
	A	B
Resin mesh size	20–50	20–50
Resin weight, g	11	11
Bed diameter, cm	2.25	1.08
Bed height, cm	8.5	36
C_0, meq/ml	0.0083	0.00756
V, ml/min	20	77

[1] H. Bliss, Course Problem, Yale University, 1955.

Utilizing the above data, we are asked to submit a design estimate for a plant ion exchanger which will process $V = 378,000$ liters/h of a solution of $C_0 = 8$ meq/liter. At least 85 percent utilization of the resin should be realized when $C/C_0 = 0.02$ at bed exit. Flow rate is not to exceed 400 liters/(min)(m²) of bed cross-sectional area.

Assuming the Klotz-Selke-Bliss model to apply, we examine the validity of this assumption by plotting the breakthrough data in accordance with Eq. (7-122). A semilog plot of C/C_0 versus y is shown in Fig. 7-22. We note that between C/C_0 values of about 0.02 and 0.9 both sets of data exhibit the linearity demanded of the assumed model. From Fig. 7-22 we obtain Table 7-3.

Derived values of saturated resin capacity a check nicely. Superficial velocities are 5 and 84 cm/min, as computed from V and column diameters for runs A and B, respectively. There should be a definite influence of fluid velocity upon the $k_L s$ values, i.e.,

$$\frac{(k_L s)_B}{(k_L s)_A} = \left(\frac{u_B}{u_A}\right)^n \qquad \frac{1.86}{0.665} = \left(\frac{84.4}{5}\right)^n$$

or $n = 0.37$, a bit low for fluid-film mass-transfer velocity dependence in a fixed bed, and without doubt the system studied is somewhat solid-diffusion-affected.

The design equation is

$$\ln \frac{C}{C_0} = \frac{k_L s}{aV} C_0 y - \left(\frac{k_L s x}{V} + 1\right) \qquad (7\text{-}124)$$

Since the desired goal is $C/C_0 = 0.02$, all is known in the design equation except x and y. By a material balance at breakthrough C_{bt}

$$y(C_0 - C_{bt}) = ax \text{ (fraction utilized, } f)$$

Fraction utilized f is specified; C_0 and C_{bt} are known, as is a, so that x is readily expressed in terms of y.

Table 7-2

Run A		Run B	
y, liters	C/C_0	y, liters	C/C_0
0.29	0.0122	0.77	0.012
0.58	0.0156	1.16	0.032
1.18	0.0227	1.55	0.06
1.48	0.027	1.9	0.084
1.8	0.05	2	0.133
2.1	0.105	2.17	0.16
2.4	0.232	2.5	0.3
2.7	0.47	2.8	0.5
3	0.69	3.1	0.7
		3.4	0.8

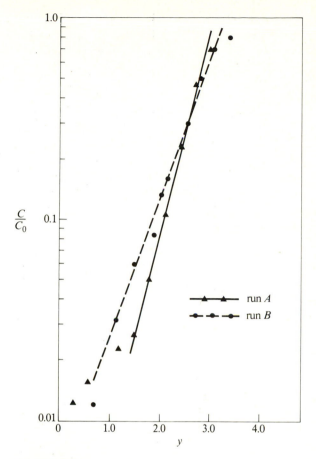

FIGURE 7-22
Experimental ion-exchange breakthrough data plotted
according to Eq. (7-122).

Having specified a maximum flow rate per unit area, we compute column
diameter

$$\frac{378,000}{60(400)} = \text{cross-sectional area of column}$$

$$\text{Area} = 15.75 \text{ m}^2 \qquad \text{Column diameter} = 4.5 \text{ m}$$

Table 7-3

Run	Slope	$\dfrac{k_{L}s}{a}$	Intercept	$k_{L}s$	a
A	2.37	0.34	0.0008	0.665	1.95
B	1.7	1.04	0.0044	1.86	1.8

Therefore superficial fluid velocity in the plant exchanger is 40 cm/min, from which the plant-scale value of $k_L s/a$ is

$$\frac{k_L s}{a} = 0.34 \left(\frac{40}{5}\right)^{0.37} = 0.734$$

$$k_L s = 0.734(1.9) = 1.4 \qquad y = \frac{ax(f)}{C_0 - C_{bt}}$$

Let $f = 0.85$; substituting all appropriate values in the design equation (7-124) and solving for x at $C/C_0 = 0.02$, we obtain

$$x = 9.85 \times 10^4 \text{ kg of resin required}$$

Resin density is computed from its weight in the laboratory column, 11 g, and volume occupied, $\rho_r = 0.33$ g/cm^3. Thus plant-exchanger resin volume is easily computed to be

$$\text{Volume of bed} = 298 \text{ m}^3$$

and bed height is

$$h = \frac{298}{15.75} = 19 \text{ m}$$

The volume issued at breakthrough is

$$y = \frac{ax(f)}{C_0(1 - 0.02)} = 20 \times 10^6 \text{ liters}$$

The cycle time is therefore

$$\theta = \frac{y}{V} = \frac{20 \times 10^6}{3.78 \times 10^5} = 54 \text{ h}$$

Far more complex modes of solution are required when a linear or unfavorable equilibrium marks the ion-exchange process. Relevant literature references are provided below, including treatments of solid-diffusion-governed exchange rate.

SUMMARY

Command of material presented in this chapter should equip the student to analyze gas-solid noncatalytic rate and conversion-time data for both isothermal and noniso-thermal environments so long as the SPM is admissible. Design estimates can be negotiated for both the constant and variable bulk-gas-composition cases. Under certain limiting conditions, the simple flat-plate approximation may be employed, and the isothermal instance of varying pellet size can also be treated.

While the case of fluid-solid noncatalytic reaction (ion exchange) has been only superficially treated, the more difficult systems lend themselves to resolution when one is armed with representative literature sources.

ADDITIONAL REFERENCES

The most comprehensive treatment of fixed- and moving-bed fluid-solid interactions is to be found in the classic papers of N. R. Amundson, *Ind. Eng. Chem.*, **48**: 26–50 (1956), while noncatalytic fluid-solid reactions are analyzed in detail by C. Wen, *Ind. Eng. Chem.*, **60**: 34 (1968), **62**: 30 (1970). A comparison between the kinetic and diffusional models is found in M. Ishida and C. Wen, *AIChE J.*, **14**: 311 (1968). The moving-bed cracker-regenerator system is discussed, among other topics, by J. Wei, *Chem. Eng. Prog. Monogr. Ser.*, **6**: 65 (1969).

An interesting case of gas-solid reaction involving surface autocatalysis is provided in J. Bandrowski, C. Bickling, K. Yang, and O. Hougen, *Chem. Eng. Sci.*, **17**: 379 (1962), in their study of NiO reduction in H_2. J. Feinman and associates, *AIChE J.*, **7**: 584 (1961), **10**: 653 (1964), present data and analyses of the H_2 reduction of ferrous oxide in a batch fluid bed. Application of kinetics to a multibed continuous process is also reported [*Ind. Eng. Chem. Process Des. Dev.*, **3**: 241 (1964)].

Theory and data for the calcination of limestone in a transport-line reactor are revealed by E. Kehat and A. Markin, *Can. J. Chem. Eng.*, **45**: 40 (1967), while rates of hydrogasification of coal char are reported by H. L. Feldkirchner and J. Huebler, *Ind. Eng. Chem. Process Des. Dev.*, **4**: 134 (1965). Data and theory for turbulent flow of gas-solid suspensions are provided by H. E. McCarthy and J. H. Olson, *Ind. Eng. Chem. Fundam.*, **7**: 471 (1968).

A detailed kinetic study of the nonisothermal hydrofluorination of uranium dioxide is set forth by E. C. Costa and J. M. Smith, *AIChE J.*, **17**: 947 (1971). Gas-solid noncatalytic reaction in a continuous fluid-bed system is discussed and a design illustration provided by P. N. Rowe, *Chem. Eng. Prog.*, **60**(3): 75 (1964). The reactions of NH_3 with carbon are analyzed by T. K. Sherwood and R. O. Maak, *Ind. Eng. Chem. Fundam.*, **1**(2): 111 (1962).

The use of the pulse-chromatographic technique to secure rate models in gas-solid reactions is detailed by N. Giordano, A. Bossi, and A. Paratella, *Chem. Eng. Sci.*, **21**: 621 (1966). Transition between regimes of rate control in gas-solid reactions is precisely analyzed by R. Aris, *Ind. Eng. Chem. Fundam.*, **6**: 315 (1967), in the light of the work of J. Shen and J. M. Smith, *Ind. Eng. Chem. Fundam.*, **4**: 293 (1965).

R. V. Mattern, O. Bilous, and E. L. Piret, *AIChE J.*, **3**: 497 (1957), present a general treatment of solid-liquid reactions occurring in a chain of CSTRs. For a dissolution process, measured specific areas and size distributions compare well with predictions. Liquid-liquid diffusion-affected reactions are analyzed by B. B. Williams et al., *Ind. Eng. Chem. Fundam.*, **9**: 589 (1970).

A general mathematical treatment of ion exchange for a broad spectrum of equilibria and kinetic relationships is to be found in the work of H. Thomas, *J. Am. Chem. Soc.*, **66**: 1664 (1944); *Ann. N.Y. Acad. Sci.*, **49**: 161 (1948); while analyses phrased in terms of both fluid- and solid-phase diffusional intrusions are provided by R. F. Baddour, E. Gilliland, N. K. Hiester, T. Vermeulen, et al., *Chem. Eng. Prog.*, **48**: 505 (1952); *Ind. Eng. Chem.*, **44**: 636 (1952); *AIChE J.*, **2**: 404 (1956). Finally R. Moison and H. O'Hern, *Chem. Eng. Prog. Symp. Ser.* (24), **55**: 71 (1959), neatly set forth negotiable ion-exchange design equations for a range of practical circumstances.

PROBLEMS

7-1   Cite several gas-solid and liquid-solid noncatalytic reactions.

7-2   Indicate the general nature of reactor types within which fluid-solid noncatalytic reactions occur.

7-3 Discuss the appropriateness of invoking the SPM for the following situations, i.e., the critical value of Da in Eq. (7-11):

(a) Combustion (regeneration) of coke deposited upon an α-alumina pellet of 3 mm diameter; total (BET) area is 4 m^2/g.

(b) Same situation but for fluidizable α-alumina of particle size of 50 μm.

(c) Repeat (a) and (b) above for a γ-alumina of total (BET) area of 200 m^2/g.

7-4 For reaction-diffusion events known to occur in series, the respective time constants give an indication of the locale of rate control, i.e.,

Reaction: $\qquad\qquad \tau_r = \dfrac{1}{k} \quad$ first order

Intraphase diffusion: $\tau_D = \dfrac{x^2}{\mathcal{D}}$

Interphase diffusion: $\tau_g = \dfrac{1}{k_g a}$

Identify the regime or regimes of rate control for the following circumstances. At 200°C

$$k_0 = 0.01 \qquad \text{and} \qquad \mathcal{D}_0 = 5 \times 10^{-2}$$

$$\text{Bulk fluid} = \text{air} \qquad \text{Particle} = \text{sphere}$$

d_p, mm	u, cm/s	E_{react}, kcal	E_{diffus}, kcal
1	1	10	2
4	100	10	2
5	100	30	2

The spheres are situated in a fixed bed. Temperature levels are 200, 300, and 500°C.

7-5 For a given fluid velocity and activation energy for reaction and intraphase diffusion as itemized in Prob. 7-4, how should d_p be varied to induce

(a) Chemical reaction control?

(b) Intraphase diffusion control?

(c) Interphase diffusion control?

7-6 An ore is to be reduced in a raining-solids reactor (countercurrent flow of gas and solids). What should be the solids contact time to achieve 40, 80, and 95 percent reduction given the following data:

(a) $\quad z = 0.6 \qquad$ particle shrinkage
$\quad R_0 = 5$ mm $\qquad \rho_B' = 0.2 \qquad b = 1 \qquad \mathcal{D} = 5 \times 10^{-2}$ cm^2/s

$\quad p_0 = 80$ Torr, $H_2 \qquad Re = \dfrac{d_p u}{\nu} = 10$

Reducing gas $= H_2$ in N_2 at 1 atm

(b) $\quad z = 1.3$

(c) $\quad z = 1.0$

What error attends the employ of the flat-plate-geometry assumption in each case at each conversion level? In all cases p_0 is constant.

7-7   A virtually ash-free coal is oxidized at 5 atm in air in a batch-fluidized bed at such an airflow rate that oxygen partial pressure is constant. The coal particles are of initial radius $R_0 = 1$ mm, of density of 2.5 g/cm³; CO_2 is the gaseous product. Assuming $n = 0.2$ in Eq. (7-43), an initial Sherwood number of 10, and a diffusivity of 0.1 cm²/s; what batch fluidization time is required to convert 10, 40, 80, and 95 percent of the coal particles? At the prevailing gas velocity, at what degree of conversion will the particles be entrained? The intrinsic rate constant is 0.1 cm/s, the mass-transfer coefficient k_g is 0.5 cm/s.

7-8   Repeat Prob. 7-7 for $n = 1$ and $n = 0$.

7-9   Repeat Probs. 7-7 and 7-8 for a flat-plate geometry.

7-10   From the Weisz-Goodwin coke burnoff data presented in Fig. 7-11, extract the intraphase diffusivity.

7-11   Create a pocket-computer program designed to generate gas-solid noncatalytic heat-generation–heat-removal relationships as a function of solids conversion.

7-12   Plot the heat-generation–heat-removal behavior of a noncatalytic gas-solid reaction as a function of conversion for the following cases:

Case	Da₀	$\bar{\beta}$	Bi$_m$	Bi$_h$	ϵ
(a)	1	0.5	1000	10	20
(b)	1	1.0	1000	10	20
(c)	1	0.5	1000	100	20
(d)	1	0.5	1000	100	40
(e)	0.1	0.5	1000	100	40
(f)	0.1	−0.5	1000	10	30

7-13   From the results secured in response to Prob. 7-12 plot solids-core temperature vs. conversion in each case. What is the value of solids external-surface temperature in each case? Assume a bulk-gas temperature of 400°C.

7-14   Compute the isothermal effectiveness factors in Prob. 7-12 and then the nonisothermal values as a function of conversion.

7-15   Obtain an analytical solution for the isothermal moving bed operated in cocurrent and countercurrent fashion when fluid-phase composition varies with bed length.

7-16   Utilizing the pyrite reduction data of Yannopoulos, Themelis, and Gauvin (*Can. J. Chem. Eng.*, **August 1966**: 231) conduct a design comparison between moving bed cocurrent and countercurrent, variable H_2 concentration, the fixed-H_2-composition moving bed, the batch fluidized bed (the simple constant-gas-composition CSTR model), and the authors' pneumatic-transport reactor. Assume a production goal of 1 metric ton/h of 90 percent reduced ore.

7-17   Wen and Huebler [*Ind. Eng. Chem. Process Des. Dev.*, **4**: 142 (1965)] report kinetic data for coal-char hydrogasification and apply their results to experimental data secured in a raining (free-fall) solids reactor. Use their data to design (a) a cocurrent transport-line and (b) batch fluid-bed reactors for 70 percent conversion of char to methane (their first-phase conversion).

7-18   Find how many transfer units are required to completely regenerate a coked catalyst in a moving bed at constant O_2 partial pressure for:

Case	Da	Bi_m
(a)	0.1	1000
(b)	1	1000
(c)	10	1000
(d)	10	100
(e)	10	10
(f)	0.1	10
(g)	100	1000
(h)	100	100
(i)	100	10

7-19 In Prob. 7-18, how serious is the assumption of isothermality in each case?

7-20 (a) Compute the moving-bed height required to burn off 100 percent of totally coked catalyst spheres at 500°C in air, given the following data

$$d_p = 3 \text{ mm} \qquad \rho = 0.1 \text{ g/cm}^3 \qquad \mathcal{D} = 0.01 \text{ cm}^2/\text{s}$$

$$\text{Da} = 1000 \qquad \text{Re} = 100 \qquad M = 14$$

$$U_s = 1 \text{ cm/s} \qquad P_{O_2} = 0.21 \text{ atm}$$

Assume the oxygen pressure to be constant.

(b) By what percentage can the catalyst throughput be increased if 90 percent burnoff can be tolerated?

(c) Justify the assumption of virtual constancy of oxygen pressure.

(e) If coked catalyst beads of $d = 1.5$ mm are to be regenerated in the same moving-bed unit, by what percentage and how is the throughput changed for 90 and 100 percent regeneration?

(f) Repeat (a), (b), and (e) for variable O_2 pressure, that is, $F_s/bu = 0.5$ and 0.1 in Eq. (7-99).

8

HETEROGENEOUS CATALYSIS AND CATALYTIC KINETICS

" We know in part, and we prophesy in part."
1 Corinthians 13:9

Introduction

Concepts of chemical kinetics (Chap. 2), reactor behavior (Chaps. 3 and 4), and diffusion and reaction (Chaps. 5 to 7) are compounded by heterogeneous catalysis, a series of surface events of immense physicochemical complexity, the detection of which now largely escapes direct observation. This chapter is an introduction to the diverse physical characteristics of common solid-catalyst types, with particular emphasis upon methods by which these characteristics can be quantified. The formulation of catalytic kinetic models in the light of well-defined postulates will be set forth, including a phenomenological description of deactivation. Appropriate comments are offered with respect to data procurement in diverse laboratory-reactor types. Finally, the known nature of catalytic action will be documented for some selected catalytic systems.

Transition-state theory teaches that chemical-reaction velocity is determined by the free energy of formation of the transition complex postulated to exist between reactants and products. Catalysis is that process in which the catalytic agent causes a reduction in the free energy of the transition-complex formation and thus a reaction-velocity increase relative to the uncatalyzed, thermodynamically permitted reaction. When the reaction network is a complex, multipathed one, the catalytic agent may alter the individual steps to differing degrees, with the consequence that overall reaction yield or selectivity is affected.

Catalysis may be realized both homogeneously and heterogeneously, the most abundant examples being heterogeneous catalysis by a solid in contact with the reactant-bearing fluid (usually a gas, though important instances of liquid-solid and gas-liquid-solid catalysis exist).

8-1 GENERAL DEFINITION OF CATALYSIS

For the overall reaction

$$A + B \rightleftharpoons C + D \qquad (8\text{-}1)$$

catalyzed by catalytic, or active, sites X_1 and X_2, fractionally poisoned by P and deactivated thermally, we can write a mechanism, i.e., a sequence of elementary steps,

Chemisorption:	$A + X_1 \rightleftharpoons AX_1$	(8-2a)
Chemisorption:	$B + X_2 \rightleftharpoons BX_2$	(8-2b)
Surface reaction and desorption:	$AX_1 + BX_2 \rightleftharpoons C + D + X_1 + X_2$	(8-2c)
Poisoning:	$P + fX_{1,2} \rightleftharpoons 0$	(8-2d)
Thermal deactivation:	$f'X_{1,2} \longrightarrow \overline{X}_{1,2}$	(8-2e)

where f is the fraction of sites poisoned and f' the fraction thermally deactivated.

Equations (8-2a) to (8-2e) neatly illustrate the key characteristics of catalysis:

1 In the absence of poisoning and/or thermal deactivation, the catalytic concentrations of sites X_1 and X_2 do not suffer consumption but are regenerated in each reaction cycle (8-2a) to (8-2c). Consequently, the catalyst does not appear explicitly in the overall reaction statement (8-1).

2 At least one reactant must be chemisorbed in any heterogeneous catalytic event (8-2a) or (8-2b). Should only one reactant be chemisorbed and the coreactant react from the gas phase, we have an Eley-Rideal mechanism

$$A + X_1 \rightleftharpoons AX_1$$
$$B + AX_1 \rightleftharpoons C + D + X_1$$

3 Various forms of deactivation [poisoning, thermal deactivation (8-2d) and (8-2e)] conspire to deny catalyst immortality. For while X_1 and X_2 are regenerated in the catalytic cycle [(8-2a) to (8-2c)] ultimately the potency of X_1 and X_2 in most instances is reduced via deactivation over periods of from several minutes to years, depending upon the particular process. In any event, the life of X_1 and X_2 is great indeed relative to reaction-cycle time.

4 Insofar as X_1 and/or X_2 can intervene to increase the rate of reaction in (8-1), in a complex, multipathed network, such as

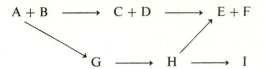

$$A + B \longrightarrow C + D \longrightarrow E + F$$
$$G \longrightarrow H \longrightarrow I$$

the yield of particular intermediates can be significantly altered relative to homogeneous reaction by the use of the catalyst which may stimulate or arrest particular steps to the benefit of desired-species generation.

5 Deactivation processes [(8-2*d*) and (8-2*e*)] demand recognition, and insights into their nature invite the possibility of devising means whereby such events can be arrested, e.g., by physicochemical promotion, discussed below.

8-2 ILLUSTRATION OF THE CATALYTIC PROCESS

By way of illustration consider the oxidation of SO_2 to SO_3, a catalytic step in the production of sulfuric acid. In the classic chamber process, catalysis is homogeneous, via NO, the possible mechanism being

Step 1: $NO + \frac{1}{2}O_2 \rightleftharpoons NO_2$

Step 2: $NO_2 + SO_2 \rightleftharpoons SO_3 + NO$

Overall $SO_2 + \frac{1}{2}O_2 \rightleftharpoons SO_3$

thus $NO \equiv X$ and $NO_2 \equiv AX$ in Eq. (8-2).

In the modern sulfuric acid process, SO_2 is oxidized over a solid catalyst, V_2O_5, presumably as follows:

Step 1: $V_2O_5 + SO_2 \rightleftharpoons SO_3 + V_2O_4$

Step 2: $V_2O_4 + \frac{1}{2}O_2 \rightleftharpoons V_2O_5$

Overall $SO_2 + \frac{1}{2}O_2 \rightleftharpoons SO_3$

hence $V_2O_5 \equiv X$ and $V_2O_4 \equiv AX$

8-3 COMPONENTS OF THE CATALYST FORMULATION

In general, the total catalyst formulation (pellet, extrusion, or particle) consists of (1) the support, usually porous Al_2O_3, SiO_2, or SiO_2-Al_2O_3, upon which is deposited (2) the catalyst agent X, and (3), in some instances, promoters.

Supports

Since catalytic activity may be proportional to the concentration of active sites X, supports of high area are commonly employed. Porous solids provide areas ranging from one to several hundred square meters per gram. For example,

α-alumina areas are about 1 to 10 m^2/g, while the η- and γ-aluminas are about 100 to 300 m^2/g. If one deposits, say, $\frac{1}{2}$ percent by weight Pt upon α-Al_2O_3 and then upon γ-Al_2O_3, the Pt atoms will clearly be better dispersed upon the higher-area γ-Al_2O_3. Thermal deactivation is a sintering process; i.e., the deposited crystallites grow in size, with the consequence that the number of surface atoms (X) per weight of deposited metal decreases. This sintering, or agglomeration process is minimized with high dispersion, hence the advantage of high-area supports.

However, a high-area support is characterized by pores of small diameter, which invites the diffusional intrusions discussed in Chap. 5. In general, a porous support is desirable, but the absolute magnitude of total area sought depends upon the particular reaction and operating conditions. For example, in the supported-silver-catalyzed oxidation of ethylene to ethylene oxide and by-product CO_2, a low-area porous support, about 1 m^2/g, is employed since intraphase heat- and mass-diffusional resistances are detrimental to yield of ethylene oxide.

The support should not be assumed to be inert in the catalytic process for it may endow the deposited catalyst with certain properties (state of oxidation, valence) and may exhibit the ability to adsorb reactants and/or atomic species dissociated by the deposited catalytic agents.[1] Finally, the higher-area supports are susceptible to sintering themselves.

Total support area (BET area) and pore-size distribution are secured by physical adsorption. These issues will be discussed shortly.

Catalytic Agents

Deposited agents are metals and semiconductors, while insulator catalysts are unsupported, for example, Al_2O_3, SiO_2-Al_2O_3. Detailed discussion of catalyst types will be set forth later. The supported catalytic ingredients are generally imposed upon the support from solution. Preparative techniques are illustrated below. It is to be noted that total impregnation of the support by the catalyst-bearing solution is not readily realized, nor, in fact, is it always desirable.

Promoters

Promoters are classified as physical or chemical. Additives which serve to maintain the physical integrity of the support and/or deposited catalytic agent are termed *physical promoters*. For example, addition of small quantities of Al_2O_3 to an iron catalyst employed in NH_3 synthesis prevents sintering of the iron crystallites. The addition of K_2O to the same catalyst seems to increase the intrinsic activity of the iron crystallites, and so it is termed a chemical promoter. In the oxidation of ethylene parts per billion of ethylene dichloride are added to the feed stream to inhibit CO_2 formation. The dichloride inhibitor is a negative chemical promoter. Thus promoters can be added during catalyst preparation or during reaction.

[1] F. Solymosi, *Catal. Rev.*, **1**: 233 (1968).

8-4 EXAMPLES OF CATALYST PREPARATION[1]

Silica-alumina cracking catalyst (89 % SiO_2, 11 % Al_2O_3)[2] A 25°Bé sodium silicate solution is stirred into an equal volume of 23°Bé sulfuric acid solution. The mixture sets as a gel and is broken up and washed. The hydrogel is soaked for 12 h in aluminum nitrate solution of the proper concentration to give an SiO_2/Al_2O_3 ratio of 8:1. The drained gel is dried and heated for 3 h at 400°C. It should be noted that Al_2O_3 is a promoter for SiO_2, the acidity of the mixture being enhanced by such promotion.

Methanol-synthesis catalyst (Zn_5Cu; Zn, Cu + Cr)[3] Five parts by weight of Zn and one part of Cu are dissolved in nitric acid and diluted with water to 15 g of metal per liter. Sodium carbonate is added to precipitate the metals. The precipitate is washed, filtered, compressed, dried, and broken into granules. To activate, granules are heated slowly from 100 to 300°C with H_2 and CO. When catalyst contains Cr, chromic acid is added to washed precipitated carbonates before drying. In this case Cu and Cr inhibit the growth of Zn crystallites, thus maintaining activity (physical promotion).

UOP platinum hydroforming catalyst (0.1 to 1 % Pt on Al_2O_3)[4] Al_2O_3 is prepared by adding ammonium hydroxide to aluminum chloride hexahydrate solution. The precipitate is washed to reduce Cl content below 0.1 percent by weight. Six washes are used with a large volume of water containing a small amount of ammonium hydroxide; a final wash is made with distilled water. Enough 4.8 % solution of HF is added to obtain 1.5 percent by weight fluorine equivalent based on dry catalyst. The wet fluorided base is then intimately mixed with colloidal platinum solution prepared by bubbling H_2S at room temperature through an aqueous solution of chloroplatinic acid. Composite is dried 17 h at 270°C and reduced in H_2 for 3 h at 500°C.

Steam-methane reaction catalyst (95-98 % MgO, 2.5 % Ni)[5] Magnesia lumps which have been calcined at high temperature are soaked in nickel nitrate solution of concentration sufficient to add 2 to 5 % Ni to magnesia. The product is dried and calcined at 800°C. Nickel can also be deposited from the vapor phase by decomposition of nickel carbonyl.

Ethylene oxide catalyst (Ag on Al_2O_3 tablets)[6] In a beaker 70 g of 8-mesh alumina tablets are placed in 21.98 g of silver oxide and 2.24 g of barium peroxide. After mixing with 100 ml of distilled water, the mixture is dried over a steam bath with

[1] See C. Thomas, "Catalytic Processes and Proven Catalysts," Academic, New York, 1970.
[2] U.S. Patent 2,363,231.
[3] U.S. Patent 2,014,883.
[4] U.S. Patent 2,479,109.
[5] U.S. Patent 1,128,804.
[6] *Ind. Eng. Chem.*, **37**: 432 (1945).

frequent stirring. Again it is dried at 115°C for 10 h. Greater selectivity ($C_2H_4 \rightarrow$ ethylene oxide) results when ethylene chloride is used as a promoter.

SO_2 oxidation catalyst (supported V_2O_5)[1] After 8.1 parts of finely powdered V_2O_5 are dissolved in 5.15 parts of KOH in 300 parts of water, dilute HCl is added to neutralize the solution and it is then diluted with 1200 parts of water. To this is added 140 parts of potassium silicate solution (21% SiO_2, 9.5% K_2O) together with 60 parts of kieselguhr and 10 parts of fibrous asbestos. The mixture is warmed to 60 to 70°C; dilute HCl is added until the solution is weakly alkaline. The mixture of zeolite and carrier is pressed lightly to remove excess liquid, particularly dried, and compressed into granules. It is then heated in air at 300 to 400°C.

Calco SO_2-oxidation catalyst (supported V_2O_5)[2] Lumps of 6-mesh kieselguhr fired at 1000°C are sprayed with a solution of Na_3VO_4 (0.4 g/g support). The mass is dried and heated to 500°C for 1 h. Heating in SO_2 is recommended.

 Although in a casual reading, some of these recipes may seem to rival that in Macbeth, it must be borne in mind that the chemistry of catalyst formulation is extremely complex and in the absence of staggering advances in inorganic and catalytic chemistry and solid-state physics, preparation of catalysts will remain primarily an art.

8-5 CATALYTIC, PROMOTER, AND TOTAL AREA

As catalytic activity and selectivity will be functions of the dispersion of X, surface concentration of X, and promoter level, it is evident that knowledge of total area and intrinsic values of catalyst and promoter surface-atom concentrations are of signal importance for intelligent interpretation of catalytic behavior.

 Total area of the catalyst formulation is routinely obtained by resolution of physical-adsorption data in terms of the Brunauer-Emmett-Teller (BET) equation. As physisorption is indiscriminate, specific area of X and that of promoters cannot be inferred from the BET analysis except with unsupported catalysts, for example, Pt-wire gauzes used in NH_3 oxidation.

 Chemisorption [(8-2a) and (8-2b)] does provide, in principle, a means whereby specific surface concentrations of catalytic agents and promoters can be ascertained. Chemisorption phenomena are of signal importance insofar as chemisorption is a step in the catalytic surface sequence and a potential tool for determining specific catalytic component concentration. The catalytic steps therefore deserve elaboration.

---

[1] British Patent 266,007.
[2] U.S. Patent 1,862,865.

8-6   STEPS INVOLVED IN THE GLOBAL CATALYTIC RATE

In general, a heterogeneous catalyst offers to reactants *active centers* situated on a solid exposed to the reactant phase.   The reactant must then (1) be transported to within the field of surface forces, a diffusional step; (2) be brought to combine with exposed *centers*, a step termed *chemisorption*; and (3) be so concentrated in the *active-center* complex as to provide reasonable rates of surface reaction to generate products.   These products must then be (4) uncoupled, desorbed, and finally (5) transported from the surface into the bulk-fluid phase surrounding the solid catalyst.   Steps 1 and 5, diffusional processes, were noted in Chap. 5 and will be considered in detail in Chap. 9.   Consideration of steps 3 to 5 indicates that two broad possibilities exist with respect to the controlling step in catalysis *at the surface*.   Assuming for the moment that the diffusive steps are very rapid, one may visualize that (1) either step 2 or step 4 is slow relative to actual surface reaction (step 3) or (2) adsorption and desorption are rapid relative to reaction between adsorbed surface species.   Under circumstances (1), the rates of adsorption and desorption become of interest, while if these rates are rapid and condition 2 exists, we then need tools for determining the surface concentration of the equilibrium adsorbed species.   In any case, we wish to know something of the accessibility of the catalytic centers, i.e., the area of the matrix (support) on which the centers of catalysis reside.   What then is the area per gram available to reactants?   Of that area, what fraction represents the active catalytic centers?   How rapidly will reactants establish themselves on these centers, and at what rate is the product molecule released from the catalytic site?   Once on the surface, what relationship exists between the surface concentration of species and the measurable bulk-phase concentrations?   We are thus brought to the question of adsorption and catalysis.

8-7   ADSORPTION ON SOLID SURFACES: QUALITATIVE DISCUSSION

A schematic representation of a typical porous-solid catalyst particle is shown in Fig. 8-1.   Sites X are dispersed throughout the porous matrix.   If the particle is nonporous, fewer sites will be accommodated; hence the use of a porous carrier

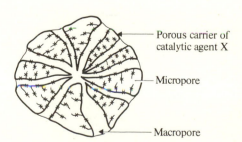

Porous carrier of catalytic agent X

Micropore

Macropore

FIGURE 8-1
Schematic of a microporous-macro-porous carrier of catalytic agent X.

provides a large area upon which the catalytic agents of a certain site concentration can be situated.

Types of Adsorption

Under suitable conditions of temperature and pressure a gas will adsorb upon a solid, ultimately covering its solid surface entirely. If the binding forces are of the magnitude of van der Waals forces, physical adsorption is involved. Physical adsorption resembles liquefaction. Referring to Fig. 8-1, a physically adsorbable gas can cover the entire surface of the porous carrier, including the area occupied by the catalytic agents X. So, in principle, if a method and model are developed whereby the amount of physically adsorbed gas corresponding to monolayer coverage can be determined, an estimate of *total* area can be realized.

Certain gases combine quite energetically with particular solids, the binding forces being comparable to those characterizing chemical compound formation. Such a process is appropriately termed *chemical adsorption*, or *chemisorption*. The process, unlike physisorption, is specific in that not all solids will chemisorb a given gas or class of gases. As noted earlier, chemisorption of at least one reactant gas is a primary prerequisite in the catalytic reaction sequence. As the process is specific, chemisorption of a gas upon catalytic sites provides, in principle, a means whereby *catalytic* (as opposed to *total*) area may be estimated.

Physisorption is therefore important as a means whereby total particle area can be determined, and, as we shall see, it also provides a basis for estimating pore volume and size and distribution of pore radii.

Chemisorption is of importance in that it constitutes a key step in the catalytic sequence and promises a means whereby actual catalytic area may be determined. Distinct aspects of chemical and physical adsorption are summarized in Table 8-1.

As an example which illustrates the characteristics cited in Table 8-1, consider the gas N_2 exposed to a porous alumina at $-195°C$. Physical adsorption occurs rapidly (negligible activation energy) and reversibly, with an adsorption heat roughly equal to that of liquefaction. On the other hand, N_2 exposed to a porous sample of reduced iron at 200°C is slowly adsorbed on the iron surface (activated chemisorption), the adsorption heat being roughly equivalent to chemical-reaction enthalpy change. At $-195°C$ the reduced iron will adsorb N_2 both physically and chemically.

The Langmuir Model: Quantitative Treatment

The concept which is fundamental to quantitative treatments of both physical and chemical adsorption is that formulated by Irving Langmuir. Although Langmuir's concern was with chemisorption, Brunauer, Emmett, and Teller utilized the Langmuir concept to derive an immensely useful relationship between the volume of a gas physically adsorbed and total surface area of the adsorbent. After treating physical adsorption, it will be shown that the Langmuir model can be extended to develop chemisorption equilibrium and rate relations even for surfaces excluded by the basic Langmuir postulates.

Langmuir postulated that the solid surface upon which adsorption occurs is homogeneous; i.e., the surface is energetically uniform, so that the free-energy change accompanying the adsorption of the first molecule or atom is identical to the change associated with subsequent adsorption of that molecular or atomic species. Further, upon adsorption, the adsorbed species do not interact with previously adsorbed neighbors. The homogeneous surface simply fills without interaction (concentration) effects, and the heat of adsorption is constant throughout the range of surface occupancy (fractional monolayer coverage θ of from 0 to 1).

Adsorption is assumed to be proportional to that fraction of the surface which is unoccupied, $1 - \theta$, and the partial pressure p of the adsorbing gas

$$\vec{r} = k_1 p(1 - \theta) \qquad (8\text{-}3)$$

Desorption is assumed to be proportional to the concentration of the adsorbed gas (or fraction of the surface occupied θ)

$$\vec{r} = k_{-1}\theta \qquad (8\text{-}4)$$

The proportionality coefficients k_1 and k_{-1} are assumed to be of the Arrhenius form, i.e..

$$k_1 = \mathscr{A}_{ads} \exp\left(-\frac{E_a}{RT}\right) \qquad \text{and} \qquad k_{-1} = \mathscr{A}_{des} \exp\left(-\frac{E_d}{RT}\right)$$

where the subscripts ads and des refer to adsorption and desorption, respectively. In accord with the homogeneous-surface postulates, $\mathscr{A}$ and E are presumed to be

Table 8-1†

	Physical adsorption	Chemisorption
Adsorbent	All solids	Some solids
Adsorbate	All gases below critical point	Some chemically reactive gases
Temperature range	Low T	Generally high temperature
Heat of adsorption	Low, $\sim \Delta H_{liq}$	High order, about enthalpy of reaction
Rate and activation energy	Very rapid, low E	Nonactivated, low E; activated, high E
Coverage	Multilayer	Monolayer and less
Reversibility	Highly reversible	Often irreversible
Importance	For determination of surface area and pore size	For determination of surface concentration, rates of adsorption and desorption, estimates of active center area, and elucidation of surface-reaction kinetics

† P. H. Emmett, Course notes in catalysis, Johns Hopkins University, 1959.

independent of surface coverage θ.   At equilibrium $\vec{r} = \bar{r}$, and so equating (8-3) with (8-4) and solving for θ

$$\theta = \frac{(k_1/k_{-1})p}{1 + (k_1/k_{-1})p} = \frac{Kp}{1 + Kp}$$

By definition k_1/k_{-1} is the adsorption equilibrium coefficient K, which by implication is independent of coverage.   Therefore adsorption enthalpy ΔH and entropy ΔS are constant, independent of surface coverage.   In the terminology of catalysis, $-\Delta H = q$, the heat of adsorption (physical or chemical).

The physical meaning of θ deserves comment.   In physical adsorption, we are not limited to monolayer adsorption since a physically adsorbed gas can be localized in multilayers and θ can be defined as equal to V/V_∞, the volume of gas physically adsorbed relative to that volume adsorbed at multilayer saturation. However, the above Equation, derived for monolayer saturation, is not applicable to multilayer coverage.   Its extension to physical adsorption will be noted shortly.

In chemisorption, θ is not necessarily equal to the volume adsorbed relative to that at complete monolayer adsorption.   Cases exist, e.g., chemisorption on semiconductors, in which chemisorption saturation occurs in monolayers at total coverage far less than unity.   So θ is best defined for chemisorption as the volume adsorbed relative to that volume adsorbed at surface monolayer saturation, whether that saturation value is equivalent to total or partial surface coverage.

Though rooted in patently idealized postulates (the homogeneous-surface implications), the Langmuir model of adsorption-desorption kinetics and equilibrium is central to the powerful relationships for chemisorption kinetics, equilibria, and physical-adsorption equilibrium derived from it.   Since the question of total surface area and pore character is primary, and since under proper conditions physical adsorption occurs on all solids, the extension of the simple Langmuir model to describe physical adsorption will be described next, followed by a discussion of chemisorption phenomena.

8-8   PHYSICAL-ADSORPTION MODEL

Procurement and Display of Physisorption Data

Obviously physisorption equilibria data can be presented in a number of convenient forms, e.g., volume adsorbed vs. pressure or temperature.   The dependent variable, the volume or mass of gas adsorbed, can be measured either volumetrically or gravimetrically.   The direct measurement of volume adsorbed is, comparatively speaking, the simpler task, as gravimetric means involve considerable refinement of apparatus due to the low mass values associated with an adsorbed layer of gas.   In each case evacuation of the sample is necessary.   In a volumetric determination, a void-space measurement in the sample chamber is required (generally with helium), and the net change in system volume is accurately measured upon introduction of the physically adsorbable gas.   Such isothermal volume-change determinations

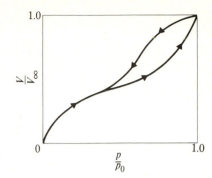

FIGURE 8-2
Typical physical absorption-desorption
isotherm for N_2 at $-195°C$.

are made at equilibrium for various adsorbate pressures, with results displayed as
shown in Fig. 8-2.   There the volume adsorbed (STP) is plotted against a reduced
pressure (pressure of adsorbate divided by that pressure corresponding to boiling
at the temperature of the isotherm).   Gravimetric determination of an isotherm is
in principle direct; however, as noted, considerable sophistication in equipment is
required and a buoyancy correction is involved.

An alternative technique for isotherm and (specifically) surface-area measurement merits comment.   This is the flow-adsorption method, described by Shell
Development Company research workers.[1]   Referring to Fig. 8-3, a gas stream
of fixed adsorbate partial pressure is passed continuously over the adsorbent
sample.   A conductivity-cell–recorder adjunct provides notice of zero base line
and subsequent events.   When steady flow and composition are realized, the
adsorbent sample tube is immersed in a suitable temperature bath, say, liquid N_2,
causing physical adsorption of the flowing component carried in an inert or non-
adsorbable carrier gas.   The adsorption and ultimate saturation at the chosen
partial pressure cause a time-compositional change in the effluent stream passing
through the conductivity cell.   Integration of the recorded effluent concentration-
vs.-time curve yields the volume of gas adsorbed at the prevailing adsorbate pressure.   By varying adsorbate partial pressure, a series of volume-vs.-pressure data
are obtained to yield an isotherm over a limited pressure range.   The limitation is
obvious, in that higher concentration of adsorbate creates flow disturbances upon
adsorption.   A technique involving operation at high total pressures in the sample
section with subsequent letdown and composition measurement at 1 atm has been
cited as a means whereby data at high values of p/p_0 can be obtained via the flow
technique.[2]

Types of Physisorption Isotherms

Brunauer, Deming, Deming, and Teller[3] cite and discuss five types of adsorption
isotherms, three of which are displayed in Fig. 8-4.   The significance of an isotherm
will be discussed in our treatment of the mathematical models of adsorption.

[1] F. M. Nelson and F. T. Eggertsen, *Anal. Chem.*, **30**: 1387 (1958).
[2] A. J. Haley, *J. Appl. Chem.*, **13**: 392 (1963).
[3] S. Brunauer, L. S. Deming, W. E. Deming, and E. Teller, *J. Am. Chem. Soc.*, **62**: 1723 (1940).

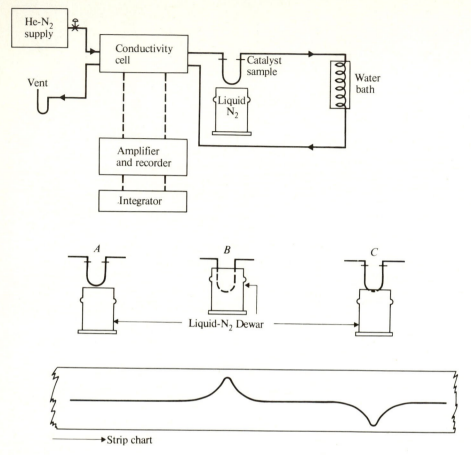

FIGURE 8-3
Flow BET apparatus and technique, also applicable to chemisorption at any temperature.

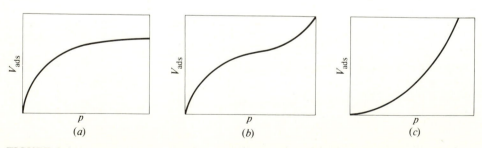

FIGURE 8-4
Three types of physisorption isotherms: (a) Langmuir, type I, (b) type II (c) type III.

Types I and II are rather common, while others arise under somewhat exceptional circumstances.

8-9 THE MULTILAYER-ADSORPTION THEORY (BET EQUATION)

In 1938 Brunauer, Emmett, and Teller[1] published the paper Adsorption of Gases in Multimolecular Layers, in which a generalized theory of physical adsorption was set forth in some detail. A signal feature of this development was a clear specification of the volume of gas adsorbed at the point of monolayer completion. In consequence, the means became available for computing with rigor the total surface area of a porous body, e.g., catalyst, from physical-adsorption equilibrium data. Adsorption equilibrium models proposed before the BET theory failed to yield an unequivocal means for determining monolayer coverage. As Sir Hugh Taylor put it, "The technique [BET] is a standardized one which can be reproduced by different workers in different laboratories with considerable accuracy and reproducibility."

It is perhaps safe to say that the BET development was sponsored more by concern for catalyst-surface-area determination than for an adsorption-process rationale. Obviously a model of physical adsorption was required to facilitate a meaningful relationship between surface area and measured quantities. Earlier Emmett and Brunauer had described the use of adsorption isotherms for the determination of surface areas of various iron catalysts in ammonia synthesis. For Benton[2] had pointed out that in systems exhibiting type II isotherms the point of monolayer completion might correspond to that region of the isotherm where linearity starts. It was reasoned that the linear portion represented second-layer development; in consequence this point of initial multilayer formation would correspond to monolayer completion. This point was termed *point B* by Emmett and Brunauer. Other criteria had been suggested; however, point B proved to be the most reliable of proposed candidates.

In systems not characterized by a type II isotherm, the point-B method obviously fails. The physical adsorption of *n*-butane on silica gel is such an example. From 0 to about 300 mm linearity is manifest, thus making it impossible to specify point B. It was clear to Emmett and Brunauer that a generalized theory for physical adsorption was required which would prove more powerful, less ambiguous, and of broader scope in the determination of surface area.

The BET equation can be developed, first, for the case of adsorption upon a free surface so that no limit upon the number of layers of adsorbed gas is imposed. The integration can also be carried out for *n* layers, this corresponding to bounded surfaces (cracks, crevices, pores). In principle the derivation amounts to an extension of Langmuir's theory (monolayer adsorption) to multilayer adsorption.

---

[1] S. Brunauer, P. H. Emmett, and E. Teller, *J. Am. Chem. Soc.*, **60**: 1309 (1938).
[2] See Emmett, loc. cit.

Let s_0, s_1, s_2, ..., s_i be the surface area covered by only 0, 1, 2, 3, ..., i layers of adsorbed molecules. At equilibrium the rate of condensation upon s_0 is equal to the rate of evaporation from s_1;

$$a_1 p s_0 = b_1 s_1 \exp\left(-\frac{E_1}{RT}\right) \qquad (8\text{-}5a)$$

which is Langmuir's equation for unimolecular adsorption. The first assumption of the general theory is clear: a, b, and E are independent of the number of adsorbed molecules in the first layer.

At equilibrium, application of the principle of microscopic reversibility demands that all the processes which occur on the surface (evaporation, condensation, surface migration) be balanced by reverse processes of equal frequency. We can then imagine s_1 changing via condensation on a bare surface, evaporation from the first layer, and evaporation from the second layer. At equilibrium, the net result for the first layer is that the rate of condensation upon the first layer must equal the evaporation rate from the second layer

$$a_2 p s_1 = b_2 s_2 \exp\left(-\frac{E_2}{RT}\right)$$

The argument is extended to i layers

$$a_3 p s_2 = b_3 s_3 \exp\left(-\frac{E_3}{RT}\right)$$

$$\cdots \cdots \cdots \cdots \cdots \cdots \cdots \qquad (8\text{-}5b)$$

$$a_i p s_{i-1} = b_i s_i \exp\left(-\frac{E_i}{RT}\right)$$

The total area S of the catalyst is the sum of all areas covered by layers of gas, i.e., the summation of s_i values

$$S = \sum_{i=0}^{i=\infty} s_i \qquad (8\text{-}6)$$

while the total volume of gas adsorbed is

$$V = \frac{V_m}{S} \sum_{i=0}^{i=\infty} i s_i \qquad (8\text{-}7)$$

where V_m is the volume corresponding to monolayer coverage of the entire surface of area S. Therefore

$$\frac{V}{V_m} = \frac{\sum_0^\infty i s_i}{\sum_0^\infty s_i} \qquad (8\text{-}8)$$

The indicated summations can be carried out by making certain simplifying assumptions. First, it is assumed that the heats of adsorption of the second, third, and ith layers are equal to each other and to the heat of liquefaction of the adsorbate $E_2 = E_3 = \cdots = E_i = E_L$. Further, let

$$\frac{b_2}{a_2} = \frac{b_3}{a_3} = \cdots = \frac{b_i}{a_i} = g = \text{const} \quad \text{and} \quad \epsilon_i = \frac{E_i}{RT} \qquad (8\text{-}9)$$

Now let $s_1, s_2, s_3, \ldots$ be expressed in terms of s_0:

$$s_1 = \left(\frac{a}{b}\right)_1 p s_0 \exp \epsilon_1$$

$$s_2 = \frac{p}{g} s_1 \exp \epsilon_L \qquad (8\text{-}10)$$

$$s_i = s_0 \frac{a_1}{b_1} g \left[\exp (\epsilon_1 - \epsilon_L)\right] \left(\frac{p}{g}\right)^i \exp \epsilon_i$$

Letting

$$y = \frac{a_1}{b_1} p \exp \epsilon_1 \qquad x = \frac{p}{g} \exp \epsilon_L \quad \text{and} \quad c = \frac{y}{x} \qquad (8\text{-}11)$$

means that s_i can be substituted into Eq. (8-8) to obtain

$$\frac{V}{V_m} = \frac{c s_0 \sum\limits_{1}^{\infty} i x^i}{s_0 \left(1 + c \sum\limits_{1}^{\infty} x_i\right)} \qquad (8\text{-}12)$$

The summations are carried out by noting that

$$\sum_{i=1}^{i=\infty} x^i = \frac{x}{1-x}$$

and

$$\sum_{1}^{\infty} i x^i = x \frac{d}{dx} \sum_{1}^{\infty} x^i = \frac{x}{(1-x)^2} \qquad (8\text{-}13)$$

Hence

$$\frac{V}{V_m} = \frac{cx}{(1-x)(1-x+cx)} \qquad (8\text{-}14)$$

As noted, this free-surface model tolerates an infinite number of adsorbed layers. Thus at $p = p_0$, V must be infinite. That is, x must equal unity at $p = p_0$ or $p_0/g \exp \epsilon_L = 1$, so that $x = p/p_0$. Equation (8-14) in terms of p and p_0 is

$$\frac{V}{V_m} = \frac{cp}{(p_0 - p)[1 + (c-1)p/p_0]} \qquad (8\text{-}15)$$

Equation (8-15) does indeed describe type II isotherms qualitatively, for as c is much larger than unity, at low pressure

$$\frac{V}{V_m} = \frac{c(p/p_0)}{1 + cp/p_0} \qquad (8\text{-}16)$$

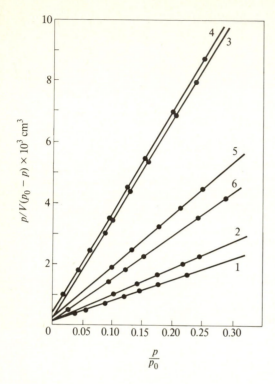

FIGURE 8-5
Linearization of physisorption data according to the BET equation (8-17). Adsorption of N_2 at 90.1 K on unpromoted Fe catalyst 973 per 489.0 g (*curve 1*); Al_2O_3-promoted Fe catalyst 424 per 49.8 g (*curve 2*); Al_2O_3-K_2O-promoted Fe catalyst 958 per 54.4 g (*curve 3*); fused-copper catalyst per 550.0 g (*curve 4*); chromium oxide gel per 1.09 g (*curve 5*); and silica gel per 0.606 g (*curve 6*). [*S. Brunauer, P. H. Emmett, and E. Teller, J. Am. Chem. Soc.*, **60**: *309 (1938)*.]

which is the Langmuir isotherm, while as p approaches p_0, the volume adsorbed versus p/p_0 displays the convex character typical of type II isotherms at high pressures.

Equation (8-15) is readily cast into a form convenient for the abstraction of V_m

$$\frac{p}{V(p_0 - p)} = \frac{1}{V_m c} + \frac{c-1}{V_m c}\frac{p}{p_0} \qquad (8\text{-}17)$$

A plot of the left-hand side (experimental data) against p/p_0 yields a straight line from which both V_m and c can be obtained by employing the slope and intercept (Fig. 8-5).

The free-surface model is an idealization, as most solid catalysts contain cracks, crevices, and pores, within which the major fraction of the surface area resides. Only n layers of gas can be adsorbed within a pore or crevice, so that the upper limit of the summations in Eq. (8-13) should be specified as n rather than infinity. In this instance

$$\frac{V}{V_m} = \frac{xc}{1-x}\frac{1 - (n+1)x^n + nx^{n+1}}{1 + (c-1)x + cx^{n+1}} \qquad (8\text{-}18)$$

For $n = 1$

$$\frac{V}{V_m} = \frac{xc}{1-x}\frac{1 - 2x + x^2}{1 + (c-1)x - cx^2} = \frac{xc}{1-x}\frac{(1-x)^2}{(1+cx)(1-x)} \qquad (8\text{-}19)$$

or
$$\frac{V}{V_m} = \theta = \frac{cx}{1 + cx}$$

which is the Langmuir isotherm. Since x is less than unity, it is apparent that as n approaches infinity, Eq. (8-18) reduces, as expected, to Eq. (8-15).

Further Remarks on the BET Equation

Given data on volume adsorbed vs. pressure, the BET equation (8-17) provides a reliable means whereby total surface area can be determined. The method of data procurement is irrelevant with respect to application of Eq. (8-17), though it is certainly significant that more data are obtained in a given period by the Shell flow method, cited earlier. Whether static or flow techniques are employed, inspection of the BET relation suggests that since the constant c is of the order of 10^2 for N_2, one low-pressure point on the isotherm may be sufficient for a rather accurate surface-area determination. Thus a single adsorption equilibrium point in the range of p/p_0 of 0.2 to 0.3, when extrapolated to zero V at $p/p_0 = 0$, will yield a slope not significantly different from the true value obtained using several relative pressures. This follows from the fact that the intercept term, $1/V_m c$, will approach zero for most systems displaying a type II isotherm with a nitrogen adsorbate.

Pore Size and Its Distribution

In Fig. 8-2 a typical physical adsorption-desorption isotherm is shown for a porous solid. Hysteresis is evident, i.e., the desorption curve lies above that of adsorption equilibrium from p/p_0 of from unity to some intermediate value. In effect, then, adsorbate is retained during desorption and released at a p/p_0 value less than that required to cause adsorption. This behavior and its rationalization provides the basis for estimating pore size and its distribution. The phenomenon is due to capillary condensation-evaporation processes.

Lord Kelvin recognized that the vapor pressure of a liquid contained in a small-diameter capillary is less than the normal value predicted for a free surface. Consider a capillary partially filled with a liquid of surface tension σ. The change in free energy due to evaporation of a differential volume of liquid $n \, \Delta G$ equals the change in surface times the surface tension, where $n = dv/v_m$ and v_m is molar volume:

$$n \, \Delta G = +(2\pi r \, d\ell)\sigma \cos \theta = -nRT \left(\ln \frac{p}{p_0} \right)$$

$$n = \frac{dv}{v_m} = \frac{\pi r^2 \, d\ell}{v_m}$$

Then
$$\ln \frac{p}{p_0} = v_m \frac{2\pi r \, d\ell}{(\pi r^2 \, d\ell)RT} \qquad \cos \theta \approx 1$$

or

$$\frac{p}{p_0} = \exp \frac{-2\sigma v_m}{rRT} \qquad (8\text{-}20)$$

which is the common form of the Kelvin equation, assuming a wetting angle of zero. This equation states that the pressure at which condensation or evaporation will occur is always less than the free-surface vapor pressure *if* the capillary radius is small enough for a given liquid. We see then that if a nonporous solid or one containing very large pores is subjected to physical adsorption-desorption equilibrium experiments, actual condensation of adsorbate will not occur until the gas pressure equals the vapor pressure at the prevailing temperature, i.e., at $p/p_0 = 1$ in Fig. 8-2. On the other hand, if pores of appropriate radius r exist, condensation will occur before $p/p_0 = 1$; thus liquid N_2 will form in the pores at $p/p_0 < 1$. On adsorption, this would account for the rapid increase in V with p/p_0 typical of adsorption isotherms for porous solids. However, this mechanism would not account for hysteresis if filling of pores on adsorption and their emptying on desorption followed the same mechanism. This physisorption mechanism is, of course, vertical filling and emptying of the pores. While various theories have been set forth to rationalize hysteresis, a very reasonable one due to Cohan[1] will be discussed here with the understanding that the matter is not yet totally resolved. Cohan argues that on *adsorption* the pores do not fill vertically but *radially*. So we see then that with condensation of the first layer, r is decreased, thus causing further condensation at a fixed p/p_0. In other words pores of a radius r, corresponding to a given p/p_0, fill instantaneously. The Kelvin equation can be derived for this situation, where now the change in volume dv is $2\pi r\ell\, dr$ and that of surface is $2\pi\ell\, dr$. Balancing the free-energy change as above, we find

$$\frac{p}{p_0} = \exp \frac{-\sigma v_m}{rRT} \qquad (8\text{-}21)$$

Following Cohan, for a given pore radius r, adsorption with radial capillary condensation occurs at

$$p_{\text{ads}} = p_0 \exp \frac{-\sigma v_m}{rRT} \qquad (8\text{-}22)$$

while *vertical* emptying of the pores occurs during desorption at about

$$p_{\text{des}} = p_0 \exp \frac{-2\sigma v_m}{rRT} \qquad \text{implies wetting angle of zero} \qquad (8\text{-}23)$$

It follows then that the adsorption pressure p_{ads} is related to that of desorption p_{des} by

$$p_{\text{ads}}^{\,2} = p_0\, p_{\text{des}} \qquad (8\text{-}22a)$$

or the reduced pressure required to empty the capillary is equal to the square of that necessary to fill it. A pore of radius r which fills at p/p_0 of say 0.7 is desorbed at

[1] L. H. Cohan, *J. Am. Chem. Soc.*, **60**: 433 (1938), **66**: 98 (1944).

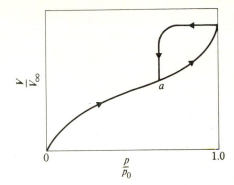

FIGURE 8-6
Isotherm for a porous solid of uniform pore radius. Point a is that of step change in V/V_∞.

p/p_0 of about 0.5; hence hysteresis.   Some data support this quantitative prediction.

Given the adsorption-desorption hysteresis loop, pore size and its distribution can, in principle, be obtained.   If, for example, a porous catalyst sample contained pores of one diameter only, the hysteresis loop would resemble that shown in Fig. 8-6.   When we know σ and v_m and that the desorption value of p/p_0 occurs at point a in Fig. 8-6, the pore radius can be calculated from the Kelvin equation. Rarely does the hysteresis loop prove to be as simple as that shown in Fig. 8-6; a less abrupt desorption curve is common, as in Fig. 8-2, suggesting a distribution of pore radii.   The problem of determining the exact distribution is not unlike that encountered earlier, namely, finding an analytical residence-time distribution of flowing molecules which fits an experimental residence-time response curve for a vessel.

Wheeler[1] and Shull[2] set forth the general relationships which permit the evaluation of pore-size distribution from physical-desorption isotherms.   The volume of gas not adsorbed at a pressure p is $V_s - V$, where V_s is total pore volume, i.e., volume of gas adsorbed at p_0.   This volume $V_s - V$ must equal the summation of volumes of unfilled pores of radii which may be distributed.   Before a pore is filled, its effective radius is less than the true radius by the thickness t of the physically adsorbed multilayer.   On emptying, of course, all liquid is evaporated except the physically adsorbed layer of thickness t.   In the Kelvin equation the radius of interest is not r but $r - t$.   Wheeler's reasoning suggests

$$V_s - V = \pi \int_{r_c}^{\infty} (r - t)^2 L(r)\, dr \qquad (8\text{-}24)$$

i.e., the summation of all pore volumes composed of volumes $\sum \pi(r_i - t)^2 L(r)\, dr$, where $L(r)\, dr$ is the total length of pores of radii between r and $r\, dr$. r_c is the critical pore radius (corrected for physical adsorption), which means that all pores of radii less than r_c are filled and those of greater radii are empty.   The

[1] A. Wheeler, in P. H. Emmett (ed.), " Catalysis," vol. 2, Reinhold, New York, 1955.
[2] C. G. Shull, *J. Am. Chem. Soc.*, **70**: 1405 (1948).

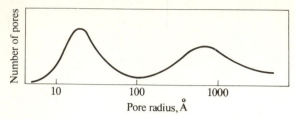

FIGURE 8-7
Bimodal pore-size distribution.

key to the use of Eq. (8-24) is the proper choice of $L(r)\,dr$.   From the Kelvin equation

$$r_c = t + \frac{2\sigma v_m}{RT \ln (p_0/p)} \qquad (8\text{-}25)$$

so that once $L(r)\,dr$ is specified, the Wheeler-Shull relation can be solved.   In general for an assumed $L(r)\,dr$ function the V-versus-p/p_0 data and prediction are compared and $L(r)\,dr$ adjusted until conformity is realized between model and data.   Barrett, Joyner, and Helenda[1] provide a detailed procedure for pore-size determination.   In any case, t must be secured from experimental isotherms for non-porous surfaces.   Anderson[2] provides an improvement of the method.   Typical pore-size distribution is presented in Fig. 8-7 for a microporous-macroporous catalyst (Fig. 8-1).   Note the bimodal distribution.

A Simplified Model of Average Pore Size

In contrast to Fig. 8-1, if only macropores characterize the catalyst, an estimation of *average* pore radius can be made by assuming a cylindrical pore of length L and radius r.   The internal surface-to-volume ratio of this idealized pore is

$$\frac{S}{V_p} = \frac{2rL}{r^2L} = \frac{2}{r} \qquad (8\text{-}26)$$

Hence a value of pore volume V_p and BET area S permits a rough estimate of average pore radius r by

$$r = \frac{2V_p}{S} \qquad (8\text{-}27)$$

Obviously if this simplified model is applied to a macroporous-microporous catalyst sample exhibiting a bimodal distribution, a totally meaningless value of r will result.   Values of S, V_p, and r for several catalysts are given in Table 8-2.

---

[1] E. P. Barrett, L. G. Joyner, and P. P. Helenda, *J. Am. Chem. Soc.*, **73**: 373 (1941).
[2] R. B. Anderson, *J. Catal.*, **3**: 50 (1964).

Table 8-2   SURFACE AREA, PORE VOLUME, AND MEAN PORE RADII FOR TYPICAL SOLID CATALYSTS†

Catalyst	Surface area S_g, m²/g	Pore volume V_g, cm³/g	Mean pore radius, Å
Activated carbons	500–1500	0.6–0.8	10–20
Silica gels	200–600	0.4	15–100
SiO_2-Al_2O_3 cracking catalysts	200–500	0.2–0.7	33–150
Activated clays	150–225	0.4–0.52	100
Activated alumina	175	0.39	45
Celite (kieselguhr)	4.2	1.1	11,000
Synthetic ammonia catalysts, Fe	4–11	0.12	200–1000
Pumice	0.38		
Fused copper	0.23		

† From J. M. Smith, "Chemical Engineering Kinetics," 1st ed., McGraw-Hill, New York, 1956. Some values from A. Wheeler, *Adv. Catal.*, **3**: 250-326 (1950).

8-10   RELEVANCE OF CHEMISORPTION TO CATALYSIS

Attention was just directed to a physical process (physisorption) which is germane to catalysis insofar as knowledge of the physical area and pore distribution of catalyst particles can be derived from physical-adsorption isotherms. The relevance of chemisorption to catalysis is perhaps best stated in a general historical context.

When we recognize that catalysis developed as a branch of chemical kinetics, it is hardly surprising that early catalytic studies were conducted as were homogeneous chemical-kinetic studies. The procedure followed to realize an understanding of, say, homogeneous gas-phase reactions is today clearly recognized. As noted by Boudart[1] and Halsey,[2] the steps are as follows: (1) the overall reaction must be dissected into its most elemental steps in a sequence which yields the net rate then (2) analyzed theoretically and if possible experimentally. By way of illustration, the first-order decomposition of N_2O_5, long considered a simple process, was eventually shown (Ogg[3]) to be the result of a series of elemental steps which yield the overall result.

Analyses of catalytic reactions must follow this general procedure. However, difficulties become legion in the light of early catalytic kinetic studies. For overall catalytic rates could not be simply expressed in terms of measured gas-phase concentrations. As surface concentrations of reactants clearly determined surface (catalytic) rates, an additional body of data was required to establish the relationship between surface and fluid concentrations.

[1] M. Boudart, *Chem. Eng. Prog.*, **58**: 73 (1962).
[2] G. Halsey, *J. Phys. Chem.*, **67**: 2038 (1963).
[3] R. A. Ogg, *J. Chem. Phys.*, **15**: 337, 613 (1947).

That body of data is the chemisorption isotherm for the system when surface reaction is rate-controlling, thereby establishing a relationship between surface concentration and the observable gas-phase concentration. When, however, adsorption is rate-controlling, the desirability of chemisorption *rate* data becomes apparent. In terms of catalytic rate phenomena, then, chemisorption equilibria and rate studies are required for a rational formulation of surface kinetics. These kinetic and thermodynamic data then permit a clear definition of the overall catalytic rate and thus invite, in principle, the dissection of overall rate into likely elementary steps. Given the fact the surface is an active participant in the sequence of elementary steps, it is not difficult to appreciate why our understanding of surface catalysis lags behind that of homogeneous reactions. For while notable progress has been made in the physics of the solid state, crucial property and energetic differences are recognized as existing between bulk and surface regimes in the solid state. As catalysis involves primarily solid surfaces, we can appreciate the profound complexities involved in characterizing the surface of a solid even when bulk properties are well known or measurable.

In addition to its kinetic significance, chemisorption, by its specificity with respect to particular surfaces as opposed to total surface (determined by non-specific BET physisorption) permits, in principle, the measurement of catalytic sites existing upon a support matrix which itself does not invite chemisorption. Specific examples of catalytic (as opposed to total-surface-area) determination by chemisorption and/or titration will be cited later in this chapter.

8-11 CHEMISORPTION EQUILIBRIA AND KINETICS

At chemisorption equilibrium, Langmuir asserted that ideal surface coverage is

$$\theta = \frac{Kp}{1 + Kp} \qquad (8\text{-}28)$$

where

$$K = \frac{k_{\text{ads}}}{k_{\text{des}}} \qquad (8\text{-}29)$$

The adsorption and desorption rate coefficients, k_{ads} and k_{des}, are assumed to obey an Arrhenius functionality. In essence, activation energies and adsorption heats are presumed to be constant, independent of θ. These assertions imply surface ideality; i.e., heat of adsorption q is independent of coverage θ, as are the activation energies E_{ads} and E_{des}.

In point of fact, few gas-solid catalyst systems conform to the Langmuir postulates. Measured values of q as a function of coverage θ, indicate a definite decline in q with θ, from which one is justified in concluding that either (1) catalytic surfaces are energetically nonideal (heterogeneous) and/or (2) interaction forces are manifested as a function of coverage. In fact Langmuir anticipated non-ideality and suggested that in such circumstances

$$\theta = \sum_{i=1}^{i=n} \frac{K_i p}{1 + K_i p} \qquad (8\text{-}30)$$

That is, the real surface might be considered as a summation of small discrete ideal surfaces. This suggestion proved to be seminal.

Real Surfaces

Chemisorption equilibria are often described by the Freundlich isotherm

$$\theta = Kp^{1/n} \qquad n > 1 \qquad (8\text{-}31)$$

or the isotherm of Temkin

$$\theta = \frac{1}{f} \ln \bar{K}p \qquad (8\text{-}32)$$

while in numerous instances the kinetics of chemisorption follow the Elovich equation

$$\vec{r}_{ads} = k_{ads}\, p \exp\left(-g\theta\right) \qquad (8\text{-}33)$$

and desorption rates tend to conform to the relation

$$\vec{r}_{des} = k_{des} \exp h\theta \qquad (8\text{-}34)$$

Thus observations of chemisorption equilibria and kinetics on real surfaces stand in radical contrast to the ideal-surface models. We note that upon equating (8-33) with (8-34), the equilibrium relation conforms to that of Temkin; this relation, however, fails to predict values of θ of zero and unity in the limits of p, and while the Freundlich isotherm predicts a zero value of θ at p equal to zero, values of θ exceed unity with increasing p.

Real-Surface Models

Brunauer, Love, and Keenan[1] have profitably analyzed chemisorption kinetics and equilibria upon real, heterogeneous surfaces. They assume adsorption and desorption activation energies (and hence equilibrium heats of adsorption) to be linear functions of coverage θ:

$$E_{ads} = E_{ads}{}^0 + \gamma\theta$$
$$E_{des} = E_{des}{}^0 - \beta\theta \qquad (8\text{-}35)$$
$$q = q_0 - \alpha\theta \qquad \text{where } \alpha = \gamma + \beta$$

The physical bases for their assumption can be (1) surface heterogeneity, (2) adsorbed-species interaction, or (3) a combination of them.

For the rate of adsorption, assuming a continuum of ideal patches, we have

$$\vec{r}_{ads} = \int_0^\theta k_{ads}\, p(1 - \theta)\, d\theta \qquad (8\text{-}36)$$

where

$$k_{ads} = \mathscr{A} \exp \frac{-(E_{ads}^0 + \gamma\theta)}{RT}$$

[1] S. Brunauer, K. S. Love, and R. G. Keenan, *J. Am. Chem. Soc.*, 64: 751 (1942).

Then upon integration of (8-36) for $1 - \theta \rightarrow 1$ we have

$$\vec{r}_{ads} = \mathscr{A} p \frac{RT}{\gamma} \exp \frac{-E_{ads}^{0}}{RT} \left(\exp \frac{-\gamma\theta}{RT} - \exp \frac{-\gamma}{RT} \right) \qquad (8\text{-}37)$$

which properly reduces to zero at $\theta = 1$ and achieves a maximum at $\theta = 0$. For large values of γ, (8-37) becomes

$$\vec{r}_{ads} = k_{ads} p \exp(-g\theta) \qquad \text{where } g = \frac{\gamma}{RT} \qquad (8\text{-}38)$$

the Elovich relation. By similar reasoning the desorption rate

$$\vec{r}_{des} = \int_{0}^{\theta} k_{des}\, \theta\, d\theta \qquad (8\text{-}39)$$

becomes upon integration, assuming a linear $E_{des} - \theta$ function (at $\theta \rightarrow 1$) and a large value of β,

$$\vec{r}_{des} = k_{des} \exp h\theta \qquad \text{where } h = \frac{\beta}{RT} \qquad (8\text{-}40)$$

Chemisorption equilibrium for the real surface characterized by a linear q-versus-θ relation is

$$\theta = \int_{0}^{1} \frac{p\mathscr{A} \exp(q_0/RT - \alpha\theta/RT)\, d\theta}{1 + p\mathscr{A} \exp[q_0/RT - (-\alpha\theta/RT)]} = \int_{0}^{1} \frac{pa_0 \exp(-\alpha\theta/RT)\, d\theta}{1 + pa_0 \exp(-\alpha\theta/RT)} \qquad (8\text{-}41)$$

which, upon integration, yields

$$\theta = \frac{RT}{\alpha} \ln \frac{1 + a_0 p}{1 + a_0 p \exp(-\alpha/RT)} \qquad (8\text{-}42)$$

Two limiting cases are of interest.

CASE 1 The change in q with θ is large; that is, α is large and $a_0\, p \exp(-\alpha/RT) \ll 1$. Then

$$\theta = \frac{RT}{\alpha} \ln(1 + a_0\, p) \qquad (8\text{-}43)$$

If α is quite small, q varies negligibly with θ and the ideal Langmuir isotherm applies.

CASE 2 If $a_0 p \gg 1$,

$$\theta = \frac{RT}{\alpha} \ln a_0 p \qquad (8\text{-}44)$$

which is the celebrated Temkin isotherm.

A significant consequence of (8-42) results upon expanding the logarithm and excluding all but the first term

$$\ln x \approx 2\,\frac{x-1}{x+1}$$

Thus

$$\theta = \frac{RT}{\alpha}\ln\frac{1+a_0 p}{1+a_0 p\,\exp\left(-\alpha/RT\right)} \approx \frac{RT2}{\alpha}\,\frac{a_0 p[1-\exp\left(-\alpha/RT\right)]}{2+a_0 p[1+\exp\left(-\alpha/RT\right)]}$$

when α is large, $\exp\left(-\alpha/RT\right) \ll 1$; then

$$\theta = \frac{RT}{\alpha}\,\frac{a_0 p}{1+a_0 p/2} \equiv \text{Langmuir isotherm form} \qquad (8\text{-}45)$$

It should be noted that the error attending the approximation of the logarithm is only 10 percent at a value of $x = 3$.

We see then that an adsorption isotherm rooted in real-surface physics readily reduces to an ideal (Langmuir) form. If we had assumed an exponential q-versus-θ functionality, then, as shown by Halsey and Taylor,[1] a Freundlich isotherm would result, that is, $\theta = K(p)^{1/n}$.   In fact since[2]

$$\frac{ax}{1+ax} \approx kx^{1/n} \qquad n > 1 \qquad (8\text{-}46)$$

the equivalence of the Langmuir, Freundlich, and Brunauer-Love-Keenan isotherms is demonstrated.   The situation is displayed in Fig. 8-8.

While noting that diverse chemisorption kinetic and equilibria functionalities are, in fact, well approximated by the Langmuir model rooted in ideality, it is to be observed that distinct advantages reside in the Langmuir formulation when

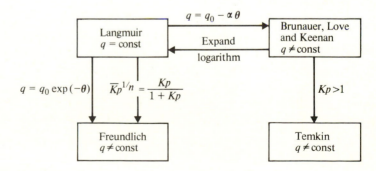

FIGURE 8-8
Equivalence of diverse adsorption equilibria.

[1] G. D. Halsey and H. S. Taylor, *J. Chem. Phys.*, **15**: 624 (1947).
[2] M. Boudart, *AIChE J.*, **2**: 62 (1956).

multicomponent systems are encountered.   For while the Brunauer-Love-Keenan isotherm

$$\theta = \frac{1}{f} \ln (1 + Kp) \qquad (8\text{-}47)$$

and its limiting form, the Temkin isotherm,

$$\theta = \frac{1}{f} \ln Kp \qquad (8\text{-}48)$$

rather adequately describe equilibrium adsorption of N_2 upon reduced-iron ammonia-synthesis catalyst, upon which coreactant H_2 is sparsely adsorbed, the nonideal isotherms above do not anticipate adsorption of other species.   The Langmuir treatment readily lends itself to multicomponent adsorption.   When one recalls (8-2a) and (8-2b) and retains the implicit assumption of two sites X_1 and X_2, the Langmuir argument leads to

$$\theta_A = \frac{K_1 A}{1 + K_1 A} \qquad \text{and} \qquad \theta_B = \frac{K_2 B}{1 + K_2 B} \qquad (8\text{-}49)$$

If $X_1 = X_2$, that is, one site type is involved, the Langmuir thesis permits us to write

$$\theta_A = \frac{K_1 A}{1 + K_1 A + K_2 B} \qquad (8\text{-}50)$$

We know of no a priori rational means whereby such a situation might be described in terms of the Brunauer-Love-Keenan model.

We observe that despite the realities set forth in heat-of-adsorption-vs.-coverage data for gas-catalytic solid systems, the ideal-surface model of Langmuir is a limiting form of more complex isotherms derived from the acknowledged non-idealities.   By inference, one might then expect that chemisorption kinetic relationships rooted in nonideality of the surface and/or interaction will, in the limit, reduce to simple Langmuir chemisorption kinetics.   As the rate of a diffusion-uninfluenced catalytic reaction can be determined only by (1) chemisorption rates, (2) surface reaction between equilibrium chemisorbed species, (3) desorption, or (4) an Eley-Rideal process, the topic of catalytic kinetics can now be discussed in the light of surface realities and the ideal approximations of them.

8-12   CATALYTIC-REACTION KINETIC MODELS

When we speak of *surface-reaction* kinetic models, it is to be understood that rate phenomena of an exclusively chemical nature are under consideration, i.e., chemisorption, desorption, or surface chemical reaction.   Excluded by definition are all limiting physical-transport steps, such as mass transfer of species from the bulk-fluid phase to the external surface of the catalyst particle (interphase transport) and those diffusive events which convey the species to within the pores of the catalyst where reaction occurs simultaneously with diffusion (intraparticle transport).   It is assumed at this juncture that interphase and intraparticle transport processes are very rapid relative to the chemical rate processes occurring on and within the catalyst particle.   Also implicit in this chapter is the additional assumption of a zero gradient (interphase or intraphase) with respect to the diffusion of heat; thus isothermality exists about and within the catalyst.   Extensive treatment of both heat and mass transport to and within porous catalysts will be provided in Chap. 9, where appropriate chemical kinetics will be combined (when possible) with those of heat and mass transfer to fashion overall rate expressions.

Catalytic reaction-rate expressions can be derived for ideal surfaces in two ways: (1) by expressing the rate in terms of surface coverage θ and then employing the Langmuir isotherm to relate θ to fluid concentrations.   This is the approach employed by Hinshelwood[1] and is commonly termed the *Langmuir-Hinshelwood formulation*.   (2) A somewhat more explicit approach was established by Hougen and Watson,[2] who derived rate equations in terms of surface concentrations of adsorbed species and free sites and then expressed these concentrations in terms of the Langmuir isotherm.

While it may appear that little difference in form exists between Langmuir-Hinshelwood rate expressions and those of Hougen-Watson, it must be emphasized that the developments of Watson and Hougen prove to be far more comprehensive.   For in addition to surface reaction per se, the Hougen-Watson formulations include explicit terms for catalyst activity (sites), catalyst effectiveness due to diffusion, and provision for activity decay.   Further, multisite as well as adjacent-site models were anticipated.   The Hougen-Watson catalytic rate equations represent not merely a modification of the Langmuir-Hinshelwood formulations but highly instructive extensions and refinements, the unique utility of which is demonstrated in the analysis and design of catalytic reactors.

Insofar as the ideal-surface postulate yields useful results, the Hougen-Watson formulations, being more explicit in their derivation, are to be preferred so long as it is clearly understood that models result, not mechanistic descriptions.   To do justice to the signal contributions of both schools, such ideal-surface models will be referred to as Langmuir-Hinshelwood–Hougen-Watson (LHHW) formulations.[3]

Ideal Surface Occupancy

Following Hougen and Watson, we designate the total concentration of active catalytic centers or sites as S_0, a constant for a given catalyst.[4]   The concentration of unoccupied sites is S, and that occupied by various adsorbed species is S_a, S_b, S_c, etc.   It follows then that

$$S_0 = S + S_a + S_b + S_c + \cdots \qquad (8\text{-}51)$$

[1] C. N. Hinshelwood, "The Kinetics of Chemical Change," Clarendon, Oxford, 1940.
[2] O. A. Hougen and K. M. Watson, "Chemical Process Principles," vol. III, Wiley, New York, 1943.
[3] As we are distributing justice in the most equitable fashion, it must be noted that the precise form of the Langmuir isotherm and resulting chemical rate expression (hyperbotic law) was first put forth by Henri in 1902 to describe enzyme-catalysis kinetics [V. C. R. Henri, *Acad. Sci. Paris*, **135**: 916 (1902)].
[4] Another assumption of ideality: sites may indeed be created in situ during the catalytic process, as first suggested by H. S. Taylor.

In order that occupied-site concentrations be specified in terms of measurable gas-phase concentration, some sequence of steps will be assumed for, say, a bimolecular reversible reaction of the general type

$$A + B \rightleftharpoons C + D \qquad (8\text{-}52)$$

The assumed sequence is

Step 1: $A + S = (AS)$

adsorption

Step 2: $B + S = (BS)$

Step 3: $(AS) + (BS) = (CS) + (DS)$ surface reaction

Step 4: $(CS) = C + S$

desorption

Step 5: $(DS) = D + S$

Surface-reaction control model   Suppose that a bimolecular surface reaction is rate-determining (step 3)

$$r = k_3(AS)(BS) - k_{-3}(CS)(DS) \qquad (8\text{-}53)$$

Implicit in the concept of a single rate-determining step within a sequence of several steps is the understanding that all steps except that controlling the overall rate exist in equilibrium, or, perhaps more accurately, a steady state prevails, so that a ratio of reactant and products of constant value exists for all except the rate-controlling step.

Step 1: $\dfrac{(AS)}{AS} = K_1$

Step 2: $\dfrac{(BS)}{BS} = K_2$

$(8\text{-}54)$

Step 4: $\dfrac{(CS)}{CS} = K_4$

Step 5: $\dfrac{(DS)}{DS} = K_5$

Substituting for chemisorbed species concentrations in (8-53), we have

$$r = k_3 K_1 K_2 \left(AB - \frac{K_4 K_5}{K_e K_1 K_2} CD \right) S^2 \qquad (8\text{-}55)$$

where A, B, etc., are gas-phase concentrations or partial pressures and the overall experimental equilibrium constant K_{eq} is defined as $K_{eq} = k_3/k_{-3}$.   Specification of free-site concentration S follows from (8-51) and the steady-state ratios, where, for example, $S_a = (AS)$:

$$S_0 = S + (AS) + (BS) + (CS) + (DS) + (IS) \qquad (8\text{-}56)$$

where (IS) accounts for site occupancy by inert (nominally nonreactive) components, such as nitrogen or water often used as diluents in feed streams. Substituting in terms of the ratios given in (8-54), we find

$$S_0 = S + SK_1 A + SK_2 B + SK_4 C + SK_5 D + SK_6 I$$

hence

$$S = \frac{S_0}{1 + K_1 A + K_2 B + K_4 C + K_5 D + K_6 I} \qquad (8\text{-}57)$$

or

$$\frac{S}{S_0} = \frac{1}{1 + \sum K_i(X)_i} \qquad (8\text{-}58)$$

where S/S_0 represents the fraction of total sites or centers which are unoccupied. The sites occupied by, say, species A are, since $SK_1 A = (AS)$,

$$\frac{(AS)}{S_0} = \frac{SK_1 A}{S_0} = \frac{K_1 A}{1 + \sum K_i X_i} \qquad (8\text{-}59)$$

and so on for other species. The relationship between these Hougen-Watson forms and the Langmuir-Hinshelwood formulations is evidently

$$\frac{S}{S_0} = 1 - \theta_{\text{total}} \qquad \theta_{\text{total}} = \text{total coverage}$$

$$\theta_A = \frac{(AS)}{S_0} = \frac{K_1 A}{1 + \sum K_i X_i}$$

and

$$\theta_{\text{total}} = \frac{K_1 A + K_2 B + K_4 C + K_5 D + K_6 I}{1 + K_i A + K_2 B + K_4 C + K_5 D + K_6 I}$$

or

$$\theta_{\text{total}} = \frac{\sum K_i X_i}{1 + \sum K_i X_i} = \theta_a + \theta_b + \cdots \qquad (8\text{-}60)$$

In Langmuir-Hinshelwood terms, Eq. (8-53) could be written

$$r = k'_3 \theta_a \theta_b - k'_{-3} \theta_c \theta_d \qquad k'_3 \neq k_3 \qquad (8\text{-}61)$$

Substituting for θ's, we obtain

$$r = k'_3 K_1 K_2 \frac{AB - CD/K_0}{(1 + K_1 A + K_2 B + K_4 C + K_5 D + K_6 I)^2} \qquad (8\text{-}62)$$

where

$$K'_0 = \frac{K_{\text{eq}} K_1 K_2}{K_4 K_5} \qquad K_{\text{eq}} = \frac{k'_3}{k'_{-3}}$$

Substituting for S in (8-55) results in

$$r = k_3 K_1 K_2 S_0{}^2 \frac{AB - CD/K_0}{(1 + K_1 A + K_2 B + K_4 C + K_5 D + K_6 I)^2} \qquad (8\text{-}63)$$

where $K_0 = K_{eq} K_1/K_2 K_4 K_5$. Note that in the ideal-surface formulation the exponent in the denominator (2 in this case) indicates the number of sites involved per reaction cycle. In the reaction scheme illustrated here all sites are identical and two are involved per reaction sequence, as A and B are both assumed to be chemisorbed. If, however, A adsorbs on one type of site S_1 and B upon another, say S_{11}, and if for the sake of illustration we assume irreversible reaction without adsorption of products, C and D, expressed by the sequence

$$A + S_1 \quad = (AS_1) \tag{8-64a}$$

$$B + S_{11} \quad = (BS_{11}) \tag{8-64b}$$

$$(AS_1) + (BS_{11}) = C + D + S_1 + S_{11} \tag{8-64c}$$

then again assuming that step 3 controls and expressing the adsorbed-species concentrations in terms of the equilibrium in (8-64a) and (8-64b) gives

$$r = k_3 K_1 K_2 [A \cdot B] S_1 S_{11} \tag{8-65}$$

Now recognizing that we have two types of total sites S_{1_0} and S_{11_0}, we find

$$S_1 = \frac{S_{1_0}}{1 + K_1 A} \quad \text{and} \quad S_{11} = \frac{S_{11_0}}{1 + K_2 B}$$

Thus

$$r = k_3 K_1 K_2 S_{1_0} S_{11_0} \frac{AB}{(1 + K_1 A)(1 + K_2 A)} \tag{8-66}$$

An interesting lesson is conveyed by considering Eq. (8-63) in the case where K_4, K_5, and K_6 are zero and K_0 large:

$$r = k_3 K_1 K_2 S_0{}^2 \frac{AB}{(1 + K_1 A + K_2 B)^2} \tag{8-67}$$

Carrying out the indicated operation on the denominators of both (8-66) and (8-67), we find

$$r = \frac{KAB}{1 + K_1 A + K_2 B + K_1 K_2 AB} \tag{8-66a}$$

and

$$r = \frac{K_3' AB}{1 + 2KA + 2KB + 2KKAB + KA^2 + KB^2} \tag{8-67a}$$

From the comparison of (8-66a) and (8-67a) we note that in the likely instance where $\overline{K_1 A^2} + K_2 B^2$ is small relative to other terms, no functional difference can be found between single- and dual-site models.

Adsorption-control model   To continue with reaction scheme (8-64), assume that chemisorption of A is the rate-determining step. Then the mass-action law demands that the rate be proportional to A and the free-site concentration

$$r_1 = k_1 A S_1 = k_1 A (1 - \theta)$$

Since

$$S_1 = \frac{S_{1_0}}{1 + K_1 A}$$

we have

$$r_1 = \frac{k_1 S_{1_0} A}{1 + K_1 A} \qquad (8\text{-}68)$$

Suppose, however, that A is a diatomic molecule such as H_2 and capable of dissociating upon adsorption. If dissociation occurs so that an atom is adsorbed upon one site, then designating the molecule A_2 for a reaction $A_2 \rightarrow C$

Step 1: $\quad A_2 + 2S \rightarrow 2(AS)$

Step 2: $\quad 2(AS) \rightarrow C + 2S$ $\qquad\qquad\qquad (8\text{-}69)$

Then

$$S_0 = S + 2(AS)$$

and since

$$K_1 = \frac{(AS)^2}{A_2 S^2} \qquad (8\text{-}70)$$

we have

$$S_0 = S + 2S\sqrt{K_1 A_2}$$

or

$$S = \frac{S_0}{1 + K_1'\sqrt{A_2}} \qquad (8\text{-}71)$$

For chemisorption of A_2 controlling, $r_1 = k_1 A_2 S^2$

$$r_1 = \frac{S_0^2 k_1 A_2}{(1 + K_1'\sqrt{A_2})^2} \qquad (8\text{-}72)$$

Comparing (8-72) with (8-68), it might be supposed that kinetic evidence of dissociation adsorption could be found by a comparison of data with these two differing functions. Such is not the case. Consider the situation where A_2 chemisorbs with dissociation upon adjacent site pairs, designated S_2:

Step 1: $\quad A_2 + S_2 \rightarrow (AS_2A)$

Step 2: $\quad (AS_2A) \rightarrow C + S_2$ $\qquad\qquad\qquad (8\text{-}73)$

Clearly then

$$r_1 = \frac{k_1 A}{1 + K_1 A}$$

Thus we have dissociative adsorption resulting in a kinetic model identical to that fashioned for molecular, nondissociative adsorption.

Desorption rate-control model   For the scheme, $A \rightarrow C$, where

Step 1: $\quad A + S \longrightarrow (AS) \qquad (AS) = K_1 SA$

Step 2: $\quad (AS) \longrightarrow (CS) \qquad (CS) = K_2(AS) = K_2 K_1 SA$

Step 3: $\quad (CS) \longrightarrow C + S$

$$r_3 = k_3(CS) = k_3' \theta_c \qquad (8\text{-}74)$$

we have

$$S_0 = S + (AS) + (CS) = S + SK_1 A + SK_2 K_1 A$$

Therefore
$$(CS) = \frac{K_1 K_2 A}{1 + (K_1 + K_2 K_1)A}$$

or
$$r_3 = \frac{k_3 K_1 K_2 A}{1 + K'A} \qquad (8\text{-}75)$$

a form identical to (8-68) (adsorption-of-A control).

8-13   DATA ANALYSIS IN TERMS OF VARIOUS MODELS

Following their development and presentation, LHHW methods of catalytic rate data analysis and correlation gained wide adoption.   Despite the caution urged by those responsible for the development of these models, numerous enthusiasts flooded the literature with mechanistic assertions derived from agreement between their rate data and particular LHHW models.   Well aware of the inherent limitations of any kinetic model, Weller[1] initiated a fruitful dialogue by pointing out that insofar as LHHW rate equations cannot reflect a unique mechanistic physicochemical reality, far simpler methods of rate-data correlation demand attention. Specifically, Weller suggested that rates of heterogeneous catalytic reaction be expressed in terms of a simple power law

$$r = kA^\alpha B^\beta P^\gamma \cdots \qquad (8\text{-}76)$$

In addition to simplicity of form and, in many cases, ease of integration (8-76) clearly rests upon a nonideal, real-surface premise, namely, one characterized by a logarithmic q-versus-θ functionality, a basis of the Freundlich isotherm

$$\theta = kA^{1/n} \qquad (8\text{-}77)$$

So if $r = k_1 \theta_A \theta_B \theta_P$, then

$$r = k_1 K_A K_B K_P A^{1/n} B^{1/m} P^{1/P} \cdots$$

Weller supported his thesis by analyzing data for a number of reaction systems previously correlated in terms of complex LHHW expression.   One such study, reanalyzed to advantage by Weller, concerned the kinetics of the reaction

$$CO + 3H_2 \; \rightleftharpoons \; CH_4 + H_2O \qquad (8\text{-}78)$$

conducted over a nickel-kieselguhr catalyst.   The correlating rate equation offered by Akers and White[2] was

$$r = \frac{(CO)(H_2)^3}{[A + B(CO) + D(CO_2) + E(CH_4)]^4} \qquad (8\text{-}79)$$

[1] S. Weller, *AIChE J.*, **2**: 59 (1956).
[2] W. W. Akers and R. R. White, *Chem. Eng. Prog.*, **44**: 553 (1948).

with the assertion that the "mechanism" involved is surface reaction between three adsorbed H_2 molecules and one adsorbed CO molecule, a contention certainly implied by (8-79) if its form is taken seriously. Quite aside from the deceptive advantages residing in a four-constant-power function, it is to be noted that neither CO_2, a by-product, nor CH_4 is likely to be anything but weakly adsorbed on the catalyst employed. Indeed, on reanalyzing the raw data, Weller found that correlation could be realized with equal accuracy by an expression containing only one coefficient, namely,

$$r = k(CO)(H_2)^{1/2} \qquad (8\text{-}80)$$

As we noted earlier, the Freundlich isotherm, though derivable on an assumption of logarithmic q-versus-θ behavior, is a limited isotherm and indeed is profitably considered to be an approximation to the hyperbolic function

$$\frac{K_1 A}{1 + K_1 A} \approx K_2 (A)^{1/n} \qquad (8\text{-}81)$$

where K_2 and n will in all likelihood be temperature- and pressure-dependent. In consequence, over a limited range of pressure and temperature

$$\frac{k_1 K_1 K_2 AB}{1 + KA + K_2 B + K_3 P} \approx k_2 A^z B^\beta P^{-\gamma} \qquad (8\text{-}82)$$

Weller justly focused attention upon the folly of inferring a mechanism from a model and the utility of a simpler model when data do not demand excessive elegance. However, as Boudart asserted in a companion paper,[1] some telling advantages are to be found in the proper use of LHHW rate equations. Primarily, a judiciously selected LHHW form conveys very useful information about the effects of component partial pressures, total pressure, and temperature upon reaction rate, as we shall illustrate shortly. This sort of information can be conveyed by the organizing equation in spite of the well-known fact that the actual mechanism of the reaction remains obscure. Between the LHHW model and its equivalent, the Freundlich-based power-law model, the LHHW form is to be preferred, Boudart notes, because greater understanding and control of reaction conditions are provided by this, the more flexible model. Here then the merits of the LHHW form relative to the Freundlich function are established. In the previous section, these merits relative to the Brunauer-Love-Keenan-based rate equations were stated. The demonstrated equivalence of these apparently diverse relationships suggests that distinctions between ideal and real surfaces, from an applied kinetic point of view, are meaningless.

Furthermore, adsorption coefficients extracted from LHHW rate equations bear little relationship to those found directly in chemisorption equilibrium studies. For example, in correlating the temperature dependency of adsorption coefficients in a rate equation of the form

$$r = \frac{kK_1 A}{1 + K_1 A + K_2 B} \qquad (8\text{-}83)$$

[1] M. Boudart, *AIChE J.*, 2: 62 (1956).

the derived *heat of chemisorption* is often found to be small, about 2 to 10 kcal/mol, although independent calorimetric studies might suggest values in the range of 10 to 50 kcal/mol, depending upon coverage. It is tempting to assert that if steady-state reaction occurs at high coverage, a low chemisorption heat would be expected on analysis of the rate-equation K values. Such an assumption is not necessary when one recalls that while surface occupancy is a function of pressure, it is more drastically affected by temperature. With increasing temperature, coverage decreases, and in consequence the chemisorption heat increases. As an approximation Boudart postulates

$$q = q_0 + aT \qquad (8\text{-}84)$$

where q_0 is the chemisorption heat at high coverage (low temperature). Assuming the usual adsorption equilibrium coefficient-temperature relation and applying (8-84), we have

$$K = K_0 \exp \frac{q}{RT} = K_0\, e^a \exp \frac{q_0}{RT} \qquad (8\text{-}85)$$

with the result that for any coverage, a plot of $\ln K$ versus $1/T$ will reveal a slope equivalent to q_0, the lowest value of q corresponding to high coverage. The adsorption heats derived from kinetic models, as well as those inferred by pulse-flow techniques, are not likely to be of great meaning so long as nonideal surfaces are involved.[1]

Adsorption Enhancement

In all the models considered so far, molecules were assumed (1) to compete for sites of one kind or (2) to adsorb independently on sites of different kinds; e.g., two site types, one for A the other for B, lead to a rate equation of the form

$$r = \frac{k_0\, AB}{(1 + K_1 A)(1 + K_2 B)}$$

while a competitive, one-site formulation gives

$$r = \frac{k_0\, AB}{(1 + K_1 A + K_2 B)^2}$$

The LHHW models developed earlier predict that the addition of gas A to the system reduces the amount of B previously adsorbed; i.e.,

$$\theta_B = \frac{K_2 B}{1 + K_2 B} \quad \xrightarrow{+A} \quad \frac{K_2 B}{1 + K_2 B + K_1 A} \qquad (8\text{-}86)$$

As Weller remarks in his critique of LHHW rate formulations, though few data are available, there are important instances where the addition of A actually *increases* the adsorption coverage of B, a situation described by

$$\theta_B = \frac{K_2 B}{1 + K_2 B} \quad \xrightarrow{+A} \quad \frac{K_2 B}{1 + K_2 B - K_1 A} \qquad (8\text{-}87)$$

[1] J. J. Carberry, *Nature (Lond.)*, **189**: 391 (1961).

In the procedures followed in testing data with various LHHW-model candidates, a negative adsorption coefficient is generally considered a physical impossibility, and rate models yielding such negative coefficients are discarded.

However, such instances of enhancement of adsorption of one gas by another can in fact be treated by LHHW techniques without resorting to negative adsorption coefficients. We need only invoke the notion of chemisorption of a gas upon a surface complex consisting of a site and the other adsorbed species. For example

Step 1:	$A + S$	$\longrightarrow$	(AS)
Step 2:	$B + S$	$\longrightarrow$	(BS)
Step 3:	$(AS) + B$	$\longrightarrow$	(ASB)
Step 4:	$(BS) + A$	$\longrightarrow$	(ASB)
Step 5:	$2ASB$	$\longrightarrow$	$P + 2S$
Overall:	$A + B$	$\rightleftharpoons$	P

(8-88)

If adsorption upon an adsorbed complex (steps 3 and 4) does not occur, coverage is given by the usual expressions

$$\theta_A = \frac{K_1 A}{1 + K_1 A + K_2 B} \qquad \theta_B = \frac{K_2 B}{1 + K_1 A + K_2 B} \qquad (8\text{-}89)$$

Assume that steps 3 and 4 do take place; in the usual manner

$$S_0 = S(1 + K_1 A + K_2 B + K_1 K_3 AB + K_2 K_4 BA) \qquad (8\text{-}90)$$

$$A_{ads} = (K_1 A + K_1 K_3 AB + K_2 K_4 AB)S = K_1 A\left[1 + K_2 B\left(\frac{K_3}{K_2} + \frac{K_4}{K_1}\right)\right]S$$

$$B_{ads} = (K_2 B + K_2 K_4 AB + K_1 K_3 AB)S = K_2 B\left[1 + K_1 A\left(\frac{K_3}{K_2} + \frac{K_4}{K_1}\right)\right]S$$

hence
$$\theta_A = \frac{A_{ads}}{S_0} = \frac{K_1 A[1 + K_2 B(K_3/K_2 + K_4/K_1)]}{1 + K_1 A + K_2 B + K_1 K_3 AB + K_2 K_4 AB}$$

or
$$\theta_A = \frac{K_1 A}{\dfrac{1 + K_2 B}{1 + K_2 B(K_3/K_2 + K_4/K_1)} + K_1 A} \qquad (8\text{-}91a)$$

Similarly
$$\theta_B = \frac{K_2 B}{\dfrac{1 + K_1 A}{1 + K_1 A(K_3/K_2 + K_4/K_1)} + K_2 B} \qquad (8\text{-}91b)$$

Clearly as long as $K_3 > K_2$ or $K_4 > K_1$, then for θ_B, for example,

$$\frac{1 + K_1 A}{1 + K_1 A(K_3/K_2 + K_4/K_1)} < 1$$

On comparison with (8-89) we see that adsorption of A enhances the coverage of B and that of B increases θ_A. Clearly when the term $K_3/K_2 + K_4/K_1$ is small, the usual expressions result. Once again the flexibility of LHHW formulations is clear. An excellent example of catalysis involving adsorption enhancement is analyzed by Powers.[1]

Multistep Rate Control

In the previous section catalytic-reaction-rate models were developed which rest upon the assumption that only *one* step in a series of supposed elementary events is the rate-determining step. This supposition, though nearly universally employed in fashioning rate models, is easily abandoned with interesting and instructive consequences, to be illustrated shortly. In fact, since in general each step in a mechanistic sequence can be expected to be characterized by a unique activation energy and, indeed, a unique pressure (π) dependency, there are few bases upon which to contend that if step 1 is the rate-determining step at T_1 and π_1, that same step should be the rate-determining step at T_2 and/or π_2. Further if step 1 is rate-determining at T_1, π_2, surely an intermediate range of T and π must exist within which *both* steps become rate-determining.

Consider the very simple catalytic reaction A $\rightleftarrows$ B. The rate of adsorption of A is

$$r_1 = k_A \left[A(1 - \theta_A - \theta_B) - \frac{\theta_A}{K_A} \right] \tag{8-92}$$

Surface-reaction rate is

$$r_2 = k_r \left(\theta_A - \frac{\theta_B}{K_r} \right) \tag{8-93}$$

The rate of desorption of B is

$$r_3 = k_B \left[\frac{\theta_B}{K_B} - B(1 - \theta_A - \theta_B) \right] \tag{8-94}$$

We assume *not* that overall rate $\mathcal{R}_0 = r_1$ or r_2 or r_3 but that $\mathcal{R}_0 = r_1 = r_2 = r_3$; so on solving for $\mathcal{R}_0$, we obtain a nonunique rate-control model

$$\mathcal{R}_0 = \frac{K_r K_A A - K_B B}{\left(\dfrac{K_r}{k_r} + \dfrac{K_r K_A}{k_A} + \dfrac{K_B}{k_B} \right) + \left[\dfrac{K_r}{k_r} + \dfrac{K_B(1 + K_r)}{k_B} \right] K_A A + \left[\dfrac{K_r}{k_r} + \dfrac{K_A(1 + K_r)}{k_A} \right] K_B B} \tag{8-95}$$

At a fixed temperature, Eq. (8-95) assumes the phenomenological form

$$\mathcal{R}_0 = \frac{K_1 A - K_2 B}{1 + \bar{K}_A A + \bar{K}_B B} \tag{8-96}$$

[1] J. E. Powers, *J. Phys. Chem.*, **63**: 1219 (1959).

which, of course, bears an amazing resemblance to developments rooted in the unique rate-determining step.

An example of nonunique surface rate control has been cited,[1] and Bischoff and Froment[2] reveal the hazards encountered when a unique rate-determining-step model is applied to data reflecting dual rate-determining-step character.

Nonequilibrium Kinetics

If the assumption of adsorption equilibrium of A is made, the traditional LHHW treatment of the catalytic reaction $A(g) \xrightarrow{\text{cat}} B(g)$ leads to the plausible rate-determining step as that of decomposition of adsorbed A, that is,

$$r = k_1 \theta_A \qquad (8\text{-}97)$$

Invoking the Langmuir isotherms to describe θ_A, we find

$$\mathcal{R}_0 = \frac{k_1 KA}{1 + KA} \qquad (8\text{-}98)$$

It can be reasonably supposed that k_1 will exhibit Arrhenius temperature dependency while the kinetic adsorption coefficient K for anticipated exothermic chemisorption will decrease with temperature.

As Sinfelt[3] shows, one must not think solely in terms of LHHW chemisorption equilibrium-rooted rate formulations.   Given the reaction

$$A(g) \xrightarrow[\text{cat}]{} B(g)$$

it may be supposed that the process involves

$$A(g) \xrightarrow{\ 1\ } B_{ads} \xrightarrow{\ 2\ } B(g) \qquad (8\text{-}99)$$

i.e., adsorption equilibrium is *not* established between $A(g)$ and A_{ads}, which is the thesis leading to Eq. (8-97) and thus (8-98).   Instead, Sinfelt argues, each step in (8-99) is irreversible, and if it is postulated that the rate of desorption of adsorbed B_{ads} is the rate-determining step, then, at steady state

$$r_1 = k_1 A(1 - B_{ads}) = k_2 B_{ads} \qquad (8\text{-}100)$$

or

$$B_{ads} = \frac{k_1/k_2 \, A}{1 + k_1/k_2 \, A} \qquad (8\text{-}101)$$

Since it is assumed that the rate-determining step is $r_2 = k_2 B_{ads}$, we have

$$\mathcal{R}_0 = r_2 = \frac{k_2(k_1/k_2)A}{1 + (k_1/k_2)A} \qquad (8\text{-}102)$$

or

$$\mathcal{R}_0 = \frac{kKA}{1 + KA}$$

[1] L. H. Thaller and G. Thodos, *AIChE J.*, 6: 369 (1960).
[2] K. B. Bischoff and G. Froment, *Ind. Eng. Chem. Fundam.*, 1: 195 (1962).
[3] J. H. Sinfelt, H. Hurwitz, and R. A. Shulman, *J. Phys. Chem.*, 64: 1559 (1969).

Although this form is identical to that issuing from LHHW theses [Eq. (8-98)], the significance and expected behavior of the respective values of K can be markedly different. To wit, in Eq. (8-102), the nonequilibrium model

$$K = \frac{k_1}{k_2} = K_0 \exp\left[-(\epsilon_1 - \epsilon_2)\left(\frac{1}{t} - 1\right)\right] \qquad (8\text{-}103)$$

and so K may increase, decrease, or remain constant with temperature, its behavior depending solely on the value of and sign of $\epsilon_1 - \epsilon_2$ and not upon the exothermicity or endothermicity of chemisorption equilibrium, as required in the LHHW formulations. The moral would seem to be that the K-versus-T behavior should not necessarily be interpreted in terms of chemisorption equilibrium laws.

Significances of the Dual Rate-Determining-Step and Nonequilibrium Kinetic Models

The standard LHHW catalytic kinetic models yield expressions of the form

$$r = \frac{k'AB}{1 + K_1 A + K_2 B}$$

where, in principle, K_1 and K_2, properly termed kinetic adsorption coefficients, are expected to obey trends exhibited by equilibrium constants for exothermic processes (though in principle endothermic chemisorption is possible).

Note, however, that in the simplest example of dual-step rate control, we secure a rate functionality identical in form to the LHHW model rooted in single-step rate control [Eq. (8-95)]. So too does the nonequilibrium kinetic model yield a form equivalent to the LHHW model. Patently the coefficients found in the denominators of these models, which are alien to the LHHW postulates, reflect a reality and an anticipated behavior hardly in conformity with a thermodynamic equilibrium constant.

In sum, the so-called kinetic adsorption coefficients might decrease or increase with temperature; indeed in the two-step rate-control model, a maximum or minimum in the apparent $\ln K$-versus-$1/T$ behavior could be manifested for a particular combination of activation and adsorption energies which characterize the several terms of which an apparent K is composed.

8-14 CATALYST DEACTIVATION

Our general definition of catalytic action, set forth in the early stages of this chapter, anticipates the fact that man-made catalytic agents suffer a decline in potency with time on-stream, which, while great indeed relative to the time cycle of the transformation of a molecule of reactant to product, is brief in contrast to nature's enzymes. The life-span of a cracking catalyst is of the order of seconds, because of coking. The promoted-iron catalyst employed in NH_3 synthesis enjoys a lifetime of years.   To be sure, regeneration processes can restore catalyst activity, as in the instance of coked cracking catalysts, which regain their vigor when an oxygen-bearing stream is utilized to burn the coke deposits from that catalyst; yet, other instances of deactivation do not admit to regeneration, catalyst sintering and/or irreversible phase transformations being notable examples.   A catalyst is caused to deactivate in three ways.

Poisoning   Poisoning is chemisorption of reactants, products, or impurities (singly or in combination) found in the reactor feed, which occupy sites otherwise available for catalysis.   Chemisorption of poisoning agents may be reversible or irreversible. In the first case, elimination of the poison precursor from the feed restores vitality. If the reversibly poisoning species is a reactant, its elimination is meaningless; if it is a product, a remedy might consist in a low-conversion, high-recycle reactor network with product removal in the recycle loop.   Poisoning, insofar as it is a chemisorptive event, constitutes chemical deactivation.

Fouling   Fouling is caused by species in the fluid phase being physically deposited upon the surface, thereby covering or blocking sites otherwise disposed to catalysis. Fouling can also be a result of surface reactions yielding products that foul the surface, e.g., coke produced by overcracking of hydrocarbons.   Particulate Pb, in an automotive-engine exhaust gas, will deactivate an oxidation catalyst by chemically reacting with catalytic sites and by physically plugging pores, thus denying reactants access to the sites within the pores.

Sintering or phase transformations[1]   As a consequence of local high temperature and in some instances the nature of the oxidizing or reducing atmosphere, the catalyst per se and/or its support suffer a reduction in specific surface area or the chemical nature of the catalytic agent is so altered as to render it catalytically mortal.

In general, it can be cautiously suggested that whereas poisoning and fouling rates will depend upon reactant and/or product concentrations, sintering and phase transformations may be assumed to be independent of fluid-phase composition for a given oxidizing or reducing atmosphere.

How then does one fashion a model of deactivation and catalytic reaction given the diverse character of intrinsic catalytic reactions, diffusional intrusions, and the poorly comprehended detailed nature of the three general classes of deactivation cited above?   The problem can be partitioned at this point.   First, the general classification of catalysts and their poisons are given in Table 8-3. Second, we shall postpone discussion of the problem of deactivation in diffusion-affected catalysis until Chap. 9.   Therefore the kinetics of deactivation-reaction treated in this chapter will be rooted in the presumption that all deactivating sites are equally accessible, a postulate not too severe if proper consideration of laboratory-reactor design criteria is granted, as set forth later in this chapter.

[1] S. E. Wanke and P. C. Flynn, *Catal. Rev.*, **12**(1): 93 (1975).

Deactivation-Reaction Models

Levenspiel and his students[1] have fruitfully analyzed the deactivation-reaction problem in simple yet negotiable terms which encompass a wide diversity of poisoning-fouling precursor networks as well as the important case of species-independent deactivation. Since their analysis assumes nth-order catalytic reaction and dth-order activity decline, it is a phenomenological treatment of a problem which, although of immense intrinsic sophistication, is best attacked from a point of view devoid of detailed mechanistic speculation. Insofar as many LHHW hyperbolic rate models can be gainfully phrased in terms of a power-law nth-order model, the Levenspiel-Szepe approach retains potency particularly with respect to the uncoupling of the deactivation-reaction events.

Following Levenspiel, the catalyzed-reaction rate is given by

$$r_A = -\frac{dA}{d\theta} = kA^n a \qquad (8\text{-}104)$$

where

$$a = \frac{\text{rate at any } t}{\text{rate at } t = 0} \qquad (8\text{-}105)$$

Then a is an activity which is initially equal to unity and declines with time. Now the kinetic model of activity decline is presumed to be

$$r_p = -\frac{da}{d\theta} = \bar{k}(A, B, P)^m a^d \qquad (8\text{-}106)$$

Table 8-3 POISONS FOR VARIOUS CATALYSTS†

Catalyst	Reaction	Type of poisoning	Poisons
Silica-alumina	Cracking	Chemisorption Deposition Stability Selectivity	Organic bases Carbon, hydrocarbons Water Heavy metals
Nickel, platinum, copper	Hydrogenation Dehydrogenation	Chemisorption	Compounds of S, Se, Te, P, As, Zn, halides, Hg, Pb, NH_3, C_2H_2, H_2S, Fe_2O_3, etc.
Cobalt	Hydrocracking	Chemisorption	NH_3, S, Se, Te, P
Silver	$C_2H_4 + O \rightarrow C_2H_4O$	Selectivity	CH_4, C_2H_6
Vanadium oxide	Oxidation	Chemisorption	As
Iron	Ammonia synthesis Hydrogenation Oxidation	Chemisorption Chemisorption Chemisorption	O_2, H_2O, CO, S, C_2H_2 Bi, Se, Te, P, H_2O VSO_4, Bi

† In part from W. B. Innes in P. H. Emmett (ed.), "Catalysis," vol. 1, chap. 7, p. 306, Reinhold, New York, 1954.

[1] O. Levenspiel, *J. Catal.*, **25**: 265 (1972).

where A = reactant concentration

$\quad\quad$ B = product or intermediate concentration

$\quad\quad$ P = poison or fouling concentration

$\quad$ m, d = empirical orders of deactivation with respect to species concentrations and activity, respectively.

The precise functionality which Eq. (8-106) will assume depends upon the reaction-deactivation network. Several networks suggest themselves:

1 Simultaneous deactivation:

$$A \longrightarrow B \quad\quad r_A = k(A)^n a$$
$$\searrow$$
$$P\downarrow \quad\quad r_P = \bar{k}(A)^m a^d$$

2 Consecutive deactivation:

$$A \longrightarrow B \longrightarrow P\downarrow \quad\quad r_A = kA^n a$$
$$r_P = \bar{k}(B)^m a^d$$

3 Parallel deactivation:

$$A \longrightarrow B \quad\quad r_A = kA^n a$$
$$P \longrightarrow P\downarrow \quad\quad r_P = \bar{k}P^m a^d$$

4 Independent deactivation:

$$A \longrightarrow B \quad\quad r_A = kA^n a$$
$$S \text{ sites} \longrightarrow (S\text{-}s) \text{ sites} \quad\quad r_P = \bar{k}a^d$$

5 Simultaneous-consecutive deactivation:

$$A \xrightarrow{1} B \xrightarrow{3} P\downarrow \quad\quad r_A = (k_1 + k_2)A^n a$$
$$\searrow^{2}$$
$$P\downarrow \quad\quad\quad\quad r_P = \bar{k}(A + B)^m a^d$$

Case 5 becomes equivalent to independent deactivation since $A + B$ is a constant for a fixed feed composition.

These phenomenological reaction-deactivation functions have the merit of adequately describing a number of deactivation-time-on-stream observations such as exponential, hyperbolic, and power-law decay, as Levenspiel notes. Use of these models to determine rate coefficients and orders from laboratory data will be discussed in Sec. 8-15. Here we extend Levenspiel's technique to illustrate the possibilities of drawing inferences from the models with respect to the nature of deactivation per se.

Let it be supposed that reaction rate is observed to vary with real time in accordance with the relationship

$$r_A = \frac{kA^n}{1 + Kt} \quad\quad (8\text{-}107)$$

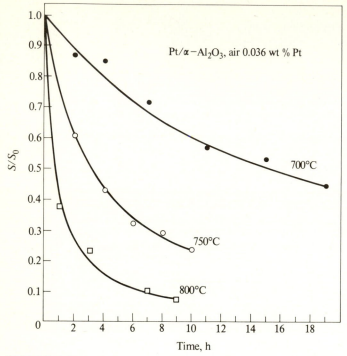

FIGURE 8-9
Supported-Pt surface area vs. time of sintering. [*E. McCarthy, J. Zahradnik, G. C. Kuczynki, and J. J. Carberry, J. Catal.*, **39**: 29 (*1975*).]

for a supported-noble-metal–catalyzed aromatic-reforming reaction, for which it is well known that coke is formed.   Thus deactivation network 5 becomes a reasonable candidate as the cause of catalyst mortality.   Since $A + B = $ constant, integration of the deactivation rate, r_P for $d = 1$, yields

$$\frac{a}{a_0} = \exp(-\bar{K}t) \qquad \text{where} \quad \bar{K} = \bar{k}(A + B) \qquad (8\text{-}108)$$

which is not in accord with the observed r_A-versus-time behavior [Eq. (8-107)]. However, integration of r_P (case 5) for $d = 2$ yields

$$a = \frac{1}{1 + Kt} \qquad (8\text{-}109)$$

The observed behavior is

$$r_A = \frac{kA^n}{1 + Kt} \qquad (8\text{-}101)$$

But it is hardly likely that the rate of deactivation due to coke laydown should be second order with respect to residual activity; the first-order law would seem the more reasonable postulate. An alternative mechanism suggests itself, namely, sintering, which is a complex phenomenon whereby supported-metal crystallites grow, or agglomerate, thus reducing the specific surface per weight of deposited metal. In Fig. 8-9 the normalized surface area of a highly dispersed Pt on Al_2O_3 is plotted vs. time of exposure to sintering conditions.

These data are roughly correlated by the rate model

$$\frac{dS/S_0}{dt} = -K\left(\frac{S}{S_0}\right)^2 \qquad (8\text{-}111)$$

which in terms of activity corresponds to independent deactivation, case 4, above. On integration $a = 1/(1 + Kt)$ and so $r_A = kA^n/(1 + Kt)$, the observed reaction-deactivation functionality.

Further discrimination between cases 4 and 5 is possible since in case 5 the experimental deactivation constant K is a function of pressure, in contrast to case 4.

In actual fact a catalyst may deactivate by reason of both coking and sintering. The coking-rate activity is

$$\frac{-da_c}{dt} = \frac{k_c a}{1 + Kt} \qquad (8\text{-}112)$$

and reaction-activity-decline rate becomes

$$\frac{-da}{dt} = \frac{k_c a}{1 + a_0 Kt} + Ka^2 \qquad (8\text{-}113)$$

As an exercise derive Eq. (8-113) and attempt to solve it.

Wojciechowski[1] presents an alternative description of deactivation, the time-on-stream theory of catalyst decay, which is rooted in the model

$$r_P = k_P a^m \qquad (8\text{-}114)$$

Hence
$$a = \left[\frac{1}{1 + (m - 1)k_P t}\right]^{1/(m-1)} \qquad (8\text{-}115)$$

where k_P contains deactivation-precursor concentrations. We see that for $m = 0$ linear deactivation results; for $m \to 1$, $1/(m - 1)$ becomes large, and exponential decay is approximated, while hyperbolic deactivation behavior obtains for $m = 2$.

Potential Deactivation Remedies

Poisoning or fouling Obviously if the source of poisoning is a species impurity present in the reactor feed, its elimination solves the problem. It is not inconceivable that a poison might be present in the feed at so low but potent a concentration that traditional separation techniques would fail to remove it. In such a case, this independent poison must be considered equivalent to a poison which

[1] B. W. Wojciechowski, *Catal. Rev.*, **9**: 79 (1974).

finds its source in the reactant and/or product, e.g., coke precursors. Assuming, then, that the poison cannot be denied entrance to the reactor, what remedies might prevent or at least minimize the deleterious influence of this species upon activity or yield?

In the case of coking, increasing H_2 partial pressure is known to inhibit that mode of physical deactivation. Obviously restoration of the coked catalyst is achieved by burning the coke deposits under modest conditions of combustion to avoid local high-temperature sintering of the deposited catalyst and/or its support.

If poisoning is of a purely chemical nature, promotion of the catalyst would seem to be a feasible remedy; however, such a remedy is not readily realized. In fact, experience teaches that a change of catalyst type proves to be a more effective mode of relief. For example, in the supported-Pt-catalyzed oxidation of SO_2 to SO_3, arsenic compounds naturally found in the products of sulfur burning and removed with great difficulty poison the platinum. By changing the catalyst from an arsenic-sensitive Pt to promoted V_2O_5, the ever-present arsenic compounds become innocuous. As another example, CO and hydrocarbon species which issue from the automotive engine are readily oxidized to CO_2 and H_2O over diverse metal oxides in the *absence of* SO_2, which speedily poisons these metal-oxide catalyst candidates. However, the noble metals prove to be impervious to the SO_2 poisoning and are therefore prime catalyst candidates for oxidation of CO and hydrocarbons. Note that these noble metals are insensitive to SO_2 simply because they catalyze SO_2 to SO_3, which in the presence of water vapor creates H_2SO_4, a species whose environmental importance surely rivals that of CO and hydrocarbons.

In not a few situations a change in catalyst type may not be a reasonable or possible solution. When promotion appears to be the only means of arresting poisoning or reducing its rate to tolerable levels, the problem can be solved by (1) intuition, (2) art, (3) profound knowledge of *surface* chemistry and physics, or (4) good fortune. On issue 3 more will be said below.

Sintering and/or phase transformations   These forms of deactivation are generally promoted by severe local values of high temperature. We term these modes of decay *environmentally sponsored* insofar as only temperature and the net nature of the fluid-phase oxidizing or reducing atmosphere determines the rate of deactivation, commonly termed *independent deactivation*, since it is independent of particular species concentrations (case 4).

As shown in Fig. 8-9, a sample of 0.036 percent by weight Pt on α-Al_2O_3 suffers average crystallite growth of from about 30 to 50 Å diameter to 1000 to 2000 Å upon exposure to air at several hundred degrees Celsius for only 10 to 12 h. While the precise sintering mechanism escapes our understanding, the global result is obviously that of crystallite agglomeration or growth such that while fresh Pt *spheres* display a surface area, and therefore Pt-surface-site availability of

$$s_i = \frac{6}{d_p} \propto \frac{6}{50} \text{ Å} \quad (8\text{-}116a)$$

upon sintering under conditions of specific time and temperature cited in Fig. 8-9

$$s_0 = \frac{6}{d_p} \propto \frac{6}{2000} \text{ Å} \quad (8\text{-}116b)$$

Catalytic surface-site availability clearly is drastically reduced as $s_i \rightarrow s_0$. But what of the nature of these sites s_i as opposed to those available to reactants when $s = s_0$? If the sites s_i are identical chemically and physically with those of s_0, we are permitted to surmise that our observed activity or selectivity will suffer with the advent of sintering. This postulate assumes, however, that the physicochemical nature of the surface is independent of crystallite size (sintering). Conceivably this nature will itself depend upon crystallite size; hence if reaction is morphologically sensitive, i.e., dependent upon surface structure, there is no reason to expect that specific rate (moles converted per time × surface atom) will be constant as crystallite size changes due to sintering. Further, if more than one catalytic agent is supported by a particular support, the chemical differences between small and large crystallites must be respected.

Sintering phenomena may be deleterious or beneficial to specific activity and/or yield. We postpone comment upon this issue for a moment and address ourselves to the matter of how one might control the sintering process to whatever level proves of benefit. As noted earlier, we can physically promote the catalyst to maintain a desired crystallite size by incorporating noncatalytic agents which bring such an end about. For example, the growth of Fe crystallites in the ammonia-synthesis catalyst is drastically retarded by incorporating small amounts of Al_2O_3 into the catalyst. This remedy constitutes physical promotion.

For a given level of physical promotion, specific activity of available catalytic surface can be favorably altered with respect to activity or yield by the judicious addition of promoters which chemically alter surface disposition. In principle, therefore, promotion by alien agents can render the catalytic species immune to poisoning, fouling, and sintering.

Again citing the NH_3 synthesis catalyst, addition of potassium oxides is now known to increase its chemical activity significantly. Such an agent is termed a *chemical promoter*. The specific details of physicochemical promotion will be given later in this chapter.

In general, it can be declared that catalytic activity or selectivity or yield as a function of time on-stream will depend upon average crystallite size, its chemical and physical promotion, and the environment which bathes these agents.

In sum one can anticipate that catalytic activity and specificity (yield or selectivity) will be governed by

1 Reactant- and product-species concentrations and local temperature
2 Time on-stream
3 The nature and size of specific, usually supported, catalytic agents
4 Physicochemical promoters
5 The support, whose chemical nature and physical structure dictate catalyst, and promoter size, dispersion, and simultaneous reaction-intraphase transport phenomena

Items 1 and 2 have been detailed, while 3 and 4 have been touched upon and will be the subject of some elaboration in forthcoming sections. Item 5 has been generally treated in Chap. 5 and will be detailed in Chap. 9. Given an understanding of items 1 to 3, it is now appropriate to discuss the nature of devices yielding those kinetic data which are the roots of both negotiable rate models and insights into catalytic mechanisms interpretable in terms of items 4 and 5.

8-15 DATA PROCUREMENT AND ANALYSIS

While model formulation rooted in diverse specifications of one or more rate-determining steps has been developed in the above sections, there remains the crucial problem of securing the experimental data from which rate models are to be formulated and evaluated under conditions typical of heterogeneous catalysis. These conditions are:

1 Catalysis in an environment within which interparticulate (long-range) temperature and concentration gradients persist
2 Catalysis affected by interphase and intraphase (short-range) gradients of concentration and temperature
3 Catalysis in an environment which prompts deactivation with a consequent time-on-stream dependency of activity and yield, or selectivity[1]

It is evident that one extracts chemical-kinetic information with great difficulty when one or more of the above conditions prevails in a laboratory catalytic study. The problem of laboratory catalytic-reactor design then becomes that of cleverly circumventing those conditions which give rise to, as J. Wei phrases it, "disguised kinetics."

For each of these intrusions, certain laboratory-scale remedies can, in principle, be visualized.

Condition 1   Long-range gradients are eliminated by use of:
     a   A differential reactor.
     b   A CSTR.
     c   A recycle reactor.

Condition 2   Intraphase (internal) gradients are minimized by:
     a   Utilizing small pellets to guarantee that the Thiele modulus is small, $\phi < 1.0$.
     b   Interphase (external) gradients obviously become negligible for small pellets (small $L = 1/a$) and high Reynolds numbers (large velocities).

---

[1] In defiance of the Proclamations of the Lucrezia Borgia Catalysis Society (*Chem. Technol.*, February 1974, p. 124) some catalysts exhibit an *increase* in activity with time on-stream. However, we and catalyst manufacturers rejoice in the fact that even these agents ultimately succumb.

Condition 3   This condition is simply a fact of catalyst life.   One can:

    a   Extrapolate rate data to initial fresh catalyst activity.

    b   Live with the reality and attempt to uncouple activity and deactivation by a clever choice of laboratory reactor type and data analyses.

Hence the realities of long- and short-range gradients and deactivation admit to circumvention and/or uncoupling in terms of reactor type, reaction conditions, and mathematical analyses.

Laboratory Catalytic Reactors

Borgia communicant Weekman addressed himself to the general problem,[1] namely, a gas-liquid-powdered solid-catalyst system which naturally suffers deactivation. The diverse laboratory reactors are sketched in Fig. 8-10.   Weekman has rated each reactor type with respect to characteristics of importance in construction, sampling, isothermality, contact-time definition, and activity-selectivity disguise due to decay for both constant- and variable-activity catalysis.   His judgments are summarized in Tables 8-4 and 8-5.

It is clear that the differential and integral fixed-bed laboratory reactors are poor devices compared with other candidates.   The ratings with respect to activity-selectivity disguise assume, it would seem, rather rapid deactivation, in which case sophisticated instantaneous methods of sampling and analyses would be required.   In such rapidly decaying networks, the continuous solids-fluid-flow reactors are of extraordinary merit since steady state is achieved.

If one comtemplates somewhat less demanding reaction systems, the choice of laboratory catalytic reactor accordingly becomes simpler.

Gas-solid catalysis is so prevalent that it merits special consideration.   Assuming that deactivation is not very rapid or (if rapid) that instantaneous analysis is realizable, the continuous-stirred-tank catalytic reactor (CSTCR)[2] or its equivalent, the open-loop recycle reaction, is of signal merit.[3]   As has been noted, a CSTCR is unique insofar as integral conversions are obtained under uniform conditions of bulk-fluid concentrations and temperature.   Both types are illustrated in Fig. 8-11.   A variation on the CSTCR is the fixed-bed internal recycle reactor described by Bennett et al.,[4] Berty,[5] and Livbjerg.[6]

While it is evident that CSTR conditions eliminate long-range gradients (condition 1), short-range fluid- and catalyst-phase gradients (condition 2) can persist in a CSTCR.   The external film-transport coefficients have been determined

[1] V. W. Weekman, *AIChE J.*, **20**: 833 (1974).
[2] J. J. Carberry, *Ind. Eng. Chem.*, **56**: 39 (1964); D. G. Tajbl, J. B. Simons, and J. J. Carberry, *Ind. Eng. Chem. Fundam.*, **5**: 171 (1966).
[3] F. V. Hanson and J. E. Benson, *J. Catal.*, **31**: 471 (1973).
[4] C. O. Bennett, M. B. Cutlip, and C. C. Yang, *Chem. Eng. Sci.*, **27**: 2255 (1972).
[5] J. Berty, *Chem. Eng. Prog.*, **70**(5): 78 (1974).
[6] H. Livbjerg and J. Villadsen, *Chem. Eng. Sci.*, **26**: 1495 (1971).

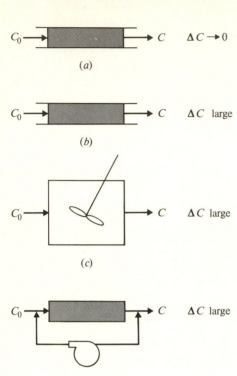

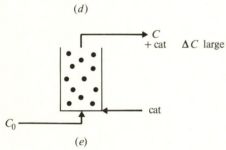

FIGURE 8-10
Laboratory catalytic reactor types: (a) differential, (b) integral, (c) CSTR, (d) recycle, (e) transport. The rate is determined by a simple material balance in (a), (c), and (d).

for different spinning-basket CSTCRs,[1] and, as discussed below, these intrusions can be uncoupled in terms of observable parameters, as can the intrapellet diffusional influences.

The transients imposed by in situ deactivation of the catalyst complicates analyses of data secured on the laboratory scale; however, it is evident that a transient CSTR system is far simpler to analyze than the fixed-integral-bed reactor, as shown in Chap. 4. So the complexities visited upon the analyst by condition 3, while not circumvented, can be uncoupled more easily by use of the CSTCR type.

In Sec. 8-14 simple kinetic models of deactivation and catalytic reaction were

[1] L. K. Doraiswamy and D. G. Tajbl, *Catal. Rev.*, **10**: 177 (1974); J. Periera and P. H. Calderbank, *Chem. Eng. Sci.*, **30**: 167 (1975).

Table 8-4 SUMMARY OF REACTOR RATING†, §

Reactor type	Sampling and analysis	Isother-mality	Residence-contact time	Selectivity disguise-decay	Construction problems
Gas-liquid, powdered-catalyst, decaying-catalyst system					
Differential	P–F	F–G	F	P	G
Fixed bed	G	P–F	F	P	G
Stirred batch	F	G	G	P	G
Stirred-contained solids	G	G	F–G	P	F–G
Continuous stirred tank	F	G	F–G	G	P–F
Straight-through transport	F–G	P–F	F–G	G	F–G
Recirculating transport	F–G	G	G	G	P–F
Pulse	G	F–G	P	F–G	G
Gas-liquid, powdered-catalyst, nondecaying-catalyst system					
Differential	P–F	F–G	F	G	G
Fixed bed	G	P–F	F	G	G
Stirred batch	F	G	G	G	G
Stirred-contained solids	G	G	F–G	G	F–G
Continuous stirred tank	F	G	F–G	G	P–F
Straight-through transport	F–G	P–F	F–G	G	F–G
Recirculating transport	F–G	G	G	G	P–F
Pulse	G	F–G	P	G	G

† G=good, F=fair, P=poor.
§ V. W. Weekman, *AIChE J.*, **20**: 833 (1974).

Table 8-5 SUMMARY OF REACTOR RATINGS†, §

Reactor type	Sampling and analysis	Isother-mality	Residence-contact time	Selectivity disguise-decay	Construction problems
Single fluid phase, powdered catalyst, nondecaying catalyst					
Differential	P–F	F–G	F	G	G
Fixed bed	G	P–F	F–G	G	G
Stirred batch	F	G	G	G	G
Stirred-contained solids	G	G	G	G	F–G
Continuous stirred tank	F	G	G	G	F–G
Straight-through transport	F–G	P–F	F–G	G	F–G
Recirculating transport	F–G	G	G	G	P–F
Pulse	G	F–G	P	G	G
Single fluid phase, non-diffusion-limited (pelleted catalyst), nondecaying catalyst, low heat-release system					
Differential	P–F	G	F	G	G
Fixed bed	G	G	F–G	G	G
Stirred batch	F	G	G	G	G
Stirred-contained solids	G	G	G	G	G
Continuous stirred tank	F	G	G	G	F–G
Straight-through transport	F–G	G	F–G	G	F–G
Recirculating transport	F–G	G	G	G	P–F
Pulse	G	G	P	G	G

† G=good, F=fair, P=poor.
§ V. W. Weekman, *AIChE J.*, **20**: 833 (1974).

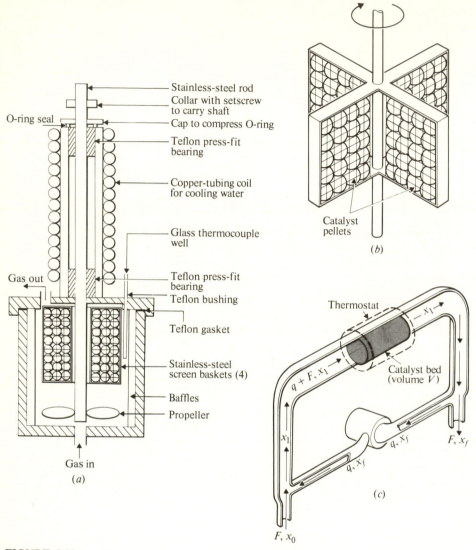

Stainless-steel rod

Collar with setscrew
to carry shaft

O-ring seal

Cap to compress O-ring

Teflon press-fit
bearing

Copper-tubing coil
for cooling water

Glass thermocouple
well

Gas out

Teflon press-fit
bearing

Teflon bushing

Teflon gasket

Stainless-steel
screen baskets (4)

Baffles

Propeller

Gas in

(a)

Catalyst
pellets

(b)

Thermostat

x_1

$q + F, x_1$

Catalyst bed
(volume V)

x_1

q, x_f

q, x_f

F, x_f

(c)

F, x_0

FIGURE 8-11
(a) The Notre Dame CSTCR for gas–solid and gas–liquid–solid studies; (b) the catalyst disposition;
(c) catalytic recycle reactor.

set forth.   In Chap. 5, external and internal effectiveness in terms of observable
parameters were derived.   In the following section, these developments will be
utilized for the detection of interphase or intraphase gradients in a CSTCR and, for
the mortal catalyst, means whereby catalytic and deactivation kinetics can be uncoupled will be noted.

Criteria for Detection of Short-Range Gradients

Suppose that catalytic-reaction-rate data are being secured in a CSTCR. Assuming that deactivation is either not apparent or slow, the rate by CSTR material balance for a particular component A is

$$\frac{A_0 - A}{\theta} = \mathscr{R} \qquad (8\text{-}117)$$

While the CSTCR is macroscopically isothermal, local gradients in A and T can exist. For a CSTCR a value of k_g, the external mass-transport coefficient, should be on hand or predictable in the light of literature revelations.[1] Catalyst pellet dimensions establish a; hence the observable $\bar{\eta}\mathrm{Da}$ is

$$\bar{\eta}\mathrm{Da} = \frac{\mathscr{R}}{k_g a A_0} \qquad (8\text{-}118)$$

Therefore the difference between external catalyst-surface concentration and the uniform bulk value in the CSTCR is (as shown in Chap. 5)

$$\frac{A_s}{A_0} = 1 - \bar{\eta}\mathrm{Da} \qquad (8\text{-}119)$$

The external ΔT (again as shown in Chap. 5) is

$$\frac{T_s}{T_0} = 1 + \bar{\beta}\bar{\eta}\mathrm{Da} \qquad \text{where } \bar{\beta} = \frac{-\Delta H\, A_0}{\rho C_p T_0\, \mathrm{Le}^{2/3}} \qquad (8\text{-}120)$$

We may assert that interphase gradients exert no influence upon catalytic rate if

$$\bar{\eta}\mathrm{Da} < 0.01 \qquad \text{and} \qquad \bar{\beta}\bar{\eta}\mathrm{Da} < 0.01 \qquad (8\text{-}121)$$

Internal catalytic effectiveness is readily phrased in terms of the internal observable, the Wheeler-Weisz number

$$\Phi = \frac{\mathscr{R}}{\mathscr{D} A_0\, a^2} = \bar{\eta}\mathrm{Da}\ \mathrm{Bi}_m \qquad (8\text{-}122)$$

As Chap. 5 indicates, in the isothermal-pellet case, internal gradients are insignificant ($\eta = 1$) when

$$\Phi < 1.0 \qquad (8\text{-}123)$$

The existence of an internal ΔT is ascertained by an a priori computation of the Prater temperature (with β from Chap. 5):

$$\frac{T}{T_s} = 1 + \beta \qquad \text{where } \beta = \frac{-\Delta H\, \mathscr{D} A_s}{\lambda T_s} \qquad (8\text{-}124)$$

If then

$$\beta < 0.01 \qquad (8\text{-}125)$$

---

[1] Ibid.

internal gradients of temperature do not affect our observed catalytic rate *for normal reaction kinetics*. This qualification is necessary since, as will be illustrated in Chap. 9, negative, abnormal, kinetic models prove to be more sensitive to β and in fact exhibit values of $\eta > 1$, even when the Wheeler-Weisz number Φ is less than 1.

In terms of both interphase and intraphase overall ΔT_0, the fraction of ΔT_0 which is confined within the external fluid film bathing the catalyst pellet (as derived in Chap. 5) is

$$\frac{\Delta T_x}{\Delta T_0} = \frac{r\bar{\eta}\text{Da}}{1 + \bar{\eta}\text{Da}(r - 1)} \qquad (8\text{-}126)$$

where

$$r = \frac{\text{Bi}_m}{\text{Bi}_h} = \frac{\lambda}{\mathscr{D}\rho C_p \text{Le}^{2/3}} \qquad (8\text{-}127)$$

As noted in Chap. 5, Bi_m is usually a few or more orders of magnitude greater than Bi_h for gas-porous solid systems; hence by Eq. (8-126) and its display (Fig. 5-14) the major ΔT can be expected to reside in the external gas film. Should ΔT_0 be finite, while $\Delta T_x/\Delta T_0$ approaches unity, our rate is unaffected so long as it is correlated with the catalyst temperature T_s and not with bulk-gas temperature.

In the light of Eqs. (8-119) to (8-122) and (8-126), it is evident that the key observables which govern the magnitude of short-range gradients of concentration and the locale of the major ΔT are $\bar{\eta}\text{Da} = \mathscr{R}/k_g a A_0$ and $\bar{\beta}$.

Integral Catalytic-Reactor Criteria

The CSTCR lends itself quite naturally to specification of the observable, $\bar{\eta}\text{Da}$, since by virtue of its uniform bulk concentration and temperature, $\mathscr{R}$ is obtained by a simple material balance. Quite often, however, catalytic reactions are conducted in small laboratory fixed beds (usually a U tube packed with catalyst particles). When they are operated at other than differential conversion levels, one can generally anticipate that both long- and short-range gradients of concentration and temperature will exist, according to the general continuity equations established in Chap. 4. If the laboratory fixed-bed reactor is characterized by axial and radial temperature gradients as well as local interpellet and intrapellet gradients of temperature and concentration, the task of rescuing chemical-kinetic parameters from data for conversion and/or yield vs. contact time is surely hopeless.

Given laboratory-scale fixed-bed data and operating conditions of flow, particle size, etc., what criteria might one employ to ascertain whether long- and short-range gradients are masking the intrinsic catalytic dispositions?

Insofar as the integral reactor cannot provide rate per se but merely conversion and/or yield vs. contact time, the fashioning of the observable $\bar{\eta}\text{Da}$ or Φ is not possible except for differential conversion. In addition, a criterion must be developed to assess the existence of radial temperature gradients in the integral catalytic-bed unit.

A summary of criteria for the fixed bed is presented by Doraiswamy and

Tajbl[1] and should be utilized if a point rate at bed entrance (where rate is a maximum for normal kinetics) can be extracted from conversion-vs.-contact-time data and, in the expectation of thermal gradients, an activation energy is at hand. The bases and limitations of these criteria are discussed by Mears.[2]

Too often packed-bed laboratory reactors are operated at quite low particle Reynolds numbers $Re = d_p u/v$ since powdered catalysts of small d_p are employed, whether for convenience or a wish to minimize intraparticle diffusional gradients. This strategy, however, not only reduces Re and consequently interphase values of h and k_g, but the tube-to-catalyst-diameter ratio is increased, thereby inviting interparticulate (long-range) radial gradients. The use of a thermowell with a traveling thermocouple can yield the illusion that radial and axial temperature gradients are negligible since the thermowell axial thermal conductivity is usually far larger than that of the packed bed.

In utilizing small catalyst particles to reduce the Thiele modulus, and thus internal gradients, velocity should be increased to maintain a high Reynolds number, thus increasing h and k_g to reduce external (interphase) gradients. However, tube-to-particle-diameter ratio should not exceed a value of about 5 or 6 to minimize radial temperature gradients. Finally, heat release per unit volume of bed can be reduced by diluting the catalyst bed with inert particles. The packed-bed catalytic reactor on the laboratory level of inquiry will then consist of a bed of large length-to-tube-diameter ratio and small tube-to-particle-diameter ratio operated at high Reynolds numbers.

In the commonly suggested tactic of varying particle size in order to detect internal gradient intrusions, some care must be exercised. For an increase in d_p may reveal no rate taxation, in spite of an increase in internal concentration gradients since the large catalyst may be at a locally higher temperature due to an external heat-transport limitation in the exothermic-reaction case.

In sum, the fixed-bed catalytic reactor in general operates in an environment characterized by long-range (interparticle) and short-range (interphase and intraphase) gradients, the temperature gradients being the most telling with respect to activity and selectivity disguise. In order of importance, the heat effects can be ranked as

$$\text{Interparticulate} > \text{interphase} > \text{intraphase}$$

8-16 DATA REDUCTION

Steady State: $A + B \rightarrow P \cdots$

If the catalytic rate model assumes the suspected form

$$r = \frac{kAB}{(1 + K_A A + K_B B + K_P P)^2} \tag{8-128}$$

---

[1] Loc. cit.

[2] D. E. Mears, *Ind. Eng. Chem. Process Deg. Dev.*, **10**: 541 (1971); *J. Catal.*, **20**: 127 (1971).

then in general the equation can be transformed to yield the relationship

$$\sqrt{\frac{AB}{r}} = \frac{1}{\sqrt{k}} + \frac{K_A A}{\sqrt{k}} + \frac{K_B B}{\sqrt{k}} + \frac{K_P P}{\sqrt{k}} \qquad (8\text{-}129)$$

When $B \gg A$ and P or $K_P = 0$, a plot of the measured value $\sqrt{A/r}$ versus A is linear, thus providing values of $K_A/\sqrt{k}$ and $1/\sqrt{k}$ from slope and intercept, respectively. Similar reasoning guides analysis when $A \gg B$. The effect of P, product, can, in principle, be ascertained by design; i.e., let P in the feed be much greater than A and B.

So much for idealizations. In fact, in complex catalytic reaction networks the most prudent procedure involves organization of the rate and or yield as a power function of effective species concentrations:

$$r = \bar{k}A^\alpha B^\beta p^p \qquad \text{and/or} \qquad Y = \bar{K}A^{\alpha'} B^{\beta'} P^{p'} \qquad (8\text{-}130)$$

Based upon estimated values of α, β, p and α', β', p', a more explicit LHHW form or its equivalent should be fashioned, for example, $r = \bar{k}AB^{0.2}P^{-1}$ may suggest

$$r = \frac{\bar{k}AB}{(1 + K_B B + K_P P)^n} \qquad (8\text{-}131)$$

Data procurement, that is, r versus A, B, ... or conversion versus A, B, ... at various values of contact time, is realized by use of one or more of the laboratory catalytic-reactor types cited earlier.

In sum, for the *integral reactor*, one obtains conversion, overall yields versus θ, where

$$\theta \equiv \frac{\text{catalyst weight}}{\text{flow rate}} \equiv \frac{\text{reactor volume}}{\text{flow rate}}$$

As with homogeneous reactions, one can analyze integral data by differentiating x-versus-θ results to secure $dx/d\theta$ versus x and thus α, β etc., or assume a rate model, integrate it, and attempt to fit x-versus-θ data to the integrated equations.

For the *differential, CSTCR* or *recycle reactor*, rate r is secured directly by a material balance, from which the rate-vs.-concentration functionality is ascertained.

Not uncommonly, a vast corpus of data confront the analyst (some reliable), and the problem reduces to one of model discrimination, i.e., determining that model which best reflects the phenomenological behavior of the catalytic system. (It should now be apparent that no amount of statistical manipulation of kinetic data will yield a mechanism.)

Armed with conversion-contact time or directly or indirectly secured rate-vs.-concentration data, the methodology detailed by Bard and Lapidus[1] proves to be quite potent in the analysis of LHHW kinetic formulations for complex catalytic reaction networks.

[1] Y. Bard and L. Lapidus, *Catal. Rev.*, **2**: 67 (1969).

Reaction-Deactivation

In this unsteady-state circumstance, one must devise a laboratory reactor type and manner of operation which permits uncoupling of the reaction-deactivation kinetic parameters. Power-law reaction-deactivation rate equations were set forth earlier.

Let us assume that for a deactivating catalyst, a CSTCR is used on the laboratory scale, and the gas phase is therefore perfectly mixed

$$\frac{A_0 - A}{\theta} = kA^n a \qquad (8\text{-}132)$$

while the catalyst is in the batch state, e.g., a spinning-basket reactor,

$$-\frac{da}{dt} = k_d(C)^m a^d \qquad (8\text{-}133)$$

where C refers to reactant A, intermediate B, or feed-laden poison P. Levenspiel[1] suggests a novel tactic by which the kinetics of reaction and deactivation can be uncoupled, namely, varying the contact time to maintain concentration at a constant level. This may not be so easily realized inasmuch as the investigator must adjust flow in a manner precisely commensurate with the unknown activity-time behavior. An alternative is to be found in certain limiting analytical solutions for Eqs. (8-132) and (8-133).

Simultaneous network

By a CSTR balance one finds for A for linear kinetics,

$$\mathcal{R} = \frac{A_0 - A}{k\theta a} \qquad k = k_1 + k_2 \qquad \text{or} \qquad \frac{A}{A_0} = \frac{1}{1 + k\theta a}$$

Substituting into the batch deactivation function (8-133) gives

$$-\frac{da}{dt} = \frac{k_d a^d A_0{}^m}{(1 + k\theta a)^m} \qquad (8\text{-}134)$$

or

$$\int \frac{(1 + k\theta a)^m}{a^d} \, da = -A_0{}^m k_d t \qquad (8\text{-}135)$$

For $m = 1$:

$$\int \frac{da}{a^d} + k\theta \int a^{1-d} \, da = -A_0 k_d t \qquad (8\text{-}136)$$

For $m = 2$:

$$\int \frac{1 + 2k\theta a + \overline{k\theta a^2}}{a^d} \, da = -A_0{}^2 k_d t \qquad (8\text{-}137)$$

[1] O. Levenspiel, *J. Catal.*, **25**: 265 (1972).

Both these equations can be integrated for diverse values of d. Values of d between 1 and 3 can be anticipated. For $d = 1$ we find for $m = 1$

$$\ln a + k\theta a = -\bar{k}_d t \qquad (8\text{-}138)$$

but for $n = 1$

$$k\theta a = \frac{x}{1 - x}$$

and so

$$\ln \frac{x}{k\theta(1 - x)} + \frac{x}{1 - x} = -\bar{k}_d t \qquad (8\text{-}139)$$

If θ is varied to maintain a constant conversion x, then as Levenspiel suggests, $\ln \theta$ versus t will be linear. More commonly x is observed to vary with t at constant θ, in which case one assumes various values of k in plotting the log of Eq. (8-139) vs. t to establish k_d and k.

As an exercise, develop integrated, negotiable expressions for (a) $m = 1$, $d = 3$ and (b) $m = 0$, $d = 1$, where $n = 1$.

Note that for the perfectly mixed gas phase of equal order in each step, n

$$k\theta a = \frac{x}{A_0^{n-1}(1 - x)^n} \qquad (8\text{-}140)$$

so that observaton of a at constant temperature k and θ as a function of feed composition A_0 gives an indication of reaction order n.

Parallel network

$$A \longrightarrow B$$
$$P \longrightarrow P\downarrow$$

For $m = 1$ and $n = 1$:

$$P = \frac{P_0}{1 + k\theta a}$$

and

$$-\frac{da}{dt} = \frac{k_d P_0 a^d}{1 + k\theta a} \qquad (8\text{-}141)$$

which is readily integrated, as is the simultaneous case.

Consecutive network   Again assuming linear kinetics for the principal reactions, we have for $m = 1$

$$-\frac{da}{dt} = \frac{k_d A_0 k_1 \theta a^{d+1}}{(1 + k_1 \theta a)(1 + k_2 \theta a)} \qquad (8\text{-}142)$$

which is integratable for anticipated values of d.

Independent network   In the case of sintering

$$-\frac{da}{dt} = ka^2 \qquad (8\text{-}143)$$

or

$$a = \frac{1}{1 + Kt}$$

For nth-order reaction in a CSTR we find

$$\frac{(1 - x)^n}{x} = \frac{1}{\overline{K}} + K_0 t \qquad (8\text{-}144)$$

As an exercise, derive the above equation.

Should deactivation-reaction events be observed in other than a CSTR environment, uncoupling becomes more difficult and the varying flow-rate–constant-composition strategy recommended by Levenspiel must be utilized.

Chemical-Kinetic Criteria

If one is persuaded that the kinetic model, found to fit catalytic conversion and/or selectivity data, conforms to the LHHW postulate of single-step rate control and equilibrium chemisorption for other than the rate-determining step, then certain criteria can be invoked to test, as it were, the legitimacy of derived kinetic adsorption coefficients in terms of the adsorption enthalpy and entropy, ΔH and ΔS, respectively.

Firstly, evidence of endothermic chemisorption as might be inferred should K_a *increase* with temperature for a rate model of the form $r = kA/(1 + K_a A)$ must be greeted with suspicion. While endothermic chemisorption is a possibility, the probability of its occurrence is so rare as to alert us to the possibility that that "adsorption" coefficient is, in fact, either a ratio of kinetic coefficients as in the instance of nonequilibrium kinetics (Sinfelt) or multiple-step rate control.

Boudart has noted that too little attention has been paid to the entropy of chemisorption.[1] For insofar as chemisorption from the three-dimensional gas phase to the two-dimensional surface-adsorbed phase is implicit in the notion of chemisorption equilibrium, then ΔS values derived from LHHW rate formulations should exhibit characteristics in agreement with the physics of the situation, whether the catalytic surface is ideal or nonideal in the sense discussed earlier. Boudart, Mears, and Vannice argue in this spirit[2] and have established a few guidelines with respect to derived values of ΔS of chemisorption which give the kineticist a basis whereby foolish rate models can be separated from those which are, at least, thermodynamically consistent. Their thesis is as follows. The standard entropy of chemisorption is $\Delta S = S_{ads} - S_g$, where S_{ads} and S_g are entropies of the adsorbed and fluid (gaseous) states. Insofar as the adsorbed state is the more ordered one ($S_{ads} < S_g$), we have

Rule 1: $\Delta S < 0$ (8-145)

Further, the negative value of ΔS must be less than S_g in absolute value, an obvious deduction since a molecule cannot lose more entropy than it possesses.

Rule 2: $|\Delta S| < S_g$ (8-146)

[1] M. Boudart, "Surface Chemistry of Metals and Semi Conductors," p. 409, Wiley, New York, 1960.
[2] M. Boudart, D. E. Mears, and M. A. Vannice, *Ind. Chim. Belg.*, **32**: 281 (1967).

Boudart and colleagues then go on to propose "less rigorous" but nevertheless prudent guidelines:

Rule 3: $\quad |-\Delta S| = 10$ entropy units $\hfill$ (8-147)

Rule 4: $\quad |-\Delta S| < 12.2 - 0.0014\Delta H$

where $\Delta H =$ the chemisorption enthalpy

In general, Boudart et al, assert that

$$0 < -\Delta S < S_g \qquad \text{and} \qquad 10 < -\Delta S < 12.2 - 0.0014\Delta H \qquad (8\text{-}148)$$

Boudart et al. assessed some 129 literature values of ΔS of adsorption in the light of the above rules, and a significant number of these kinetic studies from which ΔS values were extracted must be suspect insofar as rules 1 and 2 were violated; nor were the approximate guidelines obeyed.

It is obvious that not a few of those studies which yield ΔS values in violation of rules 1 and 2 may well reflect nonequilibrium adsorption kinetics or multiple-step rate control, in either of which cases the "apparent" adsorption coefficient(s) should not be expected to obey thermodynamic dogma.

We now offer an illustration of catalytic-rate-data analyses. The steps in the analysis are as follows:

1   Express the rate in terms of a power-law formulation.
2   Translate the power-law result into a LHHW model and evaluate the parameters and their temperature dependencies.
3   Assume a sequence of elementary steps and a rate-determining step which yields a rate law agreeing with that found to organize the data.
4   Test the adsorption parameters in the light of the Boudart-Mears-Vannice rules.
5   If possible, evaluate the sequence of assumed elementary steps in terms of microscopic fundamentals, e.g., transition-state theory. In following this procedure, full use of all physicochemical information should be made; e.g., the influence species foreign to the reaction per se should be noted, as shown in the following example.

Example   Utilizing a differential reactor, Sinfelt et al.[1] obtained initial rate data for the dehydrogenation of methylcyclohexane over a 0.3% Pt/Al_2O_3 catalyst in the presence of H_2 (to reduce coking). The data are given in Table 8-6. The reaction is

$$M \xrightarrow[\text{Pt}]{\text{H}_2} T \qquad (a)$$

STEP 1   Following the procedure cited above, one first analyzes the data in terms of a power-law rate expression

$$\mathscr{R} = kM^a(\text{H}_2)^b \qquad (b)$$

[1] J. H. Sinfelt, H. Hurwitz, and R. A. Shulman, *J. Phys. Chem.*, **64**: 1559 (1960).

Note that Sinfelt et al. secured rate data at constant M (runs a and b at three temperatures) and at constant H_2 pressures (runs c to e). Runs a and b at three temperature levels indicate that the reaction rate is independent of H_2 partial pressure, therefore

$$\mathscr{R} = k'M^a \qquad (c)$$

In general at any temperature

$$\left(\frac{\mathscr{R}_1}{\mathscr{R}_2}\right)_T = \left(\frac{M_1}{M_2}\right)^a \qquad (d)$$

Utilizing the data of runs c to e at each temperature we find $a = 0.2, 0.15, 0.18$ at 315, 344, and 372°C, respectively, indicating an inhibiting influence of M on the rate, which suggests a more formal LHHW type of rate model of the following form:

STEP 2

$$\mathscr{R} = \frac{kM^m}{(1 + KM)^n} \qquad (e)$$

This model is then tested, first in its most primitive form, $m = n = 1$. Linearizing gives

$$\frac{M}{\mathscr{R}} = \frac{MK}{k} + \frac{1}{k} \qquad (f)$$

so that a plot of the experiment rates $M/\mathscr{R}$ versus M should be linear for each temperature if the assumed model is valid.

Table 8-6[†]

Run, $T°C$	M, atm	H_2, atm	$\mathscr{R}$, mol toluene/(h)(g cat)
Aa 315	0.36	1.1	1.2
b	0.36	3	1.2
c	0.07	1.4	0.86
d	0.24	1.4	1.1
e	0.72	1.4	1.3
Ba 344	0.36	1.1	3
b	0.36	3.1	3.2
c	0.08	1.4	2
d	0.24	1.4	3.4
e	0.68	1.4	3.4
Ca 372	0.36	1.1	7.6
b	0.36	4.1	8
c	0.36	4.1	12.4
d	0.22	4.1	13.1

† J. G. Sinfelt, H. Hurwitz, and R. A. Shulman, *J. Phys. Chem.*, **64**: 1559 (1960).

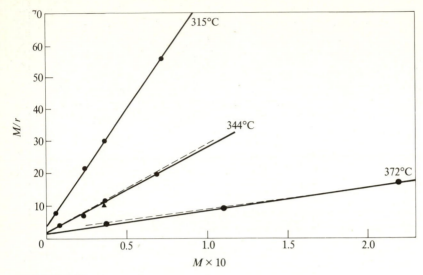

FIGURE 8-12
Experimental data plotted according to Eq. (f), Sec. 8-16.

In Fig. 8-12 the experimental data are so plotted.   Indeed the data conform to the model.   The slope K/k and intercept $1/k$ give both coefficients at each temperature.   The results are shown in Table 8-7.

STEP 3   The LHHW model is now examined in more detail by supposing a sequence of elementary steps, specifying a rate-determining step so that significance can be assigned the coefficients which emerge from the organization of the data. Let us assume the mechanism to be

Step 1: $X + M \rightleftharpoons MX$ $\quad\quad\quad\quad\quad\quad\quad\quad\quad\quad\quad\quad\quad$ (g)

Step 2: $MX \xrightarrow{k_0} T + X$

which is the classic Langmuir-Hinshelwood model of equilibrium chemisorption of reactant followed by the rate-determining decomposition of the adsorbed complex (MX) into product toluene (T).   Therefore

$$\mathscr{R} = k_0(MX) \quad\quad \text{but} \quad\quad (MX) = MX \quad\quad\quad (h)$$

TABLE 8-7

$T, °C$	k	K	k_0
315	0.3	21.2	0.00141
344	0.44	10	0.044
372	0.67	5	0.134

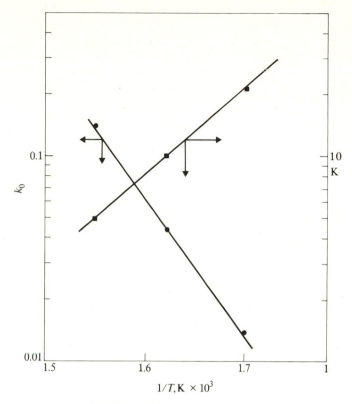

FIGURE 8-13
Temperature dependencies of K_1 and k_0 (Table 8-7).

A site balance yields

$$X_0 = X + (MX) = X(1 + KM) \qquad (i)$$

Thus, normalizing by total sites X_0, we find

$$\mathscr{R} = \frac{k_0 K_1 M}{1 + K_1 M} \qquad (j)$$

so that $k_0 = k/K$ in terms of the extracted coefficients. k_0 is listed at each temperature in the last column of Table 8-7.

This agreement between a model rooted in a mechanistic assumption and the experimental data does not prove the mechanism, of course. Next we must examine the coefficient-versus-temperature behavior. In Fig. 8-13 K_1 and k_0 are plotted in the manner of the Arrhenius law, from which it is evident that the behavior of k_0 and K_1 is normal; that is, k_0 increases with temperature, the activation energy being $E_0 = 30,000$.

The kinetic adsorption coefficent decreases with temperature, as would be expected for exothermic chemisorption. The value of the apparent adsorption enthalpy is $-\Delta H = 19,000$.   The entropy of adsorption is now readily computed from

$$K = \exp\left(-\frac{\Delta H}{RT} + \frac{\Delta S}{R}\right) \qquad (k)$$

and is found to be $\Delta S = -26$ eu.

STEP 4   We can now test this result in accord with the criteria of Boudart, Mears, and Vannice, cited earlier

Rule 1 ΔS must be negative.   This rule is obeyed.
Rule 2 ΔS in absolute value must be less than the gas-phase value of S.
Rule 3 $|-\Delta S|$ must probably be greater than 10 eu.   This rule is obeyed.
Rule 4 $-\Delta S$ must probably be smaller than $12.2 - 0.0014\Delta H$.   As $\Delta H = -19,000$, this rule is obeyed.

All would seem to be in accord with respect to the assumed mechanism and the behavior of the kinetic parameters.   *However*, Sinfelt and colleagues conducted additional experiments in which aromatics, known to chemisorb strongly and preferentially on the catalyst, were added to the methylcyclohexane feed.   Even an equimolar feed of benzene or *m*-xylene and methylcyclohexane reduced the rate by a mere 20 percent.   If the methylcyclohexane is in fact in a state of chemisorbed equilibrium, this modest rate inhibition in the presence of aromatics is inexplicable.

This difficulty (and, incidentally, the rather large value of the heat of chemisorption of methylcyclohexane, 19 kcal) is circumvented if it is assumed that chemisorption equilibrium is not achieved (see Sec. 8-13).   Sinfelt et al. postulate

Step 1: $M \longrightarrow M_{ads}$

Step 2: $M_{ads} \longrightarrow T_{ads}$ $\qquad (l)$

Step 3: $T_{ads} \longrightarrow T$

Equating the three nonequilibrated steps, where θ is the toluene coverage, gives

$$k_1 M(1 - \theta) = k_2 M_{ads} = k_3 \theta \qquad (m)$$

Assuming M_{ads} to be small in deference to the observation that aromatics affect rate only modestly, we have

$$k_1 M(1 - \theta) = k_3 \theta \qquad (n)$$

so that

$$\theta = \frac{k_1 M}{k_3 + k_1 M} \qquad (o)$$

If desorption of toluene (step 3) is the rate-determining step,

$$\mathscr{R} = k_3 \theta$$

or

$$\mathscr{R} = \frac{k_3(k_1/k_3)M}{1 + (k_1/k_3)M} \qquad (p)$$

a form identical that LHHW model found to correlate kinetic observations faithfully.

   With this more reasonable interpretation, the activation energy of $E_3 = 30$ kcal is that of desorption of the strongly bound unsaturate, toluene.   Further

$$\ln \frac{k_1}{k_3} = -\frac{E_1 - E_3}{RT} + \ln \frac{\mathscr{A}_1}{\mathscr{A}_3} \qquad (q)$$

or

$$19{,}000 = -(E_1 - 30{,}000)$$

and the activation energy E_1 for chemisorption of methylcyclohexane is 11 kcal.

   In this nonequilibrium adsorption model, instead of entropy, we have the natural logarithm of the ratio of preexponential factors

$$\ln \frac{\mathscr{A}_1}{\mathscr{A}_3} = -13 = -\frac{\Delta S}{R} \qquad \text{or} \qquad \frac{\mathscr{A}_1}{\mathscr{A}_3} = \exp(-13) = 10^{-6}$$

STEP 5   The significance and reasonability or absurdity of such a value can be *very* approximately assessed by invoking transition-state theory.   Sinfelt et al. did so in order to indicate the reasonableness of the observed desorption rate.   We do so to verify the observed magnitude of the numerical ratio $\mathscr{A}_1/\mathscr{A}_3$.

   As Laidler shows,[1] the rate of chemisorption is

$$r_1 = C_M C_S \frac{k_B T}{h} \frac{f_*}{F_g f_s} \exp(-\epsilon_1) \qquad (r)$$

and that of desorption

$$r_3 = C_T \frac{k_B T}{h} \frac{\tilde{f}_{\pm}}{f_a} \exp(-\epsilon_3)$$

where f and F are partition functions.   We may approximate the preexponential factor ratio $\mathscr{A}_1/\mathscr{A}_3$ by

$$\frac{\mathscr{A}_1}{\mathscr{A}_3} = \frac{C_S f_*}{F_g f_s} \qquad \text{since} \frac{\tilde{f}_{\pm}}{f_a} \approx 1 \qquad (s)$$

As Laidler indicates, the adsorption partition functions are rendered explicit in terms of fundamental constants; thus

$$\frac{\mathscr{A}_1}{\mathscr{A}_3} = \frac{C_S}{[(2\pi m k_B T)^{3/2}/h^3]8\pi^2 I k_B T/h^2} \qquad (t)$$

C_S is the concentration of sites per square centimeter and equal to about 10^{15}. On page 73 of Laidler's text, order-of-magnitude values of the terms in the denominator are given, i.e.,

$$\frac{(2\pi m k_B T)^{3/2}}{h^3} = 10^{24} \text{ to } 10^{25} \qquad \frac{8\pi I k_B T}{h^2} = 10 \text{ to } 10^2$$

Therefore

$$\frac{\mathscr{A}_1}{\mathscr{A}_3} \approx 10^{-10}$$

[1] K. J. Laidler, "Chemical Kinetics," 2d ed., pp. 73, 286, McGraw-Hill, New York, 1965.

In view of the vast number of simplifications involved in this type of analysis, the result is rather satisfactory.

This example is indeed a fascinating one insofar as the criteria for LHHW adsorption parameters are fulfilled, as are the magnitudes of the preexponential factors for the nonequilibrium model. The one observation which prompts us to give preference to the nonequilibrium mechanism of Sinfelt et al. is that of the modest influence of aromatics upon observed rates.

In an exhaustive kinetic study of the Deacon reaction, $2HCl + \frac{1}{2}O_2 \rightarrow Cl_2 + H_2O$, Jones, Bliss, and Walker[1] found the rate far from equilibrium to be described in terms of partial pressure by

$$\mathcal{R} = \frac{k(O_2)^{1/2}}{1 + K(HCl)}$$

However, as the authors noted, since K increased with temperature, it must not be considered an adsorption equilbrium coefficient. As an exercise, utilize the original reference and endeavor to fashion a sequence of elementary steps which will yield a model in accord with observations.

8-17 CLASSIFICATION OF CATALYSTS

Broadly speaking, catalytic agents are (1) metal conductors (Fe, Pt, Ag, etc.), (2) insulators (metal oxides, Al_2O_3, MgO), and (3) semiconductors (NiO, ZnO). As we shall see in the next section, seldom does a commercial catalyst fall into one of the above exclusive categories, as the total formulation is generally a complex mixture of components from one or more of the above classes. Nevertheless, the above classification, broad as it is, provides a general basis for some understanding of catalytic reactions.

Metallic catalysts Generally found as oxides capable of reduction, these metals effectively chemisorb hydrogen as well as oxygen and function rather effectively as oxidation catalysts as well as hydrogenation-dehydrogenation catalysts. Since they are conductors, the essential chemisorptive-catalytic step seems to be an exchange of electrons between metal and adsorbate. That the above classification is not definitive is apparent from consideration of NiO, a semiconductor, yet upon reduction nickel becomes a conductor, with all the properties of class 1. The catalytic metals are generally found in the transition- and noble-metal groups.

Insulators Insulators of catalytic character function quite differently from conductors or semiconductors. The absence of conductivity excludes the conductor mechanism of chemisorption and catalysis. The insulators are generally characterized by acidity, which renders them effective in reactions requiring a carbonium-ion mechanism. Thus cracking, polymerization, alkylation, isomerization, and

[1] A. M. Jones, H. Bliss, and C. A. Walker, *AIChE J.*, **12**: 260 (1966).

hydration-dehydration reactions are catalyzed by the acidic insulators.   The key importance of the insulator acidity is evident in the fact that the above reactions are often catalyzed homogeneously by H_2SO_4, HF, etc.   The insulators are generally irreducible.

Semiconductors   These are characterized by conductivity which, while low or negligible at low temperature, increases dramatically with temperature.   Semi-conductivity may be intrinsic or (more often) induced by the creation of a cationic or anionic vacancy by addition of foreign ions.   Capable of electron exchange with adsorbed species, the semiconductor catalysts function effectively in reactions of the same type catalyzed by the transition and noble metals.

Based upon this simple classification and very brief characterization, some general conclusions concerning the selection of catalysts for a particular reaction may be stated: namely, bases are likely poisons in acidic catalysis; the metal catalysts, being strong electron-exchange agents, tend to chemisorb strongly, and thus are freed of oxygen, CO, etc., with difficulty.   Less susceptible to such poisoning are the metal-oxide semiconductors.   As high temperature is likely to sinter the metals, the semiconductors, while less active at moderate temperatures, often prove more durable at high temperature.

To illustrate, the commercially important catalytic oxidation of SO_2 to SO_3 can be effected by a Pt-on-alumina catalyst and also by a supported semiconductor V_2O_5.   The Pt catalyst is more active at about $400°C$, but because of its suscepti-bility to poisons and sensitivity to high temperature, the platinum catalyst has largely been replaced by V_2O_5.

Finally, features of two catalyst types can be combined to produce multi-function catalysts; for example, Pt/SiO_2-Al_2O_3 is capable of hydrogenation and isomerization activity.

While it is hardly possible to detail what is known of the nature of catalysis for representatives of each catalyst class, it will prove instructive to discuss catalytic (as opposed to total catalyst-surface) area; then the nature of supported-metal catalysis will be treated, followed by illustrations of catalytic rate modeling and mechanistic findings for the NH_3 synthesis reaction and SO_2 oxidation as catalyzed by V_2O_5 and Pt.   We conclude the chapter with brief comments on dual function and zeolite catalysis.

Catalytic and Total Surface Area

Physical adsorption, being indiscriminate, provides, via the BET method, total sur-face area of the pellet.   To determine what fraction of total area is catalytic and what portion is covered with promoter requires a selective adsorption process, namely, chemisorption.

A classic case worthy of discussion is that of the iron catalyst used in the high-pressure synthesis of NH_3 from N_2 and H_2.   Many of our definitions will become meaningful in the light of this case history, and the practical virtues of chemisorp-tion-equilibrium studies will become clear.

The reduced-iron catalyst used to synthesize NH_3 was one of some 20,000 catalyst candidates evaluated in the search for an efficient catalyst carried out in Germany some decades ago. This fact suggests the empirical nature of catalyst selection for a given reaction duty.

The earliest iron catalyst consisted of reduced iron, unpromoted. However, catalyst life proved to be too brief, and a search for promoters led to a formulation consisting of iron promoted with small quantities of Al_2O_3, alumina. This singly promoted catalyst then gave way to doubly and triply promoted formulations containing K_2O and CaO with Al_2O_3. An extensive investigation into the nature of the promoted-iron catalyst started after World War I, and one of the most fruitful inquiries was undertaken by Emmett and his colleagues at the Fixed Nitrogen Laboratory in the 1930s. Armed with the powerful BET method, Emmett and coworkers estimated the total catalytic and promoted-surface areas of the NH_3 catalyst in the following fashion:

1 Total catalyst area was determined by physical adsorption of N_2 at $-195°C$. These results showed that under synthesis conditions unpromoted iron rapidly sintered, with consequent loss of surface area and therefore activity. Promoted iron retained its surface area, suggesting that the alkali promoters prevented sintering of the iron crystallites.

2 The actual, active surface area of reduced iron in a promoted catalyst was estimated by measuring the equilibrium chemisorption of CO, correcting for physical adsorption, and showing that CO did not chemisorb on the alkali.

3 The actual area occupied by promoter was estimated by meauring the equilibrium chemisorption of CO_2, which does not chemisorb upon iron itself at low temperature. Of course, correction for physically adsorbed CO_2 was made.

4 CO chemisorption on "pure" iron catalyst suggested that all surface atoms of iron chemisorbed CO. Finally, the sum of areas obtained in steps 2 and 3 agreed nicely with the total BET area. The results of this study are summarized in Table 8-8, summarizing the results of Emmett and Brunauer.

Note that in spite of the small percentage by weight of promoters added to the catalyst formulation, about 60 percent of the total area is occupied by promoter, a figure verified by the CO_2 chemisorption studies (3, above). Also, the most

Table 8-8 AMMONIA-SYNTHESIS CATALYST ACTIVITY†

Catalyst	Relative activity	Total BET area, m^2/g	Fe area, m^2/g
973 Fe + 0.15% Al_2O_3 as impurity	1	1§	1
954 Fe + 10% Al_2O_3	2.5	13.2	5.1
931 Fe + 1.6% K_2O + 1.3% Al_2O_3	3.75	3.7	1.5

† A. Nielson, *Adv. Catal.*, **51**: (1953); see also *Cat. Rev.* (1969).
§ Reference value.

active catalyst is not that of the highest total and iron-surface area.   Other studies have shown that the promoters form a " varnish " about the iron crystallites, thus arresting sintering and accounting for the high promoter area found relative to the actual amount of promoters added.   Therefore, one function of the promoter is *physical* in nature in that surface area is stabilized.

The alkali promoters also function *chemically*, for it is clear that the doubly promoted iron catalyst containing less exposed iron is more active than the singly promoted formulation of higher iron-surface area.   Evidently multiple promotion causes surface electronic and/or energetic alterations which effectively render a few iron sites more active than the greater number available in the singly promoted sample.

8-18   THE NATURE OF SUPPORTED-METAL CATALYSTS

The state of metal dispersion upon the support is of signal import.   Dispersion is

$$D = \frac{\text{no. surface metal atoms}}{\text{total no. of metal atoms}}$$

(While "metal" atoms will be discussed, it is evident that catalytic compounds of metals are understood to be embraced by this entire discussion.)

For a given level of dispersion, what is the physicochemical nature of the supported metal?   Do small crystallites of Pt, for example, exhibit the same metallic properties as the bulk metal?

If more than one metal is dispersed upon a carrier, do they exist as alloys or separate distinct entities?   If alloyed, what is the surface composition at the alloy crystallite surface in contrast with overall global composition?   For a given mono-metallic or polymetallic system, how does the nature of the support affect dispersion and nature of the crystallites?   A comprehensive discussion of any one of these issues is beyond the scope of this text.   However, each topic can be touched upon to provide the reader with a brief introduction to each issue.

Dispersion and Its Determination

As surface catalysis involves, by definition, only those sites exposed to reactants a dispersion D of unity would be generally considered the catalytic ideal, while a value of D of near zero (typical of a bulk specimen, such as a Pt wire or rod) is a catalytic undesirable.   This is a generalization, for dispersions between zero and unity can prove of greater interest since the properties of highly dispersed metal are not necessarily those of its bulk form.   Let us illustrate by defining a specific catalytic rate in terms of area of the dispersed supported metal M

$$\mathcal{R}_s = \frac{\text{molecules reacted}}{(\text{time})(\text{m}^2 \text{ of M})} \qquad (8\text{-}149)$$

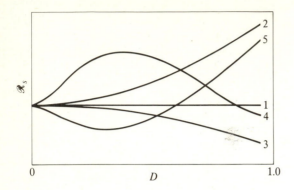

FIGURE 8-14
Specific rate vs. dispersion. [See, for example, *G. C. Bond, 4th Int. Cong. Catal., Moscow, 1968, pap. 67.*]

In Fig. 8-14 the behavior of $\mathscr{R}_s$ as a function of dispersion is schematically set forth. Five possibilities exist: $\mathscr{R}_s$ is independent of absolute catalytic metal area, (curve 1). As the surface structure of large and small crystallites are known to differ, this behavior is characteristic of a reaction which is structure-insensitive or, in Boudart's phrase, a *facile reaction*. $\mathscr{R}_s$ increases (curve 2), decreases (curve 3), or passes through a maximum (curve 4) or a minimum (curve 5), with increasing dispersion. Evidently reactions exhibiting one of these forms of behavior are, indeed, structure sensitive, or in Boudart's terminology, *demanding reactions*. Not only will activity $\mathscr{R}_s$ be either sensitive or insensitive to structure, but reaction yield will fall into one of these two categories.

It is important to bear in mind that the global rate $\mathscr{R}$ [molecules reacted/(t) (grams of total catalyst)] is

$$\mathscr{R} = \mathscr{R}_s \times \frac{\text{m}^2 \text{ of metal}}{\text{g of total catalyst}}$$

so that a large value of $\mathscr{R}_s$ may prevail for a large average crystallite size (small dispersion), e.g., curve 3, Fig. 8-14, but with the result that $\mathscr{R}$ may be smaller than its value at higher dispersion levels.

Dispersion is measured by chemisorption, e.g.,

$$H_2 + 2Pt \longrightarrow 2PtH \quad \text{or} \quad O_2 + 2Pt \longrightarrow 2PtO$$

or titration,[1] e.g.,

$$PtO + \tfrac{3}{2}H_2 \longrightarrow PtH + H_2O \quad \text{or}$$

$$PtO + 2CO \longrightarrow PtCO + CO_2$$

When crystallite diameter is greater than about 50 Å, scanning electron microscopy, electron microscopy, and (for diameters greater than 100 Å) x-ray diffraction

[1] J. E. Benson and M. Boudart, *J. Catal.*, **4**: 704 (1965).

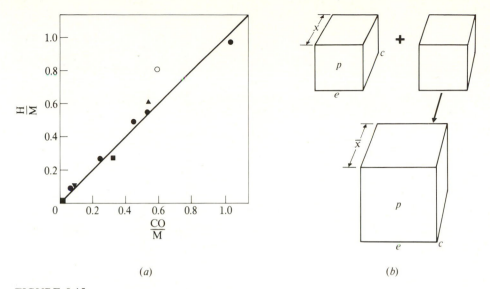

(a)

(b)

FIGURE 8-15

(a) Comparison of hydrogen and carbon monoxide chemisorption results on metals at room temperature: ●, rhodium samples of varying dispersion, including bulk rhodium and silica-supported rhodium; ▼, 10% rhenium on silica; ■, 5% ruthenium on silica; ▲, 10% osmium on silica; ○, 10% iridium on silica. (b) Schematic of coalescence of two cubes, where p, e, and c stand for planar edge, and corner atoms, respectively. (Fig. 8-15a only), [*J. H. Sinfelt, Catal. Rev.,* **3**: *175 (1970).*]

can be employed to determine crystallite sizes and possibly their distribution.[1] It must be emphasized that chemisorption and titration techniques presuppose agreement between the assumed stoichiometry, set forth above, and reality. Further, crystallite-size distribution cannot be inferred by chemisorption or titration; only the totality of surface-atom population is secured, on the assumption, of course, that subsurface atoms are not participants.

While these methods are not absolute, they do give a measure of relative dispersion or dispersed-metal catalytic-surface area when the area per surface atom is at hand. In Fig. 8-15(a), H_2 chemisorption data are compared with CO chemisorption results for various dispersions of rhodium and iridium. A value of CO/metal and hydrogen/metal ratios of unity corresponds to 100 percent dispersion. Near-zero values are for the bulk metal.[2]

In sum, knowledge of supported-metal dispersion proves of importance with respect to specific activity-selectivity and as a barometer of deactivation by sintering, chemical-surface poisoning, and fouling.

Example Consider two cubic crystallites (Fig. 8-15b) of linear dimension x. The reactant sees planar sites p, edge atoms e, and corner atoms c. The total number of

[1] T. E. Whyte, *Catal. Rev.,* **8**: 117 (1974).
[2] J. H. Sinfelt, *Catal. Rev.,* **3**: 175 (1970).

surface atoms M_s per total number of metal atoms on the surface and within the cube, M_0 (as known from the weight of metal deposited) is $M_s/M_0 = 5x^2/x^3$ since the plane at the crystallite-supported interface is not exposed to the reactant. If, now, the two cubes sinter to form one cube of dimension $\bar{x}$, then $(\overline{M}/M)_s = x/\bar{x}$.

If the catalytic reaction depends simply upon the number of surface atoms available and not on whether the catalytic sites must be corner and/or edge atoms, it follows that specific rate $\mathscr{R}_s$ will be independent of crystallite size; i.e., the reaction is facile. Suppose, however, that reaction (or selectivity) is dictated only by corner atoms insofar as the energetics of chemisorption may be expected to differ on a corner (or edge) atom in contrast to a planar atom. Then assuming for the sake of illustration that the two cubes sinter to produce one cubic crystal, it is evident that we are witness to a twofold reduction in corner sites. Total surface-atom availability is reduced, however, by $2^{2/3}$. Our reaction is consequently demanding since $\mathscr{R}_s$ is no longer independent of M_s/M_0. Should the two cubic crystals sinter to form a single sphere, corner and edge atoms disappear in this idealized illustration. In reality corners, edges, steps, and dislocations in general will vary in their population with the severity and nature of the sintering environment.

As an exercise, assume that the two cubes sinter to form a single sphere. What is the surface-atom availability ratio?

Physicochemical Properties of Dispersion

It is well recognized that surface properties, e.g., Gibbs free energy, of a bulk metal differ from those of the bulk. In the case of a catalytic wire or rod the number of bulk atoms is virtually infinite relative to the surface-atom population, and therefore property measurements will predominantly reflect bulk character. If, in a thought experiment, one were to shave from the surface of a bulk specimen several atomic layers and highly disperse them upon a high-area support, it could be reasonably expected that property measurements would reveal characteristics unlike those of the parent bulk entity. However, these dispersed crystallites upon growth (sintering) would achieve a certain size beyond which bulk properties would become manifest. Note, however, that even when the dispersed crystallites exhibit bulk characteristics, their surface properties will not necessarily be those of the bulk.

Under conditions of NH_3 synthesis from N_2 and H_2, thermodynamics teaches that the iron nitride cannot be formed, while kinetic and allied data support the contention that nitrogen chemisorption occurs. Discuss this issue as an exercise.

The data of Carter and Sinfelt[1] on the magnetic properties of bulk and dispersed Ni are reproduced in Fig. 8-16. The chemical nature of dispersed catalytic agents can be inferred from selective chemisorption data, as noted earlier, the specificity of the catalyst with respect to certain reactions, and direct detection of surface as opposed to bulk compositions.

The most tantalizing systems are those involving bimetallic or polymetallic

---

[1] J. L. Carter and J. H. Sinfelt, *J. Catal.*, **10**: 134 (1968).

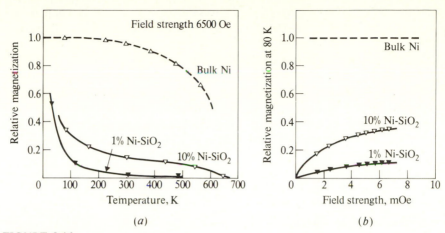

FIGURE 8-16

(a) Thermomagnetic curves for, and (b) field dependence of the magnetization of, Ni catalysts. $\triangledown$ = bulk Ni, $\triangle$ = 10% Ni on silica, $\blacktriangledown$ = 1% Ni on silica. The Ni-silica catalysts were prepared by impregnating silica with nickel nitrate solution. All samples were reduced in flowing hydrogen for 16 h at 370°C. After reduction the samples were evacuated at 370°C before making the magnetic measurements. [*J. L. Carter and J. H. Sinfelt, 1st Middle Atl. Reg. Meet. Am. Chem. Soc., Philadelphia, Feb. 3–4, 1966.*]

dispersions. Earlier we alluded to the question of the state of two dispersed metals upon a support: Are they alloyed, or do they exist in a noninteracting, independent state? If alloyed, what is the surface composition relative to the overall average composition? Can two metals which exhibit bulk immiscibility become miscible (alloy) at high levels of dispersion, where bulk thermodynamics no longer apply?

An excellent example of the utilization of selective chemisorption and reaction specificity in elucidating the nature of supported metals is to be found in the work of Sinfelt and associates as summarized in two reviews[1] dealing with the reactions of hydrocarbons catalyzed by supported metals. The essential elements of these studies will be discussed insofar as they illustrate the key points touched upon earlier in this chapter.

Hydrogenolysis-Dehydrogenation over Dispersed Metals

As Sinfelt shows, hydrogenolysis involves a class of catalytic reactions characterized by rupture of chemical bonds via interaction with hydrogen. The severed bond can be C—C, C—N, or C—halogen. Dehydrogenation consists of hydrogen removal from a saturated hydrocarbon. Two reactions catalyzed by metals typify each reaction type:

$$C_2H_6 + H_2 \longrightarrow 2CH_4 \qquad C_4H_{12} \longrightarrow C_6H_6 + 3H_2$$

Hydrogenolysis of ethane   Dehydrogenation of cyclohexanes

[1] J. H. Sinfelt, *Catal. Rev.*, **3**: 175 (1970), **9**: 147 (1974).

We have previously alluded to the kinetics of dehydrogenation of methyl-cyclohexane by way of illustrating irreversible adsorption-kinetic sequences which yield a rate equation identical in form to that generated by the LHHW assumption of surface coverage dictated by reversible chemisorption.

We shall address ourselves to hydrogenolysis of ethane, its kinetics, inferred mechanism, specificity with respect to a broad family of catalytic elements, solid-state correlations, dispersion effects, the role of the support, and dehydrogenation-hydrogenolysis catalytic selectivity for dispersed bimetallics.   In the last instance, some provocative insights into the state of dispersed alloys are revealed.

Kinetics and mechanism of ethane hydrogenolysis   The kinetics of ethane hydro-genolysis is found to conform to a phenomenological rate model

$$\mathscr{R} = k(E)^n (H_2)^m \qquad (8\text{-}150)$$

where E and H_2 are partial pressures of ethane and hydrogen, respectively.   The reaction has been studied for a number of group VIII metals supported on silica gel, with results summarized in Table 8-9.   Armed with a rate model and the rather provocative values of n and m itemized in Table 8-9, the analyst next postulates a sequence of elementary steps (a mechanism) in an attempt to lend a rationale to the

Table 8-9   SUMMARY OF KINETIC PARAMETERS FOR ETHANE HYDROGENOLYSIS OVER SILICA-SUPPORTED METALS †

Metal	Temper-ature range, °C	Reaction order With respect to ethane, n	With respect to H_2, m	Temper-ature at which reaction orders deter-mined, °C	Apparent-activa-tion energy E, kcal/mol	r_0' §
Fe	239–376	0.6	+0.5	270		
Co	219–259	1.0	−0.8	219	30	3.0×10^{25}
Ni	177–219	1.0	−2.4	177	40.6	4.9×10^{31}
Cu	288–330	1.0	−0.4	330	21.4	4.5×10^{20}
Ru	177–210	0.8	−1.3	188	32	1.3×10^{28}
Rh	190–224	0.8	−2.2	214	42	5.8×10^{31}
Pd	343–377	0.9	−2.5	354	58	3.7×10^{33}
Re	229–265	0.5	+0.3	250	31	1.8×10^{26}
Os	125–161	0.6	−1.2	152	35	7.0×10^{30}
Ir	177–210	0.7	−1.6	210	36	5.2×10^{28}
Pt	344–385	0.9	−2.5	357	54	5.9×10^{31}

† J. H. Sinfelt, *Catal. Rev.*, **3**: 175 (1970).
§ Preexponential factor, molecules/(s) (cm²), in the equation

$$r_0 = r_0' \exp(-E/RT)$$

observed kinetic model. Following Sinfelt, suppose these elementary steps to be:

Step 1: $\qquad C_2H_6 \underset{k_{-1}}{\overset{k_1}{\rightleftarrows}} (C_2H_5) + (H)$

Step 2: $\quad (C_2H_5) + (H) \overset{k_2}{\rightleftarrows} (C_2H_x) + aH_2$

Step 3: $\quad (C_2H_x) + H_2 \overset{k_3}{\longrightarrow} (CH_y) + (CH) \overset{H_2}{\longrightarrow} CH_4$

where parentheses signify the adsorbed state.

Now the stoichiometric coefficient a is $a = (6-x)/2$. Let it now be assumed that coverage of C_2H_x is governed by equilibriation of steps 1 and 2; hence

$$\theta_x = \frac{K(E)/(H_2)^a}{1 + K(E)/(H_2)^a} \qquad (8\text{-}151)$$

where $K = K_1 K_2$. As noted earlier, this hyperbolic form can be reexpressed by the simpler relationship

$$\theta_x = \left[\frac{K(E)}{(H_2)^a}\right]^n \qquad (8\text{-}152)$$

As step 3 is the rate-determining step,

$$r = k_3(H_2)\theta_x = k(E)^n(H_2)^{1-na} \qquad \text{where } k = k_3 K^n \qquad (8\text{-}153)$$

It will be recalled that the observed rate over diverse catalysts conforms to the expression

$$\mathscr{R} = k(E)^n(H_2)^m \qquad (8\text{-}154)$$

Table 8-10 KINETICS OF ETHANE HYDRO-
GENOLYSIS: COMPARISON OF
OBSERVED AND CALCULATED
EXPONENTS ON HYDROGEN
PRESSURE †

		Exponent on hydrogen	
Catalyst	a	Obs. m	Calc $1-na$
Fe	1	+0.5	+0.4
Co	2	−0.8	−1.0
Ni	3	−2.4	−2.0
Cu	1	−0.4	0.0
Ru	3	−1.3	−1.4
Rh	3	−2.2	−1.4
Pd	3	−2.5	−1.7
Re	1	+0.3	+0.5
Os	3	−1.2	−0.8
Ir	3	−1.6	−1.1
Pt	3	−2.5	−1.7

† J. H. Sinfelt, *Catal. Rev.*, 3: 175 (1970).

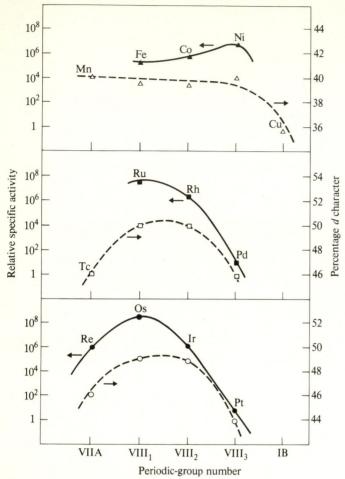

FIGURE 8-17

Catalytic activities of metals for ethane hydrogenolysis in relation to the percentage d character of the metallic bond. The closed points represent activities compared at a temperature of 205°C and ethane and hydrogen pressures of 0.030 and 0.20 atm, respectively, and the open points represent percentage d character. Three separate fields are shown in the figure to distinguish the metals in the different long periods of the periodic table. [*J. H. Sinfelt, Adv. Catal.*, **23**: 91 (1973).]

where the observed orders are listed in Table 8-9; hence $n = n$ and $m = 1 - na$, from which it follows that an assumption of a value for a for the observed, experimental value of n fixes $m = 1 - na$. Further, a is uniquely related to the nature of adsorbed C_2H_x by $a = (6 - x)/2$.

Logically a should assume integer values, 1, 2, etc. For each of the hydrogenolysis catalyst listed in Table 8-9 a value of a is selected such that with the observed order in ethane n, the order in hydrogen is predicted. These results are shown in Table 8-10, where observed (m) and predicted ($1 - na$) order in H_2 are contrasted. The agreement is satisfactory and constitutes a fine defense of the merits of intelligent use of kinetics in elucidating mechanism.

Catalytic Specificity and Its Correlation

As is indicated in Table 8-9, the specific rates of hydrogenolysis of ethane over diverse supported metals vary significantly. Sinfelt has established these rates at fixed values of H_2 and ethane partial pressure and a temperature of 205°C; his results are reproduced in Fig. 8-17, where the rate is shown for group VIII and 1B metals in the order of increasing atomic number. Figure 8-17 represents metals of the first, second, and third transition series, and the indicated rates are specific value, i.e., per unit surface area of supported metal as inferred by H_2 chemisorption.

Superimposed upon Fig. 8-17 is the value of percentage of *d* character for each catalyst. Basically percentage *d* character is a measure of the strength of the metallic bond. By inference, it is speculated that the higher the percentage of *d* character, the more effectively will gaseous species be chemisorbed as precursors of catalytic transformation. An apparent correlation exists between catalytic activity and percentage of *d* character. It is to be noted that metallic radius, a geometric factor, will be a function of *d* character.

Dispersion and Support Effects

Again we cite the results of Sinfelt and coworkers with respect to crystallite size and support-induced influences upon specific rate (Tables 8-11 and 8-12) and selectivity (Table 8-13). As the data of Table 8-11 indicate, the *specific* rate of ethane hydro-

Table 8-11 EFFECT OF SINTERING TEMPERATURE ON THE CRYSTALLITE SIZE AND SPECIFIC CATALYTIC ACTIVITY OF SUPPORTED NICKEL[a]
Catalyst composition 10% Ni on SiO_2-Al_2O_3

Sintering temperature,[b] °C	Crystallite size, Å	Specific catalytic activity for ethane hydrogenolysis[c]
370	29[d]	1070
450	. . .	500
500	57[e]	90
580	71[e]	53
700	88[e]	57

[a] J. L. Carter, J. A. Cusumano, and J. H. Sinfelt, *J. Phys. Chem.*, **70**: 2257(1966).
[b] The catalysts were sintered in flowing hydrogen.
[c] Micromoles of ethane converted to methane per hour per square meter of nickel; conditions 267°C, hydrogen pressure 0.20 atm, ethane pressure 0.030 atm.
[d] Determined from field dependence of magnetization.
[e] Determined by x-ray-diffraction line broadening.

genolysis decreases by a factor of 20 as crystallite size increases from 30 to 90 Å. Somewhat less demanding behavior has been observed for Ni supported on silica,[1] in contrast with the Si-Al$_2$O$_3$ support utilized in the study reported in Table 8-11. As Sinfelt observes, sintering of the Si-Al$_2$O$_3$ supported Ni occurred in H$_2$, that of the silica-supported Ni, in air; thus the sintering atmosphere (reducing or oxidizing) may well affect the average size, distribution, and nature of dispersed crystallites.

Table 8-12 ETHANE HYDROGENOLYSIS ACTIVITY OF RHODIUM IN RELATION TO ITS STATE OF DISPERSION†

State of dispersion	Crystallite size, Å	Specific activity§
Very low:		
Bulk Rh	2560	0.79
5% Rh-SiO$_2$ sintered	127	0.41
Intermediate:		
1–10% Rh on SiO$_2$	12–41	8–16
Very high:		
0.1–0.3% Rh on SiO$_2$	<12	4.4

† Millimoles ethane converted to methane per hour per square meter of rhodium. Conditions 253°C, hydrogen pressure 0.20 atm, ethane pressure 0.030 atm.
§ D. J. C. Yates and J. H. Sinfelt, *J. Catal.*, **8**: 348 (1967).

Table 8-13 SUPPORT EFFECT IN THE REACTION OF CYCLOPROPANE WITH HYDROGEN OVER NICKEL CATALYSTS^a

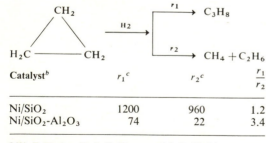

Catalyst^b	$r_1{}^c$	$r_2{}^c$	$\dfrac{r_1}{r_2}$
Ni/SiO$_2$	1200	960	1.2
Ni/SiO$_2$-Al$_2$O$_3$	74	22	3.4

^{a} W. F. Taylor, D. J. C. Yates, and J. H. Sinfelt, *J. Catal.*, **4**: 374 (1965).
^{b} Catalysts contained 10 percent nickel by weight.
^{c} Reaction rates, micromoles per hour, per square meter of Ni; conditions 42°C, cyclopropane and hydrogen pressures of 0.030 and 0.20 atm, respectively.

[1] D. J. C. Yates, W. F. Taylor, and J. H. Sinfelt, *J. Am. Chem. Soc.*, **86**: 2996 (1964).

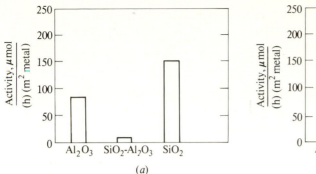

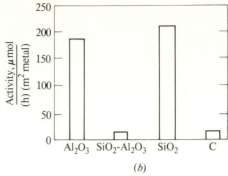

(a)

(b)

FIGURE 8-18
The effect of carrier on the specific catalytic activity of (a) Ni at 191°C and (b) Co at 233°C for ethane hydrogenolysis. Activities were measures at respective ethane and hydrogen partial pressures of 0.030 and 0.20 atm. The metal concentration was 10 percent by weight in all cases. [*J. H. Sinfelt, Catal. Rev.*, **3**: *175 (1970)*.]

Data set forth in Table 8-12 for ethane hydrogenolysis over bulk and dispersed (supported) rhodium reveals that the *specific* rate passes through a maximum with increasing dispersion. The influence of dispersed-metal support is revealed in Fig. 8-18 and Table 8-13. Note, in particular the selectivity variation with support (Table 8-13).

A comment on dispersion effects   Insofar as chemisorption is logically presumed to be an essential step for at least one reactant in surface-catalyzed events, the noted effects with respect to crystallite size and specific rate might well be profitably viewed in the light of Sabatier's teachings regarding catalysis. In essence, Sabatier argued that for a reactant-solid catalyst bond that is (1) very weak, (2) modest, or (3) very strong, we should witness poor catalysis in case 1, since adsorbed bond distortion is weak, and no catalysis in case 3, since irreversible reactant-solid compound formation will occur. In case 2, bond distortion is appropriate to maintain the catalytic cycle of adsorption-reaction-desorption. The Sabatier principle is nicely displayed by Boudart[1] for formic acid decomposition over diverse metals. The plot (Fig. 8-19), known as the volcano, consists of a correlation of the *reciprocal* of the temperature at which the reaction rate achieves a specific, common value for each metal versus the heat of formation of the metal formate (this heat is presumably proportional to the formic acid–metal chemisorptive bond strength).

While formic acid decomposition may not capture industrial attention, surely ethylene oxidation to generate ethylene oxide in preference to CO_2 and H_2O is an example of yield governed by the Sabatier principle. For it is generally agreed that ethylene oxide yield is governed by the surface concentration of the O_2^- species, while total combustion is catalyzed by O^-. Gold hardly chemisorbs O_2 in any form, platinum chemisorbs O_2 to form the atomic species, while only silver sig-

[1] M. Boudart, *Chem. Eng. Prog.*, **57**: 33 (1961).

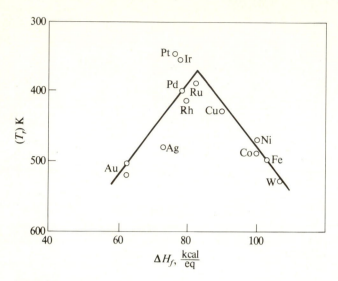

FIGURE 8-19
Isokinetic temperature of formic acid decomposition and heat of
formation of metal formate. [*M. Boudart, Chem. Eng. Prog.*, **57**:
33 (1961).]

nificantly accommodates the molecular O_2^- ion. Silver proves to be the unique
catalyst for the selective oxidation of ethylene to its oxides, gold is virtually inactive,
and platinum catalyzes the nearly total oxidation of ethylene to CO_2 and H_2O.

For a given metal catalyst, Sabatier's principle may well apply for demanding
(structure-sensitive) reactions. Quite small crystallites may chemisorb too strongly
and thus retard the catalytic cycle; large crystallites chemisorb only weakly, to the
detriment of specific rate, while an intermediate size may be optimum with respect
to specific activity, the differences in crystallite sizes reflecting differences in
morphology.

Supported Bimetallic Catalysts

For the bimetallic alloy Cu-Ni, Figs. 8-20 and 8-21 reveal the distinct difference
between surface and bulk alloy composition (Fig. 8-20) and the yield- or selectivity-
pattern alteration with alloying (Fig. 8-21). The alloy of Cu-Ni consisted of fine
unsupported granules. Since H_2 does not strongly chemisorb upon Cu but does
so on Ni, Fig. 8-20 indicates that with only a few percent Cu alloyed with Ni, H_2
chemisorption is drastically reduced, suggesting that the surface of the alloy is
largely populated with Cu in spite of its small overall, average presence.

For this same alloy, the activity for ethane hydrogenolysis and then cyclohexane dehydrogenation as set forth in Fig. 8-21, again suggests the dominant
presence of Cu on the alloy surface. Recall that the percentage of d character of

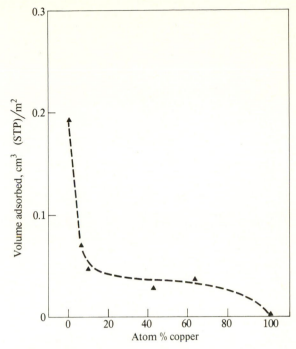

FIGURE 8-20
The chemisorption of hydrogen on Cu-Ni catalysts at room temperature as a function of composition. The chemisorbed hydrogen refers to the amount retained on the surface after 10-min evacuation at room temperature following completion of an adsorption isotherm. [*J. H. Sinfelt, J. L. Carter, and D. J. C. Yates, J. Catal.*, **24:** *283 (1972).*]

Cu is below that of Ni and that a correlation has been demonstrated to exist between *d* character and rate of hydrogenolysis (Fig. 8-17). As Sinfelt observes, carbon-carbon bond rupture involved in hydrogenolysis would be sensitive to electronic and geometric surface factors, while cyclohexane dehydrogenation is comparatively insensitive to such factors. Hence the marked effect of surface-concentrated Cu on selectivity which has been observed for other alloys of a group 1B metal with one of group VIII.

Such selectivity alterations have also been noted for bimetallics highly dispersed upon a support, surely permitting the inference that despite the high degree of dilution of both metals upon the support, alloying nevertheless is evident. Sinfelt terms these highly dispersed alloys *bimetallic clusters*, a term to be preferred since evidence of alloying of microcrystals exists for systems known to be *immiscible* at the bulk scale. Again a thought experiment may shed light on the matter. Consider the interface between two highly immiscible metals. From an overall bulk-compositional point of view, immiscibility may be evident; however, at the

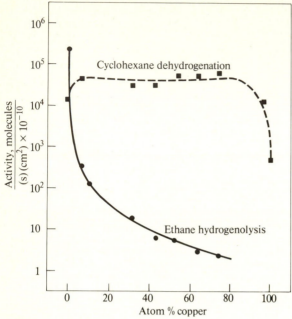

FIGURE 8-21

Activities of Cu-Ni alloy catalysts for the hydrogenolysis of
ethane to methane and the dehydrogenation of cyclohexane
to benzene. The activities refer to reaction rates at 316°C.
Ethane hydrogenolysis activities were obtained at ethane and
hydrogen pressures of 0.030 and 0.20 atm, respectively.
Cyclohexane dehydrogenation activities were obtained at
cyclohexane and hydrogen pressures of 0.17 and 0.83 atm,
respectively. [*J. H. Sinfelt, J. L. Carter, and D. J. C. Yates,
J. Catal.*, **24**: *283 (1972).*]

interface within the dimensional range of angstroms, miscibility prevails. If then,
we could shave away this narrow interfacial zone and disperse it upon a support,
miscible clusters would persist until sintering caused the clusters to grow to such size
that phase separation would occur.

The concept of bimetallic clusters is best illustrated by reference to the essen-
tials of the research findings of Sinfelt and colleagues.[1] The bimetallic systems
Cu-Ru and Cu-Os were highly dispersed upon silica. *Both systems exhibit very
low miscibility in the bulk.* The following studies were made to characterize each
bimetallic catalyst formulation.

Chemisorption of H_2 and CO From a knowledge of the weight of each group
VIII metal deposited, the hydrogen/metal and CO/metal ratios were determined

---

[1] J. H. Sinfelt, *Catal. Rev.*, **9**: 147 (1974).

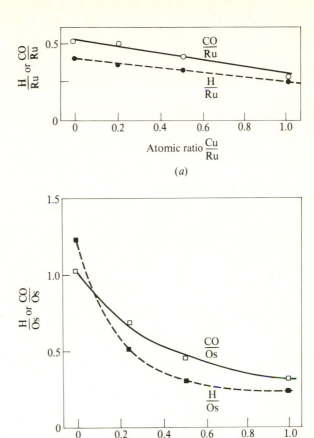

FIGURE 8-22
The chemisorption of hydrogen and carbon monoxide at room temperature on silica-supported (*a*) Ru-Cu catalysts and (*b*) Os-Cu catalysts.   The catalysts all contain 1 percent by weight Ru or Os with varying amounts of copper.   The adsorption data are expressed by the quantities H/Ru, CO/Ru, H/Os, and CO/Os, which represent the number of hydrogen atoms or carbon monoxide molecules chemisorbed per atom of Ru or Os in the catalyst.   [*J. H. Sinfelt, J. Catal.*, **29**: *308 (1973)*.]

as a function of the deposited Cu/metal ratio.   These results are shown in Fig. 8-22 for the Cu-Ru and Cu-Os systems.   Evidently for these otherwise highly immiscible systems, microcrystalline alloys or bimetallic clusters are formed, for if Ru and Os were independently dispersed, the hydrogen/metal and CO/metal ratios would be independent of the deposited Cu/metal ratio.   Recall that neither

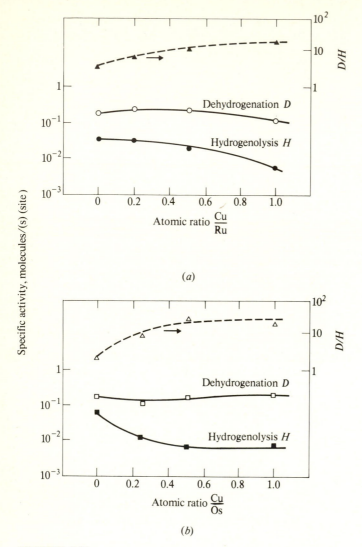

(a)

(b)

FIGURE 8-23

The specific activities of silica-supported Ru-Cu catalysts (a) and Os-Cu catalysts (b) for the dehydrogenation and hydrogenolysis of cyclohexane.   Activities are compared at 316°C and cyclohexane and hydrogen partial pressures of 0.17 and 0.83 atm, respectively.   The catalysts are the same as those in Figs. 8-22.   Specific activity is defined as in Fig. 8-21.   Selectivity, defined as the ratio of dehydrogenation activity D, to that of hydrogenolysis, H, is also shown as a function of catalyst composition.   [*J. H. Sinfelt, J. Catal., 29: 308 (1973).*]

H_2 nor CO is strongly chemisorbed on Cu. Hence by clustering, the number of Ru and Os surface atoms decreases by a factor of from 2 to 4 with increasing Cu content.

X-ray-diffraction scans These revealed no lines due to metals, indicating a high degree of dispersion of the clusters. Significantly, Cu dispersed in the absence of a cometal exhibits a diffraction pattern for 200-Å crystallites. So the interaction of Cu with Ru or Os arrests the creation of large Cu crystallites. (It might be suggested, therefore, that the creation of bimetallic clusters prevents or retards sintering.)

Catalytic reactions These reactions were conducted over both bimetallics for the conversion of cyclohexane, which in H_2 can be catalyzed via dehydrogenation and hydrogenolysis; i.e., schematically,

$$C_6H_{12} \longrightarrow \text{benzene} \qquad r_D \equiv D$$

$$C_6H_{12} \xrightarrow{H_2} CH_4 \cdots \qquad r_H \equiv H$$

The results, specific activity and selectivity, D/H, are displayed as a function of Cu/metal ratio for both bimetallic cluster systems in Fig. 8-23. Note that specific activities and selectivities are expressed on a logarithmic scale, so that the efficacy of bimetallic clusters in affecting selectivity is signal indeed.

The results of the research summarized above are of more than pure intellectual interest, for while knowledge of the existence of bimetallic clusters as found dispersed upon supports contributes to our fundamental understanding of the chemistry and physics of microcrystalline assemblies, potent practical consequences issue from such comprehension. In catalytic reforming (dehydrogenation) of saturated hydrocrabons, hydrocracking, e.g., hydrogenolysis, proves to be undesirable, as CH_4 is of little value in upgrading gasoline. Reformed cyclohydrocarbons enhance octane value. Evidently, the improvement in selectivity, r_D/r_H, found with bimetallics is then of signal commercial merit. Furthermore, deactivation by coking is a natural consequence of reforming conditions, and it is not inconceivable that a judicious choice of bimetallic or, indeed, polymetallic dispersed clusters may not only improve reaction selectivity and diminish sintering but also control coking. Some evidence in the realization of the latter exists in the case of the Chevron bimetallic Pt-Re-supported reforming catalyst, although it cannot be asserted at this time that that bimetallic network is a microscopic alloy (cluster). An ancillary benefit of Sinfelt's work was the development of a polymetallic cluster catalyst now utilized in plant-scale reformers.

8-19 FURTHER ILLUSTRATIONS OF CATALYTIC RATE MODELS AND MECHANISMS

We have attempted to convey the message that neat agreement between an assumed kinetic model and laboratory data does not prove a mechanism. Tentative mechanisms can nevertheless be inferred when kinetic data are supported by other inquiries. The key to mechanistic inference is the determination of the rate-controlling step in the catalytic sequence. The concept of stoichiometric number proves to be a powerful tool in this regard.

The Stoichiometric Number

Rarely does a chemical reaction, in particular a catalytic reaction, proceed as written, i.e.

$$A + B \underset{k_2}{\overset{k_1}{\rightleftharpoons}} C + D \qquad (8\text{-}155)$$

As has been suggested, (8-155) represents the net summation of a number of elementary steps, some of which are rapid, others (with luck, one) are relatively slow. The measured rate represents the slowest or rate-controlling step. An insight into this phenomenon is gained by considering the relationship between the ratio of *observed* rate coefficients and the equilibrium constant.

If, indeed, reaction (8-155) is elementary, i.e., occurs as written, then

$$\frac{k_1}{k_2} = K_{eq} \qquad (8\text{-}156)$$

When, however, the reaction is not elementary as written, (8-156) is not necessarily true.

Consider the general case of $R \rightleftharpoons P$. At equilibrium

$$K_{eq} = \frac{P_{eq}}{R_{eq}} \qquad (8\text{-}157)$$

Affinity is defined as the distance *from* equilibrium, or

$$\Delta G_0 = \Delta G^0 \text{ (equilibrium)} - \text{(nonequilibrium)}$$

or
$$\Delta G_0 = RT \ln K_{eq} - RT \ln \frac{P}{R} \qquad (8\text{-}158)$$

so that at equilibrium $P/R = P_{eq}/R_{eq}$ and $\Delta G_0 = 0$.

Let us suppose that a particular overall reaction can be dissected into several elementary steps, for example, $2A + B \rightleftharpoons P$, the steps being

Step 1: $\qquad A \rightleftharpoons A^*$

Step 2: $\qquad A^* + B \rightleftharpoons Y^* \qquad\qquad\qquad (8\text{-}159)$

Step 3: $\qquad A^* + Y^* \rightleftharpoons P$

To produce 1 mol of P, step 1 must occur twice while steps 2 and 3 must occur but once. Overall reaction affinity is in terms of the affinity of each elementary step $\Delta G_0 = 2\,\Delta G_1 + \Delta G_2 + \Delta G_3$ or, in general,

$$\Delta G_0 = \Sigma v_i\,\Delta G_i \qquad (8\text{-}160)$$

Now the net rate of reaction r is

$$r_{net} = \vec{r} - \overleftarrow{r} = \vec{r}\left(1 - \exp\frac{\Delta G_r}{RT}\right) \qquad (8\text{-}161)$$

where ΔG_r is the affinity of the rate-controlling step in a sequence of several elementary steps.

Suppose now that all but one step in a several-step sequence are in equilibrium; then in view of (8-160)

$$\Delta G_0 = v_r \Delta G_r \qquad (8\text{-}162)$$

and (8-161) becomes

$$r_{net} = \vec{r} - \overleftarrow{r} = \vec{r}\left(1 - \exp\frac{\Delta G_0}{v_r RT}\right) \qquad (8\text{-}163)$$

By (8-158)

$$r_{net} = \vec{r}\left[1 - \left(\frac{P}{K_{eq} R}\right)^{1/v_r}\right] \qquad (8\text{-}164)$$

or

$$\frac{k_1}{k_{-1}} = K_{eq}^{1/v_r} \qquad (8\text{-}165)$$

since k_1/k_{-1} must be independent of P and R and

$$\left(\frac{P}{K_{eq} R}\right)^{1/v_r} = \frac{\overleftarrow{r}}{\vec{r}} = \frac{k_{-1}}{k_1}\frac{P^p}{R^r}$$

Note that equilibrium exponents, which are equal to stoichiometric coefficients, are related to observed orders of the rate-determining step by

$$\frac{P^b}{R^a} = \left(\frac{P^p}{R^r}\right)^{v_r} \qquad (8\text{-}166)$$

for the overall reaction $aR \to bP$. When the reaction is sufficiently close to equilibrium, (8-161) becomes

$$r_{net} = \vec{r}\frac{\Delta G_r}{RT} = r_{eq}\frac{\Delta G_r}{RT} \qquad (8\text{-}167)$$

since $\vec{r} = \overleftarrow{r} = r_{eq}$, where r_{eq} is the forward rate at equilibrium. In view of (8-162)

$$r_{net} = \frac{r_{eq}}{v_r RT}\Delta G_0 \qquad (8\text{-}168)$$

Therefore a measurement of r_{net} close to equilibrium, concentrations close to and at equilibrium (ΔG_0), and the forward rate at equilibrium establishes v_r, the stoichiometric number. The rate at equilibrium can be established at equilibrium by the use of radioactively tagged molecules. We shall illustrate this manner of analysis for NH_3 synthesis over promoted-iron catalyst.

It should be noted, however, that the analysis set forth above presumes that there is but one rate-determining step in the mechanistic sequence. Yet it is not inconceivable that two or more steps within a detailed reaction sequence may be of comparable rate. Furthermore, whereas each elementary step is characterized by a unique activation energy, except in the rather rare circumstance in which activation energies of all elementary steps are equal, one can anticipate the probability that the actual rate-determining step may well change with temperature and, incidentally, total pressure.

Kinetics and Mechanism of NH_3 Synthesis

While in principle a kinetic model is derivable a priori from a detailed and verified mechanistic understanding, in fact kinetic models invariably are rooted in empirical observation and correlation. In NH_3 synthesis from its elements $N_2 + 3H_2 \rightleftharpoons 2NH_3$ the Temkin-Pyzhev equation adequately describes the net rate of synthesis over promoted-iron catalyst

$$r_{net} = k_1(N_2)\left(\frac{H_2^3}{NH_3^2}\right)^a - k_{-1}\left(\frac{NH_3^2}{H_2^3}\right)^b \qquad (8\text{-}169)$$

where a and b are empirically determined exponents, each varying with reaction conditions of temperature, pressure, and conversion. Surely it is evident that (8-169) does not describe the initial rate at zero partial pressure of NH_3.

A rationalization of (8-169) was naturally sought. As has been noted above, the first step in the rationalization process is that of postulating a mechanism or sequence of steps:

Step 1: $N_2 + 2S \rightleftharpoons 2NS$ N_2 chemisorption

Step 2: $3H_2 + 6S \rightleftharpoons 6HS$ H_2 chemisorption   (8-170)

Step 3: $2NS + 6HS \rightleftharpoons 2NH_3 + 8S$ surface reaction

which, as Boudart notes,[1] can be written

$$N_2 + 2S \rightleftharpoons 2NS$$

$$2NS + 3H_2 \rightleftharpoons 2NH_3 + 2S$$

Sequence (8-170) does not represent elementary steps, except that of N_2 chemisorption. Step 2, for example, is surely the sum of three elementary steps, while step 3 can be dissected. It does indicate the three possibilities of rate control.

Both NH_3 synthesis and N_2 chemisorption rates were measured by Emmett and Brunauer,[2] who found the N_2 chemisorption rate to be slow and of the same order of magnitude as that of synthesis. More precise comparison was not possible then, as the extent of N_2 coverage during synthesis was not known. However, the data suggested that step 1 is rate-controlling.

[1] M. Boudart, *AIChE J.*, **18**: 465 (1973).
[2] P. H. Emmett and S. Brunauer, *J. Am. Chem. Soc.*, **62**: 1732 (1940).

If indeed that is the case, in terms of the Elovich equation and Temkin isotherm

$$r_{ads} = k_{ads}(N_2) \exp(-g\theta) \qquad (8\text{-}171)$$

$$r_{des} = k_{des} \exp h\theta \qquad (8\text{-}172)$$

and

$$\theta = \frac{2}{f} \ln [KNS] \qquad (8\text{-}173)$$

where

$$NS = K_e^{-1} \left(\frac{NH_3}{H_3^{3/2}} \right) \qquad (8\text{-}174)$$

since steps 2 and 3 are at equilibrium.

When we substitute, the net rate of synthesis becomes

$$r_{net} = r_{ads} - r_{des} = k_1(N_2) \left(\frac{H_2^3}{NH_3^2} \right)^{g/f} - k_{-1} \left(\frac{NH_3^2}{H_2^3} \right)^{h/f} \qquad (8\text{-}175)$$

which is to be compared with (8-169). The significance of the exponents a and b becomes clear and their variation rationalized since the Temkin isotherm is but an approximation to the Brunauer-Love-Keenan isotherm.

Following Ozaki, Taylor, and Boudart,[1] we see that the Langmuir formulation for N_2 adsorption rate control upon two sites is

$$r_{net} = k_{ads} N_2 (1 - \theta)^2 - k_{des} \theta^2 \qquad (8\text{-}176)$$

where

$$\theta = \frac{K_1 NS}{1 + K_1 NS} \qquad (8\text{-}177)$$

Substituting (8-174) into (8-177) and the resulting θ function into (8-176) gives

$$r_{net} = \frac{k_{ads}(N_2) - k_{des} K_0(NH_3^2/H_2^3)}{[1 + K_0(NH_3/H_2^{3/2})]^2} \qquad (8\text{-}178)$$

which is adequately rephrased in power-law form

$$r_{net} = \bar{k}_1(N_2) \left(\frac{H_2^{3/2}}{NH_3} \right)^{2m} - \bar{k}_{-1} \left(\frac{NH_3}{H_2^{3/2}} \right)^{2n} \qquad (8\text{-}179)$$

where, of course, m and n are empirically specified and are equivalent to a and b in (8-169). Thus the usefulness of the LHHW formulation is demonstrated. As noted earlier, this mode of correlation even anticipates poisoning processes, as Boudart[2] illustrates in his treatment of water-vapor poisoning of the iron catalyst during NH_3 synthesis. The issue of mechanism will now be discussed (with caution) for those few catalytic processes which admit to such speculation.

[1] A. Ozaki, H. S. Taylor, and M. Boudart, *Proc. R. Soc.*, A258: 47 (1960).
[2] M. Boudart, *AIChE J.*, 2: 62 (1956).

Mechanism In very careful determinations, Mars, Scholten, and Zwietering[1] measured the kinetics of N_2 chemisorption and synthesis during which the weight of N_2 adsorbed was recorded gravimetrically. The N_2 adsorption velocity at the coverage corresponding to steady-state synthesis did indeed equal the rate of synthesis. (H_2, while adsorbed, contributes negligibly to the total weight of adsorbed species.)

Synthesis-rate relations were developed in terms of surface coverage, where the velocity of adsorption and desorption is expressed in terms of surface occupancy by N_2. The data indicate that the activation entropies are functions of θ, as is the activation energy. Additional evidence for N_2 adsorption control was secured by a determination of the stoichiometric number.

In NH_3 synthesis, an isotope-exchange experiment can be carried out to determine r_{eq} providing the exchange reaction passes exclusively through the same path and has the same rate-determining step as that of synthesis. Mars et al. measured the exchange rate for

$$^{30}N_2 + {}^{14}NH_3 \underset{\longrightarrow}{\overset{cat}{\longleftarrow}} {}^{29}N_2 + {}^{15}NH_3$$

and

$$^{30}N_2 + {}^{28}N_2 \longrightarrow 2{}^{29}N_2$$

(8-180)

and found r_{eq}. Rate measurements close to equilibrium were also made, and the stoichiometric number was found to be unity ($v_r = 1$), suggesting N_2 adsorption as rate-controlling. In contrast, Horiuti[2] determined $v_r = 2$, suggestive of another step as rate-controlling. This discrepancy has yet to be resolved.

More recent studies reported by Ozaki, Taylor, and Boudart,[3] while lending further support to the contention of N_2 absorption as rate-determining, reveal, by isotope experiments, the possibility that the surface of the double-promoted-Fe catalyst is covered with NH rather than N atoms. Ozaki's experiments covered a 300-fold range of conversion over a pressure range of from $\frac{1}{3}$ to 1 atm (218 to 302°C). As noted above, it is possible that the nature of a rate-determining step changes with operating conditions (pressure and temperature) and indeed two steps in the sequence may proceed at comparable rates at a particular operating condition. Furthermore, the role of the promoters in altering both the nature of the adsorbed species and the relative velocities of various surface steps is obscure. It is clear, in view of the fact that the velocity constant for synthesis is the same for H_2 and D_2, that over the wide range of conditions explored by Ozaki et al. a step involving H_2 cannot be rate-controlling. Otherwise the isotope effect would alter the rate constant as one experimented with N_2-H_2 and then N_2-D_2 feed streams.

Kinetics and Mechanism of SO_2 Oxidation

Commercially SO_2 is oxidized by vanadium pentoxides or supported platinum. Generally the platinum catalyst is active at a lower temperature than the semi-

[1] P. Mars, J. J. F. Scholten, and P. Zwietering, "The Mechanism of Heterogeneous Catalysis," p. 66, Elsevier, Amsterdam, 1960.
[2] J. Horiuti, S. Enomoto, and H. Kobayasi, *Hokkaido Univ., J. Res. Inst. Catal.*, **3**: 185 (1955).
[3] Loc cit.

conducting V_2O_5; however, the latter proves less sensitive to poisoning. The reaction clearly involves electron exchange between catalyst and adsorbed species. The conducting catalyst (Pt) will consequently readily chemisorb oxygen, probably covalently, while the n-type semiconductor, V_2O_5, presumably adsorbs oxygen less readily, electrons passing from the catalyst to the adsorbed species. The p-type semiconductors extract electrons *from* the adsorbed phase. V_2O_5 can be promoted by incorporation of foreign compounds which increase the electron density. These promoted V_2O_5 catalysts may actually exist as melts upon the support at reaction temperature (500 to 600°C). On the semiconductor the mechanism may involve

Step 1: $\quad \frac{1}{2}O_2 + 2e \longrightarrow O^{-2}$

Step 2: $\quad SO_2 + 2e \longrightarrow SO_2^{-2}$ $\qquad\qquad$ (8-181)

Step 3: $\quad SO_2^{-2} + O^{-2} \longrightarrow SO_3 + 4e$

the electrons e passing from the catalyst. In terms of the stoichiometric number, if oxygen adsorption is rate-determining, $v_r = 1$ and the surface equilibrium prevails $SO_2 + O_a \rightleftharpoons SO_3$. We can fashion a rate equation in terms of the Langmuir isotherm,

$$\theta = \frac{K(SO_3/SO_2)}{1 + K(SO_3/SO_2)} \qquad (8\text{-}182)$$

The forward rate is the adsorption rate

$$r_{ads} = k(O_2)(1 - \theta)^2 \qquad \text{for two sites}$$

or $\qquad \vec{r} = k(O_2)\left[\dfrac{1}{1 + K(SO_3/SO_2)}\right]^2 = \bar{k}(O_2)\left(\dfrac{SO_2}{SO_3}\right)^{2m} \qquad (8\text{-}183)$

But for the reversible reaction, where $v_r = 1$ and

$$\Delta G_0 = RT \ln \frac{K_{eq}}{(SO_3)^2/(SO_2)^2(O_2)}$$

we have $\qquad r = \vec{r}\left[1 - \dfrac{SO_3^{\;2}}{K_{eq}(SO_2)^2(O_2)}\right]$

Hence $\qquad r = k_1(O_2)\left(\dfrac{SO_2}{SO_3}\right)^{2m} - k_2\left(\dfrac{SO_3}{SO_2}\right)^{2(1-m)} \qquad (8\text{-}184)$

which nicely fits experimental data for $m = 0.4$.[†]

On platinum the rate-determining step is thought to be reaction of gaseous SO_2 with adsorbed atomic oxygen

Step 1: $\quad O_2 + 2S \longrightarrow 2OS$

Step 2: $\quad SO_2 + OS \longrightarrow SO_3S$ $\qquad\qquad$ (8-185)

Step 3: $\quad SO_3S \longrightarrow SO_3 + S$

[†] M. Boudart, *Ind. Chim. Belg.*, **23**: 383 (1958).

For the overall reaction $2SO_2 + O_2 \rightarrow 2SO_3$. If then step 2 is rate-controlling, $v_r = 2$ and

$$r = k(SO_2)\theta_0$$

where

$$\theta_0 = \frac{K_1\sqrt{O_2}}{1 + K_1\sqrt{O_2} + K_2(SO_3)} \approx K\left(\frac{\sqrt{O_2}}{SO_3}\right)^m \qquad (8\text{-}186)$$

Then

$$\vec{r} = k\frac{SO_2}{(SO_3)^m}(O_2)^{m/2}$$

and

$$r = k_1\frac{SO_2}{(SO_3)^m}(O_2)^{m/2} - \frac{k_1}{K_{eq}^{1/2}}\frac{(SO_3)^{1-m}}{(O_2)^{(1-m)/2}} \qquad (8\text{-}187)$$

Note that $k_1/k_2 = K_{eq}^{1/2}$. For Pt, where $m = \frac{1}{2}$,

$$r = k_1(O_2)^{0.25}\frac{SO_2}{\sqrt{SO_3}} - k_2\left(\frac{\sqrt{SO_3}}{O_2}\right)^{0.25} \qquad (8\text{-}188)$$

which agrees nicely with experimental data. Again while agreement between data and a rate model does not prove the validity of the mechanism supposed, in the case of SO_2 oxidation auxiliary data tend to support the assumed mechanism.

Not surprisingly, oxidation of organic compounds proceeds readily over V_2O_5 and other *n*-type semiconductors. Dixon and Longfield[1] suggest that in some cases the organic reduces the V_2O_5, which is then reoxidized by the ambient O_2:

$$\begin{aligned} R + O^{2-} &\longrightarrow P + 2e \\ \tfrac{1}{2}O_2 + 2e &\longrightarrow O^{2-} \end{aligned} \qquad (8\text{-}189)$$

o-Xylene oxidation to phthalic anhydride would appear to be controlled by O_2 chemisorption, the solid donating electrons.

$$\begin{aligned} \tfrac{1}{2}O_2 + e &\longrightarrow O^- \quad \text{slow} \\ o\text{-Xylene} + O^- &\longrightarrow P + e \end{aligned} \qquad (8\text{-}190)$$

Organic-compound oxidations are complex in many ways, as a number of products generally result, due in part to surface catalysis per se, but also due to heat and mass-diffusion effects which often accompany such exothermic reactions.

Dual-Function Catalysts

If a metal, say Pt, is deposited upon a silica-alumina cracking catalyst, a dual-functioning catalyst results in that Pt sites are capable of hydrogenation-dehydrogenation catalysis while the acid sites of the alumina-silica support can function to foster reactions involving the carbonium ion.

Consider isomerization. With an olefin the acid sites easily catalyze direct

---

[1] J. K. Dixon and J. E. Longfield, in P. H. Emmett (ed.), "Catalysis," vol. 7, Reinhold, New York, 1960.

isomerization, the carbonium ion being formed at the double bond. Direct iso-merization of saturated hydrocarbons is not readily realized with an acidic catalyst. However, the incorporation of small quantities of Pt into the acidic oxide causes isomerization, presumably via the steps

$$n\text{-Saturate} \xrightleftharpoons{\text{Pt}} n\text{-olefin} + H_2 \qquad \text{fast}$$

$$n\text{-Olefin} \xrightleftharpoons{\text{acid}} \text{isoolefin} \qquad \text{slow}$$

$$H_2 + \text{isoolefin} \xrightleftharpoons{\text{Pt}} \text{isosaturate} \qquad \text{fast}$$

In general it can be realized that the distinctly separate metal sites in intimate contact with acid sites serve to provide the acid site with a reactant, e.g., n-olefin, and rapidly remove the product of olefin isomerization in spite of its low partial pressure dictated by equilibrium. The role of diffusion and/or intermediate migration between acid and metal site, and therefore the necessity of intimate contact of the two catalyst types, was nicely demonstrated by Weisz and Swegler,[1] who compared rates of n-hexane isomerization on a Pt-impregnated silica-alumina (high degree of acid-metal intimacy) with physical mixtures of Pt and silica (negligible acidity) and silica-alumina, each of various particle size. As the particle size of the physical mixture was decreased from 1000 μm (10 percent conversion to isohexane) down to several micrometers, conversion approached that of the impregnated Pt/Si-Al particles (40 to 50 percent). These clever experiments and the results clearly demonstrate (1) the existence of intermediate olefins in the gas phase, (2) the fact that metal and acidic sites act as physically independent catalysts, and (3) that gas-phase diffusion of intermediates plays a role in sustaining the consecutive reaction scheme. Weisz and Swegler also measured the n-hexane concentration after exposure of n-heptane to Pt and found it to be 3×10^{-2} atm, in good agreement with attainable thermodynamic equilibrium.

Molecular-Sieve Catalysts

Crystalline alumina-silicates, termed *molecular sieves*, are solids consisting of microcrystals found in nature (zeolites) but also created synthetically. Upon dehydration, a significant intracrystalline volume is formed of uniform chambers interconnected by ports of uniform size. In consequence, the intracrystalline volume is accessible only to molecules which are of a shape and size commensurate with the port or sieve-hole size. If catalytic centers are available within the cages (intracrystalline volume), it is clear that by tailoring port size, certain molecules will be denied entrance while others will pass into the cages to be catalyzed. The original synthetic sieve (Linde sieves) were created and developed as selective adsorbents. Later Weisz and Frilette[2] revealed unexpected intrinsic catalytic activity and shape-selective properties of some synthetic zeolite salts. In the first detailed

[1] P. B. Weisz and E. Swegler, *Science*, **126**: 31 (1957).
[2] P. B. Weisz, V. J. Frilette, R. W. Maatman, and E. B. Mower, *J. Phys. Chem.*, **64**: 382 (1960); *J. Catal.*, **1**: 301, 307 (1962).

report the cracking activity of synthetic zeolites (sodium and calcium X zeolites) is disclosed in a case where port size is large enough (10 Å) for the sieve effect to be negligible. The NaX crystals exhibit high cracking activity for paraffin and olefin hydrocarbons, and the products are virtually free of branched-chain hydrocarbons. However, NaX is ineffective in dealkylating cumene. By ion exchange, the CaX crystal is prepared and the catalytic properties are significantly altered. Indeed CaX approaches conventional silica-alumina cracking-catalyst behavior, including dealkylating cumene. These results suggest that the catalytic sites are located in the interstitial spaces of dehydrated crystals. Changing the mobile cation from Ca to Na in the crystal alters its role from that of an acid catalyst to one which involves a radical mechanism not unlike that successfully postulated in thermal cracking.

In a second report[1] studies with smaller port sizes (about 5 Å) revealed the selectivity to be found when a port or sieve opening is tailored to exclude certain reactants and/or products. For example, normal and isobutyl alcohols were catalytically dehydrated over, first, a non-shape-selective (10 Å) and then a shape-selective zeolite (5 Å). Over 10 Å both alcohols are rapidly dehydrated at 230 to 260°C, the isobutyl alcohol being somewhat more reactive. With 5 Å, however, the isobutyl alcohol is hardly converted because of its inability to gain entrance through the small 5 Å sieve ports which do admit n-butanol. Weisz et al. also induced hydrogenation activity in a shape-selective zeolite by incorporating Pt within the cages. 1-Butene, isobutene, and propylene were then subjected to hydrogenation over this Pt-containing molecular sieve, with the result that 70 percent of 1-butene, less than 2 percent isobutene, and over 50 percent propylene were converted. In a feed mixture of equimolar amounts, propylene is hydrogenated exclusively, the isobutene being effectively screened from the hydrogenation sites within the sieve cage.

As the three olefins have comparable reactivities for hydrogenation, the selectivity evident in these studies can be attributed to molecular size and shape. Crystalline alumina-silicate catalysts of both the shape-selective and non-shape-selective structure open a fruitful avenue to catalytic investigators. An excellent review is provided by Weisz.[2]

SUMMARY

The fascinating subject of heterogeneous catalysis and catalytic kinetics has only been touched upon in this chapter; however, this introductory exposition should convey an insight into the nature of catalysis, catalyst components, modes of preparation, and physicohemical characterization, as well as the unique problem associated with data procurement and kinetic-model development free of diffusional disguises.

[1] Ibid.
[2] P. B. Weisz, *Chem. Technol.*, **3**: 498 (1973).

ADDITIONAL REFERENCES

Signal insights into kinetics and catalysis are set forth in M. Boudart, "Kinetics of Chemical Processes," Prentice-Hall, Englewood Cliffs, N.J., 1968. Other instructive treatises in catalysis, per se, are P. Ashmore, "Catalysis and Inhibition of Chemical Reactions," Butterworth, London, 1963; *Advances in Catalysis and Related Subjects*, an annual volume, Academic, New York, 1950 to date; G. Bond, "Catalysis by Metals," Academic, New York, 1962; G. Bond, "Heterogeneous Catalysis: Principles and Applications," Clarendon, Oxford, 1974; A. Clark, "Theory of Adsorption and Catalysis," Academic, New York, 1970; P. H. Emmett (ed.), "Catalysis," 7 vols., Reinhold, New York, 1954–1960; C. L. Thomas, "Catalytic Processes and Proven Catalysts," Academic, New York, 1970; and J. Thomas and W. Thomas, "Introduction to the Principles of Heterogeneous Catalysis," Academic, New York, 1967. An insight into Soviet contributions to catalysis is provided by A. A. Balandin et al., "Catalysis and Chemical Kinetics," Academic, New York, 1964.

S. J. Gregg and K. S. W. Sing, "Adsorption: Surface Area and Porosity," Academic, New York, 1967, is an excellent compendium of theory and data. See also D. O. Hayward and B. Trapnell, "Chemisorption," Butterworth, London, 1964.

An extremely valuable reference is R. B. Anderson (ed.), "Experimental Methods in Catalytic Research," Academic, New York, 1968, while the chemistry of catalytic processes is nicely outlined in B. Gates, J. Katzer, J. H. Olson, and G. Schuit, *Chem. Eng. Educ.*, Fall **1974**: 172. A rich and fascinating treatment of catalysis from a historical point of view is that of A. J. B. Robertson, "Catalysis of Gas Reactions," Logos, London, 1970.

The preparation by ion exchange and resulting catalytic activity of platinum on an yttrium zeolite is described by R. Dalla Betta and M. Boudart, *5th Int. Congr. Catal.*, North Holland, Amsterdam, 1972. Catalysis by molten salts is discussed by C. N. Kenny, *4th Int. Congr. Catal., Moscow, 1968*. In the same proceedings, see the contribution of D. A. Dowden et al. on the design of complex catalysts and the work of I. Fux and I. Ioffe on model-building techniques for the estimation of constants from complex catalytic-reaction data. Engineering aspects of catalysis are neatly set forth in a seminal paper by O. A. Hougen, *Ind. Eng. Chem.*, **53**(7); 509 (1961), which should be required reading for all students.

The matters of supported-crystallite size, coordination numbers, and structure are discussed in English by O. Poltorak and V. Boronin, *Russ. J. Phys. Chem.*, **40**: 1436 (1966). Isomerization-hydrogenolysis selectivity effects of platinum catalysts are reported by M. Boudart, A. Aldag, L. Ptak, and J. Benson, *J. Catal.*, **11**: 25 (1968).

Surface science and catalysis are discussed by M. Boudart, *Chem. Technol.*, December **1974**: 748. See also his Catalysis by Supported Metals, *Adv. Catal.*, **20**: 153 (1969), and Two Step Catalytic Reactions, *AIChE J.*, **18**: 465 (1972). The telling influence of ionizing radiation upon catalytic selectivity is summarized by J. J. Carberry and G. C. Kuczynski, *Chem. Technol.*, April **1973**: 237.

Hydrocarbon oxidation catalysis is discussed in terms of a surface selectivity factor by J. Callahan and R. K. Grasselli, *AIChE J.*, **9**: 755 (1963), while provocative details of olefin oxidation are to be found in a series of papers by W. K. Hall and colleagues, *J. Catal.*, **16**: 204, 220 (1970), **22**: 310 (1971).

The merits of a CSTCR are nicely illustrated in the studies of hydrogenolysis in D. Tajbl, *Can. J. Chem. Eng.*, **47**: 154 (1969); *Ind. Eng. Chem.*, (PDD) **8**: 364 (1969). The utilization of the microcatalytic reactor (pulse dosing) pioneered by R. Kokes, J. Tobin, and P. H. Emmett, *J. Am. Chem. Soc.*, **77**: 5860 (1955), is discussed by J. Hightower, *Can. J. Chem. Eng.*, **48**: 151 (1970).

As a guide to further, more advanced treatments of diverse aspects of heterogeneous catalysis, the contents of 10 volumes of *Catalysis Reviews* are itemized below.

Volume 1

Zeolites as Catalysts, I. J. Turkevich
Reactions Catalyzed by Pentacyanocobaltate(II), J. Kwiatek
Reactions of Unsaturated Ligands in Pd(II) Complexes, E. W. Stern
Application of Computers in Chemical Reaction Systems, S. S. Grover
Electronic Surface States of Ionic Lattices, P. Mark
Reflectance Spectroscopy as a Tool for Investigating Dispersed Solids and Their Surfaces, K. Klier
Importance of the Electric Properties of Supports in the Carrier Effect, F. Solymosi
Static Volumetric Methods for Determination of Adsorbed Amount of Gases on Clean Solid
    Surfaces, Z. Knor

Volume 2

The Band Picture in the Electronic Theories of Chemisorption on Semiconductors, F. Garcia-
    Molinar
Kinetics Analysis by Digital Parameter Estimation, Y. Bard and L. Lapidus
The Surface Properties of Zeolites as Studied by Infrared Spectroscopy, D. Yates
Application of Mössbauer Spectroscopy to the Study of Adsorption and Catalysis, W. Delgass
    and M. Boudart
Corrosion of Platinum Metals and Chemisorption, J. Llopis
Inhibition Processes in the Gas Phase, Z. Szabó
Determination of Heat of Adsorption on Clean Solid Surfaces, S. Černý and V. Ponec
Prediction of Catalytic Action as Presented in Papers before the Fourth International Congress
    on Catalysis, V. Kazansky
United States–Japan Conference on Catalysis, J. Turkevich

Volume 3

Electrochemical Methods for Investigating Catalysis by Semiconductors, T. Freund and W. Gomes
Olefin Disproportionation, G. Bailey
Heat and Mass Diffusional Instrusions in Catalytic Reactor Behavior, J. Carberry
The Chemical Adsorption of Sulfur on Metals: Thermodynamics and Structure, J. Bénard
Fixation of Molecular Nitrogen, K. Kuchynka
Olefin Polymerization on Supported Chromium Oxide Catalysis, A. Clark
Catalytic Hydrogenolysis over Supported Metals, J. Sinfelt
Chemical Relaxation of Surface Reactions, G. Parravano
Identification of Rate Models for Solid Catalyzed Gaseous Reactions, R. Mezaki and J. Happel

Volume 4

Review of Ammonia Catalysis, A. Nielsen
The Mechanism of the Catalytic Oxidation of Some Organic Molecules, W. M. H. Sachtler
Equilibrium Oxygen Transfer at Metal Oxide Surfaces, G. Parravano
Isotopic Exchange of Oxygen ^{18}O between the Gaseous Phase and Oxide Catalysts, J. Nováková
The Use of Molecular Beams in the Study of Catalytic Surfaces, R. P. Merrill
Heterogeneous Catalysis by Electron Donor-Acceptor Complexes of Alkali Metals, K. Tamaru
X-ray Photoelectron Spectroscopy: A Tool for Research in Catalysis, W. N. Delgass, T. R.
    Hughes, and C. S. Fadley
Electrocatalysis and Fuel Cells, A. J. Appleby
Hydrodesulfurization, S. C. Schuman and H. Shalit

PROBLEMS

8-1 Dash off to your library and secure a list of 25 gas-solid, liquid-solid, gas-liquid-solid catalyzed reactions of current importance.

8-2 In the light of transition-state theory, sketch the energy–extent of reaction profile for (a) a homogeneously and (b) heterogeneously catalyzed reaction.

8-3 In the oxidation mechanism postulated for SO_2 resketch the energy-vs.-extent-of-reaction diagram to include the intermediate (complexing) steps.

8-4 It was stated in Sec. 8-3 that high-area supports are characterized by small pore diameters. Justify that assertion by assuming a flat-plate pellet of 3 mm thickness containing pores of (a) 1000 Å and (b) 10 Å. Assume the pores to be cylindrical of uniform diameter and of length equal to that of the pellet. Compare the external pellet area to that of the total (internal plus external) area. A pore-volume fraction of 0.5 can be reasonably assumed and a unit volume of 1 cm³.

8-5 Catalysts are prepared by (a) impregnation, (b) coprecipitation, and (c) vapor deposition. Cite examples of each. Further, (d) ion exchange can be used in the preparation of a catalyst. Find at least one example.

8-6 NH_3 is oxidized to NO at 7 atm and about 900°C over a Pt-Rh-alloy wire gauze. In Chap. 5, it was noted that that process is virtually bulk-mass-transport–determined. Compute the specific rate of NH_3 oxidation, i.e., the turnover number, molecules of NH_3 reacted per square meter of Pt per second, for (a) 30 layers of 80-mesh Pt gauze 1 m² in total cross section and (b) the same gauze which by reason of surface reconstruction and Pt vaporization exhibits a 100 percent increase in Pt-Rh external area. (Refer to Prob. 5-12, for specific details.)

8-7 For the roughened catalytic wire cited in part (b) of Prob. 8-6, how can its total area be determined? Given that datum in square centimeters per cubic centimeter and the controlling mass-transport coefficient, we secure a turnover number. Can we assume that all surface Pt-Rh atoms are active sites? In effect, for a nonporous catalytic surface such as is found in the Pt-Rh wire gauze, is the area inferred by physisorption the same as that which might be inferred by chemisorption? Discuss this issue.

8-8 Given the following data for physical adsorption of N_2 on an alumina at $-195°C$, compute the total (BET) surface area. Note that the BET area, in square meters per gram, is $4.4V_m$ for N_2.

$\dfrac{P}{P_0}$	V, cm³/g at NTP
0	0
0.01	75
0.08	100
0.15	120
0.3	140
0.45	160

8-9 Assuming that in the simple Langmuir isotherm $\theta = V/V_m$, where V_m is monolayer coverage, analyze the data given in Prob. 8-8 in accord with the Langmuir isotherm and estimate V_m. Compare the area so obtained with that found in response to Prob. 8-8.

8-10 For the system cited in Prob. 8-8, data for adsorption and desorption are:

$\dfrac{P}{P_0}$	V, cm³/g at NTP	
	Adsorption	Desorption
0.6	190	200
0.72	235	260
0.78	260	305
0.8	280	375
0.86	360	545
0.88	440	580
0.92	505	620
1	700	

From these desorption data *estimate* the average pore size by applying the Kelvin equation. Assuming a pore volume of 0.4 cm³/g, compute the total surface area utilizing the simple cylindrical-pore model and contrast that result with the BET area determined in Prob. 8-8.

8-11 A simple tube contains 10 g of a porous NH_3 synthesis catalyst. Hg is admitted at 1 atm, and the volume required to bathe the bed of catalyst is 4 cm³, the unpacked volume being 10 cm³. The Hg is removed and He gas is admitted to the evacuated bed. The volume of He admitted at NTP is 8.2 cm³. Compute the void fraction of the bed, the pore volume per gram of the catalyst, the catalyst-pellet density, and the intrinsic density of the iron-crystallite structure of the catalyst.

8-12 The quite simple catalytic reaction $A \rightarrow B$ is catalyzed by a catalyst composed of the remaining elements of the alphabetic periodic table. Utilize arbitrary but meaningful values of the rate and equilibrium coefficients to fashion plots of rate vs. partial pressure of A assuming:

(*a*) Adsorption of A

(*b*) Surface reaction

(*c*) Desorption of B to be rate-controlling

8-13 It is asserted that a power-law rate formulation can be found to be equivalent to a LHHW formulation. However, reflect upon the data of McCarthy (Ph.D. thesis, University of Notre Dame, Notre Dame, Ind., 1974) obtained in a CSTCR for CO oxidation in excess O_2 over a $Pt/\alpha-Al_2O_3$ catalyst, where $T = 246°C$:

Rate, mol(h)/(g cat)	CO, %
2.5	0.1
4.5	0.2
5.65	0.3
7.5	0.48
6	.1.0
5	1.25
4	1.6

Attempt to organize these data by (*a*) a power law, (*b*) the LHHW model.

8-14 Digest Powers' analysis of enhanced adsorption-catalysis [*J. Chem. Phys.*, **63**: 1219 (1959)]. Comment upon that model in the light of Boudart's observations summarized in *Chem. Techol.*, December 1974, p. 748. For the catalyst involved in Powers' analysis, how would you determine the population of real "sites"?

8-15 McCarthy (Ph.D. thesis, Notre Dame, Notre Dame, Ind., 1974) and colleagues [*J. Catal.*, **39**: 29 (1975)] report Pt surface-area-vs.-time data for a 0.036% $Pt/\alpha-Al_2O_3$ catalyst at three temperatures in air, where S/S_0 is the normalized area:

T, °C	$\dfrac{S}{S_0}$	Time, h
700	0.9	2
	0.81	4
	0.72	7
	0.59	11
750	0.6	2
	0.42	4
	0.32	6
	0.23	8
800	0.38	1
	0.23	3
	0.1	7
	0.07	9

From these data determine deactivation order and the activation energy.

8-16 Attempt to organize the above data in terms of a Voorhies correlation, that is, S/S_0 versus $\sqrt{\text{time}}$.

8-17 Assume that you are in receipt of data for conversion yield vs. contact time for the V_2O_5-catalyzed partial oxidation of a substituted aromatic. (Name one.) These data were secured in a 0.5-cm ID U tube packed with 50-μm catalyst particles immersed in a molten-salt bath. Conversions reside between 20 and 80 percent. Superficial contact times (reactor-tube volume per volumetric flow rate of air plus organic) range from 1 to 30 s at STP. These data were secured at 1 atm and a bath temperature of 400°C in excess air. Assume the Socratic role and with appropriate calculations suggest all possible nonchemical kinetic factors which might disguise chemical kinetic inferences. Be guided by criteria set forth by Mears [*Ind. Eng. Chem. Process Des. Dev.*, **10:** 541 (1971); *J. Catal.*, **20:** 127 (1971)].

8-18 With respect to the not unusual situation described in Prob. 8-17, fashion a laboratory-reactor-operating strategy which will minimize the extrachemical kinetic intrusions.

8-19 Redirect your attention to Prob. 8-18 in the case in which the catalyst activity declines with time such that one-half the activity is lost in (*a*) 2 min, (*b*) 2 h.

8-20 Devise an analytical justification for procuring the activation energy and order of a solid-catalyzed isothermal reaction in a fixed integral-bed reactor by varying flow rate vs. temperature at constant conversion. Repeat your analysis for a CSTCR.

8-21 Suppose that you are monitoring the principal-reactant conversion in a CSTCR or its equivalent. The conversion declines with time on-stream. How would you analyze these data to secure some notion of (*a*) principal reaction kinetics and (*b*) the nature of deactivation and its kinetics?

8-22 In the application of the Boudart-Mears-Vannice criteria to the data of Sinfelt et al. on dehydrogenation of methylcyclohexane, rule 2 is cited but not verified. Verify it.

8-23 Fashion a table of catalyst types and corresponding reactions which are known to be catalyzed by each type.

8-24 Cite industrially important catalysts which *do not* appear among those shown in Fig. 8-17 and compounds derivative of those periodic groups.

8-25 (*a*) The reaction $CO + Cl_2 \rightarrow COCl_2$ has been studied over an activated-carbon catalyst. Rate control appears to be that of surface reaction between adsorbed reactants. Assuming LHHW kinetics, derive a rate expression rooted in the postulate that only Cl_2 and $COCl_2$ are strongly adsorbed.

   (*b*) Use the data given below to refashion the rate model:

Rate$\times 10^3$	P_{CO}	P_{Cl_2}	P_{COCl_2}
4.41	0.406	0.352	0.226
4.4	0.396	0.363	0.231
2.41	0.31	0.32	0.356
2.45	0.287	0.333	0.376
1.57	0.253	0.218	0.522
3.9	0.61	0.113	0.231
2.0	0.179	0.608	0.206

8-26 Utilizing a CSTCR, McCarthy (Ph.D. thesis, University of Notre Dame, Notre Dame, Ind., 1974) obtained the following rate data vs. percentage of CO at three levels of catalyst sintering. $T = 180°C$; catalyst is 0.036% Pt/α-Al$_2$O$_3$, and d is the average Pt crystallite diameter.

d, Å	Rate, mol CO$_2$/(g cat)(h)	CO, %
28	1	0.08
	1.8	0.15
	2	0.2
	0.8	1.1
	0.7	1.3
	0.65	1.7
78	0.8	0.1
	1.2	0.12
	1.75	0.2
	0.4	1.25
	0.35	1.7
1000	0.2	0.04
	0.28	0.05
	0.1	0.35
	0.08	0.65
	0.07	1.25
	0.05	1.5

Assuming spherical platinum crystallites, compute the turnover number.   Repeat assuming cubic crystallites; total catalyst used was 9 g (Pt plus support in these experiments). Average crystallite size was inferred from H$_2$ titration data.

8-27 The kinetics of *n*-pentane isomerization is reported by Sinfelt et al. [*J. Phys. Chem.*, **64**: 892 (1960)].   Test the rates reported in that work for possible transport intrusions and evaluate the selectivity pattern.

8-28 Summarize the diverse techniques whereby both physical and chemical characteristics of a porous-solid-supported catalyst can be ascertained. If a catalyst suffers deactivation what physicochemical tools would you utilize to gain an insight into the cause of its demise?

9

DIFFUSION AND HETEROGENEOUS CATALYSIS

" Some circumstantial evidence is very strong, as when you find a trout in the milk."
Thoreau,   Miscellanies

Introduction

General concepts of diffusion (of heat and mass) both surrounding and within the locale of heterogeneous reaction were developed in Chap. 5.   In Chaps. 6 and 7 these diffusional intrusions were rendered explicit in instances of fluid-fluid and fluid-solid noncatalytic reactions and reactors.   At this point, a more detailed inquiry into interphase and intraphase diffusion of heat and mass is merited, as these transport phenomena affect heterogeneous catalytic activity, yield or selectivity, multiplicity of steady states, and stability.

It will be recalled that catalytic effectiveness

$$\eta = \frac{\text{diffusion affected rate}}{\text{diffusion unaffected rate}} \qquad (9\text{-}1)$$

will be governed by the dimensionless groups which characterize the equations evolved to describe the catalytic pellet and its environment.   Anticipating both interphase (external) and intraphase (internal) concentration and temperature gradients in the instance of steady-state nth-order catalytic reaction, we found that

$$\eta = f(\phi_0, \beta, \epsilon_0, \text{Bi}_m, \text{Bi}_h) \qquad (9\text{-}2)$$

where $\phi_0 = L\sqrt{\dfrac{k_0 A_0{}^{n-1}}{\mathscr{D}}}$ = Thiele modulus

$$(9\text{-}3)$$

$\epsilon_0 = \dfrac{E}{RT_0}$ = Arrhenius number

$\mathrm{Bi}_m = \dfrac{k_g L}{\mathscr{D}}$ = mass Biot number

$$(9\text{-}4)$$

$\mathrm{Bi}_h = \dfrac{hL}{\lambda}$ = thermal Biot number

$\beta = \dfrac{-\Delta H \,\mathscr{D} A_0}{\lambda T_0}$ = adiabatic $\dfrac{\Delta T}{T_0}$ within pellet

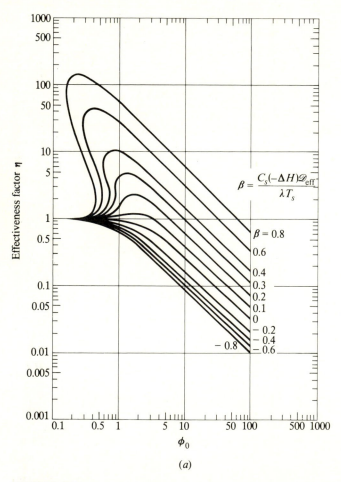

$$\beta = \frac{C_s(-\Delta H)\mathscr{D}_{\text{eff}}}{\lambda T_s}$$

(a)

FIGURE 9-1
Intraphase nonisothermal effectiveness versus (a) ϕ_0 and (b) $\eta \phi_0{}^2$ for linear kinetics; $\epsilon = E/RT = 20$, and Bi_m and $\mathrm{Bi}_h = \infty$. [*P. B. Weisz and J. S. Hicks, Chem. Eng. Sci.,* **17**: *265 (1962).*]

It was noted in Chap. 5 that an η-versus-ϕ_0 plot can be set forth for various values of β at fixed values of the Biot and Arrhenius number. However, such a display applies to only one pellet or one point in a reactor, since A_0 and T_0 appear in key dimensionless groups which dictate η-versus-ϕ_0 behavior. Such displays are therefore of greater qualitative than quantitative value in overall reactor design. A point assessment of η is facilitated by revealing the behavior of η not as a function of the usually unobservable ϕ_0 but of the observable

$$\eta{\phi_0}^2 = \frac{\mathcal{R}_0 L^2}{\mathcal{D} A_0} \qquad (9\text{-}5)$$

where $\mathcal{R}_0$ is the observed local rate. Both modes of correlation are shown in Fig. 9-1, for *infinite* values of the Biot numbers. The figure reveals only internal effectiveness for nonisothermal first-order reaction.

Hutchings' nonisothermal interphase and intraphase η results for Langmuir-Hinshelwood kinetics at various values of the Biot numbers are shown in Fig. 9-2.

In the following section a detailed general treatment of global (interphase and intraphase) effectiveness for CO oxidation over supported Pt is presented to illustrate effectiveness for an exothermic reaction of mixed normal and abnormal kinetic behavior. There follows a discussion of permissible simplifications of the effectiveness relationships from which various practical criteria to guide the analyst

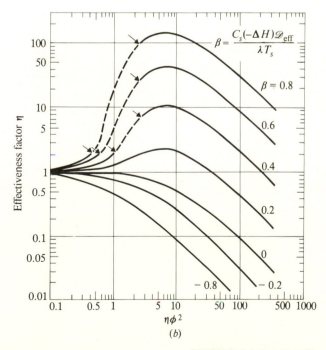

FIGURE 9-1 *(continued)*.

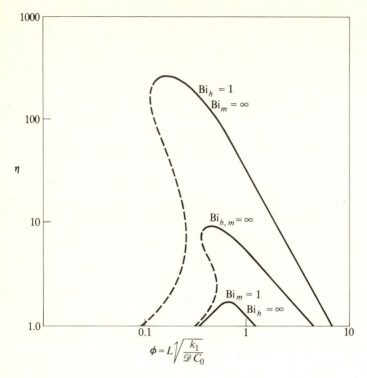

FIGURE 9-2
Interphase and intraphase nonisothermal effectiveness for LHHW kinetics for the reaction C $\rightarrow$ P; $r = k_1 K_1 C/(1 + K_1 C + K_p P)$; $\epsilon = 40$, $\beta = 0.1$, $K_1 C_0 = 10$, and $K_p C_0 = 0.1$. [*J. Hutchings and J. J. Carberry, AIChE J.*, **12**: *30 (1966)*.]

will emerge.   Selectivity alterations due to diffusion are then reviewed, followed by a brief treatment of catalysts characterized by finite external area as well as by the macropore-micropore case.   We conclude the chapter with a comment on poisoning and diffusional phenomena and steady-state multiplicity.

9-1 DIFFUSION-AFFECTED CO OXIDATION

Over supported-Pt catalyst, intrinsic CO-oxidation kinetics conforms to the rate model, in excess O_2,

$$r = \frac{k(CO)}{[1 + K(CO)]^2} \qquad (9\text{-}6)$$

The reaction is substantially exothermic and of obvious importance in combustion-gas-pollution abatement. Of signal interest is the fact that the rate model exhibits normal kinetic behavior when $K(CO) \ll 1$

$$r = k(CO)$$

However, when (as is the case at but a fraction of a percent CO) $K(CO) > 1$, abnormal (negative-order) kinetics becomes manifest

$$r = \frac{k'}{CO}$$

The kinetic parameters for the intrinsic rate equation have been measured by Voltz and coworkers[1] and are

$$k = 1.83 \times 10^{12} \exp\left(-\frac{45,000}{RT}\right) \qquad s^{-1} \qquad (9\text{-}7)$$

$$K = 0.655 \exp\frac{3460}{RT} \qquad (\text{mol }\%)^{-1}$$

$$-\Delta H = 67,000 \text{ cal/g mol}$$

The steady-state interphase and intraphase nonisothermal catalytic effectiveness is determined by numerical solution[2] of the following continuity equations for a spherical catalytic pellet, where $c = CO/CO_0$ and $\overline{K} = K_a C_0$,

$$\frac{d^2c}{dz^2} + \frac{2}{z}\frac{dc}{dz} = \phi_0^2 \frac{c}{(1 + \overline{K}c)^2} \exp\left[-\epsilon_0\left(\frac{1}{t} - 1\right)\right] \qquad (9\text{-}8)$$

$$\frac{d^2t}{dz^2} + \frac{2}{z}\frac{dt}{dz} = -\beta\phi_0^2 \frac{c}{(1 + \overline{K}c)^2} \exp\left[-\epsilon_0\left(\frac{1}{t} - 1\right)\right]$$

where at $z = 0$
$$\frac{dc}{dz} \qquad \text{and} \qquad \frac{dt}{dz} = 0$$

and at $z = 1$
$$c_s = 1 - \frac{1}{\text{Bi}_m}\frac{dc}{dz}\bigg|_s$$
$$\qquad (9\text{-}9)$$
$$t_s = 1 - \frac{1}{\text{Bi}_h}\frac{dt}{dz}\bigg|_s$$

The effect of temperature on the kinetic absorption coefficient $\overline{K}$ is ignored in view of the comparatively small value of adsorption heat (7 percent of the value of the activation energy). The governing equations are solved by the modified Picard method. One could also use collocation techniques, as described by Finlayson[3] or techniques revealed earlier.[4] The computed global effectiveness factors are displayed for diverse circumstances in Figs. 9-3 to 9-6, where $\phi \equiv \phi_0$.

[1] S. E. Voltz, C. R. Morgan, D. Liederman, and S. M. Jacob, *Ind. Eng. Chem. Process Des. Dev.*, **12**: 294 (1973).
[2] T. G. Smith, J. Zahradnik, and J. J. Carberry, *Chem. Eng. Sci.*, **30**: 763 (1975).
[3] B. Finlayson, *Catal. Rev.*, **10**: 69 (1975).
[4] J. J. Carberry and M. Wendel, *AIChE J.*, **9**: 129 (1962).

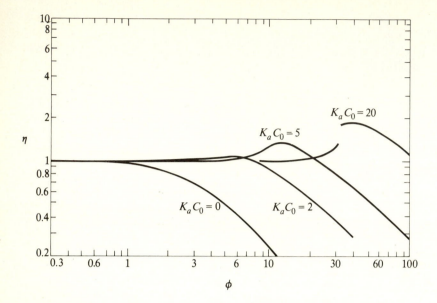

FIGURE 9-3
Spherical isothermal effectiveness; $Bi_m = \infty$, $r = kC/(1 + K_aC_0C)^2$, LHHW kinetics.
[*T. G. Smith, J. Zahradnik, and J. J. Carberry, Chem. Eng. Sci.,* **30**: *763 (1975)*.]

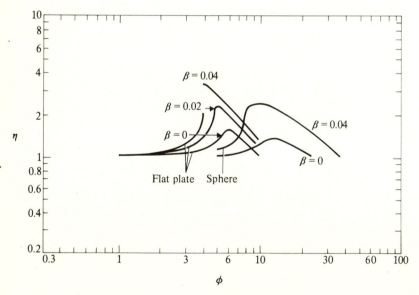

FIGURE 9-4
Comparison of nonisothermal effectiveness for a flat plate and a sphere; $Bi_{h,m} = 200$,
$\epsilon = 30$, LHHW kinetics. [*T. G. Smith, J. Zahradnik, and J. J. Carberry, Chem. Eng.
Sci.,* **30**: *763 (1975)*.]

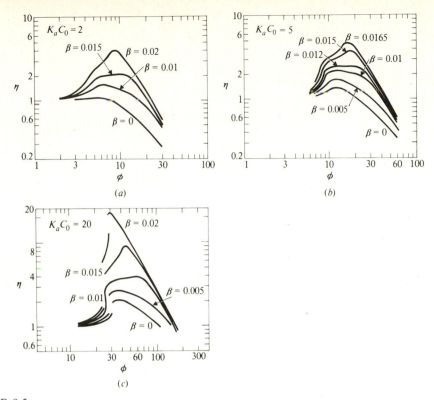

(a)

(b)

(c)

FIGURE 9-5
Nonisothermal effectiveness for diverse values of $K_a C_0$; $Bi_m = 20$, $Bi_h = 2$, $\epsilon = 30$, LHHW kinetics. [*T. G. Smith, J. Zahradnik, and J. J. Carberry, Chem. Eng. Sci.,* **30**: *763 (1975).*]

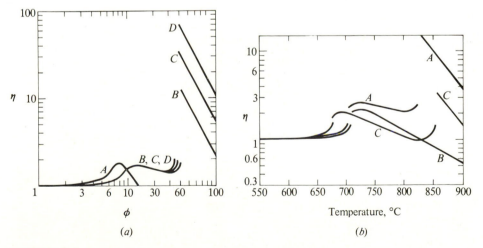

(a)

(b)

FIGURE 9-6
(a) Effect of r ($=Bi_m/Bi_h$) on spherical nonisothermal effectiveness factor; $\beta = 0.005$, $K_a C_0 = 5$, $\epsilon = 30$. Curve A: $r = 1$, $Bi_m = 2$, curve B: $r = 100$, $Bi_m = 200$; curve C: $r = 250$, $Bi_m - 500$, curve D: $r - 500$, $Bi_m - 1000$. (b) Nonisothermal effectiveness vs. ambient temperature for CO oxidation (see Table 9-1). [*T. G. Smith, J. Zahradnik, and J. J. Carberry, Chem. Eng. Sci.,* **30**: *763 (1975).*]

Isothermal Intraphase η

The internal effectiveness ($\text{Bi}_m = \infty$) under isothermal condition ($\beta = 0$) is shown in Fig. 9-3 for various values of the adsorption coefficient; the curve for $K_a C_0 = 0$ corresponds, of course, to the simple first-order isothermal case, derived and solved analytically in Chap. 5. Note that as $K_a C_0$ increases, the regime of abnormal kinetic behavior commands and *isothermal* values of η greater than unity are found. In particular for $K_a C_0 = 20$ a very narrow region of isothermal multiplicity appears. Values of η greater than unity occur with the onset of negative-order kinetics and strong intraphase diffusional limitations simply because, unlike what happens in normal kinetics, the rate within the pellet actually *increases* with a diffusion-sponsored concentration drop. For particular values of $K_a C_0$ and ϕ_0, three solutions apparently exist, giving rise to the noted multiplicity (similar to the isothermal multiplicity illustrated in Chap. 5) for external mass transfer in series with a rate of the form identical to Eq. (9-6).

Nonisothermal Intraphase η

For specific values of $K_a C_0$ and ϵ_0 and negligible external film resistances, the internal nonisothermal effectiveness is given in Fig. 9-4 for modest values of β. A comparison with the flat-plate solution is also shown, as is the isothermal case for both geometries ($\beta = 0$). β, the Prater number. is a measure of the maximum possible internal temperature gradient, as shown in Chap. 5. A modest value of T (internal)/T_0 (surface) $= 1.04$ dramatically increases η to a value of about 2 in this mixed normal-abnormal kinetic case. By contrast we note that in the simple linear kinetic case, a Prater number of 0.04 yields an η value about identical to the isothermal value (Fig. 9-1a).

Nonisothermal Interphase and Intraphase η

For $\text{Bi}_m = 20$, $\text{Bi}_h = 2$, and $\epsilon_0 = 30$ the global nonisothermal η-versus-ϕ_0 behavior is shown in Fig. 9-5 in order of increasing $K_a C_0$, spanning the spectrum of from normal to abnormal kinetic character. Again rather modest values of β are involved. This figure dramatically reveals the telling influence of the external temperature gradient upon η. For example, in Fig. 9-4 at $\phi_0 = 10$ and $\beta = 0.04$ the spherical effectiveness is about 2; in Fig. 9-5b (for the same value of $K_a C_0$) at $\phi_0 = 10$, η is 2 for a β value of merely 0.01, a signal indication of the governing role of $\bar{\beta}$, the external adiabatic ΔT parameter discussed in Chap. 5. Finally, note the existence of multiple multiplicity in Fig. 9-5c, in the kinetic region of negative order ($K_a C_0 = 20$). Recall (Chap. 5) that the internal Prater number β and its external value $\bar{\beta}$ are related to the Biot-number ratio by

$$\frac{\bar{\beta}}{\beta} = \frac{\text{Bi}_m}{\text{Bi}_h} \qquad (9\text{-}10)$$

In Fig. 9-5 this ratio is 10. The influence of this ratio and therefore the startling effect of external-film temperature gradient upon η is shown in Fig. 9-6a for the conditions noted.   Referring to Fig. 5-14, we note that a Biot ratio of 100 suggests that practically the entire temperature gradient resides in the external fluid film for even modest values of the observable.

Effectiveness-Temperature Behavior

Insofar as real reactor environments are nonisothermal with respect to spatial dimensions, it is of interest to note how η varies over a range of temperature. From start-up to full cruising speed an automotive-exhaust catalytic reactor operates over a temperature range of from less than 100°C to near 800°C.   The behavior of the nonisothermal interphase and intraphase effectiveness factor for CO oxidation over Pt, as a function of bulk-fluid-temperature environment, is indicated in Fig. 9-6b for diverse conditions cited in Table 9-1.   Again we are witness to multiple multiplicity in the steady state, a conspiracy of interphase and intraphase nonisothermality and mixed normal and abnormal kinetic character.

This illustration of global-effectiveness behavior for an actual reaction, as opposed to one involving the celebrated universal reactants-products A, B, C, ... of the alphabetic periodic table, draws attention to certain features of η-versus-ϕ_0 behavior touched upon in Chap. 5:

1 The existence of abnormal kinetic behavior gives rise to isothermal η values greater than unity.
2 For abnormal kinetics, even a modest intrapellet temperature rise β amplifies the enhancement of η.
3 A temperature gradient in the external fluid film significantly affects η for exothermic reactions.
4 An inspection of expected values of the Prater number β and its external analog $\bar{\beta}$ compels one to anticipate the conclusion noted in 3.
5 Steady-state multiplicities can exist under both isothermal and nonisothermal circumstances.

9-2 IMPLICIT LIMITATIONS OF THE GENERAL MODEL

While it might appear that the above model described by the continuity equations and their boundary conditions is comprehensive for steady-state nonisothermal global effectiveness, it is worth noting that reality presents a few physical situations

Table 9-1   PARAMETERS EMPLOYED IN GENERATING FIGURE 9-6b

Curve A:	$d_p = 1$ cm	$\mathscr{D}_{\text{eff}} = 5 \times 10^{-2}$ cm²/s
	$C_0 = 5\%$	$\lambda = 5 \times 10^{-4}$ cal/(°C)(cm)(s)
	$\text{Bi}_m = 200$	$\text{Bi}_h = 2$

Curve B: Same as curve A, but $\text{Bi}_h = 10$
Curve C: Same as curve A, but $C_0 = 3\%$

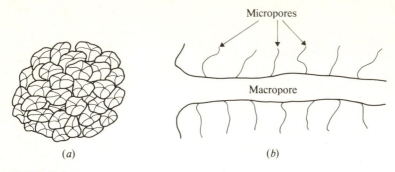

Micropores

Macropore

(a)   (b)

FIGURE 9-7
Schematic of microporous-macroporous catalyst.

which are not embraced by the seemingly general mathematical descriptions of global effectiveness. The assumptions implied in our earlier treatments of global effectiveness for porous-solid catalysis are now discussed in detail.

A simple effective diffusivity $\mathcal{D}$ is assumed to sufficiently characterize intrapellet diffusive transport. As noted in Chap. 8, a pellet composed of macropore admits such a simplification. However, as schematized in Fig. 9-7a, the pellet, being composed of microporous microspheres, gives rise to a pellet model which is best termed a macropore-micropore structure. In essence such a structure is equivalent to a macropore network from the macropores of which branched micropores extend, as the simplified cylindrical macropore-micropore model set forth in Fig. 9-7b suggests. Consequently pore effectiveness η is a function of a macropore populated by micropores of effectiveness $\bar{\eta}$. This macropore-micropore diffusion-reaction case will be treated subsequently.

A uniform distribution of catalytic sites throughout the pellet has been assumed so far. A notable exception to this postulate is found in the partially impregnated catalyst pellet. As illustrated in Chap. 8, the deposition of a catalyst-bearing solution upon a porous support involves the familiar competition of diffusion into the pores with simultaneous adsorption. Rapid adsorption of the catalytic species relative to its intrapellet diffusive transport invites preferential accumulation of the catalytic agent onto the pore mouth and neighboring region. Partial impregnation results. A support pellet thus partially impregnated with catalytic species will be properly characterized not by a Thiele modulus phrased in terms of total pore length but by the actual catalytically active pore length, about which more will be said here (Sec. 9-8) and in Chap. 10.

A catalytic pellet in which the major locale of catalytic activity resides upon the internal (pore) area has been implicitly postulated. As most porous-solid catalyst pellet or extrusion formulations are typically characterized by total BET areas of from several to 100 to 200 m^2/g, it is apparent that their external, superficial areas are relatively insignificant in magnitude, being of the order of $a = 6/d_p$ for a sphere, for example. So a 5-mm sphere of silica-alumina has an external area per gram of several square centimeters while its total BET (internal plus external) area is of

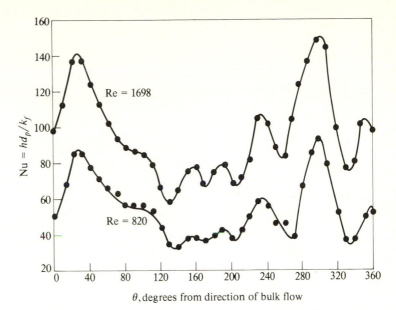

FIGURE 9-8
Local distribution of Nusselt number for heat transfer as a function of position about a single sphere in a fixed bed. [*B. M. Gillespie, E. D. Crandall, and J. J. Carberry, AIChE J.*, **14**: *483 (1968)*.]

the order of 100 to 200 m²/g. Catalytic effectiveness and yield analyses rooted solely in terms of an internal-area contribution are therefore justified. However, a few examples of quite low BET-area catalyst formulations exist (highly calcined α-alumina, catalytic wires and gauzes). In these instances, external area becomes a finite fraction of total catalytic area. Further, for any porous catalyst, as the Thiele modulus increases with reaction vigor, the zone of catalytic action retreats toward the pore-mouth entrance or external surface. In such instances, detailed below, due account must be given to effectiveness and yield in terms of the available external pellet surface.

An average global value of both the interphase heat- and mass-transport coefficients, h and k_g, has been assumed to be sufficient in the description of external transport about a single pellet and, for that matter, throughout the catalytic-reactor confines, as suggested in Chap. 10. Detailed measurements of the distribution of values of the local interphase heat-transfer coefficient about a single sphere immersed in a packed bed indicate a rather wide distribution of h values, as revealed by the data of Gillespie[1] set forth in Fig. 9-8. These detailed data suggest that when conditions persist in which global effectiveness or yield becomes dependent upon the Biot numbers, asymmetry within a single pellet may be anticipated. However, in the total formulation of catalytic-reactor models, it is unlikely that a just account of local Biot-number variations can be easily fashioned.

[1] B. M. Gillespie, E. D. Crandall, and J. J. Carberry, *AIChE J.*, **14**: 483 (1968).

9-3 SOME LIMITING CASES

Not all heterogeneous catalytic systems which confront the analyst require so detailed a consideration as our treatment of CO oxidation, nor will ignorance of all of the implicit assumptions itemized above necessarily frustrate the realization of insights into the problem. Therefore a few limiting cases will be detailed, specifically:

1   Isothermal effectiveness and yield in terms of interphase and intraphase mass-diffusional intrusions for finite external catalyst area
2   Interphase nonisothermal effectiveness in conjunction with intraphase isothermality
3   Isothermal yield for a macroporous-microporous catalyst

The general features of the isothermal interphase and intraphase effectiveness or yield relationships were discussed in Chap. 5, where it was shown that for a given value of the Thiele modulus for an nth-order reaction

$$\phi = L\sqrt{\frac{kA_0^{n-1}}{\mathscr{D}}}$$

the reaction of highest order manifests the lowest effectiveness, the limiting value of η being $\eta = 1/\phi$. In consequence, in this the limiting-internal-diffusion-affected regime, the experimental rate will assume the form

$$\mathscr{R}_0 = \frac{\sqrt{k\mathscr{D}}}{L} A_0^{(n+1)/2} \qquad (9\text{-}11)$$

so that observed order, activation energy, and pellet-size dependency will be modified relative to the diffusion-uninfluenced rate behavior.

It was also shown in Chap. 5 that the intervention of external mass-transport limitations, as determined by the mass Biot number $k_g L/\mathscr{D}$, further modifies experimental inferences insofar as, in the limit of external mass-transport control, observed order becomes unity, measured activation energy approaches zero, and a strong dependency upon pellet size and fluid velocity is shown (Table 5-1). The interphase and intraphase isothermal effectiveness for a first-order reaction for a flat plate rendered essentially equivalent to a sphere for $L = 1/a$ is

$$\eta = \frac{\tanh \phi}{\phi[1 + (\phi \tanh \phi)/\text{Bi}_m]} \qquad (9\text{-}12)$$

Chapter 5 also indicated that insofar as yield in a complex reaction network is governed by an effective-rate-coefficient ratio

$$\frac{\bar{k}_1}{\bar{k}_2} = \frac{\eta_1 k_1}{\eta_2 k_2} \qquad (9\text{-}13)$$

where k_1 and k_2 are intrinsic rate coefficients, diffusional alteration of yield is to be anticipated to the extent that ϕ and Bi_m assume values which cause η to be other than unity. As declared in Chap. 5, modifications of this nature threaten intermediate yield in a consecutive network or an independent parallel one but not in simultaneous networks of equal order. We illustrate this point later for a consecutive linear reaction network, which is of some importance, as exemplified in catalytic partial oxidation of organics, where yield of intermediate is important, while, on the other hand, in pollution abatement yield of ultimate oxidation products CO_2 and H_2O is of prime interest. Yielding to the temptation of generality (isothermal), we postulate a macropore-micropore model.

Effectiveness and Point Yield for Finite External Area[1]

We address ourselves to the catalytic circumstance in which total catalyst area consists of internal area S and external, geometric area a per unit volume. First, we consider isothermal effectiveness and then isothermal point yield of intermediate in the common consecutive-reaction network.

The issue reduces to that of specifying internal and external effectiveness and point yield. As we are dealing with two distinct areas, internal S and external a, an examination of the linear rate coefficient is justified. Generally for linear kinetics k is in reciprocal seconds. In fact, however, an heterogeneous first-order chemical-reaction-rate coefficient must be intrinsically related to catalytic area, that is k_1 is in centimeters per second; thus k_1 (area/per volume) $= k$.

For a finite external area a and internal area S (per unit volume)

$$k_1'(a + S) = k \qquad s^{-1}$$

(To be sure, actual catalytic area is only a fraction of $a + S$; however, as noted in Chap. 8, specification of true catalytic area remains to be unambiguously ascertained.)

Now we define f as the ratio of external to total (BET) area

$$f = \frac{a}{a + S} \qquad (9\text{-}14)$$

Given this definition, for the linear consecutive reaction $A \rightarrow B \rightarrow C$, we have

Internal: $\qquad \left(\dfrac{dA}{d\theta}\right)_i = \eta_i(1 - f)k_1 A_0$

External: $\qquad \left(\dfrac{dA}{d\theta}\right)_x = \eta_x f k_1 A_0$

$\qquad\qquad \left(\dfrac{dB}{dA}\right)_i = Y_i = $ internal point yield

$\qquad\qquad \left(\dfrac{dB}{dA}\right)_x = Y_x = $ external point yield

[1] W. Goldstein and J. J. Carberry, *J. Catal.*, **28**: 33 (1973).

where η_i and η_x are the internal and external effectiveness factors, each a function of diffusion-reaction and external mass-transport moduli. Global (internal, external-area) catalytic effectiveness is then

$$\eta_t = (1 - f)\eta_i + f\eta_x \qquad (9\text{-}15)$$

while point yield $(dB/dA)_t$ is

$$\frac{dB}{dA} = Y_p = \frac{[(dA/d\theta)(dB/dA)]_i + [(dA/d\theta)(dB/dA)]_x}{(dA/d\theta)_i + (dA/d\theta)_x}$$

or

$$Y_p = \frac{(1 - f)\eta_i Y_i + f\eta_x Y_x}{(1 - f)\eta_i + f\eta_x} \qquad (9\text{-}16)$$

where η_i and η_x are internal and external effectiveness factors and Y_i and Y_x the respective point yields of intermediate B. Specification of η_t and Y_p then merely depends upon identifying the internal-external η and Y functions.

As shown in Chap. 5, internal catalytic effectiveness is

$$\eta_i = \frac{\tanh \phi}{\phi[1 + (\phi \tanh \phi)/\mathrm{Bi}_m]} \qquad (9\text{-}17)$$

where

$$\phi = L\sqrt{\frac{(1 - f)k}{\mathscr{D}}} \qquad \text{and} \qquad \mathrm{Bi}_m = \frac{k_g L}{\mathscr{D}}$$

External effectiveness η_x, as derived in Chap. 5, is

$$\eta_x = \frac{1}{1 + \mathrm{Da}} \qquad (9\text{-}18)$$

where

$$\mathrm{Da} = \frac{kf}{k_g a} = \frac{\phi^2}{\mathrm{Bi}_m}\frac{f}{1 - f} \qquad (9\text{-}19)$$

Note that both internal and external η values are functions of external mass transport as reflected in the Biot and Damköhler numbers, respectively. Total effectiveness is then, by Eqs. (9-15), (9-17), and (9-18),

$$\eta_t = \frac{(1 - f)\tanh \phi}{\phi[1 + (\phi \tanh \phi)/\mathrm{Bi}_m]} + \frac{f}{1 + (\phi^2/\mathrm{Bi}_m)f/(1 - f)} \qquad (9\text{-}20)$$

This result is displayed in Figs. 9-9 and 9-10 as η_t versus ϕ for diverse values of Bi_m and f. In Figs. 9-11 and 9-12 external-internal effectiveness is set forth in terms of the internal observable $\eta\phi^2$, which is

$$\eta\phi^2 = \frac{\mathscr{R}_0 L^2}{\mathscr{D} A_0}$$

Point yield in terms of f is given by Eq. (9-16), where the internal point yield Y_i is

$$Y_i = -\gamma \frac{m_1 \phi_2 \tanh \phi_2}{m_2 \phi_1 \tanh \phi_1}\left(\frac{B_0}{A_0} + \frac{K}{\gamma K - 1}\right) + \gamma \frac{K}{\gamma K - 1} \qquad (9\text{-}21)$$

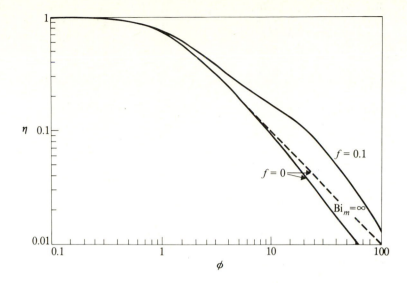

FIGURE 9-9
Isothermal intraphase and interphase effectiveness vs. Thiele modulus for finite external area, $Bi_m = 100$.

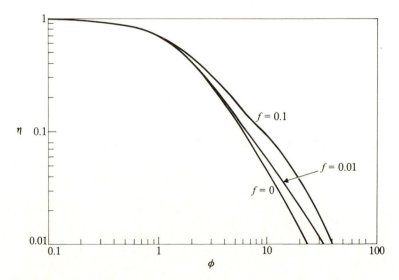

FIGURE 9-10
Isothermal intraphase and interphase effectiveness vs. Thiele modulus for finite external area, $Bi_m = 10$.

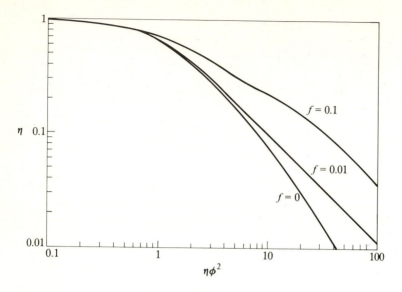

FIGURE 9-11
Finite-external-area effectiveness vs. the observable; $Bi_m = 100$.

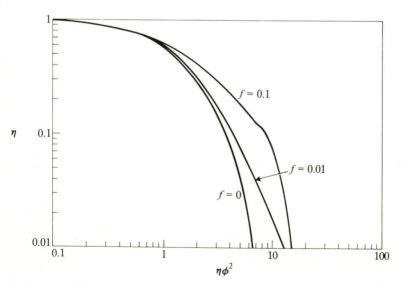

FIGURE 9-12
Finite-external-area effectiveness vs. the observable; $Bi_m = 10$.

where $K = k_1/k_2$ and $\gamma = \mathscr{D}_B/\mathscr{D}_A$ and $m = 1 + (\phi \tanh \phi)/\text{Bi}_m$. External point yield is (Chap. 5)

$$Y_x = \frac{1}{1 + \text{Da}_2} - \frac{1}{K}\frac{1 + \text{Da}_1}{1 + \text{Da}_2}\frac{B_0}{A_0} \qquad (9\text{-}22)$$

Overall point yield in terms of internal and external area is therefore

$$Y_p = \frac{(1-f)[\text{Eq. } (9\text{-}21)][\text{Eq. } (9\text{-}17)] + f[\text{Eq. } (9\text{-}18)][\text{Eq. } (9\text{-}22)]}{(1-f)[\text{Eq. } (9\text{-}17)] + f[\text{Eq. } (9\text{-}18)]} \qquad (9\text{-}23)$$

Point yield is set forth in Figs. 9-13 to 9-16 for the values of f and Bi_m cited. It is interesting to note the limiting behavior of both η_t and Y_p in terms of f and Bi_m.

Consideration of internal-external area yields a limiting value of η_t at large values of ϕ of

$$\eta_t = \frac{\text{Bi}_m{}^2}{\phi^2}(1-f) \qquad (9\text{-}24)$$

In contrast, if external mass transport is ignored, we find, $\text{Bi}_m = \infty$,

$$\eta_t = f \qquad (9\text{-}25)$$

If both external mass transfer and external area are ignored, the result, in the limit of large values of ϕ ($f = 0$, $\text{Bi}_m = \infty$), is

$$\eta_t = \frac{1}{\phi} \qquad (9\text{-}26)$$

Limiting values of point yield are sensitive to both f and Bi_m, as evident in Figs. 9-13 to 9-16. If we assume $f = 0$ and $\text{Bi}_m = \infty$, a limiting low value of Y_p is realized. If f is finite and $\text{Bi}_m = \infty$, yield passes through a minimum as a function of ϕ and then returns to a value of unity. This results from assuming an infinite value of Bi_m. By Eq. (9-23) we see that with increasing values of ϕ, the value of Da remains finite except for the rather extraordinary circumstance in which fluid velocity becomes supersonic.[1]

In most circumstances f is virtually zero inasmuch as total BET area is many orders of magnitude greater than external area a. However, even for small values of f, at high values of ϕ which are encountered at high temperatures (hot spots) the zone of catalytic activity retreats toward the external surface, at which point due account of the external surface is required, particularly in view of the fact that the implicit assumption that $f = 0$ leads to erroneous limiting values of η_t and Y_p at high values of ϕ.

---

[1] Ibid.

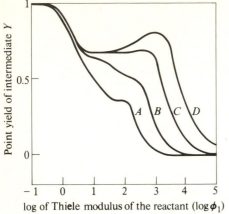

FIGURE 9-13
Point yield of B versus the Thiele modulus ϕ_1:
$f = 10^{-3}$; $B_0 = 0$; $k_1/k_2 = 4$. Values of Bi_m:
A, 10; B, 10^2; C, 10^3; D, 10^4. [*W. Goldstein and J. J. Carberry, J. Catal.,* **28:** *33 (1973).*]

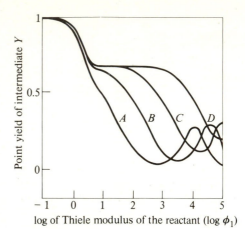

FIGURE 9-14
Point yield of B versus the Thiele modulus ϕ_1:
$f = 10^{-1}$; $B_0 = 0$; $k_1/k_2 = 4$. Values of Bi_m:
A, 10; B, 10^2; C, 10^3; D, 10^4. [*W. Goldstein and J. J. Carberry, J. Catal.,* **28:** *33 (1973).*]

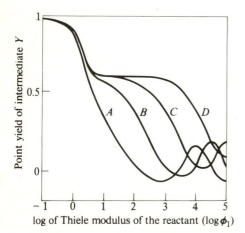

FIGURE 9-15
Point yield of B versus the Thiele modulus ϕ_1:
$f = 10^{-1}$; $B_0 = 0.1$; $k_1/k_2 = 4$. Values of Bi_m:
A, 10; B, 10^2; C, 10^3; D, 10^4. [*W. Goldstein and J. J. Carberry, J. Catal.,* **28:** *33 (1973).*]

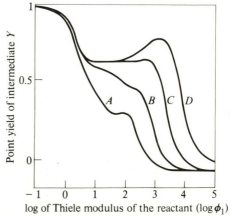

FIGURE 9-16
Point yield of B versus the Thiele modulus ϕ_1:
$f = 10^{-3}$; $B_0 = 0.1$; $k_1/k_2 = 4$. Values of Bi_m:
A, 10; B, 10^2; C, 10^3; D, 10^4. [*W. Goldstein and J. J. Carberry, J. Catal.,* **28:** *33 (1973).*]

Interphase Nonisothermality and Intraphase Isothermal Effectiveness and Point Yield[1]

As shown in Chap. 5 and illustrated in this chapter, in gas-solid catalyst systems, it is a happy fact that

$$r = \frac{Bi_m}{Bi_h} \gg 1.0 \qquad (9\text{-}27)$$

which, as Fig. 5-14 reveals, permits the prudent assumption that in general (if not universally) one may assume that since the major locale of the total interphase and intraphase temperature gradient is to be found in the external fluid film, in many instances we can treat the pellet as an isothermal entity at a temperature dictated by interphase (external) heat transport. Table 9-2 presents a comparison of experiment and theoretical prediction [Eq. (5-85)] for the ratio of the external ΔT_x relative to the total ΔT_0 in terms of r [Eq. (9-27)] and the observable $\bar{\eta}Da$

$$\frac{\Delta T_x}{\Delta T_0} = \frac{r\bar{\eta}Da}{1 + \bar{\eta}Da(r-1)} \qquad (9\text{-}28)$$

Table 9-2 COMPARISON OF PREDICTIONS OF THEORY [Eq. (9-28)] WITH EXPERIMENTAL DATA OF KEHOE AND BUTT†
$r = 20\S$

Run	$\bar{\eta}\,Da \times 10^3$	$\Delta T_0, °C$	$\Delta T_x/\Delta T_0$ Pred	Exp∓
8	2.45	8	0.05	0.07
9	3.75	9	0.07	0.11
10	2.17	0	0.04	Nil
20	12	22	0.20	0.3
21	8.25	13	0.14	0.15
22	7	10	0.10	Nil
23	17.5	37	0.26	0.27
24	7.2	30	0.13	0.13
25	47	30	0.085	0.07
26	25.6	100	0.33	0.37
27	8.7	51	0.15	0.14
28	4.35	40	0.08	0.1

† J. J. Carberry, *Ind. Eng. Chem. Fundam.*, **14:** 129 (1975).

‡ J. P. G. Kehoe and J. B. Butt, *AIChE J.*, **18:** 347 (1972).

§ For $r = 300$ all runs indicate $(\Delta T_x/\Delta T_0)_{exp}$ of about 100%, in accord with Eq. (9-28).

[1] J. J. Carberry and A. Kulkarni, *J. Catal.*, **31:** 41 (1973).

The experimental data are those of Kehoe and Butt.[1] The verification is due to Carberry.[2]

From this confrontation of data and theory we may safely conclude that for even modestly low values of the observable $\bar{\eta}$Da, the assumption of intraphase iso-thermality is tolerable for values of the Biot-number ratio r of 100 or more. Most gas-solid catalyst systems are marked by such favorable ratios, primarily because, it will be recalled [Eq. (5-79)],

$$r = \frac{\mathrm{Bi}_m}{\mathrm{Bi}_h} = \frac{\lambda}{\mathscr{D} \rho C_p \mathrm{Le}^{2/3}} = \frac{\bar{\beta}}{\beta} \qquad (9\text{-}29)$$

where, the λ pellet thermal conductivity, is of the order of 10^{-4} but ρC_p is small indeed for gaseous systems.

Assuming then that intraphase isothermality is a reasonable assumption, while an interphase temperature (and concentration) gradient is anticipated, effectiveness becomes for linear kinetics

$$\bar{\eta} = \frac{1}{L} \frac{\int_0^L kA \, dz}{k_0 A_0} = \frac{1}{L} \frac{k}{k_0} \int^L \frac{A}{A_0} \, dz \qquad (9\text{-}30)$$

or

$$\bar{\eta} = \frac{k}{k_0} \eta_{\text{iso}} \qquad (9\text{-}31)$$

and

$$\eta_{\text{iso}} = \frac{\tanh \phi}{\phi[1 + (\phi \tanh \phi)/\mathrm{Bi}_m]} \qquad (9\text{-}32)$$

where $\phi = L\sqrt{k/\mathscr{D}}$ and the rate coefficient k is determined by the surface and there-fore pellet temperature dictated by external heat transport. Now

$$\frac{k}{k_0} = \exp\left[-\epsilon_0\left(\frac{1}{t} - 1\right)\right] \qquad (9\text{-}33)$$

where $\epsilon_0 = E/RT_0$ and $t = T_s/T_0$ and, as was shown in Chap. 5,

$$t = 1 + \bar{\beta}\bar{\eta}\mathrm{Da} \qquad (9\text{-}34)$$

where

$$\bar{\beta} = \frac{-\Delta H \, A_0}{\rho C_p T_0 \mathrm{Le}^{2/3}} \qquad (9\text{-}35)$$

and the observable is

$$\bar{\eta}\mathrm{Da} = \frac{\mathscr{R}_0}{k_g a A_0} \qquad (9\text{-}36)$$

Effectiveness for the isothermal pellet at a temperature dictated by external heat transport is

$$\bar{\eta} = \frac{\tanh \phi}{\phi[1 + (\phi \tanh \phi)/\mathrm{Bi}_m]} \exp\left[-\epsilon_0\left(\frac{1}{1 + \bar{\beta}\bar{\eta}\mathrm{Da}} - 1\right)\right] \qquad (9\text{-}37)$$

[1] J. P. G. Kehoe and J. B. Butt, *AIChE J.*, **18**: 347 (1972).
[2] J. J. Carberry, *Ind. Eng. Chem. Fundam.*, **14**: 129 (1975).

$\bar{\eta}$ is readily displayed as a function of the observable $\bar{\eta}\mathrm{Da}$ since it is related to $\bar{\eta}$ and ϕ by $\bar{\eta}\mathrm{Da} = \bar{\eta}\phi^2/\mathrm{Bi}_m$.

The $\bar{\eta}$-versus-$\bar{\eta}\mathrm{Da}$ functionality is set forth in Figs. 9-17 to 9-19 for diverse values of governing parameters, $\bar{\beta}$, ϵ_0, and Bi_m. Once again we are witness to effectiveness factors greater than unity for exothermic reaction ($\bar{\beta} + $).

The point yield in complex reaction networks submits to telling analysis provided intraphase isothermality is admissible in the presence of interphase (external) temperature and concentration gradients. For insofar as point yield is determined by the effective-rate-coefficient ratio

$$\frac{\bar{k}_1}{\bar{k}_2} = \frac{\eta_1}{\eta_2}\left(\frac{k_1}{k_2}\right)_0 \qquad (9\text{-}38)$$

we have for the isothermal-intraphase–nonisothermal-interphase circumstance [by Eq. (9-37)]

$$\frac{\bar{k}_1}{\bar{k}_2} = K_0\left(\frac{m_2}{m_1}\frac{\phi_2\tanh\phi_1}{\phi_1\tanh\phi_2}\right)\exp\left[-\Delta\epsilon_0\left(\frac{1}{t}-1\right)\right] \qquad (9\text{-}39)$$

where $\Delta\epsilon_0 = \epsilon_1 - \epsilon_2$.

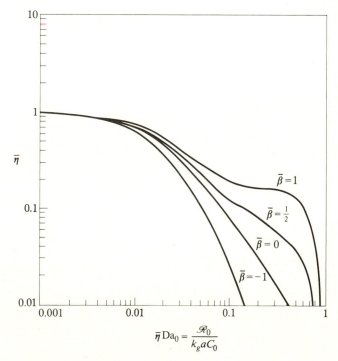

$$\bar{\eta}\,\mathrm{Da}_0 = \frac{\mathcal{R}_0}{k_g a C_0}$$

FIGURE 9-17
Interphase and intraphase effectiveness of an isothermal pellet first-order reaction in the presence of an external temperature gradient. $\mathrm{Bi}_m = 100$, and $\epsilon_0 = 10$ in terms of the external observable. [*J. J. Carberry and A. Kulkarni, J. Catal.*, **31**: *41 (1973).*]

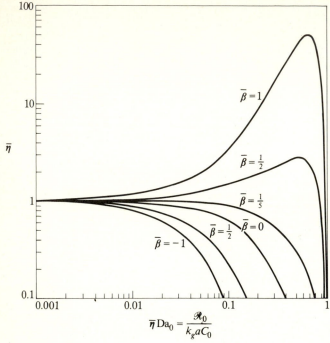

FIGURE 9-18
Interphase and intraphase effectiveness of an isothermal pellet first-order reaction in the presence of an external temperature gradient. $Bi_m = 10$, and $\epsilon_0 = 20$ in terms of the external observable. [*J. J. Carberry and A. Kulkarni, J. Catal.*, **31**: 41 (1973).]

Isothermal Yield for Macroporous-Microporous Catalyst[1]

A schematic of a macroporous-microporous catalyst pellet is given in Fig. 9-7a and an idealization of it in Fig. 9-7b. For the linear isothermal consecutive network $A \xrightarrow{k_1} B \xrightarrow{k_2} C$ we seek an analytical relationship between both the point and overall yield of B and the macropore-micropore parameters of the system. A flat-plate geometry will be assumed. For the micropores of length x,

$$\frac{d^2a}{dy^2} = x^2 \frac{k_1}{\mathscr{D}_a} a \tag{9-40}$$

$$\frac{d^2b}{dy^2} = x^2 \frac{k_2 b}{\mathscr{D}_b} - x^2 \frac{k_1 a}{\mathscr{D}_a} \tag{9-41}$$

At $y = 1$: $a = \bar{a}$ $b = \bar{b}$

At $y = 0$: $\dfrac{da}{dy} = 0$ $\dfrac{db}{dy} = 0$ (9-42)

[1] J. J. Carberry, *Chem. Eng. Sci.*, **17**: 675 (1962).

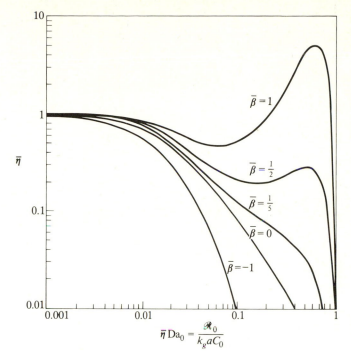

FIGURE 9-19

Interphase and intraphase effectiveness of an isothermal pellet first-order reaction in the presence of an external temperature gradient. $Bi_m = 100$ and $\epsilon_0 = 20$ in terms of the external observable. [*J. J. Carberry and A. Kulkarni, J. Catal.*, **31**: *41 (1973)*.]

Therefore

$$a = \frac{\bar{a} \cosh \phi_1 y}{\cosh \phi_1} \tag{9-43}$$

$$b = (\bar{b} + S\bar{a}) \frac{\cosh \phi_2 y}{\cosh \phi_2} - s\bar{a} \frac{\cosh \phi_1 y}{\cosh \phi_1} \tag{9-44}$$

where $\quad \phi_i = x\left(\dfrac{k_i}{\mathscr{D}_i}\right)^{1/2} \qquad \gamma = \dfrac{\mathscr{D}_b}{\mathscr{D}_a} \qquad S = \dfrac{s}{s\gamma - 1} \qquad$ and $\qquad s = \dfrac{k_1}{k_2}$

In the macropores of length L

$$\frac{d^2 \bar{a}}{dz^2} = \frac{L^2}{\mathscr{D} A} k_1 \bar{\eta}_1 \bar{a} = \Phi_1{}^2 \bar{a} \tag{9-45}$$

$$\frac{d^2 \bar{b}}{dz^2} = \Phi_2{}^2 (\bar{b} + S\bar{a}) - \Phi_1{}^2 \bar{a} S \tag{9-46}$$

where

$$\Phi = L^2 \frac{k}{\mathscr{D}} \bar{\eta}$$

and $\bar{\eta}$, the microphore effectiveness is, $\bar{\eta} = (\tanh \phi)/\phi$,   The boundary conditions are as follows:

At $z = 0$: $\dfrac{d\bar{a}}{dz} = \dfrac{d\bar{b}}{dz} = 0$

At $z = 1$: $\bar{a} = \bar{A} - \dfrac{1}{\text{Bi}_m} \dfrac{d\bar{a}}{dz}$

$$\bar{b} = \bar{B} - \dfrac{1}{\text{Bi}_m} \dfrac{d\bar{b}}{dz} \tag{9-47}$$

Solving for the flux of A and B and dividing, we see that the *point* yield of B is

$$-\frac{d\bar{B}}{d\bar{A}} = -\gamma \frac{m_1 \Phi_2 \tanh \Phi_2}{m_2 \Phi_1 \tanh \Phi_1} \left(\frac{\bar{B}}{\bar{A}} + S \right) + \gamma S \tag{9-48}$$

Integration gives the overall yield of B, where $\Phi = L\sqrt{k_1 \bar{\eta}/\mathcal{D}}$, as

$$\frac{\bar{B}}{A_0} = \frac{K - \gamma}{K - 1} S(\bar{a}^K - \bar{a}) + \frac{B_0}{A_0} \bar{a}^K \tag{9-49}$$

where $K = \dfrac{k_2}{k_1} \dfrac{\eta_{o_2}}{\eta_{o_1}} = \textit{effective}\text{-rate-coefficient ratio}$

and $\bar{a} = \dfrac{A}{A_0}$

The overall yield (for $B_0 = 0$) of B relative to feed concentration A_0 is displayed vs. conversion $(1 - \bar{a})$ in Fig. 9-20.   It is to be noted that in the total absence of diffusional influence $K = 1/s$ and Eq. (9-49) becomes identical to that expressing yield in a homogeneous consecutive reaction, as indeed must be the case.   Note also the η_o is the product of the micro, macro, and external-effectiveness factors

$$\eta_o = \bar{\eta} \eta \eta_x   \text{as}   \eta_x = \frac{1}{1 + \text{Da}} \tag{9-50}$$

Insofar as the *effective*-rate-coefficient ratio K determines yield for a given intrinsic ratio s, the limiting behavior of K is instructive.

CASE 1   Nonporous pellet

$$K = \frac{1}{s} \frac{1 + \text{Da}_1}{1 + \text{Da}_2} \tag{9-51}$$

When, of course, $k_g a \gg k$, we have surface-reaction control

$$K = \frac{1}{s}   \text{intrinsic ratio} \tag{9-51a}$$

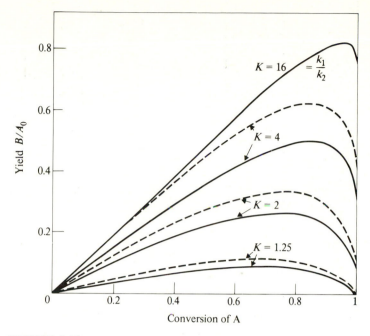

FIGURE 9-20

Yield of B versus conversion of A for A $\xrightarrow{k_1}$ B $\xrightarrow{k_2}$ C in an isothermal microporous-macroporous catalyst according to Eq. (9-49) for $B_0 = 0$, $K = \eta k_1 / \eta k_2$, $\gamma = 1$; — — — —: $k_1/k_2 = 16$, ————: $k_1/k_2 = 4$.

while if Da_1, $Da_2 > 1$, that is, bulk mass-transport control,

$$K = \frac{k_{g_2}}{k_{g_1}} \cong 1.0 \qquad (9\text{-}51b)$$

in which case yield of B becomes zero, as was shown in Chap. 5.

CASE 2   Macroporous pellet or $\bar{\eta} = 1.0$

$$K = \frac{1 + k_1\eta_1/k_g a}{1 + k_2\eta_2/k_g a} \frac{1}{s} \frac{\eta_2}{\eta_1} \qquad (9\text{-}52)$$

$k\eta \gg k_g a$:   bulk-mass-transport control and $K \to 1.0$

$k\eta \ll k_g a$:   pore diffusion-reaction control

$$K = \frac{1}{(s/\gamma)^{1/2}} \frac{\tanh \Phi_2}{\tanh \Phi_1} \qquad (9\text{-}52a)$$

for $\Phi > 3$, $\tanh \Phi \to 1$ and

$$K = \sqrt{\frac{k_2}{k_1}} \gamma \qquad (9\text{-}52b)$$

In this regime the advantage of an intrinsic rate-coefficient ratio of, say, $k_1/k_2 = 16$ is drastically reduced to an effective value of 4, as Wheeler was the first to note.[1]

CASE 3   In the regime of micropore-macropore diffusion control (ϕ and $\Phi > 3$)

$$K = \gamma^{3/4}\left(\frac{k_2}{k_1}\right)^{1/4} \qquad (9\text{-}53)$$

In such a case an intrinsic coefficient advantage of, say, 16 is reduced to 2.

It is immediately apparent that the Thiele modulus of the micropores relative to that of the macropore is an index of the real significance of a micropore-macropore diffusion-reaction model

$$\frac{\phi_1}{\Phi_1} = \frac{x}{L}\frac{k_1'}{k_1}\frac{\mathscr{D}}{\mathscr{D}'} \qquad (9\text{-}54)$$

Now $x/L \ll 1$, and the macropore-micropore diffusivity ratio can be somewhat greater than unity.   However k_1' for the micropore is not equal to k_1 for the macropore, not simply because k_1 contains $\bar{\eta}$ but because the major fraction of the surface area is to be found in the micropores.   Recall that the surface rate constant is

$$k = \frac{1}{s} = \frac{\bar{k} \text{ area}}{\text{vol}} = \frac{\text{cm}}{s}\frac{1}{\text{cm}}$$

Hence a first-order surface rate coefficient is the product of an intrinsic reaction *velocity* coefficient multiplied by a usually unknown catalytic-site area per unit volume.   Consequently a small value of micropore length relative to that of the macropore does not necessarily permit one to conclude that $\bar{\eta}$ will always be unity.

9-4   INTERPHASE-INTRAPHASE NONISOTHERMAL YIELD

We now address ourselves to the problem of catalytic yield or selectivity in instances where both interphase and intraphase gradients of temperature and concentration intrude upon the intrinsic dispositions of the catalyst.   Qualitative insights were established in Chap. 5 with respect to simultaneous, consecutive, and parallel reaction networks, where it was shown that for the diffusion-affected case of simultaneous reaction

CASE 1:

$$A \xrightarrow{k_1} B \qquad n\text{th order}$$
$$A \xrightarrow{k_2} D \qquad m\text{th order}$$

[1] A. Wheeler, in P. H. Emmett (ed.), "Catalysis," vol. 2, Reinhold, New York, 1955.

Point, and therefore overall, selectivity B/D is adversely affected if $n > m$ and in an exothermic case if $E_1 < E_2$. For the consecutive case

CASE 2: $$A \xrightarrow{k_1} B \xrightarrow{k_2} C$$

An interphase-intraphase mass-differential limitation taxes the yield of B, which nonisothermality can nullify or magnify, depending upon the sign and magnitude of the β's, and $\Delta\epsilon = (E_1 - E_2)/RT$. Finally in the instance of parallel reactions.

CASE 3: $$A \xrightarrow{k_1} B \quad n\text{th order}$$
$$X \xrightarrow{k_2} Y \quad m\text{th order}$$

The principles established for network 1 generally apply. We can now tender quantitative support of our qualitative contentions for the interesting simultaneous-consecutive network (1 and 2) in which interphase and intraphase gradients of temperature and concentrations are anticipated. Our memories of the qualitative conclusions for the simple yield systems, 1 to 3, are refreshed in Table 9-3.

Consider the solid-catalyzed network,

$$A \xrightarrow{k_1} B \xrightarrow{k_2} C \quad \text{linear}$$
$$A + A \xrightarrow{k_3} D \quad \text{second order} \tag{9-55}$$

which is recognized as the van de Vusse system analyzed in Chap. 3 with respect to CSTR-PFR and intermediate levels of gross backmixing. In that context, it was found that the choice of CSTR-PFR or the PFR with limited recycle depended, with respect to maximizing yield of B, upon the magnitudes of the respective rate coefficients *under isothermal conditions*. Fascinating possibilities exist when the van de Vusse scheme is considered to be catalyzed by a porous solid about and

Table 9-3 INFLUENCE OF INTERPHASE AND INTRAPHASE HEAT AND MASS DIFFUSION UPON YIELD OF B

Isothermal:

$$A \xrightarrow{1} B \quad n\text{th order} \qquad A \xrightarrow{1} B \qquad A \xrightarrow{1} B \xrightarrow{2} C$$
$$\searrow^{2} C \quad m\text{th order} \qquad X \xrightarrow{2} Y$$

Mass-diffusion influence on yield of B	$n = m$	None	any order
	$n > m$	Decrease	Decrease
	$n < m$	Increase	

Nonisothermal:†

	$E_1 > E_2$	$E_1 < E_2$
Exothermal	Increase	Decrease
Endothermal	Decrease	Increase

† This refers to behavior of k_1/k_2; its value is assumed to dominate under nonisothermal conditions relative to concentration-gradient effects.

within which both heat- and mass-diffusional limitations are assumed to prevail. A dilemma is presented in the isothermal case since general qualitative principles of interphase and intraphase diffusion indicate that mass-diffusional limitations tax the local yield of B in the consecutive network, while the higher, second-order reaction producing by-product D is also relatively taxed by an interphase and intraphase mass-diffusional gradient.

If interphase and intraphase local nonisothermality is assumed, the yield of B is further affected by the relative magnitudes of E_1, E_2, and E_3. It is obvious that the yield of B in this van de Vusse solid-catalyzed network will, in general, be a function of ϕ_1, ϕ_2, ϕ_3, β_1, β_2, β_3, Bi_m, Bi_h, ϵ_1, ϵ_2, and ϵ_3, the diffusion-coefficient ratios, and, in this instance, the relative orders of the simultaneous-consecutive steps.

Let it be assumed that the van de Vusse reaction network is catalyzed by a spherical porous-solid catalyst about and within which temperature and concentration gradients are anticipated. Utilizing the dimensionless groups

$$t = \frac{T}{T_0} \qquad a = \frac{A}{A_0} \qquad b = \frac{B}{A_0} \qquad Z = \frac{r}{R_0}$$

$$\epsilon_i = \frac{E_i}{RT_0} \qquad \beta_i = \frac{-\Delta H_i A_0 \mathscr{D}}{\lambda T_0} \qquad Bi_m = \frac{k_g R_0}{\mathscr{D}}$$

$$Bi_h = \frac{hR_0}{\lambda} \qquad \text{and} \qquad \phi_i = R_0 \sqrt{\frac{k_i C_0^{n-1}}{\mathscr{D}}} \qquad \text{at } T_0$$

we see that the continuity equations for A, B, and T are

$$\frac{d^2 a}{dZ^2} + \frac{2}{Z}\frac{da}{dZ} = \phi_1^2 a \exp\left[-\epsilon_1\left(\frac{1}{t} - 1\right)\right] + \phi_3^2 a^2 \exp\left[-\epsilon_3\left(\frac{1}{t} - 1\right)\right] \qquad (9\text{-}56a)$$

$$\frac{d^2 b}{dZ^2} + \frac{2}{Z}\frac{db}{dZ} = \phi_2^2 b \exp\left[-\epsilon_2\left(\frac{1}{t} - 1\right)\right] - \phi_1^2 a \exp\left[-\epsilon_1\left(\frac{1}{t} - 1\right)\right] \qquad (9\text{-}56b)$$

$$\frac{d^2 t}{dZ^2} + \frac{2}{Z}\frac{dt}{dZ} = -\sum_{i=1}^{i=3} \phi_i^2 \beta_i C_i \exp\left[-\epsilon_i\left(\frac{1}{t} - 1\right)\right] \qquad (9\text{-}56c)$$

The boundary conditions are

At $Z = 0$: $\qquad \dfrac{da}{dZ} = \dfrac{db}{dZ} = \dfrac{dt}{dZ} = 0$

At $Z = 1$: $\qquad a = 1 - \dfrac{1}{Bi_m}\dfrac{da}{dZ}$

$$b = b_0 - \frac{1}{Bi_m}\frac{db}{dZ} \qquad (9\text{-}57)$$

$$t = 1 - \frac{1}{Bi_h}\frac{dt}{dZ}$$

where $$b_0 = \frac{B_0}{A_0}$$

The governing equations have been solved numerically for a range of interesting values of the parameters. Since the simple consecutive network (case 2) is apparent for $\phi_3 = 0$, results for network 2 are presented in Figs. 9-21 to 9-23 for values of parameters cited there in terms of point yield of B, dB/dA versus ϕ_2/ϕ_1. Hutchings' results are fashioned for values of $Bi_m = \infty$, a reasonable postulate, since, as noted earlier, $Bi_m \gg Bi_h$. The point yield of B versus ϕ_2/ϕ_1 is in accord with Butt's results[1] computed for $Bi_h = Bi_m = \infty$. The telling influence of finite values of Bi_h on local yield is evident in the display of Hutchings' results, which embrace the interesting cases of mixed endothermicity and exothermicity for this important consecutive-reaction network. These quantitative findings are, of

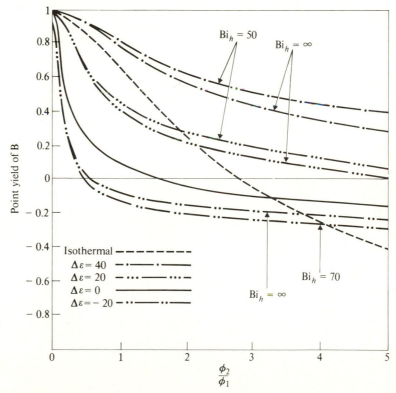

FIGURE 9-21
Point yield of B for finite and infinite thermal Biot number versus ϕ_2/ϕ_1 for the reaction A $\xrightarrow{1}$ B $\xrightarrow{2}$ C, where β_1 and $\beta_2 = 0.1$ and $Bi_m = \infty$. [*J. Hutchings, Ph.D. thesis, University of Notre Dame, Notre Dame, Ind., 1969; J. J. Carberry, Chim. Ind. (Milan), 51(9): 951 (1969).*]

[1] J. B. Butt, *Chem. Eng. Sci.*, **21:** 275 (1966).

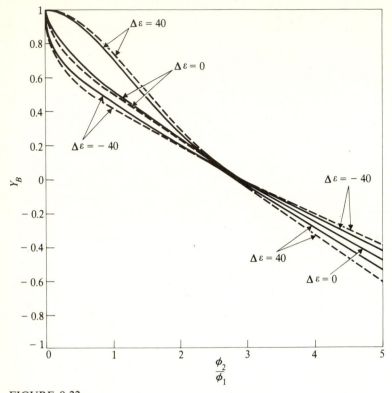

FIGURE 9-22

Point yield of B in an exothermal-endothermal reaction system $A \xrightarrow{1} B \xrightarrow{2} C$, $\beta_1 = 0.1$, $\beta_2 = -0.2$, $Bi_m = \infty$; ————: $Bi_h = \infty$, — — — —: $Bi_h = 15$. [*J. Hutchings, Ph.D. thesis, University of Notre Dame, Notre Dame, Ind., 1969; J. J. Carberry, Chim. Ind. (Milan),* **51**(9): 951 (1969).]

course, in accord with expectations rooted in the generalizations set forth earlier regarding compensation, nullification, and amplification of mass-diffusional effects in the presence of endothermal or exothermal temperature fields.

For reasons cited earlier, the van de Vusse network does not readily lend itself to qualitative a priori prediction, as is clear from the computed point-yield-versus ϕ_1 behavior patterns displayed in Figs. 9-24 to 9-28 for parameter values listed in Table 9-4.

Under isothermal conditions, the benefit to the point yield of B by reason of an intraphase gradient which suppresses the second-order reaction (9-55) is evident in Fig. 9-24, where cases *A* and *D* are compared. Note that while an expected yield superiority is evident in case *D* ($K_3 = 0.1$) relative to case *A* ($K_3 = 0.5$), with an increase in the diffusional gradient (increasing ϕ) the difference decreases markedly. For dominating values of $K_3 = 5$ we note a maximum in yield (curves *E* and *F*), again a reflection of the benefit of the intraphase gradient in suppressing step 3.

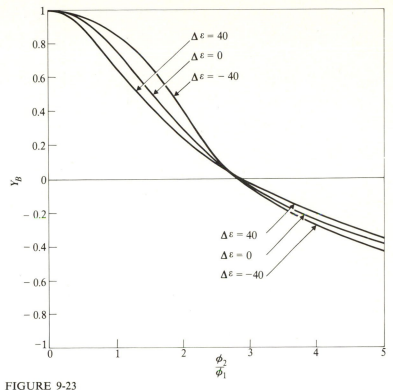

FIGURE 9-23

Point yield of B in an endothermal-exothermal reaction system $A \xrightarrow{1} B \xrightarrow{2} C$. $\beta_1 = -0.1$, $\beta_2 = 0.2$; Bi_h, $Bi_m = \infty$. [*J. Hutchings, Ph.D. thesis, University of Notre Dame, Notre Dame, Ind., 1969; J. J. Carberry, Chim. Ind. (Milan), 51(9): 951 (1969).*]

The expected deleterious consequence of increasing concentrations of product B in the gas phase is revealed in Fig. 9-25 (curves A, B, and C). Compare, however, H and B (a nonisothermal vs. isothermal case). The nonisothermal behavior proves to be rather provocative. In Fig. 9-25 the isothermal point-yield profile, curve B, is contrasted with mild and strong exothermicity (curves H and J), Fig. 9-26, for identical values of the governing parameters noted in Table 9-4. The telling effect of the thermal Biot number is evident in the contrast of curves H and I. In comparing H and I it is to be noted that $\epsilon_1 > \epsilon_2$ but $\epsilon_3 > \epsilon_2$; multiplicity of steady-state point yields is revealed and is also evident in cases M and N in Fig. 9-27, where modest activation energy differences between these two cases give rise to dramatic differences in point yield. Contrast, for example, case L, $\epsilon_1 = \epsilon_2 = \epsilon_3$, with case J, $\epsilon_1 > \epsilon_3 > \epsilon_2$. In Fig. 9-28, where cases K and J are compared, it is to be noted that in spite of a severe value of $K_3 = 5$ in case K, the favorable activation energy pattern coupled with a small value of Bi_h prompts a yield profile not too drastically lower than that evidenced in case J, for which K_3 is only 0.5, the activation-energy pattern identical to case K but Bi_h is much larger.

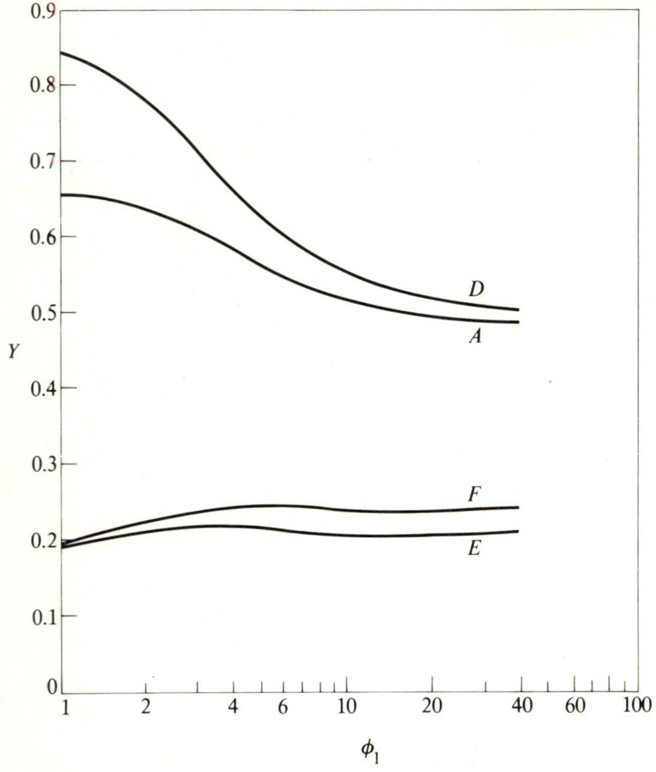

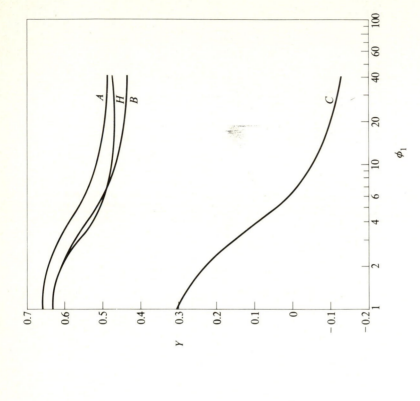

FIGURE 9-24
Point yield of B for $A \rightarrow B \rightarrow C$, $A + A \rightarrow D$ for conditions specified in Table 9-4.

FIGURE 9-25
Point yield of B for $A \rightarrow B \rightarrow C$, $A + A \rightarrow D$ for conditions specified in Table 9-4.

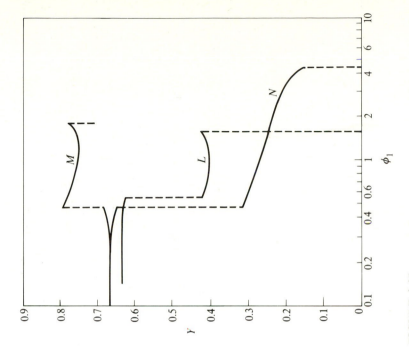

FIGURE 9-26
Point yield of B for $A \rightarrow B \rightarrow C$, $A + A \rightarrow D$ for conditions specified in Table 9-4.

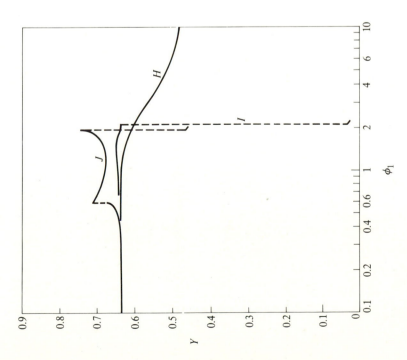

FIGURE 9-27
Point yield of B for $A \rightarrow B \rightarrow C$, $A + A \rightarrow D$ for conditions specified in Table 9-4.

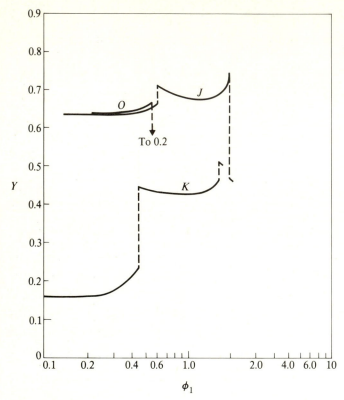

FIGURE 9-28
Point yield of B for A → B → C, A + A → D for conditions specified in Table 9-4.

Table 9-4 VALUES OF PARAMETERS USED IN FIGS. 9-24 TO 9-28

						Parameter					
Case	β_1	β_2	β_3	Bi_m	Bi_h	$\left(\dfrac{k_1}{k_2}\right)_0$	$\left(\dfrac{k_3 A}{k_1}\right)_0$	$\left(\dfrac{A}{B}\right)_0$	ε_1	ε_2	ε_3
A	0	0	0	1000	—	2	0.5	100	30	20	25
B	0	0	0	1000	—	2	0.5	10	30	20	25
C	0	0	0	1000	—	2	0.5	1	30	20	25
D	0	0	0	1000	—	2	0.1	100	30	20	25
E	0	0	0	1000	—	2	5	100	30	20	25
F	0	0	0	1000	—	5	5	100	30	20	25
H	0.02	0.01	0.02	1000	100	2	0.5	10	30	20	25
I	0.02	0.01	0.02	1000	2	2	0.5	10	30	20	25
J	0.2	0.1	0.2	1000	100	2	0.5	10	30	20	25
K	0.2	0.1	0.2	1000	100	5	5	10	30	20	25
L	0.2	0.1	0.2	1000	100	2	0.5	10	30	30	30
M	0.2	0.1	0.2	1000	100	2	0.5	100	30	20	25
N	0.2	0.1	0.2	1000	100	2	0.5	100	20	30	25
O	0.2	0.1	0.2	1000	10	2	0.5	10	30	20	25

9-5 TRANSPORT COEFFICIENTS

The evaluation of the diffusion-reaction parameters which govern interphase and intraphase diffusion-affected catalytic reaction require values of $\mathscr{D}$, λ, k_g, and h, as these transport coefficients appear in ϕ, β, Bi_m, and Bi_h.

First we consider the intraphase coefficients, $\mathscr{D}$ and λ.

Intraphase Diffusivity

Three types of diffusive transport may characterize in-pore, intraphase mass transfer.

Normal, or molecular, diffusion This takes place when pore radius is large relative to the molecular mean free path. For a porous pellet of fractional porosity ξ and a tortuosity τ, which is a measure of deviation from the straight-cylindrical-pore postulate, the effective intraphase diffusion coefficient is

$$\mathscr{D}_{\text{eff}} = \frac{\xi \mathscr{D}_m}{\tau} \qquad (9\text{-}58)$$

$\mathscr{D}_m$ is secured from reliable computation or published data. This mode of diffusion involves the predominance of intermolecular collision relative to pore-wall–molecule collision.

Intraphase transport If pore radius is of the order of or less than the mean free path, Knudsen diffusion dictates intraphase transport, in which case

$$\mathscr{D}_{\text{eff}} = \frac{\xi \mathscr{D}_K}{\tau} \qquad (9\text{-}59)$$

and $\mathscr{D}_K$ is computed from the Knudsen equation

$$\mathscr{D}_K = 9700 r_p \sqrt{\frac{T}{M}}$$

where r_p = pore radius, cm
  T = absolute temperature, K
  M = molecular weight of diffusing species

If the average pore radius admits to simple specification in terms of pore volume and BET area S, that is, as noted in Chap. 8,

$$r_p = \frac{2V_P}{S} = \frac{2\xi}{S\rho}$$

where ρ is pellet density, then

$$\mathscr{D}_{\text{eff}} = 19,400 \, \frac{\xi^2}{\tau S \rho} \sqrt{\frac{T}{M}} \qquad \text{cm}^2/\text{s} \qquad (9\text{-}60)$$

In this mechanism of diffusion, molecule–pore-wall collision dominates relative to the intermolecular event.

Surface diffusion Surface diffusion along the walls of the pores is another mode of intraphase mass transport, but little is actually known of its mechanism.   It can be expected with high-boiling species, and it will surely depend upon the nature of the surface.   Satterfield discusses the issue and sets forth a few *empirical* correlations.[1]   One of these seemingly satisfying correlations includes H_2, O_2, and other species capable of dissociation, suggesting that spillover is involved in the surface-diffusion process.   Spillover is simply dissociative chemisorption followed by atom migration.[2]

In zeolite catalysts which are characterized by " pore " diameters of 5 to 10 Å, diffusion would seem to be an inappropriate term with which to define the intraphase or intracage transport process.   Evidence suggests that transport in zeolites is a complex form of chemical creep, dependent upon stereochemical factors.

In the absence of surface diffusion, the effective intraphase diffusivity in the transition regime between molecular and Knudsen diffusion can be neatly secured by the relation

$$\frac{1}{\mathscr{D}_{eff}} = \frac{1}{\mathscr{D}_{K,eff}} + \frac{1}{\mathscr{D}_{m,eff}} \qquad (9\text{-}61)$$

as derived by Pollard and Present and others.[3]   The computation of an effective intraphase diffusion coefficient presupposes a value of ξ and τ.   Pellet porosity ξ is easily obtained—precisely by the He-Hg method and roughly by water absorption.

In the first method one measures the void volume in a tube of porous catalyst pellets with Hg, which does not penetrate the pores at atmospheric pressure.   Then total (bed void plus pore void) volume is measured with He or some other non-adsorbing gas.   The difference gives the pellet void volume.   In the second method a batch of porous catalyst is "titrated" with liquid water until external surface water becomes evident.

The determination of tortuosity τ is far less simple.   Arguing that the pore path will deviate from the straight one on an average of 45°, a value of $\tau = \sqrt{2}$ would be predicted.   In fact, comparisons of *measured* effective diffusivities with $\xi \mathscr{D}_m$ and/or $\xi \mathscr{D}_K$ reveal τ values of from $\sqrt{2}$ to as high as 10 or 12.

Given a pore-size distribution as secured by N_2-desorption hysteresis data (Chap. 8) and Hg penetration[4] one may invoke one or more of the various theoretical models of diffusion in porous pellets to predict an effective diffusivity.   In view of the imprecision inherent in the hysteresis data, the model employed to secure the pore distribution from these data, and the tentative agreement of pore

[1] C. N. Satterfield, "Mass Transfer in Heterogeneous Catalysis," MIT Press, Cambridge, Mass., 1970.
[2] P. A. Sermon and G. C. Bond, *Catal. Rev.*, **8:** 211 (1974).
[3] W. G. Pollard and R. D. Present, *Phys. Rev.*, **73:** 762 (1948); R. B. Evans, et al., *J. Chem. Phys.*, **33:** 2076 (1961).
[4] H. L. Ritter and L. C. Drake, *Ind. Eng. Chem. Anal. Ed.*, **17:** 787 (1945).

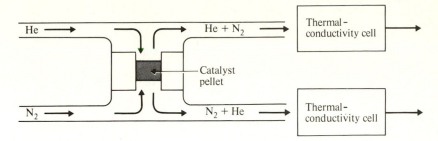

FIGURE 9-29
Schematic of apparatus for measurement of intrapellet diffusion coefficients.

model with reality, it seems risky indeed to rely upon an a priori calculation of $\mathscr{D}_{eff}$. In fact, it is far easier to measure that coefficient.

As schematically displayed in Fig. 9-29, the candidate porous catalyst is embedded between two nonadsorbing flowing gas streams. In steady state simple thermistor determination of the presence of the diffusing species in one or both streams permits the measurement of $\mathscr{D}_{eff}$ from

$$\mathscr{D}_{eff} = \frac{\text{pellet length}}{\Delta C} \times \text{rate} \qquad (9\text{-}62)$$

This method is, of course, sensitive to cracks or bypassing and insensitive to dead-ended pores. Alternatively, transient response or chromatographic techniques can be employed to obtain $\mathscr{D}_{eff}$.[1]

In sum, intraphase diffusion coefficients will range in value from 10^{-1} to 10^{-3}; surface diffusivities are in the 10^{-4} to 10^{-5} range, while diffusivities in zeolites are of a magnitude encountered in solid species, that is, 10^{-9} to 10^{-15} cm^2/s.

Intraphase Thermal Conductivity

Heat is conducted through the porous solid by conduction in an ill-defined solid matrix and via the gas phase, which permeates that porous structure. Both theoretical and experimental approaches are presently rather weak in a quantitative sense. As Satterfield notes, however, what measurements do exist reveal a happily narrow range of λ values, all of the order of 10^{-4} cal/(s)(cm)(°C). Even porous silver exhibits an effective conductivity far less than its bulk value and of a magnitude comparable to that of the porous insulator alumina. Internal solid-solid contact resistances (structure-determined) and gas-phase thermal conduction evidently dictate effective thermal conductivity of the porous catalyst.

[1] T. Furusawa and J. M. Smith, *AIChE J.*, **20**: 88 (1974); *Catal. Rev.*, **13**(1): 43 (1976).

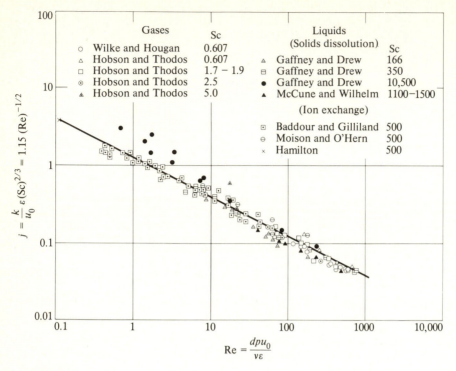

FIGURE 9-30

Fixed-bed mass-transfer data contrasted with Carberry's boundary-layer model. Data from C. R. Wilke and O. A. Hougen, *Trans. Am. Inst. Chem. Eng.*, **41**: 445 (1945); M. Hobson and G. Thodos, *Chem. Eng. Prog.*, **47**: 370 (1951); B. W. Gaffney and T. B. Drew, *Ind. Eng. Chem.*, **42**: 1120 (1950); L. K. McKune and R. H. Wilhelm, *Ind. Eng. Chem.*, **41**: 1124 (1949); R. F. Baddour and E. R. Gilliland, *Ind. Eng. Chem.*, **45**: 330 (1953); R. L. Moison and H. A. O'Hern, *Chem. Eng. Prog. Symp. Ser.*, (24)**55**: 71 (1959); P. B. Hamilton, Alfred I. du Pont Institute, private communication, 1959. [*J. J. Carberry, AIChE J.*, **6**: 460 (1960).]

Interphase Transport

We are upon more secure grounds in assessing values of k_g and h, the interphase coefficients of mass and heat transfer, respectively. To be sure, values of k_g and h extracted from familiar Nusselt, Sherwood, or j-factor correlations are average values, in contrast to local, point, values (Fig. 9-8). However, the average overall coefficients must suffice for analyses and design.

In our review of those factors which affect the form and constants in an interphase-transport-coefficient functionality, it was noted that interfacial shear, which generates a velocity gradient within the fluid-film boundary layers, gives rise to a two-thirds-power dependency upon the Schmidt and Prandtl numbers, according to boundary-layer theory. When the j-factor manner of correlation is used, k_g and h will assume the form

$$j_D = \frac{k_g}{u} \mathrm{Sc}^{2/3} \cong \frac{h}{\rho u C_p} \mathrm{Pr}^{2/3} = K \frac{1}{\mathrm{Re}^\alpha} \qquad (9\text{-}63)$$

where K and α will depend upon the geometry and nature of the fluid-particle system, e.g., whether fixed or fluid bed, a single pellet, etc.

Theory and experimental data agree for fixed-bed interphase heat and mass transport, as is evident in Fig. 9-30. From general considerations discussed in Chap. 5, a fixed-bed j factor was derived.[1] The theoretical line is

$$j_D = j_H = \frac{1.15}{\sqrt{d_p u/\nu}} \qquad (9\text{-}64)$$

where $u = u_0/\epsilon$, the effective interstitial velocity. Detailed discussion and valuable literature data are to be found in Satterfield's treatise.[2]

9-6 DIFFUSION-CATALYSIS: THEORY AND REALITY

A vast corpus of experimental data now exists to support beyond doubt the essential features of diffusion-affected porous-solid-catalyzed reaction. We cite a few key studies which demonstrate agreement between theory and fact for isothermal effectiveness, isothermal diffusion-affected yield in a consecutive network, non-isothermal effectiveness, and interphase and intraphase nonisothermal effectiveness.

Isothermal Effectiveness

Dente and Pasquon[3] studied the kinetics of methanol oxidation over Fe–molybdenum oxide catalyst in a differential reactor. Particle sizes of 3 and 0.6 mm were employed. In excess O_2 the diffusion-unaffected rate is one-half order in methanol. Global reaction rates were also measured in the regime of diffusion influence. External, interphase values of ΔT were computed from the differential heat balance and were found to vary from 2 to 30°C, while the external mass gradients proved to be rather minor. The catalysts were assumed to be isothermal at the temperature dictated by the external interphase heat transport (an assumption in accord with principles set forth in Chap. 5).

Effectiveness factors were then computed from the experimental differential rates, while the Thiele modulus is expressed in terms of the diffusion-uninfluenced rate and concentration. The experimental values of η versus ϕ are displayed in Fig. 9-31. The theoretical curve is that derived for one-half order kinetics. Theory and experiment agree rather neatly.

Diffusion-influenced Yield

An inspiring verification of diffusional modification of both catalytic activity and selectivity is offered by Weisz and Swegler[4] in the chromia-alumina-catalyzed dehydrogenation of cyclohexane

$$\underset{A}{\text{Cyclohexane}} \xrightarrow{k_1} \underset{B}{\text{cyclohexene}} \xrightarrow{k_2} \underset{C}{\text{benzene}}$$

[1] J. J. Carberry, *AIChE J.*, **6**: 460 (1960).
[2] Satterfield, op. cit.
[3] M. Dente and I. Pasquon, *Chim. Ind. (Milan)*, **47**: 359 (1965).
[4] P. B. Weisz and E. W. Swegler, *J. Phys. Chem.*, **59**: 823 (1955).

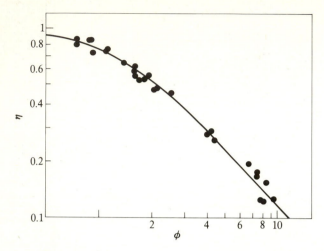

FIGURE 9-31
Comparison of experimental intraphase effectiveness with theory
in methanol oxidation. [*M. Dente and I. Pasquon, Chim. Ind.*
(*Milan*), **47**: *359 (1965)*.]

It is known that $k_2 > k_1$, and in consequence the detectable concentration of intermediate, cyclohexene, should decrease with an increase in intrapellet diffusional
resistance.

For the linear consecutive reaction as catalyzed by a spherical catalyst we
have, for point yield,

$$\frac{-dB}{dA} = \left(\frac{1}{K-1} - \frac{B_0}{A_0}\right) \frac{\phi\sqrt{K} \coth(\phi\sqrt{K}) - 1}{\phi \coth(\phi - 1)} - \frac{1}{K-1} \qquad (9\text{-}65)$$

where $K = k_2/k_1$ and $\phi = R\sqrt{k_1/\mathscr{D}}$.

Weisz and Swegler measured $-dA/dt$, dB/dt, and B_0/A_0, the bulk-gas-phase
concentration ratio, in a differential reactor for catalyst particle radii R of 0.5,
1.84, and 3.1 mm.   For experiments at two particle sizes for, say, runs 2 and 3

$$\frac{\phi_2}{\phi_3} = \frac{R_2}{R_3} \qquad (9\text{-}66)$$

The corresponding ratio of effectiveness factors η_2/η_3 is found from the rates

$$\frac{(dA/dt)_2}{(dA/dt)_3} = \frac{\eta_2}{\eta_3} \qquad (9\text{-}67)$$

Referring to Fig. 9-32, we note that η_2/η_3 and ϕ_2/ϕ_3 form a triangle on the
η-versus-ϕ ln-ln plot; thus by fitting these two ratios, ϕ_2 and ϕ_3 are estimated, from
which ϕ_1 is found since $\phi_1/\phi_3 = R_1/R_3$.

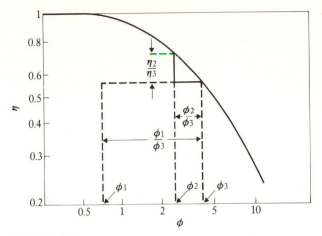

FIGURE 9-32
Functional relationship between η and ϕ, with method of obtaining ϕ's from measured η and ϕ ratios for different catalyst-bead sizes. [*P. B. Weisz and E. W. Swegler, J. Phys. Chem.,* **59:** *823 (1955).*]

We therefore have three values of the Thiele modulus, point yield dB/dA, and B_0/A_0. In Eq. (9-65) all is known except the ratio of intrinsic rate coefficients, $K = k_2/k_1$, which must be independent of catalyst size, i.e., diffusional retardation. One can, of course, find K by trial and error; however, as Weisz notes, graphical solution is the simpler technique. In Fig. 9-33 dB/dA is plotted versus B_0/A_0 for a range of K values at each of three determined values of ϕ. The experimental value of dB/dA at the observed value of B_0/A_0 is shown for each value of ϕ, thus establishing k_2/k_1. In beautiful accord with the model, k_2/k_1 is constant at a value of $k_2/k_1 = 14 \pm 2$.

If diffusional limitations had not been anticipated and the intrinsic point-yield relationship had been used,

$$-\frac{dB}{dA} = 1 - \frac{k_2}{k_1}\frac{B_0}{A_0} \qquad (9\text{-}68)$$

values of $k_2/k_1 = 50$, 125, and 230 would have resulted for the three particle sizes employed.

Finally, the individual values of k are secured, since

$$\frac{dA}{dt} = \eta k_1 A_0 \qquad (9\text{-}69)$$

It is found that $k_1 = 0.15$ s^{-1} and $k_2 = 2.1$ s^{-1}. From the values of the Thiele modulus, particle radius, and rate constant, diffusivity is determined as $\mathscr{D} = 0.8 \times 10^{-3}$ cm^2/s.

Note that the Weisz-Swegler triangulation method cannot be used at high values

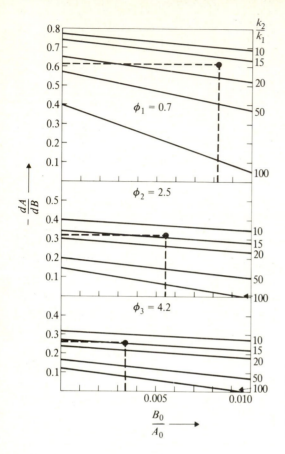

FIGURE 9-33

Graphical solutions of Eq. (9-65) for the measured values of ϕ and B_0/A_0. [*P. B. Weisz and E. W. Swegler, J. Phys. Chem.*, **59**: 823 (1955).]

of ϕ since the slope of the η-versus-ϕ relationship becomes constant on ln-ln coordinates. However, if diffusivity is measured or estimated, the η-versus-$\eta\phi^2$ (observable) plot can be utilized to determine η for each experiment and thus ϕ, and thence k.

Nonisothermal Interphase-Intraphase Effectiveness

A persuasive validation of the theory of nonisothermal effectiveness and the relative importance of interphase and intraphase temperature gradients is to be found in the work of Kehoe and Butt.[1] Single pellets of supported Ni were internally fitted with delicate thermocouples, and both internal and external temperature gradients were measured during the hydrogenation of benzene in a single-pellet reactor. By compositional changes in the support, the ratio of mass to thermal Biot number, for Le = 1,

$$r = \frac{\text{Bi}_m}{\text{Bi}_h} = \frac{\lambda}{\rho C_p \mathscr{D}}$$

---

[1] J. P. G. Kehoe and J. B. Butt, *AIChE J.*, **18**: 347 (1972).

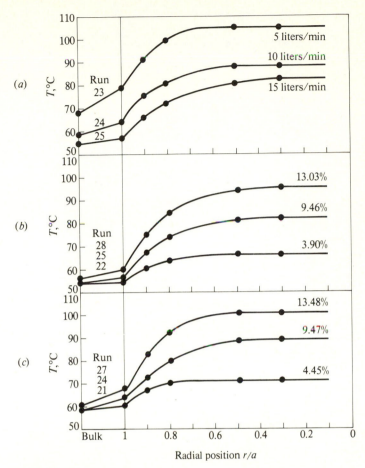

FIGURE 9-34

Measured internal and external temperature profiles for pellet 1, $r =$ $Bi_m/Bi_h = 20$ with feed temperature of 52°C. Profiles (*a*) as a function of flow rate for feed composition of $\sim 10\%$ C_6H_6, (*b*) as a function of feed composition at a high flow, 15 liters/min, (*c*) as a function of feed composition at intermediate flow, 10 liters/min. [*J. P. G. Kehoe and J. B. Butt, AIChE J.,* **18:** *347 (1972).*]

could be varied; for pellet 1, $r = 20$; for pellet 2, $r = 300$. Measured values of interphase and intraphase temperatures are shown in Figs. 9-34 and 9-35 for $r = 20$ and 300, respectively. Note that in the second case the external ΔT_x represents the total interphase and intraphase gradient, i.e., the pellet is isothermal at a temperature determined by interphase heat transport.

Nonisothermal intraphase η values, as noted earlier, will be a function of $\eta = f(\phi, \beta, \epsilon)$. Utilizing the governing kinetic and thermal parameters, Kehoe and Butt computed values of η from the theoretical model, and these are contrasted

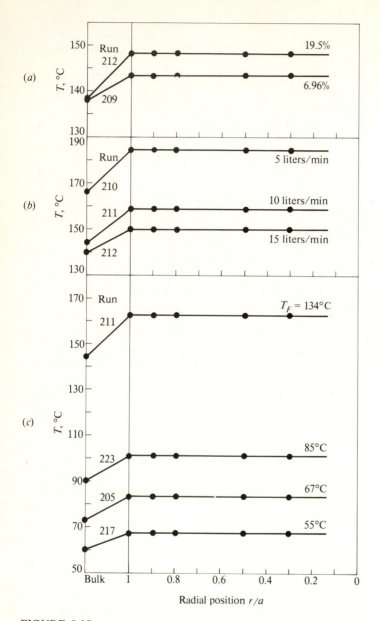

FIGURE 9-35
Measured internal and external temperature profiles for pellet 2, $r =$
$Bi_m/Bi_h = 300$. Profiles (a) as a function of feed composition, (b) as a
function of flow rate, and (c) as a function of feed temperature. [*J. P.
G. Kehoe and J. B. Butt, AIChE J.*, **18**: *347 (1972)*.]

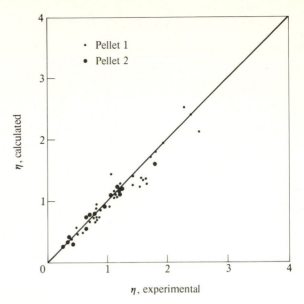

FIGURE 9-36
Comparison of computed and measured nonisothermal intra-pellet effectiveness factors. [*J. P. G. Kehoe and J. B. Butt, AIChE J.*, **18**: *347 (1972)*.]

with experimental results in Fig. 9-36. For this exothermic reaction η values in excess of unity are found, as theory predicts. The agreement between theory and reality is excellent.

Utilizing the experimental ΔT_x, ΔT_0 values in the Kehoe-Butt work, Carberry[1] compared them with his theoretical model (see Chap. 5)

$$\frac{\text{External}}{\text{Total}}\ \frac{\Delta T_x}{\Delta T_0} = \frac{r(\bar{\eta}\text{Da})}{1 + \bar{\eta}\text{Da}(r-1)} \qquad (9\text{-}70)$$

where $\bar{\eta}$Da is the observable. The contrast between theory and experiment was itemized in Table 9-2.

It is of interest to note that in spite of the modest values of the observable, $\bar{\eta}$Da, ΔT_0 values can be appreciable, and the external film contribution can be significant even for this rather low value of $r = 20$, for example, $\bar{\eta}$Da $= 0.0256$, $\Delta T_0 = 100°$C, and $\Delta T_x/\Delta T_0 = 0.33$. The observable can be considered as a measure of that fraction of the global rate which is affected by external mass transport. Other verifications of nonisothermal effectiveness are provided by Maymo and Smith[2] and Cunningham et al.[3] One may prudently conclude that the theory of steady-state catalytic activity and isothermal yield as affected by heat and/or mass transfer is verified by experiment.

[1] J. J. Carberry, *Ind. Eng. Chem. Fundam.*, **14**: 129 (1975).
[2] J. A. Maymo and J. M. Smith, *AIChE J.*, **12**: 845 (1966).
[3] R. E. Cunningham, J. J. Carberry, and J. M. Smith, *AIChE J.*, **11**: 636 (1965).

9-7   DEACTIVATION-DIFFUSION

Wheeler's Analysis

In Chap. 8 the problem of catalytic reaction-deactivation was discussed on the assumption that diffusional limitations do not intervene to affect the phenomenological deactivation rate.   We need no longer embrace such an assumption, for insofar as diffusional phenomena affect catalytic reaction per se, it can be anticipated that deactivation dependent upon the concentration of intermediate and product can, in principle, be subject to the dictates of transport processes.

Wheeler's early insights into the problem proved to be seminal.[1]   He considers a porous catalyst within which a first-order reaction and a poisoning reaction occur.   The rate of poison deposition may be low relative to its transport to within the pores, in which case catalytic sites are poisoned uniformly with time.   On the other hand, if poison deposition is rapid relative to its intraphase transport, the poison will deposit preferentially on the pore mouth initially and grow inward with time as described in Chap. 7 for the SPM of gas-solid reaction.   These limiting types of poisoning are termed *uniform* and *pore-mouth poisoning*, respectively.

Wheeler addressed himself to the question: How is catalytic activity affected as a function of the level of poisoning for these limiting and intermediate modes of poisoning?

First it is obvious that if the main reaction is of very modest vigor, in which instance reaction rate per se is low relative to intraphase diffusion, the effective rate will simply fall in proportion to the population of poisoned sites, whether uniformly or preferentially (pore-mouth) deposited.

In the general situation in which a potentially diffusion-affected main reaction is considered, the consequent activity–poison-level behavior, as Wheeler shows, can be nicely analyzed in terms of effectiveness and the Thiele modulus for the main reaction.

Uniform poisoning   The rate of the principal reaction is

$$\mathscr{R} = \eta k(1 - \alpha)A_0 \qquad (9\text{-}71)$$

where α is the fraction of the catalytic sites which are poisoned.   Obviously the effective surface-rate coefficient k_s is $\eta k(1 - \alpha)$.   Thus it is that the rate-versus-α behavior will depend upon η.   For modest reactivity ($\eta \to 1.0$) observed rate merely declines linearly with α

$$\mathscr{R} = k(1 - \alpha)A_0 \qquad (9\text{-}72)$$

For high reactivity ($\eta \to 0$) the Thiele modulus is $\phi = L\sqrt{k(1 - \alpha)/\mathscr{D}}$, and for values of $\phi > 3$, $\eta = 1/\phi$; hence the rate becomes

$$\mathscr{R} = \frac{k(1 - \alpha)A_0}{\phi} = \frac{1}{L}\sqrt{k(1 - \alpha)\mathscr{D}}\,A_0 \qquad (9\text{-}73)$$

---

[1] A. Wheeler, in P. H. Emmett (ed.), "Catalysis," vol. 2, Reinhold, New York, 1955.

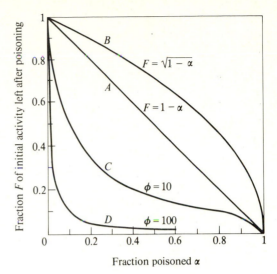

FIGURE 9-37
The ratio of the activity of a catalyst after poisoning to the activity before poisoning as a function of the fraction of the surface poisoned. Curve A: $\eta \rightarrow 1.0$, any distribution of poison; curve B: $\eta \rightarrow 0$, uniform distribution of poison; curve C: $\eta = 0.1$, pore-mouth poisoning; curve D: $\eta = 0.01$, pore-mouth poisoning. (Effectiveness factors refer to conditions before poisoning.) [A. Wheeler, in P. H. Emmett (ed.), "Catalysis," vol. 2, Reinhold, New York, 1955.]

As Wheeler suggested, the poison-affected rates in both cases can be normalized by the unpoisoned rates, respectively [$\alpha = 0$ in Eqs. (9-72) and (9-73)] to yield

For $\eta = 1$:
$$F = \frac{k(1 - \alpha)A_0}{kA_0} = 1 - \alpha \qquad (9\text{-}74)$$

For $\eta \ll 1$:
$$F = \sqrt{1 - \alpha} \qquad (9\text{-}75)$$

This behavior is set forth in Fig. 9-37 for uniform poisoning.

Pore-mouth poisoning   The SPM postulate of pseudo steady state will be invoked. With the creation of a region of total poisoning at the pore mouth, we equate reaction-free diffusion through the poisoned pore mouth to potentially diffusion-affected reaction within the interior unpoisoned pore region $(1 - \alpha)$. Hence, for first-order reaction in a sphere

$$\mathcal{R} = \mathcal{D}4\pi r^2 \frac{dA}{dr} \qquad (9\text{-}76)$$

or, upon integration,

$$\mathcal{R} = \frac{4\pi \mathcal{D} \gamma R_0}{1 - \gamma}(A_0 - A_c) \qquad (9\text{-}77)$$

where $\gamma = r_c/R_0$, subscript c and 0 referring to the poisoned-unpoisoned radial interface and sphere radius, respectively.

We equate (9-77) to reaction within the unpoisoned region of the pore

$$\frac{4\pi \mathcal{D} \gamma R_0}{1 - \gamma}(A_0 - A_c) = \tfrac{4}{3}\pi R_0^3 \gamma^3 \eta k A_c \qquad (9\text{-}78)$$

We solve for A_c, substitute it into (9-77), and obtain the rate per unit volume

$$\mathscr{R}' = \frac{kA_0}{1/\eta(1-\alpha) + 3\phi^2[1 - (1-\alpha)^{1/3}]/(1-\alpha)^{1/3}} \qquad (9\text{-}79)$$

since $\gamma^3 = 1 - \alpha$ and $\phi = (R_0/3)\sqrt{k/\mathscr{D}}$. At $\alpha = 0$ (unpoisoned state) $\mathscr{R}' = \eta k A_0$; thus the poisoned- to unpoisoned-rate ratio for pore-mouth poisoning becomes

$$F = \frac{1}{1/(1-\alpha) + 3\eta\phi^2[1 - (1-\alpha)^{1/3}]/(1-\alpha)^{1/3}} \qquad (9\text{-}80)$$

For large ($\phi = 10$ and 100) values of the Thiele modulus, the F-versus-α behavior is plotted in Fig. 9-37. It is to be observed that for a given amount of imposed poison, pore-mouth deposition proves to be more of a rate taxation than when uniform poisoning prevails.

Experimental insights into the manner of poisoning can, of course, be realized by the measurement of the poisoned- to unpoisoned-rate ratio. Petersen and his students have developed an ingenious single-pellet reactor, which permits the direct assessment of catalytic reaction-deactivation processes.[1]

Levenspiel's Analysis

Khang and Levenspiel[2] have extended the phenomenological description of reaction

$$r_A = k'A^n a \qquad (9\text{-}81)$$

and deactivation

$$r_d = k(A, B, P)^m a^d \qquad (9\text{-}82)$$

discussed in Chap. 8, to those situations where r_A and/or r_d are affected by intra-phase mass diffusion for the simultaneous, consecutive, and parallel reaction-deactivation networks detailed in Chap. 8.

When diffusion of the reactant and/or poison precursor is to be anticipated, the respective Thiele moduli for reactant and poison, ϕ_r and ϕ_p, will determine the distribution of poison within the porous catalyst and thus the activity-selectivity behavior. For simultaneous and parallel sources of reaction-decay the poison distribution as a function of time for small, moderate, and large values of ϕ is schematically shown in Fig. 9-38, while Fig. 9-39 indicates expectations for consecutive deactivation. Khang and Levenspiel solved the diffusion-reaction equations for each of these poisoning networks to establish the mean activity of the pellet and concluded that the order of deactivation d in the equation

$$-\frac{da}{dt} = kAa^d \qquad (9\text{-}83)$$

[1] L. L. Hegedus and E. E. Petersen, *Catal. Rev.*, **9**: 245 (1974).
[2] S. J. Khang and O. Levenspiel, *Ind. Chem. Eng. Fundam.*, **12**: 185 (1973).

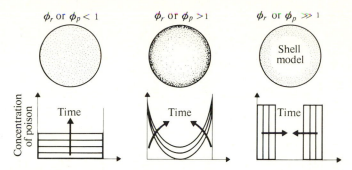

FIGURE 9-38
Resistance to pore diffusion determines where the poison deposits; for parallel or simultaneous deactivation. [*S. J. Khang and O. Levenspiel, Ind. Eng. Chem. Fundam.*, **12**: *185 (1973)*.]

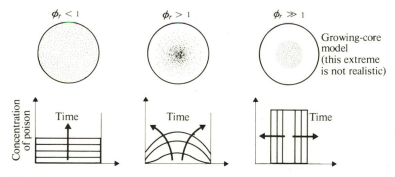

FIGURE 9-39
Resistance to pore diffusion determines where the poison deposits; for series deactivation. [*S. J. Khang and O. Levenspiel, Ind. Eng. Chem. Fundam.*, **12**: *185 (1973)*.]

assumes the following range of values:

Simultaneous deactivation d varies from 1 to 3 as ϕ increases.

Consecutive deactivation d is equal to unity so long as the bulk stream contains the intermediate.

Parallel deactivation When ϕ_r and ϕ_p are small, $d = 1$ and

$$\phi_r \begin{cases} = \phi_p = 1 & d \lessdot 1 \\ \gg \phi_p > 1 & d \to 2 \\ = \phi_p > 1 & d \to 3 \end{cases}$$

$\phi_p > \phi_r > 1$ d not const

Independent deactivation if this is due to reactants and products, $d = 1$; if it is due to sintering, the cause is thermal, and while independent of mass intraphase diffusion, it will be sensitive to intraphase temperature gradients. If the internal ΔT is negligible, $d \approx 2$, as suggested in Chap. 8.

In view of the fact that these analyses are rooted in linearity of the principal reactions and that the deactivation exponent proves to be somewhat variable, one is well advised to ascertain d from experiments as described in Chap. 8. Extra-kinetic insights are surely required if the precise mechanism of deactivation is to be established.

Masamune-Smith Analysis[1]

For the simultaneous, consecutive, and independent deactivation schemes, Masamune and Smith solved the governing differential equations, compared them with the pseudo-steady-state SPM and presented the results of the deactivating catalyst in terms of an effectiveness factor which is a function of both the Thiele moduli and time on-stream. They conclude that for consecutive and independent deactivation, the pellet of lowest intraphase resistance yields the highest activity-time profile, while for simultaneous deactivation, a pellet exhibiting an intermediate level of intraphase resistance gives the highest activity, particularly at long process times. These findings are presented in Fig. 9-40, where

$$\theta_i = \frac{k_i C_0 t}{q_0}$$

where k_i = main-reaction-rate constant
C_0 = reactant-feed concentration
q_0 = saturation concentration on solid
t = time on-stream

Carberry-Gorring Analysis[2]

In this mode of assessing reaction-deactivation, the deactivation process is viewed in terms of gas-solid noncatalytic reaction (Chap. 7) under conditions where the SPM can validly be invoked. This application of the SPM to coking, poisoning, and deactivation in general is inspired by the fact that observations of catalyst decay in petroleum processing have been correlated in terms of the Voorhies equation,[3] which states that the extent of poisoning, coking, or fouling α is related to time on-stream t by

$$\alpha = K\sqrt{t} \qquad (9\text{-}84)$$

or, in general,

$$\alpha = Kt^n \qquad (9\text{-}85)$$

[1] S. Masamune and J. M. Smith, *AIChE J.*, **12**: 384 (1966).
[2] J. J. Carberry and R. L. Gorring, *J. Catal.*, **5**: 529 (1966).
[3] A. Voorhies, *Ind. Eng. Chem.*, **38**: 318 (1945).

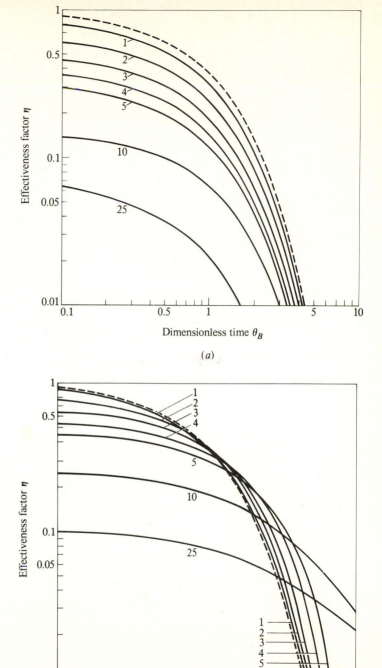

FIGURE 9-40a and b

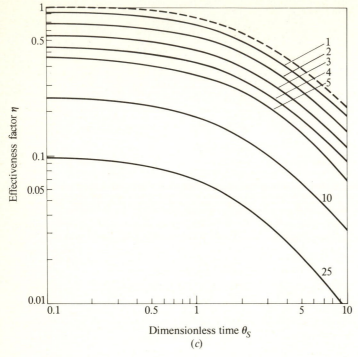

FIGURE 9-40
Effectiveness factor vs. dimensionless time for diverse deactivation schemes with parameter ϕ, reaction Thiele modulus where ϕ_p is that of deactivation process: (a) series fouling, $\phi_s = 0.1$, (b) parallel fouling, (c) independent fouling, $\phi_s = 10$. [*S. Masamune and J. M. Smith, AIChE J.*, **12**: *384* (*1966*).]

A value of $n = \frac{1}{2}$ has been interpreted as an indication of a diffusion-affected deactivation event.   In practice, therefore, one plots α versus t on ln-ln coordinates, and it is found that the order with respect to time often assumes a value of $\frac{1}{2}$, though values of $n \gtreqless \frac{1}{2}$ have been observed.

In fact, if we anticipate that interphase and intraphase diffusion and poison-solid reaction may be of comparable velocity, then, as developed in Chap. 7, the time-on-stream behavior can be expressed in terms of the three rate processes as

$$t = \frac{\rho' R^2}{b \mathscr{D} C_0}\left[\frac{1-\gamma^3}{3}\left(\frac{1}{\text{Bi}_m} - 1\right) + \frac{1-\gamma^2}{2} + \frac{1-\gamma}{\text{Da}}\right] \qquad (9\text{-}86)$$

where R = pellet radius
ρ' = molar density
$\mathscr{D}$ = intraphase diffusivity
C_0 = bulk poison-precursor concentration

b is the stoichiometric coefficient for

$$C_0 + b \text{ (solid)} \rightarrow \text{product of poisoning}$$

For a sphere $\alpha = 1 - \gamma^3$, where $\gamma = r/R$ and $\text{Bi}_m = \dfrac{k_g R}{\mathscr{D}}$ $\text{Da} = \dfrac{kR}{\mathscr{D}}$

k, the poisoning-rate coefficient [Eq. (9-86)] obviously applies to regeneration, e.g., coke burnoff by an oxygen-bearing gas stream.

The extent of poisoning, coking, or regeneration is plotted versus

$$\tau = \frac{1}{R}\sqrt{\frac{\mathscr{D}C_0 t}{\rho'}} \qquad (9\text{-}87)$$

in Fig. 9-41 for a range of Bi_m and Da values. The flat-plate solution, derived in Chap. 7, is also indicated. We see that a simple α-versus-$\sqrt{t}$ correlation can be detected over certain ranges of τ for cases in which intraphase diffusion is *not* the unique rate-determining step. Further, the flat-plate solution which is an explicit α-versus-t functionality rather accurately describes this mixed rate-control process up to about 20 to 40 percent of extent of poisoning or regeneration.

Wheeler's analysis gives insight into the limiting modes of poisoning; Levenspiel's treatment encompasses phenomenological formulation of reaction-deactivation rates; the Masamune-Smith analysis reveals effectiveness-factor behavior for diverse poisoning mechanisms, while the Carberry-Gorring commentary views the time on-stream and extent of poisoning process in the limiting circumstances where the SPM postulates are obeyed. What of the issue of experimental verification and procurement of reaction-deactivation parameters? The single-pellet reactor of Petersen[1] and his students proves to be of signal merit in response to such queries.

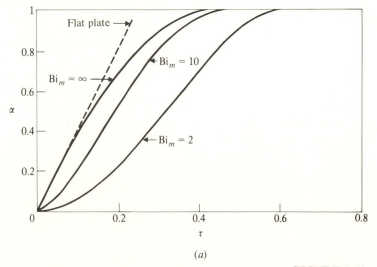

(a)

FIGURE 9-41

[1] L. L. Hegedus and E. E. Petersen, *Catal. Rev.*, 9: 245 (1974).

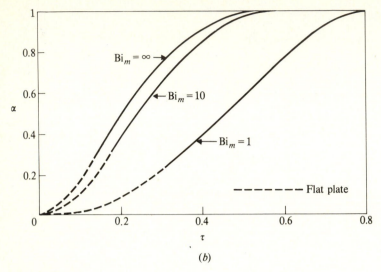

(b)

FIGURE 9-41

Fraction of sphere poisoned α versus $\tau = (1/R)\sqrt{\mathscr{D}C_0 t/\rho'}$ for (a) $Da = \infty$ and (b) $Da = 10$.   [J. J. Carberry and R. L. Gorring, J. Catal., **5**: 529 (1966).]

Petersen's Poisoned Pellet

The essential features of the Petersen single-pellet reactor are shown in Fig. 9-42. Reactants are found on one face of the pellet, while the other face is equivalent to the center plane of a pellet twice as long as the experimental pellet.   By analyzing concentrations at the back face, centerline concentrations are secured, from which, as shown below, reaction-deactivation parameters can be extracted.

First, for the steady-state nondeactivating system, one measures the reaction rate by noting the moles of reactant consumed per unit time in the recirculating reactant-bearing loop, which is equivalent to a CSTR at high recycle ratios.   Concentrations at the back face or centerline are also measured.   Reaction order is readily established by the observed variation in centerline reduced concentration C/C_0 with reactant concentration C_0.

For illustrative purposes let us assume that C/C_0 is independent of C_0, for such a first-order system

$$\eta = \frac{\tanh \phi}{\phi} \tag{9-88}$$

$$\frac{C}{C_0} = \frac{1}{\cosh \phi} \tag{9-89}$$

and the rate is

$$\overline{\mathscr{R}} = \eta k_1 C_0 V_{\text{pellet}} \quad \frac{\text{mol}}{\text{time}} \tag{9-90}$$

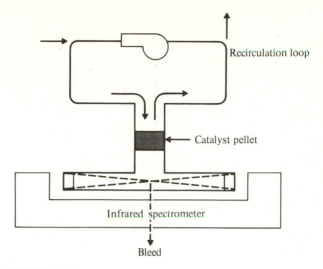

FIGURE 9-42
Schematic of Petersen's poisoned-pellet reactor. [*L. Hegedus and E. E. Petersen, Catal. Rev.*, **9**: *245 (1974).*]

In a single experiment, one measures $\bar{\mathcal{R}}$ and C/C_0. From Eq. (9-89), ϕ is secured, and then, by Eq. (9-88), η.   Then k_1 is obtained from Eq. (9-90).   Finally the effective diffusivity is extracted from ϕ since pellet length and k_1 are known. This strategy is applicable for any reaction order, though the η and C/C_0 functionalities assume more complex forms, of course, and are tabulated in the Hegedus-Petersen reference for orders of from -2 to $+2$.

When deactivation-reaction occur at rates which admit the pseudo-steady-state approximation, the Petersen reactor permits a determination of the mode of poisoning as well as the rate parameters.   (In addition to uniform and selective, or pore-mouth, poisoning discussed above, one may encounter core poisoning; i.e., the pellet poisons from its center outward, the opposite of the pore-mouth process.)

When linear kinetics is assumed, rate relative to its zero-time unpoisoned value $\mathcal{R}_0$ and the respective center-plane reduced concentration are for uniform poisoning,

$$\frac{\mathcal{R}}{\mathcal{R}_0} = \frac{\tanh \phi}{\tanh \phi_0} \frac{\phi_0}{\phi}$$

$$\frac{C}{C_0} = \frac{1}{\cosh \phi} \qquad \left(\frac{C}{C_0}\right)_0 = \frac{1}{\cosh \phi_0}$$

(9-91)

Each ratio is established in a single experiment during which poisoning occurs. For other modes of poisoning the functionalities are given in the reference work.

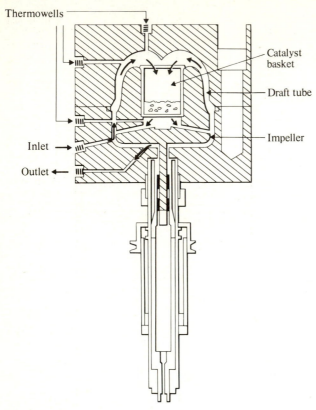

Thermowells

Catalyst basket

Draft tube

Impeller

Inlet →

Outlet ←

FIGURE 9-43
Modified Berty CSTCR. [*J. A. Mahoney, J. Catal.*, **32**: *247 (1974).*]

Example Mahoney[1] provides an excellent example of the utilization of a CSTCR to ellucidate the reaction-deactivation behavior of a complex reforming (dehydro-cyclization) process as catalyzed by a supported-Pt and a supported bimetallic catalyst.

An internal recycle CSTCR, a modified version of the Berty reactor[2] (Fig. 9-43), was employed. The basic reaction network is for reactant *n*-heptane

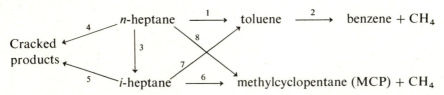

[1] J. A. Mahoney, *J. Catal.*, **32**: 247 (1974).
[2] J. Berty, *Chem. Eng. Prog.*, **70**(5): 78 (1974).

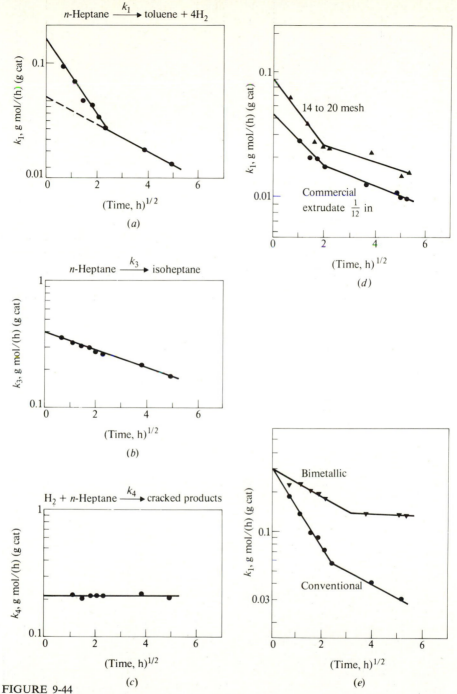

FIGURE 9-44
(a) Dehydrocyclization deactivation; (b) paraffin isomerization deactivation; (c) paraffin hydrocracking deactivation; (d) effect of pore diffusion on dehydrocyclization rate constant; (e) effect of bimetallic promoter on deactivation rate. [*J. A. Mahoney, J. Catal.*, **32**: *247* (*1974*).]

Table 9-5 MASS BALANCES

Y = mole fraction

θ = (molar space velocity)$^{-1}$ = $\dfrac{W}{F}$

$Y_{\text{toluene}} = \theta[k_1(Y_{nC70} + Y_{iC7}) - k_2 Y_{\text{toluene}}]$

$Y_{\text{benzene}} = \theta k_2 Y_{\text{toluene}}$

$Y_{nC7,i} - Y_{nC7,0} = \theta Y_{nC7,0}(k_1 + k_3 + k_4)$

$Y_{iC7} = \theta[k_3 Y_{nC7,0} - (k_1 + k_4 + k_6)Y_{iC7}]$

$Y_{\text{MCP}} = 2\theta k_6(Y_{iC7} + Y_{nC7,0})$

Because of coke formation, the catalyst deactivates with time on-stream, but since the decay time is of the order of hours, the CSTCR can be gainfully employed to determine the time (deactivation-level) dependency of each of the rate constants in the above network (Table 9-5). Measured values of k_1, k_3, and k_4 are shown versus $\sqrt{t}$ in ln-ln coordinates in Fig. 9-44a to c, a method of deactivation correlation known as the Voorhies relationship. In Fig. 9-44d the influence of intraphase diffusion on k_1 is evident, while Fig. 9-44e shows the dramatic role of a bimetallic catalyst in reducing the deactivation rate. The data in Fig. 9-44a to c were secured in a *single* experiment.

9-8 POSSIBLE REMEDIES FOR DIFFUSIONAL INTRUSIONS

Realistically, for a given catalyst, pore-size distribution, and structure, in the usually limited temperature range of interest, the pellet dimension L (or for a sphere $R/3$), would seem to be the only reasonable parameter in the Thiele modulus which can be engineered to alter the diffusion-reaction consequences set forth above. However in securing a reduction of the Thiele modulus by reducing pellet size, tube-to-pellet-diameter ratio is increased, thus inviting radial temperature gradients. Further, pressure drop increases, though this may not be important in a laboratory scale study. And, as observed in Chap. 8, an increase in intrinsic activity associated with a decrease in pellet size may well create an interphase temperature gradient associated with the consequent enhancement of local heat generation ($\pm$).

In fact, however, even a seemingly minor tailoring of average pore size, and hence intraphase diffusivity, can yield benefits, as Weisz testifies.[1] By modifying the pore structure and thus doubling the effective diffusivity of a cracking catalyst, diffusion-sponsored attrition losses in the regeneration kiln were halved.

A further remedy is to be found in partial impregnation; i.e., the support is impregnated to the extent of some fraction (<1) of its volume so that the active catalytic zone is a shell. Under such circumstances $\phi = F_1 L\sqrt{k/\mathscr{D}}$, where $F_1 < 1.0$. This simple insight assumes, of course, that k remains identical to the fully impregnated value. This cannot be so unless the catalytic species is concentrated within

---

[1] P. B. Weisz, *Science*, **179**: 433 (1973).

the shell, a measure which may, however, invite sintering. More generally one can envision a modification of k with F_1, that is, $\phi = F_1 L \sqrt{k F_2 / \mathscr{D}}$, where, of course, both F_1 and F_2 are less than unity. This manner of analysis points to the fact that benefits may well follow if both partial impregnation *and* a reduction of intrinsic specific activity are entertained when diffusional intrusions reveal taxing consequences upon activity and yield. The apparent benefits of partial impregnation will be discussed in Chap. 10.[1]

9-9 MULTIPLICITY OF THE STEADY STATE

Our earlier treatments of multiplicity in the CSTR, nonporous catalysts, gas-liquid and gas-solid systems should suffice to indicate that multiplicity in a porous-solid-catalyzed pellet is possible and in the instance of consecutive reaction, multiple multiplicities may be anticipated. A rich literature exists on this topic and that of stability as cited below.

SUMMARY

The material touched upon in this chapter should provide the reader with insights into diffusional phenomena which may significantly alter the intrinsic dispositions of the porous-catalyst formulation. Such diffusional intrusions can be of benefit in circumstances where activation energies or surface environmental requirements are so ordered in nature that the constant interphase and intraphase concentration-temperature condition is not to be sought.

ADDITIONAL REFERENCES

The seminal work first identifying and quantifying the intraphase diffusion-reaction problem is to be found in E. Thiele; *Ind. Eng. Chem.*, **31**: 916 (1939).

The most profound, cosmic treatment of the interphase-intraphase diffusion-reaction issue is set forth in R. Aris, "The Mathematical Theory of Diffusion and Reaction in Permeable Catalysts," vols. I and II, Clarendon, Oxford, 1975, a monumental work. C. N. Satterfield, "Mass Transfer in Heterogeneous Catalysis," MIT Press, Cambridge, Mass., 1970, is rich in data, correlation, and illustrative computations. Physical aspects of heterogeneous catalysis are tersely set forth by E. Wicke, *Mem. Soc. R. Sci. Liege*, 6 ser., tome 1, fasc. 4, p. 211 (1971), in English, and *Adv. Chem. Ser.*, **109**: 183 (1972). W. E. Corbett and D. Luss, *Chem. Eng. Sci.*, **29**: 1473 (1974), treat the influence of nonuniform distribution of catalytic sites, within a sphere, upon activity, selectivity, and deactivation.

Data and theory stand nicely in accord in the work of J. A. Maymo and J. M. Smith, *AIChE J.*, **12**: 845 (1966), in their study of H_2 oxidation over supported Pt in the presence of interphase and intraphase diffusion of heat and mass. A comprehensive analysis of mass transport and reaction during catalyst regeneration is presented by J. M. Ausman and C. C. Watson, *Chem. Eng. Sci.*, **17**: 323 (1962).

---

[1] The benefits of partial impregnation upon yield are examined by T. G. Smith and J. J. Carberry, *Can. J. Chem. Eng.*, **53**: 347 (1975).

Diverse aspects of catalyst poisoning are tersely rendered in a note by J. B. Butt, *Adv. Chem.*, **109**: 259 (1972). See also L. Hegedus, *Ind. Eng. Chem. Fundam.*, **13**: 190 (1974). Key insights into diffusion-affected catalytic activity, selectivity, and deactivation were first presented in terms of observables by A. Wheeler, in P. H. Emmett (ed.), "Catalysis," vol. 2, p. 123, Reinhold, New York, 1955.

The numerical method whereby diffusion-flow-reaction models of a nonlinear nature (with flux boundary conditions) can be resolved is detailed in J. J. Carberry and M. M. Wendel, *AIChE J.*, **9**: 129 (1963). The method of collocation is described in B. Finlayson, *Cat. Rev. Sci. Eng.*, **9**: 169 (1974). The Prater-Wei analysis of complex linear reaction networks is extended to include diffusional effects in J. Wei, *J. Catal.*, **1**: 526, 538 (1962). A lucid presentation of the Prater-Wei methodology is to be found in M. Boudart, "Kinetics of Chemical Processes," Prentice-Hall, Englewood Cliffs, N.J., 1968. G. Roberts and C. N. Satterfield, *Ind. Eng. Chem. Fundam.*, **5**: 317 (1966), treat isothermal intraphase effectiveness for diverse LHHW kinetic models; these computations reveal pathological behavior for strong rate inhibition by a coreactant.

A provocative study of heat transport in the Pt-wire–catalyzed decomposition of NO_2 is given in B. Ong and D. M. Mason, *Ind. Eng. Chem. Fundam.*, **11**: 169 (1972). Multiplicity and instability problems in heterogeneously catalyzed systems are analyzed in J. Cardoso and D. Luss, *Chem. Eng. Sci.*, **24**: 1699 (1969).

PROBLEMS

9-1   An isothermal bed of porous spherical catalyst pellets of 3 mm diameter exhibits a conversion commensurate with a first-order rate constant of 0.3 s^{-1}. Intrapellet diffusivity is $3 \times 10^{-2} \text{ cm}^2/\text{s}$. If a 2-m depth of the present catalyst is required to realize design conversion, what depth will be needed if because of reduced pressure drop catalyst of 6 mm diameter is to be utilized?

9-2   Interpret the following data for possible diffusional intrusions:

T_0,°C	100	150
Rate, mol/(min)(kg)	50	63

9-3   A 100-μm sample of supported-Ni benzene-hydrogenation catalyst exhibits linear kinetics with a diffusion-unaffected rate constant of 5 min^{-1} at 150°C. At 1 atm intrapellet diffusivity is $0.2 \text{ cm}^2/\text{s}$.

(*a*)   What is the largest size porous sphere of this catalyst which will exhibit an intraphase effectiveness of 0.8 at 1 atm?

(*b*)   Repeat part (*a*) for 20 atm, assuming $\mathscr{D}$ is inversely proportional to pressure.

(*c*)   If liquid-phase hydrogenation is anticipated at the same specific rate but $\mathscr{D}$ is now $10^{-5} \text{ cm}^2/\text{s}$, what catalyst size would be required for $\eta = 0.5$?

9-4   Cylinders (5 by 5 mm) of γ-alumina upon which Pt is dispersed catalyzed dehydrogenation of cyclohexane. At 700 K the intrinsic rate constant is 4 s^{-1}. What will the apparent, observed rate constant be at that temperature? Each pellet weighs 150 mg; true density is 2.5 g/cm^3, and the BET area is $50 \text{ m}^2/\text{g}$.

(*a*)   Assume Knudsen diffusion; use a tortuosity factor of $\sqrt{2}$.

(*b*)   Assume bulk diffusion of cyclohexane through H_2 to prevail.

(*c*)   What factors dictate whether intraphase diffusion is of Knudsen or bulk variety?

(*d*)   If $\mathscr{D}_K = 5 \times 10^{-2} \text{ cm}^2/\text{s}$ and $\mathscr{D}_B = 7 \times 10^{-2} \text{ cm}^2/\text{s}$, what is $\mathscr{D}_{\text{eff}}$?

9-5 Sucrose inversion as catalyzed by the acid form of an ion-exchange resin has been studied by Reed and Dranoff [*Ind. Eng. Chem. Fundam.*, **3**: 304 (1964)] at 70°C:

Average resin diameter, mm	k, min^{-1}	For the 1-mm resin particles	
		Temp, °C	k, min^{-1}
1	0.134	50	0.032
0.715	0.168	60	0.067
0.5	0.226	70	0.134

Interpret these data and determine the true activation energy for this reaction.

9-6 In the kinetic study of Pt-catalyzed oxidation of CO in a CSTCR, McCarthy [Ph.D. thesis, University of Notre Dame, Notre Dame, Ind., 1974; see also McCarthy, Zahradnik, Kuczynski, and Carberry, *J. Catal.*, **39**: 29 (1975)] reports the following rate vs. percentage CO data at 260°C:

Rate, mmol CO/(h)(cm^3)	CO %
10	0.25
15	0.37
16	0.5

The diameter of catalyst pellets was 0.2 cm. At 1600 r/min the data of Periera and Calderbank [*Chem. Eng. Sci.*, **30**: 167 (1975)] indicate that the interphase mass-transfer coefficient is 7 cm/s. Test these data for heat- and mass-diffusional intrusions.

9-7 Maymo and Smith [*AIChE J.*, **12**: 845 (1966)] report the following rate data for the supported-Pt-catalyzed oxidation of hydrogen as a function of catalyst-pellet surface temperature in excess hydrogen and constant partial pressure of oxygen.

Rate, μmol/(g)(s)	Surface temp, °C
30	93
52	130
80	172
140	209

What is your interpretation of these results?

9-8 For a macroporous catalyst we know that in the limit of severe isothermal intrapellet diffusion limitation the apparent measured activation energy will be about one-half the intrinsic value. What will be the apparent value of E in the instance of severe isothermal internal diffusional intrusions for a microporous-macroporous catalyst? Assume that the mass Biot number is large.

9-9 If the assumption of an isothermal pellet operative at a temperature dictated by external, interphase heat transport can be tolerated, one can, as shown in Sec. 9-3, fashion a simple expression for isothermal-intraphase–nonisothermal-interphase effectiveness [Eq. (9-37)]. Develop such an expression for a catalyst of finite external area relative to the total area. Plot the results for $f = 10^{-1}$, 10^{-2}, and 10^{-3}, for mass Biot numbers of 1000 and 10.

9-10   The catalytic rate of methane total oxidation is first order in methane and zero order in oxygen.   The rate coefficients for Pd/γ-Al$_2$O$_3$ and Pt/γ-Al$_2$O$_3$, respectively, are at 450°C:

For Pd: $k = 40 \exp\left[-13.7\left(\frac{1}{t} - 1\right)\right]$ s^{-1}

For Pt: $k = 10 \exp\left[-6.4\left(\frac{1}{t} - 1\right)\right]$ s^{-1}

For a bed of 6-mm-diameter spherical catalyst, compare the activities of these catalysts at 300, 400 450, 600, and 800°C.   Intraphase diffusivity is of the Knudsen type and its value at 450°C is 10^{-3} cm^2/s.   The packed-bed gas velocity is 20 cm/s, defined by Eq. (9-64).   The Schmidt number is that of methane diffusing in excess air; void fraction of the bed is 0.4.

9-11   In an integral reactor like that posited in Prob. 9-10, it is difficult to obtain the observed rate since the system is a distributed one.   However since methane-oxidation kinetics is normal, in an isothermal integral reactor the maximum rate is found at bed inlet.   For the conditions of Prob. 9-10 estimate the bulk-gas to pellet-center ΔT_0 and the fraction of that value found in the gas film at bed entrance for a feed of 3% methane in air for both catalysts at the temperature (bulk) values specified above.   The thermal conductivity of γ-Al$_2$O$_3$ is 2×10^{-4} cal/(°C)(cm)(s) and may be assumed temperature-independent.

9-12   Seek out the work of Fulton and Crosser [*AIChE J.*, **11**: 513 (1965)] and render an interpretation of their data in terms of observables.

9-13   From Mahoney's data (Fig. 9-44) attempt to fashion a deactivation model rooted in postulates other than that of Voorhies.

9-14   Kehoe and Butt (*Chem. React. Eng. 5th Eur./2d Int. Symp. Chem. React. Eng., Amsterdam, 1972*, pap. B-8-13, Elsevier) present experimental data and analyses of the transient response and stability of a diffusion-limited exothermic solid-catalyzed reaction.   Study this paper and write a short, precise, detailed report on the topic.

9-15   Consider an adiabatic, nonporous, solid-catalyzed reaction governed by bulk mass transport.   Plot surface and gas-phase temperature vs. reactor length for Le numbers of 0.5, 1.0, and 1.5.

9-16   Given results found in response to Prob. 9-15, discuss their significance in the light of thermodynamic and kinetic teachings.   See L. L. Hegedus [*AIChE J.*, **21** (5): 849 (1975)].

10

ANALYSES AND DESIGN OF HETEROGENEOUS REACTORS

" I go to seek a great perhaps...,"
Rabelais

Introduction

Assuming that the student now has a command of the elements of reaction behavior (Chap. 2), ideal and real reactor behavior (Chaps. 3 and 4), the general
principles of heterogeneous reaction (Chap. 5), and its specification in terms of
fluid-fluid reaction (Chap. 6), fluid-solid reaction (Chap. 7), heterogeneous catalysis
(Chap. 8), and diffusion-affected heterogeneous catalysis (Chap. 9), we now address
ourselves to problems of overall reactor analyses for heterogeneous catalytic reactions. It should now be apparent that a priori design rooted in first principles of
chemistry, physics, mathematics, and economics is not to be realized except in the
most trivial chemical reaction engineering circumstances.

In this, the terminal chapter, elements previously discussed will be assembled
to describe and specify the parameters which govern behavior and design of the
following catalytic reactor types:

Fixed-bed catalytic reactor
Batch fluidized-bed catalytic reactor
Slurry (gas-liquid-solid catalyst) reactor

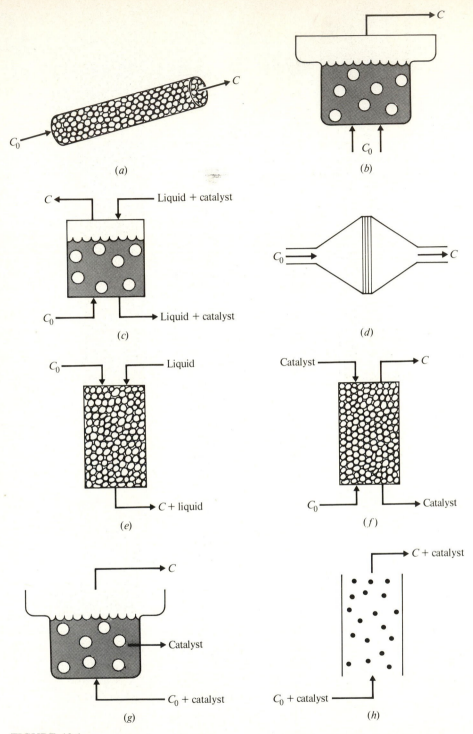

FIGURE 10-1
Heterogeneous-catalytic-reactor types: (a) fixed bed; (b) batch fluid bed; (c) slurry reactor; (d) catalytic gauze; (e) trickle bed; (f) moving bed; (g) continuous fluid bed; (h) transport line.

Catalytic-gauze (wire-mesh) reactor
Trickle-bed reactor

Steady-state catalyst activity will be assumed in the analyses of these reactor types.

Cognizant of the fact that catalysts rarely enjoy immortality and in parochial deference to the Lucrezia Borgia Society,[1] we next direct attention to catalytic reactors in which intrinsic catalytic activity is dependent on time on-stream. Reactors which accommodate catalyst deactivation are moving-bed catalytic reactors, continuous fluid-bed reactors, and transport-line reactors. Fixed-bed, fluidized, and slurry catalytic reactors, subject to deactivation, also deserve a commentary, though in these time-on-stream-dependent circumstances, analyses and design prove to be rather complex.

The diverse catalytic reactor types are schematized in Fig. 10-1. The conditions of both fluid and catalyst phase, i.e., batch, continuous, constant, variable, are itemized in Table 10-1, which also indicates dependence of catalyst activity on time on-stream and the scope of anticipated global nonisothermality. Obviously in cases where catalytic activity depends on time on-stream, temperature will be a function of that time as well as length Z and radial position.

10-1 THE FIXED-BED REACTOR

The general equations of continuity for a cylindrical fixed bed operating under nonisothermal, nonadiabatic conditions will be set forth first. Dimensionless parameters generated by these equations and the boundary conditions are then discussed, and means whereby these parameters may be determined are given. Tolerable simplifications of the general equations are noted. Solutions of two practical problems are then described, and several important features of reactor analyses and design are discussed. The first practical problem is that of adiabatic oxidation of SO_2, and the second is that of nonisothermal, nonadiabatic oxidation of naphthalene.

Table 10-1   FLUID-SOLID CATALYTIC-REACTOR CHARACTERISTICS

Type	Fluid	Solid	Activity, t	Temperature
Fixed bed	PF	Batch	No or yes	$f(Z, R)$
Fluid bed	PF	CST	No or yes	$\sim$ uniform
Slurry	CST or PF	CST	No or yes	Uniform
Gauze	Between PF and CST	Batch	No or yes	Uniform
Trickle bed	$\sim$ PF	Batch	No or yes	$f(Z, R)$
Moving bed	PF	PF	No	$f(Z, R)$
Transport line	PF	PF	No	$f(Z)$

[1] J. J. Carberry, *Chem. Technol.*, February **1974**: 124

Continuity Equations

Following the procedure explicitly set forth in Chap. 4, the mass and thermal-energy continuity equations for a fixed-bed cylindrical reactor of length Z and radius R_0 for key species concentration are

$$\frac{\partial C}{\partial t} - \mathscr{D}_a \frac{\partial^2 C}{\partial Z^2} - \mathscr{D}_r \left(\frac{\partial^2 C}{\partial R^2} + \frac{1}{R} \frac{\partial C}{\partial R} \right) + u \frac{\partial C}{\partial Z} = \mathscr{R} \qquad \text{mol/(vol)(time)} \qquad (10\text{-}1)$$

For fluid temperature, assuming u and ρC_p to be constant,

$$\rho C_p \frac{\partial T}{\partial t} - \rho C_p K_a \frac{\partial^2 T}{\partial Z^2} - \rho C_p K_r \left(\frac{\partial^2 T}{\partial R^2} + \frac{1}{R} \frac{\partial T}{\partial R} \right) + \rho C_p u \frac{\partial T}{\partial Z} = q \qquad \text{cal/(vol)(time)}$$

$$(10\text{-}2)$$

K_a, $\mathscr{D}_a$ and K_r, $\mathscr{D}_r$ are turbulent diffusivities (in square centimeters per second) for axial and radial dispersion of heat K and mass $\mathscr{D}$. Hence, $\rho C_p K_a$ and $\rho C_p K_r$ are effective thermal conductivities of the bed. The generation terms $\mathscr{R}$ and q express mass and heat exchange between the flowing fluid and the source of generation, the catalyst particle.

Reactor Parameters

As Hulbert[1] has observed, the left-hand sides of Eqs. (10-1) and (10-2) describe the reactor, while the right-hand sides describe the global reaction. Key reactor parameters emerge when Eqs. (10-1) and (10-2) are rendered dimensionless. Letting

$$n = \frac{Z}{d_p} \qquad m = \frac{R_0}{d_p} \qquad \mathscr{A} = \frac{Z_0}{R_0} \qquad \theta = \frac{Z_0}{u}$$

$$f = \frac{C}{C_0} \qquad t = \frac{T}{T_0} \qquad \bar{r} = \frac{R}{R_0} \qquad z = \frac{Z}{Z_0} \qquad \tau = \frac{t}{\theta}$$

we obtain the dimensionless continuity equations

$$\frac{\partial f}{\partial \tau} - \frac{\mathscr{D}_a}{Z_0 u} \frac{\partial^2 f}{\partial z^2} - \frac{\mathscr{D}_r}{R_0 u} \frac{Z_0}{R_0} \left(\frac{\partial^2 f}{\partial \bar{r}^2} + \frac{1}{\bar{r}} \frac{\partial f}{\partial \bar{r}} \right) + \frac{\partial f}{\partial z} = \frac{\mathscr{R}\theta}{C_0} \qquad (10\text{-}3)$$

$$\frac{\partial t}{\partial \tau} - \frac{K_a}{Z_0 u} \frac{\partial^2 t}{\partial z^2} - \frac{K_r}{R_0 u} \frac{Z_0}{R_0} \left(\frac{\partial^2 t}{\partial \bar{r}^2} + \frac{1}{\bar{r}} \frac{\partial t}{\partial \bar{r}} \right) + \frac{\partial t}{\partial z} = \frac{q\theta}{\rho C_p T_0} \qquad (10\text{-}4)$$

For a fixed bed the Peclet numbers are defined in terms of particle diameter, and so

$$\text{Pe}_a = \frac{d_p u}{\mathscr{D}_a} \qquad \text{Pe}_r = \frac{d_p u}{\mathscr{D}_r}$$

[1] H. H. Hulbert, Peter C. Reilly Lecturer, University of Notre Dame, 1967.

Therefore

$$\frac{Z_0 u}{\mathscr{D}_a} = \frac{d_p u}{\mathscr{D}_a}\frac{Z_0}{d_p} = \mathrm{Pe}_a n$$

(10-5)

and

$$\frac{R_0 u}{\mathscr{D}_r}\frac{R_0}{Z_0} = \frac{d_p u}{\mathscr{D}_r}\frac{R_0}{d_p}\frac{R_0}{Z_0} = \mathrm{Pe}_r\frac{m}{\mathscr{A}}$$

Our dimensionless equations then become

$$\frac{\partial f}{\partial \tau} - \frac{1}{\mathrm{Pe}_a n}\frac{\partial^2 f}{\partial z^2} - \frac{\mathscr{A}}{\mathrm{Pe}_r m}\left(\frac{\partial^2 f}{\partial \bar{r}^2} + \frac{1}{\bar{r}}\frac{\partial f}{\partial \bar{r}}\right) + \frac{\partial f}{\partial z} = \frac{\mathscr{R}\theta}{C_0}$$

(10-6)

and

$$\frac{\partial t}{\partial \tau} - \frac{1}{\overline{\mathrm{Pe}}_a n}\frac{\partial^2 t}{\partial z^2} - \frac{\mathscr{A}}{\overline{\mathrm{Pe}}_r m}\left(\frac{\partial^2 t}{\partial \bar{r}^2} + \frac{1}{\bar{r}}\frac{\partial t}{\partial \bar{r}}\right) + \frac{\partial t}{\partial z} = \frac{q\theta}{\rho C_p T_0}$$

(10-7)

where $\overline{\mathrm{Pe}}$ = Peclet number for heat dispersion
$n, m, \mathscr{A}$ = aspect ratios

Reactor behavior is thus governed by the following dimensionless coefficients: the Peclet numbers, aspect ratios, and dimensionless time τ for a given generation function and boundary conditions. Assuming steady-state operation ($\partial f/\partial \tau$ and $\partial t/\partial \tau = 0$), the boundary conditions are:

At $z = 0$ (reactor entrance):

$$f = 1 + \frac{1}{\mathrm{Pe}_a n}\left(\frac{\partial f}{\partial z}\right)_{z=0} \qquad t = 1 + \frac{1}{\overline{\mathrm{Pe}}_a n}\left(\frac{\partial t}{\partial z}\right)_{z=0}$$

(10-8)

At $z = 1$: $\qquad \dfrac{\partial f}{\partial z} = 0 \quad$ and $\quad \dfrac{\partial t}{\partial z} = 0$

(See Chap. 4 for the derivation of the entrance-exit boundary conditions when axial dispersion exists.) At the bed centerline, by symmetry,

$$\frac{\partial f}{\partial \bar{r}} = 0 \qquad \text{and} \qquad \frac{\partial t}{\partial \bar{r}} = 0$$

while at the wall, for mass transfer

(10-9)

$$\frac{\partial f}{\partial \bar{r}} = 0$$

and for heat transfer at the wall, in dimensional form,

$$K_r \rho C_p \frac{\partial T}{\partial R}\bigg|_{R=R_0} = U(T - T_c) \qquad (10\text{-}10)$$

or

$$\frac{\partial t}{\partial \bar{r}}\bigg|_{\bar{r}=1} = \frac{UR_0}{\rho C_p K_r}(t - t_c)$$

U is an overall wall coefficient and $UR_0/\rho C_p K_r$ is the Biot number at the wall Bi_w. In terms of radial aspect ratio m, the radial Peclet number for heat transport

$\overline{Pe}_r$, and the Stanton number for overall heat transfer at the wall $St = U/u\rho C_p$, the wall Biot number is

$$Bi_w = \frac{UR_0}{\rho C_p K_r} = \frac{U}{\rho u C_p} \frac{R_0}{d_p} \overline{Pe}_r = St\overline{Pe}_r m \qquad (10\text{-}11)$$

Reaction Parameters

The right-hand sides of Eqs. (10-1) and (10-2) express the nature and magnitude of global reaction and associated heat generation or abstraction. The term $\mathcal{R}$ defines the global rate of species generation. As noted in Chaps. 5 and 9, $\mathcal{R}$ includes diffusional (interphase and intraphase) events and the chemical-reaction steps. Recall that for first-order reaction where both external and internal mass diffusion intervenes

$$\mathcal{R} = \frac{\eta k C}{1 + \eta k / k_g a} \qquad (10\text{-}12)$$

C being the local bulk-fluid-phase concentration bathing the catalyst.

In the continuity equations $\mathcal{R}$ and q actually express the addition of mass and heat to the continuous fluid phase or removal from it, and so

$$\mathcal{R} = k_g a(C - C_s) \qquad q = ha(T - T_s) \qquad (10\text{-}13)$$

When these are normalized and put in dimensionless form as dictated by the right-hand side of Eqs. (10-6) and (10-7), we have

$$\frac{\mathcal{R}\theta}{C_0} = k_g a\theta(f - f_s) \qquad (10\text{-}14)$$

$$\frac{q\theta}{\rho C_p T_0} = \frac{ha}{\rho C_p} \theta(t - t_s) \qquad (10\text{-}15)$$

The actual reaction-internal diffusion process within the porous-catalyst pellet is governed by the local continuity equations discussed in the previous chapter. For a flat-plate geometry and power-law kinetics,

$$\mathcal{D}\frac{d^2 C}{dx^2} = kC^n \qquad \lambda\frac{d^2 T}{dx^2} = -\Delta H\, kC^n \qquad (10\text{-}16)$$

In dimensionless form, normalizing with C_b, T_b (bulk values) gives

$$\frac{d^2 f}{d\bar{z}^2} = L^2 \frac{k_b C_b^{n-1}}{\mathcal{D}} f^n \exp\left[-\epsilon\left(\frac{1}{t} - 1\right)\right] \qquad (10\text{-}17)$$

$$\frac{d^2 t}{d\bar{z}^2} = \frac{-\Delta H\, \mathcal{D} C_b}{\lambda T_b} \frac{L^2 k_b C_b^{n-1}}{\mathcal{D}} f^n \exp\left[-\epsilon\left(\frac{1}{t} - 1\right)\right] \qquad (10\text{-}18)$$

As Chap. 5 reveals, the intraphase temperature difference is secured by elimination of kC^n between Eqs. (10-17) and (10-18) to yield in dimensionless form the Prater temperature

$$t = 1 + \beta(f_s - f_c) \qquad (10\text{-}19)$$

thus defining the Prater number

$$\beta = \frac{-\Delta H \, \mathscr{D} C_b}{\lambda T_b} = \frac{\Delta T_{ad}}{T_b} \qquad (10\text{-}20)$$

The boundary conditions for solution of the intraphase continuity equations are as follows. At pellet center, by symmetry,

$$\frac{dt}{d\bar{z}} = 0 \quad \text{and} \quad \frac{df}{d\bar{z}} = 0 \qquad (10\text{-}21)$$

and at the external surface, equating fluxes between the bulk-fluid phase and the catalyst gives

$$k_g(C_b - C) = -\mathscr{D}\frac{dC}{dx}$$

$$f_s = \frac{C_s}{C_b} = f_b - \frac{1}{\text{Bi}_m}\frac{df}{d\bar{z}} \qquad (10\text{-}22)$$

and for heat

$$t_s = t_b - \frac{1}{\text{Bi}_h}\frac{dt}{d\bar{z}} \qquad (10\text{-}23)$$

where $k_g L/\mathscr{D} = $ mass Biot number Bi_m

$hL/\mathscr{D} = $ thermal Biot number Bi_h

Hence on substituting for interphase gradients [Eqs. (10-22) and (10-23)] into Eqs. (10-3) and (10-4), our local generation terms become

$$\frac{\mathscr{R}\theta_0}{C_0} = -\frac{k_g a\theta_0}{\text{Bi}_m}\frac{df}{d\bar{z}} = -\frac{\theta_0\mathscr{D}}{L^2}\frac{df}{d\bar{z}}$$

$$\frac{q\theta_0}{\rho C_p T_0} = -\frac{ha\theta_0}{\rho C_p \text{Bi}_h}\frac{dt}{d\bar{z}} = -\frac{\theta_0\lambda}{L^2}\frac{dt}{d\bar{z}} \qquad (10\text{-}24)$$

or, as expected, since $a = 1/L$,

$$\mathscr{R} = -a\mathscr{D}\frac{dC}{dx} \qquad q = -a\lambda\frac{dT}{dx} \qquad (10\text{-}25)$$

Recalling the general definition of the overall (interphase-intraphase effectiveness factor), we have

$$\eta_o = \frac{-a\mathscr{D}\, dC/dx}{f'(k, C, T)_{\text{bulk}}}$$

Hence

$$\mathscr{R} = \eta_o f'(k, C, T)_{\text{bulk}} \qquad (10\text{-}26)$$

and since

$$-a\mathscr{D}\frac{dC}{dx} = \frac{a\lambda}{-\Delta H}\frac{dT}{dx}$$

we have

$$q = \eta_o(-\Delta H)f'(k, C, T)_{\text{bulk}}$$

If we suppose that we are confronted with first-order nonisothermal reaction, where interphase and intraphase diffusional limitations may intervene, the overall effectiveness η_o is

$$\eta_o = \bar{\eta}\eta = \frac{\eta}{1 + \eta k/k_g a} \qquad (10\text{-}28)$$

and

$$\eta_o = f'(\phi, \beta, \epsilon, \text{Bi}_h, \text{Bi}_m)$$

as shown in Chap. 9.

Our reaction parameters, i.e., those dimensionless groups governing the right-hand side of the reactor continuity equations, are

$$\frac{\mathcal{R}\theta_0}{C_0} = \frac{\eta_o f'(k, C, T)_{\text{bulk}}}{C_0} \theta_0$$

$$\frac{\mathcal{R}\theta_0}{C_0} = g(\phi, \text{Da}, \beta, \epsilon, \text{Bi}_m, \text{Bi}_h) \qquad (10\text{-}29)$$

and

$$\frac{q\theta_0}{\rho C_p T_0} = \frac{\mathcal{R}\theta_0}{C_0} \bar{\beta} \qquad (10\text{-}30)$$

where $\bar{\beta}$ is the overall adiabatic temperature change.

We observe, therefore, that the local rate of global reaction can be nicely expressed in terms of overall effectiveness for the kinetics and nonisothermal environment confronting the investigator or designer.

Significance of Dimensionless Parameters

The continuity equations established dimensionless parameters important in defining the *long-range* gradients (interparticle) throughout the reactor. They are

$$\text{Pe}_a = \text{axial Peclet number} = \frac{d_p u}{\mathcal{D}_a} = \frac{\text{axial convection}}{\text{axial dispersion}}$$

$$\text{Pe}_r = \text{radial Peclet number} = \frac{d_p u}{\mathcal{D}_r} = \frac{\text{axial convection}}{\text{radial dispersion}}$$

$$n = \text{axial aspect number} = \frac{Z}{d_p} = \frac{\text{bed length}}{\text{particle diameter}}$$

$$m = \text{radial aspect number} = \frac{R_0}{d_p} = \frac{\text{bed radius}}{\text{particle diameter}}$$

$$\mathcal{A} = \text{overall reactor aspect number} = \frac{Z}{R_0} = \frac{\text{bed length}}{\text{radius}}$$

$$\tau = \text{dimensionless time} = \frac{t}{\theta} = \frac{\text{real time}}{\text{contact time}}$$

$$\text{Bi}_w = \text{wall Biot number} = \frac{UR_0}{\rho C_p K_r}$$

$$\equiv \frac{\text{temperature gradient in core of bed}}{\text{temperature gradient in wall film}}$$

$$(\text{Da})_1 = \text{Damköhler number} = k_0 \theta_0 C_0^{n-1}$$

$$\epsilon = \text{Arrhenius number} = \frac{E}{RT_0}$$

$$\bar{\beta} = \text{dimensionless adiabatic temperature change} = \frac{-\Delta H C_0}{\rho C_p T_0} = \frac{\Delta T_{ad}}{T_0}$$

Short-range, or local, gradients (interphase and intraphase) are governed by dimensionless parameters established by local continuity equations and boundary conditions:

$$\phi = \text{Thiele modulus} = \frac{\text{rate of surface reaction}}{\text{rate of intraphase diffusion}} = L \sqrt{\frac{k}{\mathscr{D}_i} C^{n-1}}$$

$$\text{Bi}_m = \text{local mass Biot number} = \frac{k_g L}{\mathscr{D}_i} = \frac{\text{interphase mass transfer}}{\text{intraphase mass transfer}}$$

$$\text{Bi}_h = \text{local thermal Biot number} = \frac{hL}{\lambda} = \frac{\text{interphase heat transfer}}{\text{intraphase heat transfer}}$$

$$\beta = \text{Prater number} = \frac{-\Delta H \, C_0 \mathscr{D}}{\lambda T_0}$$

$$= \text{normalized intraphase adiabatic temperature difference}$$

$$\epsilon = \text{Arrhenius number} = \frac{E}{RT_b}$$

The Biot numbers, containing interphase transport coefficients h and k_g, are functions of the interphase j-factor dimensionless groups; i.e.,

$$h, k_g = f(\text{Re, Pr, Sc})$$

where Re = Reynolds number, $d_p u/v$ = convective flow/viscous flow
Pr = Prandtl number, v/K_f = momentum diffusion/thermal diffusion
Sc = Schmidt number, $v/\mathscr{D}_f$ = momentum diffusion/mass diffusion

Thus a totally detailed description of the nonisothermal, nonadiabatic fixed-bed catalytic reactor involves over a dozen dimensionless parameters. Formal analytical solution of Eqs. (10-3), (10-4), (10-8) to (10-10), and (10-16) to (10-30) is obviously impossible in view of the nonlinearity imposed by the Arrhenius rate-temperature functionality. Nor is formal proof required to declare that scale-up based upon retaining precise numerical values of the dimensionless parameters in passing from laboratory or bench scale to the large-scale production reactor is also impossible. Bosworth's proof[1] suffices for those not confident in their in-tuition in this regard. Numerical solution via computer is possible. However, considerable sophistication and computing time are required to resolve the general transient nonisothermal, nonadiabatic case involving axial as well as radial dis-persion and local chemical rates affected by interphase and intraphase diffusion of heat and mass.[2] Some tolerable simplifications of the general case may be em-braced, as will be noted shortly. First, however, methods of determining or estimating dimensionless parameters will be discussed.

The dimensionless parameters governing overall and local events in the heterogeneous catalytic fixed-bed reactor can be conveniently classed as chemical and physical. The chemical parameters, intrinsic to the chemical-reaction–catalyst system of concern, are the Arrhenius number ϵ, Damköhler group Da, dimension-less overall adiabatic temperature change $\bar{\beta}$, the Prater number (local intraphase reduced adiabatic ΔT) β, and the Thiele modulus ϕ. To be sure each of the above groups is a function of extrinsic variables such as C_0, T_0, intraphase diffusivity $\mathscr{D}_i$, and (in the case of the Thiele modulus) the ratio L of the particle volume to external surface area. In contrast with these essentially chemical dimension-less parameters, those of an exclusively physical nature, i.e., determined by physical properties and rate processes as well as reactor-catalyst dimensions, are the Peclet numbers Pe, various aspect ratios n, m, and $\mathscr{A}$, and the wall and inter-phase and intraphase Biot numbers. By inference, of course, the Reynolds number is included as well as the Schmidt and Prandtl numbers (governed by the fluid properties but generally not by the nature of the chemical reaction per se).

Chemical Dimensionless Parameters

Table 10-2 lists typical values of ϵ and β for some catalytic reactions of industrial importance. Typical values of the Thiele modulus ϕ are not easily anticipated, as catalyst size and pore dimensions (physical properties) can be manipulated by the practitioner so that for a given reaction, ϕ values can be found to be low or high, depending primarily upon the size of the catalyst pellet employed in a particular plant.

The precise evaluation of ϵ, β, and/or $\bar{\beta}_0$ rests upon careful rate measure-ments free from diffusional intrusions (to secure activation energy in ϵ), diffusion-coefficient measurements or computations (for β), thermodynamic data (for β and $\bar{\beta}_0$), thermal conductivity of the catalyst pellet (for β), and volumetric heat-capacity data for the reactant-product-bearing fluid (for $\bar{\beta}_0$). Values of C_0 and T_0 are dictated by design, these values being chosen in the light of catalyst selectivity, activity, life, and plant-safety requirements.

Physical Dimensionless Parameters

We deal here with parameters totally independent of the particular catalytic reac-tion, though to be sure specific reaction systems demand that these parameters be

[1] R. C. L. Bosworth, "Transport Processes in Applied Chemistry," Wiley, New York, 1956.
[2] M. McGuire and L. Lapidus, *AIChE J.*, **2**: 82 (1965).

controlled within limits, where possible, to guarantee safe, stable, productive performance of the reactor.

The Peclet numbers for axial and radial transport of heat and mass are products of Reynolds and turbulent property groups, i.e.,

$$\text{Pe}_m = \text{Mass} = \frac{d_p u}{v}\frac{v}{\mathscr{D}} = \frac{d_p u}{\mathscr{D}} = \frac{\text{convective transport}}{\text{dispersive transport}}$$

where $d_p u/v$ is the Reynolds number and $v/\mathscr{D}$ is a turbulent Schmidt number;

$$\frac{v}{\mathscr{D}} = \frac{\text{molecular momentum transport}}{\text{dispersive mass transport}}$$

By analogy

$$\text{Pe}_h = \frac{d_p u}{v}\frac{v}{K} = \frac{d_p u}{K}$$

where

$$\frac{v}{K} = \frac{\text{molecular momentum transport}}{\text{dispersive heat transport}}$$

Unlike their molecular counterparts, both $\mathscr{D}$ and K (in square centimeters per second) are of a complex physical nature; i.e., the level of flowing-fluid turbulence, bypassing, capacitance, and conductive properties of the core (packed bed) influence their magnitude.

Experimental techniques which permit determination of axial and radial dispersion coefficients, and thus the respective Peclet numbers, were outlined in Chap. 4. For axial and radial dispersion of mass, experiment and theory suggest for the packed bed

$$\text{Pe}_{am} = 1 \text{ to } 2$$
$$\text{Pe}_{rm} = 8 \text{ to } 12 \qquad \text{Re} > 30$$

Table 10-2   RANGE OF DIFFUSION-REACTION PARAMETER VALUES FOR SOME CATALYTIC PROCESSES†

Process	β	ϵ
Ethylene hydrogenation	0.2–0.8	23–27
Dissociation of N_2O	0.09	22
SO_2 oxidation	0.01	15
Ethylene oxidation	0.13	14
Vinyl chloride synthesis	0.25	6.5
Benzene hydrogenation	0.4–0.8	10
Hydrogen oxidation	0.02–0.3	7–9

† M. C. Mercer and R. Aris, *Lat. Am. J. Chem. Eng. Appl. Chem.*, **2**: 149 (1971).

Thermal Peclet numbers are less easily secured experimentally, though some data are available which would suggest that if the sole mode of thermal transport is by the mass so dispersed, thermal Peclet numbers will be of the value suggested above for mass dispersion. However, whereas mass is essentially dispersed through axial and radial voids, heat may be transported by additional mechanisms, e.g., radiation and particulate conduction. Since these agencies are not available as modes of mass transport, we can anticipate differences in the values of mass and heat Peclet numbers.

Radial Peclet Number for Heat and Mass

Experiment and theory suggest a mass Peclet-number value for radial dispersion in fixed beds to be of the order of 10 (see Chap. 4). Singer and Wilhelm[1] and later Argo and Smith[2] devoted thought to the issue of the thermal Peclet number for radial dispersion and the following tentative mechanism for radial transfer of heat, based upon their reflections and available data, can safely be suggested (Fig. 10-2).

Heat is assumed to be transported radially through the packed-bed voids by parallel paths of conduction, convection, and radiation. In contrast, radial transport via the solid-particle phase is visualized as a series mechanism embracing fluid-film convection, pellet-phase conduction, and radiation. These assumptions yield a radial thermal Peclet number $\overline{Pe}_r$ in terms of the mass Peclet number Pe_m, the Stanton number for interphase heat transfer $St = h_p/\rho C_p u$, the thermal Biot number $Bi_h = hd_p/\lambda$, the Stefan–Boltzmann constant σ, emmissivity e, and solids absolute temperature. Smith[3] details this development, and the result, as modified

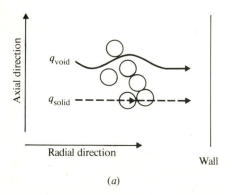
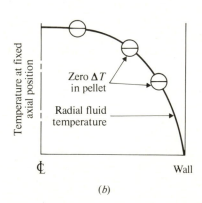

FIGURE 10-2
Fixed-bed radial-heat-transport model.

[1] E. Singer and R. H. Wilhelm, *Ind. Eng. Chem.*, **46**: 343 (1950).
[2] W. B. Argo and J. M. Smith, *Chem. Eng. Prog.*, **49**: 443 (1953).
[3] J. M. Smith, "Chemical Engineering Kinetics," 2d ed., McGraw-Hill, New York, 1970.

by Beek,[1] for radial heat dispersion in a packed bed of void fraction ϵ is

$$\frac{1}{\overline{Pe}_r} = \frac{1}{Pe_m} + \frac{\epsilon 6.3 St}{1 + 0.35 Bi_h} + \frac{2e}{1 - e} \frac{\epsilon\sigma}{\rho u C_p} T^3 \qquad (10\text{-}31)$$

Inspection of Eq. (10-31) compels us to conclude that since radial heat transport knows more paths (conduction and radiation) than are available to mass transport, the radial Peclet number for heat will be less than that of mass. One can anticipate that radial $\overline{Pe}_r$ values will lie in the range of about 5 to 10. Patently, since St and Bi and the radiation term are each Reynolds-number-dependent, the wall Biot number [Eq. (10-11)], which governs the core-wall temperature gradient, will be sensitive to Reynolds number. Note that in Eq. (10-11), St refers to the wall coefficient, while in Eq. (10-31), St refers to the interphase (particle-fluid) coefficient.

Aspect Numbers

Values of Z/d_p, R_0/d_p, and Z/R_0 are, of course, instrumental in determining the magnitudes of radial gradients and the second derivative of the axial gradient (the first derivative being quite naturally established by the generation term). To illustrate, the magnitude of the radial gradient at the point of axial temperature maximum in Eq. (10-4) ($dt/dz = 0$) is

$$\left(\frac{\partial^2 t}{\partial \bar{r}^2} + \frac{1}{\bar{r}} \frac{\partial t}{\partial \bar{r}}\right) = \overline{Pe}_r \frac{R_0}{d_p} \frac{R_0}{Z_0} \frac{q\theta_0}{C_p \rho T_0} \qquad (10\text{-}32)$$

For a given value of thermal generation and radial thermal Peclet number $\overline{Pe}_r$, the radial and overall aspect numbers determine the magnitude of the radial temperature gradient in the core of the fixed bed.

To be sure, the aspect ratios are determined a priori on the basis of factors governing pellet size (Thiele modulus, interphase surface-to-volume ratio), contact time insofar as Z is fixed for a given velocity, and heat-transport requirements as expressed in terms of tube surface-to-volume ratio ($2/R_0$ for a cylinder).

The merits of aspect-ratio manipulation are not to be underestimated. Consider Eq. (10-32); if a severe temperature gradient within the core of the fixed bed is undesirable, then since the Peclet number is fixed at Re > 30, a reduction in $m = R_0/d_p$ and increase in $\mathscr{A} = Z/R_0$ will prove of benefit because the goal is that of minimizing the left-hand side of Eq. (10-32).

In sum, if it is supposed that we are witness to a highly exothermic solid-catalyzed reaction where excessive temperature rise and backmixing prove destructive to the desired species, e.g., naphthalene oxidation over certain V_2O_5 catalyst formulations, in order to minimize radial temperature gradients we would be advised to reduce R_0/d_p, increase Z/R_0 and minimize detrimental backmixing, and to increase Z/d_p. All such manipulations of the aspect ratios are consistent.

---

[1] J. Beek, *Adv. Chem. Eng.*, **3**: 203 (1962).

For with a fixed value of d_p (dictated by local interphase and intraphase diffusion-reaction moduli), an increase in Z with a decrease in tube radius R_0 achieves the desired end (reduction of R_0/d_p and increase in Z/R_0 and Z/d_p).

Biot Numbers, Local and Overall (Wall)

Our equations and boundary conditions cast in dimensionless form yield local, or interphase and intraphase, Biot numbers

$$\mathrm{Bi}_m = \frac{k_g L}{\mathscr{D}_i} = \frac{\text{intraphase } \Delta C}{\text{interphase } \Delta C} = 10^2 \text{ to } 10^5 \qquad (10\text{-}33)$$

$$\mathrm{Bi}_h = \frac{hL}{\lambda} = \frac{\text{intraphase } \Delta T}{\text{interphase } \Delta T} = 10 \text{ to } 10^3 \qquad (10\text{-}34)$$

and a wall Biot number [see Eq. (10-11)]

$$\mathrm{Bi}_w = \frac{U R_0}{K_r \rho C_p} = \frac{\text{core } \Delta T}{\text{wall } \Delta T} = 10^{-1} \text{ to } 10^2 \qquad (10\text{-}35)$$

Some Permissible Simplifications

Rigorous application of the models set forth [Eqs. (10-1) to (10-30)] is required only if and when the real system under investigation suffers from an oversimplified description. The anticipated magnitudes of governing dimensionless parameters, as set forth above, allow one to prudently modify the rigorous model while maintaining confidence in the resulting simulation. So, for example, while the student is aware, by virtue of his tutoring in thermodynamics, that ρC_p and reaction enthalpy are temperature-dependent, the fact that the reaction-rate coefficient varies exponentially with temperature should cause the prudent student to assume, *for the sake of realistic simulation*, that ρC_p and $-\Delta H$ are sensibly constant in a nonisothermal reaction system. While, in fact, these properties are temperature-dependent, such variations with temperature are indeed minor in contrast to the acknowledged temperature dependency of the chemical rate coefficient.

Armed, then, with some notions of the reasonable magnitudes of the dimensionless parameters which govern the nonisothermal, nonadiabatic fixed-bed catalytic reactor, we can invoke some prudent simplifications.

The Peclet numbers Our comparatively rigorous model [Eqs. (10-1) to (10-11)] contains both axial and radial Peclet numbers for heat- and mass-dispersive transport. With respect to axial dispersion, it was shown in Chap. 4 that in a reactor of length Z_0 through which fluid flows at a velocity u and the axial dispersion coefficient is $\mathscr{D}_a$ the system may be considered equivalent to n CSTRs set in series, the relationship being

$$\bar{n} = \frac{Z_0 u}{2 \mathscr{D}_a} \qquad (10\text{-}36)$$

We noted further that for a packed bed of axial aspect ratio Z_0/d_p

$$\bar{n} = \frac{d_p u}{2\mathscr{D}_a} \frac{Z_0}{d_p} \qquad (10\text{-}37)$$

Since the axial Peclet number is equal to about 2 for gases flowing through packed beds at Reynolds numbers above 10, we have

$$\bar{n} = \frac{1}{2} \text{Pe}_a \frac{Z_0}{d_p} = \frac{Z_0}{d_p} = n \qquad (10\text{-}38)$$

A packed bed is therefore equivalent to Z_0/d_p perfectly mixed stirred tanks set in series. As has been shown,[1] this assures us that except for unusual reactor configurations, one can safely assume that axial dispersion exerts a very minor influence upon performance so long as bed-length-to-pellet-diameter ratio is greater than about 150.†

For example, in NH_3 oxidation as conducted commercially over some 30 layers of Pt-alloy gauze, total gauze-thickness-to-wire-diameter ratio is the axial aspect ratio, and since NH_3 conversion is 100 percent, backmixing of mass and particularly heat (via mass dispersion, radiation, and conduction through the metallic-gauze matrix) plays a significant role in that reactor system. In general, however, a typical fixed-bed reactor model can be realistically fashioned by ignoring the axial dispersion term in the continuity equations.

Radial dispersion terms cannot be dismissed so neatly. Following the mixing-cell–dispersion analogy, since the radial Peclet number is about 10, if one is permitted to extend the analogy, the number of radial mixers is

$$\bar{m} = \frac{R_0 u}{2\mathscr{D}_r} = \frac{d_p u}{2\mathscr{D}_r} \frac{R_0}{d_p} \qquad (10\text{-}39)$$

or

$$\bar{m} = \frac{1}{2} \overline{\text{Pe}}_r \frac{R_0}{d_p} \approx 5 \frac{R_0}{d_p} \approx 5 \text{ m} \qquad (10\text{-}40)$$

As the thermal Peclet number for radial transport lies in the range of 5 to 10, we can anticipate $\bar{m}$ to be in the range of 2 to 5 times the radial aspect ratio R_0/d_p. As the radial aspect ratio can be expected to be in the range of 2 to 20, we conclude that a condition of zero radial gradient is not to be anticipated; i.e., the condition $\bar{m} = 1$ cannot be realized. However, a negligible radial gradient can be secured by imposing a lower practical limit on R_0/d_p, say, by reduction in tube diameter such that perhaps three or four catalyst pellets occupy the tube cross-sectional area. This is common practice in such demanding reaction systems as naphthalene oxidation, where safety and product survival demand a negligible radial temperature gradient. Certainly Eq. (10-40) shows that a large radial aspect ratio invites only small radial mixing and hence a significant radial gradient. It must be emphasized, of course, that the locale of radial resistance to heat transfer (whether

[1] J. J. Carberry, *Can. J. Chem. Eng.*, **36**: 207 (1958).
† This figure reflects our awareness of the influence of axial dispersion of heat.

largely at the wall or within the bed per se) depends on the wall coefficient as well as upon m [see Eq. (10-11)].

In sum, axial dispersion can generally be ignored in considering a fixed bed of axial aspect ratio greater than about 150; however, it is far less prudent to ignore radial gradients unless the radial aspect ratio is small (about 2 to 4) and the major seat of radial thermal resistance is at the wall. Of course, when the system is adiabatic, a radial temperature and concentration gradient cannot exist unless a severe radial velocity variation is encountered.

Velocity Variations

Fluid velocity can be a function of both axial and radial position within the reactor. Axial variation can be prompted by a change in number of moles due to reaction, e.g., as in NH_3 synthesis or phosgene decomposition. In addition, severe axial temperature changes cause an axial gas-velocity variation. In either case, the effect is to alter contact time, and, as noted earlier, the proper formulation of the continuity equation in terms of moles flowing accounts for velocity variations in the direction of flow.

Radial variations in fluid velocity can be due to the nature of flow, e.g., laminar flow, and, particularly in fixed beds, to radial variations in void fraction. Smith and his students[1] measured radial velocity variations, and one must acknowledge their existence. However, as Beek[2] notes, it is very difficult to give analytical expression to these observed variations. Thus an average fluid velocity independent of radial position can be assumed, except in pathological cases such as flow at very low Reynolds numbers, where a parabolic profile might be anticipated.

Fixed-bed reactor simulation will now be illustrated for three cases: (1) adiabatic oxidation of SO_2, (2) nonisothermal, nonadiabatic oxidation of naphthalene, and (3) case 2 conducted in an inert-packed-bed–catalytic-wall reactor or a packed tubular-wall reactor (PTWR).

10-2 ADIABATIC FIXED BED

Since, by definition, heat is neither removed or added at the reactor wall, one may justly assume a zero radial temperature gradient. However, two issues deserve analysis: (1) the importance of axial dispersion in the presence of a severe axial temperature gradient and (2) the importance of interphase and intraphase concentration-temperature gradients as they may affect local and therefore overall reactor behavior.

If axial dispersion and interphase and intraphase diffusional factors prove *not* to be important, the adiabatic fixed-bed-design problem then becomes equivalent to its homogeneous counterpart, and this yields to simple graphical design methods.

---

[1] J. M. Smith et al., *Ind. Eng. Chem.*, **45**: 1209 (1953).
[2] Beek, loc. cit.

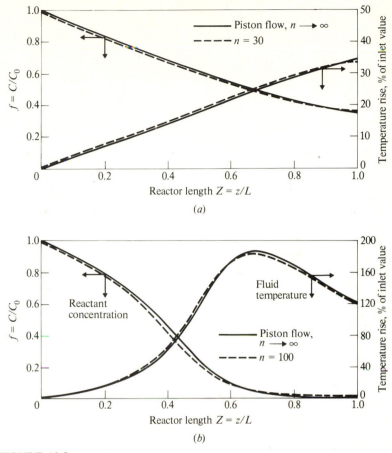

FIGURE 10-3
Computed profiles for plug flow and finite backmixing for A → B, homogeneous reactor: (a) adiabatic, $k_0\theta_0 = 1.0$, $\bar{\beta} = 0.63$; (b) nonisothermal and nonadiabatic, $k_0\theta_0 = 0.2$, $\bar{\beta} = 3.2$.

For both the adiabatic and one-dimensional fixed-bed catalytic reactor Carberry and Wendel[1] solved numerically the governing equations, which included both axial dispersion and local diffusional intrusions of both heat and mass. As Fig. 10-3 suggests for beds of particle aspect ratio Z_0/d_p greater than 100, an insignificant difference exists between the plug-flow and limited-backmixed model. More recent work, however, suggests that axial dispersion of heat is more significant than that of mass; thus a more meaningful criterion for negligible axial dispersion influence in a packed-bed reactor should be $Z_0/d_p > 150$, as is implicit in Finlayson's analyses.[2]

[1] J. J. Carberry and M. M. Wendel, *AIChE J.*, **9**: 129 (1963).
[2] B. A. Finlayson, *Catal. Rev.*, **10**: 69 (1974).

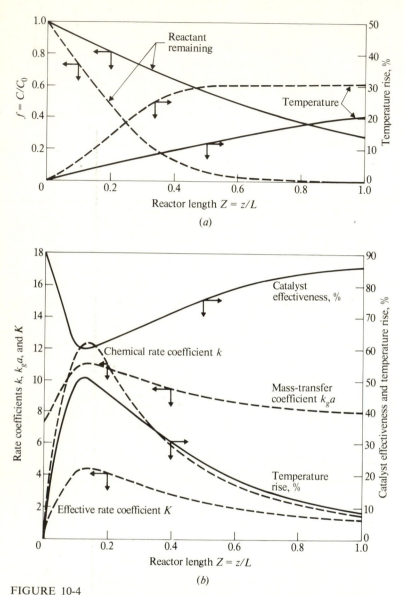

FIGURE 10-4
(a) Interphase-intraphase diffusional influence upon reactor profiles; — — —:
fine-mesh catalyst, high flow rate; ————: mass transfer and diffusion influence.
(b) Variation of individual rate coefficients; overall η and K in a nonisothermal
bed. (*J. J. Carberry and M. M. Wendel, AIChE. J., 9: 129 (1963).*]

In contrast, the signal importance of interphase and intraphase diffusion
upon local rate is evident in Fig. 10-4a and b. A drastic modification in terminal
temperature and conversion is evident when local gradients are anticipated in the
fixed-bed model. Note also the variation and changing relative importance of

the diffusion-reaction parameters along the length of the nonisothermal bed (Fig. 10-4b).

Example: Oxidation of SO$_2$†

When we assume plug flow but anticipate an interphase-intraphase concentration gradient and an interphase temperature gradient (therefore an isothermal pellet, a limiting assumption justified in Chaps. 5 and 9), the continuity equations reduce to

$$\frac{df}{dz} = \frac{\mathscr{R}\theta_0}{C_0} \qquad (10\text{-}41)$$

$$\frac{dt}{dz} = \frac{q\theta}{\rho C_p T_0} \qquad (10\text{-}42)$$

where local rates are Eqs. (10-14) and (10-15) and reactor boundary conditions are

$$z = 0 \qquad t = 1 \qquad f = 1 \qquad (10\text{-}43)$$

The governing equations are readily resolved numerically for the intrinsic kinetics, which is, for Pt-catalyzed SO$_2$ oxidation,[1]

$$r_{\text{Pt}} = k \frac{\text{SO}_2\sqrt{\text{O}_2}}{[1 + \sqrt{K_1(\text{O}_2)} + K_2(\text{SO}_3)]^2} \left[1 - \frac{\text{SO}_3}{K(\text{SO}_2)\sqrt{\text{O}_2}}\right] \qquad (10\text{-}44)$$

For V$_2$O$_5$ catalysis, the kinetic model of Mars[2] is

$$r_v = \frac{k(\text{SO}_2)(\text{O}_2)}{[\sqrt{\text{SO}_3} + \sqrt{K_1(\text{SO}_2)}]^2} \left[1 - \frac{\text{SO}_3}{K(\text{SO}_2)\sqrt{\text{O}_2}}\right] \qquad (10\text{-}45)$$

The temperature dependencies of equilibrium and adsorption-reaction coefficients were included in the models described by Minhas.

Given the LHHW form of the kinetics, the formulation of the η-versus-ϕ functionality requires compromise. In this simulation Minhas and Carberry utilized a definition of ϕ in terms of the order of limiting reaction (SO$_2$) and the equilibrium nature of the reaction SO$_2$ + $\frac{1}{2}$O$_2 \rightleftarrows$ SO$_3$. Therefore

$$\phi = L\sqrt{\frac{k(K_{\text{eq}} + 1)}{\mathscr{D}K_{\text{eq}}}} (\text{SO}_2)_b^{\,n-1} \qquad (10\text{-}46)$$

(The formulation of ϕ for equilibrium reaction has been established by Smith and Amundson[3] and by Carberry[4] for a microporous-macroporous catalyst.)

† S. Minhas and J. J. Carberry, *Br. Chem. Eng.*, **14**(6): 799 (1969).
[1] O. Hougen and K. M. Watson, "Chemical Process Principles," vol. 3, Wiley, New York, 1947.
[2] P. Mars, *3d Int. Congr. Catal., Amsterdam, July, 1964*.
[3] N. L. Smith and N. R. Amundson, *Ind. Eng. Chem.*, **43**: 2156 (1951).
[4] J. J. Carberry, *AIChE J.*, **8**: 557 (1962).

Several interesting conclusions emerge from this simulation:

1 Given the intrinsic kinetics for both catalysts, interphase-intraphase transport influence becomes insignificant only at ludicrously small values of $d_p < 0.01$ cm (Fig. 10-5).

2 For pellet sizes commensurate with reasonable pressure drop, catalytic effectiveness is small indeed (Fig. 10-6).

3 The V_2O_5 catalyst compares unfavorably with Pt/Al_2O_3 at a feed temperature of 400°C but is comparable in performance at a feed temperature of 450°C (Figs. 10-7 and 10-8).

4 An optimum mass-flow rate exists for specific conditions of feed temperature (Figs. 10-9 and 10-10); these results are the consequence of adiabatic reaction under interphase-intraphase transport influence visited upon a reaction governed by kinetic and thermodynamic equilibrium considerations.

The results set forth for catalyzed SO_2 oxidation are not at variance with plant experience. As operated commercially, the reactor network consists of multistaged adiabatic bed with interstage indirect heat exchange, for reasons discussed in Sec. 3-13.

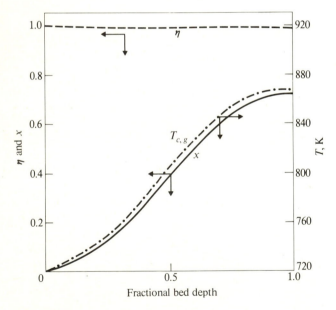

FIGURE 10-5
Adiabatic profiles in the absence of interphase-intraphase diffusional gradients; SO_2 oxidation over 0.01 cm V_2O_5, feed 2844 kg/(h)(m²), bed depth 15 cm. [*S. Minhas and J. J. Carberry, Br. Chem. Eng.*, **14**(6): *799 (1969).*]

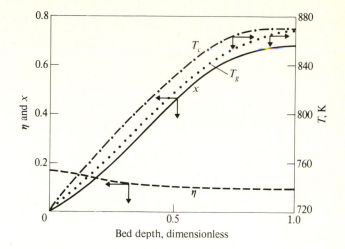

FIGURE 10-6
Realistic profiles for SO_2 oxidation over Pt/Al_2O_3, feed 5700 kg/(h)(m²), bed depth 60 cm. [*S. Minhas and J. J. Carberry, Br. Chem. Eng.*, **14**(*6*): *799* (*1969*).]

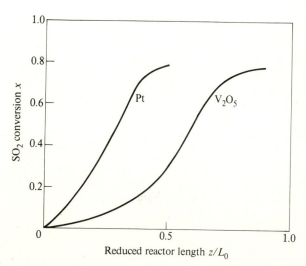

FIGURE 10-7
Comparison of activity of Pt and V_2O_5 in SO_2 oxidation at 400°C; $L_0 = 146$ cm, $d_p = 1$ cm, feed rate = 2844 kg/(h)(m²). [*S. Minhas and J. J. Carberry, Br. Chem. Eng.*, **14**(*6*): *799* (*1969*).]

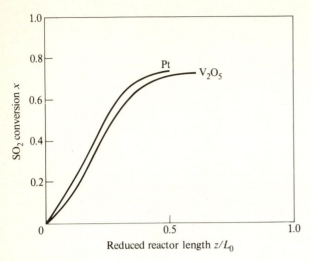

FIGURE 10-8
Activity of Pt and V_2O_5 in SO_2 oxidation for a feed temperature of 450°C; $L_0 = 73$ cm, $d_p = 1$ cm, feed rate = 2844 kg/(h)(m²). [*S. Minhas and J. J. Carberry, Br. Chem. Eng.,* **14**(*6*): *799 (1969)*.]

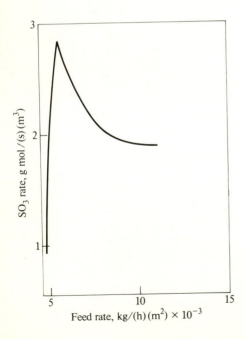

FIGURE 10-9
Space-time yield of SO_3 versus mass flow; $T_0 = 400°C$, Pt/Al_2O_3. [*S. Minhas and J. J. Carberry, Br. Chem. Eng.,* **14**(*6*): *799 (1969)*.]

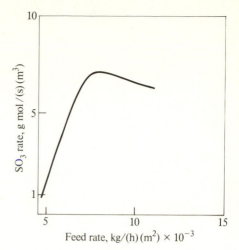

FIGURE 10-10
Space-time yield of SO_3 versus mass flow; $T_0 = 450°C$, Pt/Al_2O_3. [*S. Minhas and J. J. Carberry, Br. Chem. Eng.*, **14**(6): *799 (1969)*.]

Remedies for the deleterious influence of local temperature-concentration gradients are not obvious in such vigorous exothermic equilibrium-limited catalytic reactions. Obviously pellet-size reduction is not feasible, as Fig. 10-5 teaches. Partial impregnation does suggest itself, however. As discussed in Chap. 9, if the catalytic agent, for example, Pt, is partially imposed upon the support to a fractional penetration F_1 at a concentration level F_2, then

$$\phi = F_1 L \sqrt{\frac{F_2 k}{\mathscr{D}}} \qquad (10\text{-}47)$$

For SO_2 oxidation over supported Pt, the benefits of partial impregnation are indicated in Table 10-3, based upon computations rooted in the theses noted above for the adiabatic oxidation of SO_2.†

From the viewpoint of design, the computed two-stage results with interstage cooling are as shown in Table 10-4. Considerably more Pt catalyst is required since T_f for Pt is 400°C and, as Figs. 10-7 and 10-8 indicate, V_2O_5 must be utilized at a higher feed temperature (450°C) to compete with Pt.

This extraordinary effect of inlet temperature upon V_2O_5 activity, in contrast to that of Pt/Al_2O_3, is an interesting consequence of the physical chemistry of the V_2O_5 surface. For, in fact, the V_2O_5 actually exists upon the support as a melt at temperatures above about 420 to 440°C (depending upon promotion). Until the melting point is reached, the V_2O_5 is of rather modest activity; in the molten state its activity for SO_2 oxidation is comparable to that of Pt (Fig. 10-8).

† S. Minhas and J. J. Carberry, *J. Catal.*, **14**: 270 (1969).

Adiabatic Fixed-Bed-Reactor Yield

For a consecutive reaction A → B → C Carberry and Wendel[1] simulated an adiabatic fixed bed with results shown in Fig. 10-11. It is interesting to note that in this case diffusional interventions reduce reactor yield of intermediate B from its diffusion-unaffected value of 87 percent to 71 percent when transport factors are introduced into the model.

Table 10-3   MERITS OF PARTIAL IMPREGNATION AS COMPUTED FOR ADIABATIC OXIDATION OF SO_2 OVER SUPPORTED Pt^a

Catalyst concentration F_2	F_1, % impregnation	Conversion, %	Z, cm^b	$\eta_{in}{}^c$	$\eta_{out}{}^c$
1	100^d	81	60	0.3	0.04
	50	81	47	0.5	0.26
	10	81	30	0.96	0.84
0.5	50	81	100	0.54	0.25
	10	81	72	0.96	0.84
0.1	50	77	473	0.5	0.23
	10	77	335	0.96	0.8

^{a} S. Minhas and J. J. Carberry, *J. Catal.*, **14**: 270 (1969).
^{b} Bed length required to achieve the noted conversion.
^{c} η values refer to catalytic effectiveness at bed inlet and outlet, respectively.
^{d} Base case.

Table 10-4

Stage	x, %	T_{exit}, °C	Bed length, cm
V_2O_5, $T_f = 450°C$, 8% SO_2 in air, $d_p = 1$ cm, $G = 2840$ kg/(h)(m^2)			
1	73	602	46
2	90	493	83
Pt/Al_2O_3, $T_f = 400°C$, other conditions identical			
1	77	585	60
2	90	430	180

[1] Loc. cit.

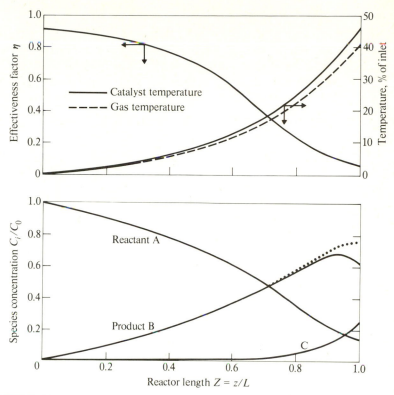

FIGURE 10-11

Influence of transport factors on yield (adiabatic reactor), for $A \xrightarrow{k_1} B \xrightarrow{k_2} C$, where B is product; $(k_1/k_2)_0 = 10, k_1\theta = 2.8, E_1 = 30\text{kcal/mol}, E_2 = 30\text{kcal/mol}, T_0 = 495°C$, $x_0(-\Delta H_1)/C_p T_0 = 0.345$, $\Delta H_2 = 30 \text{ kcal/mol}$, $\text{Re} = 160$, $d_p = 3 \text{ mm}$, $\mathscr{D}_k = 0.1$ cm²/s, $\theta_0 = 0.1$ s; $\cdots\cdots$: diffusion-free yield of B = 87 percent, ————: actual yield of B = 71 percent. [*J. J. Carberry and M. M. Wendel, AIChE J.*, **9**: *129 (1963).*]

10-3 NONISOTHERMAL, NONADIABATIC FIXED BED

Again assuming steady state and negligible axial dispersion of heat and mass, one can fashion a reasonable two-dimensional model of the nonisothermal fixed bed which must include radial as well as axial gradients of concentration and temperatures of both fluid and catalytic phases.

The following nonisothermal, nonadiabatic models will be treated: (1) V_2O_5-catalyzed oxidation of naphthalene in a steady-state fixed bed for a network which is supposed to characterize this reaction and (2) V_2O_5-wall-catalyzed oxidation of naphthalene in a steady-state tubular-wall inertly packed bed reactor for another supposed reaction network.

The steady state is presumed to prevail; however, parametric sensitivity can nevertheless be illustrated, i.e., the sensitivity of reactor profiles, conversion, and yield to minor variations in such physicochemical parameters as feed composition, wall temperature, Peclet numbers for radial transport, and reaction activation energies.

Example: Naphthalene Oxidation[1]

This complex reaction network has been conceived to be of the form[2]

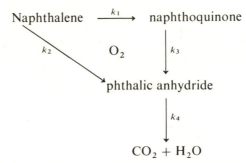

for particular forms of a V_2O_5 catalyst, designated A and B in Table 10-5, where rate constants at two temperature levels and activation energies are tabulated. Since $k_1 = k_2$ and $k_3 > k_4$ we can further simplify the network

$$\text{Napthalene} \xrightarrow{K_1} \text{anhydride} \xrightarrow{K_2} CO_2 + H_2O.$$

K_1/K_2 is also reported in Table 10-5, where $K_1/K_2 = 2k_1/k_4$.

Table 10-5 NAPHTHALENE OXIDATION KINETICS†

	Temp, °C	$k_1 = k_2$	k_3	k_4	K_1/K_2
Catalyst A					
	320	0.25	0.4	0.01	50
	370	3.5	12	0.035	200
Activation energy, kcal/mol	...	38	50	20	
Catalyst B					
	320	0.11	0.45	0.0016	140
	370	0.42	1	0.03	28
Activation energy, kcal/mol	...	20	12	43	

† J. J. Carberry and D. White, *Ind. Eng. Chem.*, **61**(7): 27 (1969), based on data from F. DeMaria, J. E. Longfield, and G. Butler, *Ind. Eng. Chem.*, **53**: 259 (1961).

[1] J. J. Carberry and D. White, *Ind. Eng. Chem.*, **61**(7): 27 (1969).
[2] F. DeMaria, J. E. Longfield, and G. Butler, *Ind. Eng. Chem.*, **53**: 259 (1961).

The catalysts exhibit quite different behavior with respect to activation energies, presumably due to differences in promotion of the V_2O_5. With catalyst A, temperature increases K_1/K_2 with obvious benefit to anhydride yield, while that ratio decreases with a temperature increase for catalyst B. Given the highly exothermic nature of this reaction, any fixed-bed operation will be conducted isothermally with some difficulty and expense; hence one would not be inclined to utilize catalyst B in a nonisothermal fixed-bed reactor. The near isothermality of a fluidized bed makes it a candidate reactor for catalyst B if backmixing and bypassing associated with fluid-bed reactors prove not too taxing upon conversion or yield and if instabilities can be avoided.[1]

The character of catalyst A invites its use in a fixed-bed unit, but, as we shall see, apparent benefit of the exothermic temperature excursions in the fixed bed of catalyst A are somewhat nullified by interphase-intraphase diffusional intrusions.

The model   For reactant A, naphthalene, Eq. (10-6) reduces to

$$\frac{\partial f}{\partial z} - \frac{\mathscr{A}}{\text{Pe}_r \, m}\left(\frac{\partial^2 f}{\partial \bar{r}^2} + \frac{1}{\bar{r}}\frac{\partial f}{\partial \bar{r}}\right) = \frac{\mathscr{R}_1 \theta_0}{A_0} \qquad (a)$$

For B, anhydride,

$$\frac{\partial b}{\partial z} - \frac{\mathscr{A}}{\text{Pe}_r \, m}\left(\frac{\partial^2 b}{\partial \bar{r}^2} + \frac{1}{\bar{r}}\frac{\partial b}{\partial \bar{r}}\right) = \frac{(\mathscr{R}_2 - \mathscr{R}_1)\theta_0}{A_0} \qquad (b)$$

For gas temperature Eq. (10-7) reduces to

$$\frac{\partial t}{\partial z} - \frac{\mathscr{A}}{\text{Pe}_r \, m}\left(\frac{\partial^2 t}{\partial \bar{r}^2} + \frac{1}{\bar{r}}\frac{\partial t}{\partial \bar{r}}\right) = \frac{q\theta}{\rho C_p T_0} \qquad (c)$$

The boundary conditions are

At $z = 0$: $f = 1$ $t = 1$

At $\bar{r} = 0$: $\dfrac{df}{d\bar{r}} = 0$ $\dfrac{dt}{d\bar{r}} = 0$

At $\bar{r} = 1$ (wall): $\dfrac{\partial f}{\partial \bar{r}} = 0$

$$\frac{\partial t}{\partial \bar{r}} = \frac{UR_0}{\rho C_p K_r}(t - t_c) = \text{St }\overline{\text{Pe}_r}\, m(t - t_c)$$

Equations (a) to (c) must be solved numerically for the stated boundary conditions. A simplification can be invoked, i.e., to ignore Eq. (b) and relate the net consumption of anhydride to that of A locally, as will be shown and justified below.

At any point in the bed for A

$$\mathscr{R}_1 = \eta_1 k_1 \bar{A} = k_g a(A - \bar{A}) \qquad q = \sum (-\Delta H)\,\mathscr{R} = ha(T - \bar{T}) \qquad (d)$$

[1] K. R. Westerterp, *Chem. Eng. Sci.*, **17**: 423 (1962).

where the bar indicates a value at the catalyst-pellet surface. For anhydride we make use of the interphase-intraphase point-yield relationship developed in Chap. 9

$$\frac{dB}{dA} = \gamma \frac{m_1}{m_2} \frac{\phi_2 \tanh \phi_2}{\phi_1 \tanh \phi_1} \left(\frac{B}{A} + \frac{K}{\gamma K - 1}\right) - \gamma \frac{K}{\gamma K - 1} \qquad (e)$$

where

$$m = 1 + \frac{\phi \tanh \phi}{\text{Bi}_m} \qquad \gamma = \frac{\mathscr{D}_B}{\mathscr{D}_A}$$

The justification for ignoring the continuity equation (b) for anhydride rests upon the observation that computed profiles proved virtually independent of the value of the radial Peclet number for *mass;* this is to assert that the behavior of the reactor is not altered by dropping the radial gradient term in Eqs. (a) and (b). The radial-temperature-gradient term *cannot* be dropped, however. The simulation is based, then, upon numerical solution of Eqs. (a) and (c) for the overall and local boundary conditions (d).

Point yield is determined by Eq. (e); thus an isothermal pellet is assumed at a temperature determined by interphase heat transport. In sum the model anticipates:

1 Radial and axial temperature gradient for both gas and solid
2 Radial and axial species distribution
3 Local interphase-intraphase concentration gradients
4 Local interphase temperature gradients
5 Constancy of physicochemical properties such as ρC_p, $-\Delta H$, E, and $\mathscr{D}$ while, of course, the rate coefficients obey the Arrhenius functionality at pellet temperature t_s

$$k = k_0 \exp \left[-\frac{E}{RT_0} \left(\frac{1}{t_s} - 1\right) \right]$$

6 Heat transport at the wall governed by the process-side film coefficient in the Yagi-Wakao[1] correlation

Results[2] In Fig. 10-12 axial (centerline) and radial catalyst temperature profiles (at two axial positions) are shown for a mass Peclet number of 10 but for thermal radial Peclet numbers of 5 and 10. Recall that a Pe_r value of 10 is in agreement with experimental data (Chap. 4); but since heat can be transported by radiation and conduction as well as by mass, a value of $\overline{Pe}_r < 10$ can be anticipated.

Note particularly that for enhanced radial heat transport ($\overline{Pe}_r = 5$, Fig. 10-13) the predicted yield of anhydride is about 10 percentage points higher than at $\overline{Pe}_r = 10$. The reason for this significant difference is found in Fig. 10-14, where the overall interphase-intraphase point effectiveness is plotted vs. bed length at the centerline and at the wall for the two values of $\overline{Pe}_r$.

[1] S. Yagi and N. Wakao, *AIChE J.*, **5**: 79 (1959).
[2] "They ought to give medals for making phthalic anhydride." *Chem. Week*, **70**: 40 (1952).

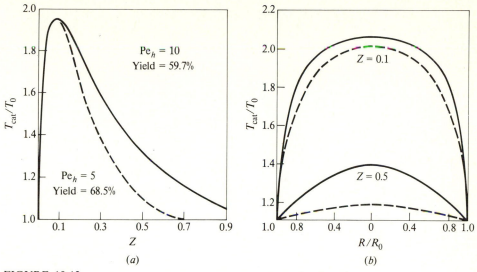

FIGURE 10-12
(a) Normalized catalyst-temperature profile.   Influence of radial Pe_h on profile; $\frac{1}{2}\%$ naphthalene in feed, radial $Pe_m = 10$.   (b) Computed radial catalyst-temperature profiles at $Z = 0.1$ and $Z = 0.5$. ———: $Pe_h = 10$; ————: $Pe_h = 5$.   Conditions for both (a) and (b) are given in Fig. 10-13.   [*J. J. Carberry and D. White, Ind. Eng. Chem.*, **61**: *27 (1969)*.]

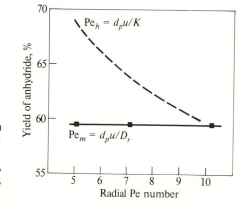

FIGURE 10-13
Influence of radial Peclet number on overall yield of anhydride; conversion = 99%. $T_0 = 320°C$, $T_w = 310°C$, $y_0 = \frac{1}{2}\%$, $Re = 184$, $d_p = \frac{1}{2}$ cm, $d_T = 5$ cm, $\theta_0 = 2$ s.   [*J. J. Carberry and D. White, Ind. Eng. Chem.*, **61**: *27 (1969)*.]

The conditions of operation are shown in the legend of Fig. 10-13, which gives yield of anhydride vs. both Pe_r and $\overline{Pe_r}$.   Reactor performance is obviously quite insensitive to the mass radial Peclet number yet very sensitive to its thermal counterpart.

Parametric sensitivity   The predicted influence of feed composition upon overall conversion and yield for various pellet sizes is shown in Table 10-6, where perfor-

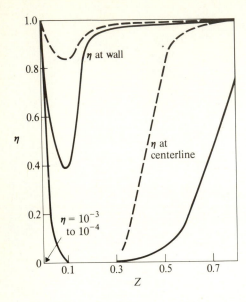

FIGURE 10-14
Interphase-intraphase catalytic effectiveness vs. reactor length at reactor wall and centerline. ———: $Pe_h = 10$; - - - - - -: $Pe_h = 5$. (Conditions are identical to those cited in Fig. 10-13.) [*J. J. Carberry and D. White, Ind. Eng. Chem.*, **61**: 27 (*1969*).]

mance of catalyst B is also noted. Also tabulated is the maximum radial temperature difference between centerline and wall ΔT_r at the hot spot and the maximum interphase ΔT_p at that point.

Table 10-6 INFLUENCE OF FEED COMPOSITION ON CONVERSION AND YIELD OF ANHYDRIDE†
Tube diameter = 5 cm

d_p, cm	Feed composition, % naphthalene in air y_0	Conversion x, %	Yield Y, %	Maximum radial temperature difference ΔT_r, °C	Interphase temperature difference ΔT_p at T_{max}, °C
		Catalyst A, Re = 360, T_w = 599 K			
0.5	0.5	9	0.7	10	1.7
	0.75	96.5	74	770	470
	1	97.6	66	1000	590
0.25	0.5	36	32	12	2.4
	1	97	80	590	240
		Catalyst B, Re = 184, T_w = 588 K			
0.25	0.5	5	0	7	0.5
	0.75	8	1.5	65	518
	1	98.5	55	650	590

† J. J. Carberry and D. White, *Ind. Eng. Chem.*, **61**(7): 27 (1969).

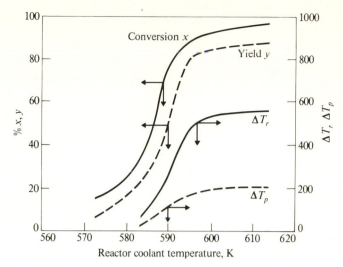

FIGURE 10-15

Fixed-bed parametric sensitivity: influence of coolant temperature on conversion x, yield y, and ΔT_r and ΔT_p; Re $= 730$, $y_0 = \frac{1}{2}$ percent. [*J. J. Carberry and D. White, Ind. Eng. Chem.*, **61**(7): *27* (*1969*).]

In Fig. 10-15 parametric sensitivity to changes in the constant coolant temperature is displayed. Note particularly the dramatic changes between 580 and 590 K. Table 10-7 indicates the reactor sensitivity to variation in activation energy of the first step in the consecutive sequence at two feed compositions. Conversion-yield results for diverse values of tube-to-pellet-diameter ratios are set forth in Table 10-8.

Table 10-7 **INFLUENCE OF ACTIVATION-ENERGY VARIATION ON** x, y, ΔT_r, **and** ΔT_p†

| | | A $\xrightarrow{\ 1\ }$ B $\xrightarrow{\ 2\ }$ C | | | |
| | Re $= 360$ | $T_w = 599$ K | $d_p = 0.25$ cm | $E_2 = 20$ kcal | |
$y_0, \%$	E_1	$x, \%$	$Y, \%$	ΔT_r	ΔT_p
0.75	42	97.4	75.6	770	470
	32	9	3.7	0	0
0.5	42	37	32	110	2.4
	32	7.4	0	0	0

† J. J. Carberry and D. White, *Ind. Eng. Chem.*, **61**(7): 27 (1969).

It is evident that computed performance is sensitive to a number of physico-chemical factors and operating parameters. In view of the limited precision which characterizes the Peclet number for radial heat transfer, the wall-coefficient correlations, and interphase-intraphase transport coefficients, it would be imprudent to suppose that precise a priori predictions of fixed-bed reactor performance could be realized, particularly for such demanding reactions as partial oxidation of hydrocarbons, aromatics, etc. Rather accurate simulation of a less vigorous reaction, dehydrogenation, has been demonstrated by Hawthorn et al.[1]

Although a priori prediction for vigorous reaction is not possible, relative simulation is not to be overlooked. As illustrated above, the qualitative effects of pellet size, tube diameters, feed composition, etc., as computed, tell the analyst-designer which design-operating variables should be manipulated to enhance performance. For example, a cause of yield loss in the naphthalene-oxidation reactor is detected by inspecting the radial profile at the hot spot.

In Table 10-9 values of gas-solid temperature, intrinsic rate-coefficient ratio, catalytic effectiveness, and the effective-rate-coefficient ratio (a product of intrinsic

Table 10-8 COMPUTED OVERALL CONVERSION AND YIELD FOR NAPHTHALENE OXIDATION†

$T_w = 583$ K, $y_0 = \frac{1}{2}\%$, $Pe_r = 10$, $\overline{Pe}_r = 7$, $\theta = 2$ s

Run	d_p, cm	D_T, cm	S/V, cm^{-1}	Re	x, %	y, %
a	1	10	0.4	365	99.9	42
b	0.5	5	0.8	365	98	74
c	0.5	2.5	1.6	184	93	74
d	0.5	5	0.8	184	99	65
e	0.5	10	0.4	184	100	47

† J. J. Carberry and D. White, *Ind. Eng. Chem.*, **61**(7): 27 (1969).

Table 10-9 COMPUTED RADIAL PROFILE AT $Z = 0.2$ FOR NAPHTHALENE OXIDATION†

$y_0 = \frac{1}{2}\%$, $d_p = \frac{1}{2}$ cm, $d_T = 5$ cm
Re = 365, $T_0 = 310°$C, $x = 99\%$, $Y = 84\%$

$\dfrac{r}{R}$	$t_g = \dfrac{T_g}{T_0}$	$t_c = \dfrac{T_c}{T_0}$	$\dfrac{k_1}{k_2}$	η_1	$\dfrac{\eta_1 k_1}{\eta_2 k_2}$
0 (centerline)	1.3	1.6	1.5×10^4	7.3×10^{-4}	13
0.6	1.25	1.53	1×10^4	1.5×10^{-3}	17
1 (wall)	1.0	1.005	50	0.98	50

† J. J. Carberry and D. White, *Ind. Eng. Chem.*, **61**(7): 27 (1969).

[1] R. D. Hawthorn, G. H. Ackerman, and A. C. Nixon, *AIChE J.*, **14**: 69 (1968).

ratio and effectiveness) are shown at three radial positions. It is immediately evident that in the absence of diffusional limitations a 100 percent yield of anhydride could be anticipated since the intrinsic rate-coefficient ratio increases from 50 at the wall to 10^4 at the bed centerline, a consequence of the fact that for this catalyst (A) $E_1 > E_2$ (38 versus 20 kcal/mol). However, in the intraphase-diffusion regime, the apparent, effective activation energy $\bar{E}_1 = E_1/2 = 19$. Hence $\bar{E}_1 < E_2$, with the result that the effective-rate-coefficient ratio actually decreases in moving from the wall to centerline. The reactor yield in this case is 84 percent, well below what one would expect from intrinsic kinetics unaltered by diffusion.

Such analyses suggest two remedial steps: (1) reduce d_p, but in so doing a respect for pressure drop and amplification of radial ΔT_r must be borne in mind. For while not included in the model, excessive ΔT_r values will prove damaging to catalyst physicochemical integrity. Thus tube diameter should be reduced when d_p is reduced. (2) Given pressure drop and radial ΔT_r constraints, partial impregnation can effectively increase η with benefit to yield.[1]

In accord with practice, fixed-bed naphthalene-oxidation reactors are characterized by low values of tube-to-pellet-diameter ratio, which calls for an expensive shell-and-tube reactor configuration. An alternative is the PTWR.

The Packed Tubular-Wall Reactor

The previous illustration shows that a significant resistance to heat removal during exothermic reaction in a fixed bed may be localized in the core or the packing per se in contrast to the wall film resistance. This situation places obvious limits on the concentration of combustible feed species, attainable yield, etc.

For a rather vigorous exothermic catalytic reaction it is evident that effectiveness factors tend to be low, so that utilization of the catalytic packing is surely less than total. In such circumstances one might fruitfully envision a tube of which the inner wall is coated with the highly active catalyst and which is packed with inert pellets, for example, Al_2O_3, to maximize radial mass transport of the reactants to the active wall. Since in such a configuration (Fig. 10-16) heat release is at the tube wall, the minimum of thermal resistance prevails. This is the PTWR.

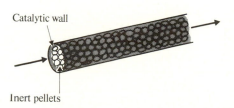

Catalytic wall

FIGURE 10-16
Packed tubular-wall reactor with inert
packing and catalytic wall.

Inert pellets

[1] T. G. Smith and J. J. Carberry, *Can. J. Chem., Eng.* **53**: 347 (1975).

Smith and Carberry[1] have compared, by numerical solution, the conventional fixed-bed reactor with the PTWR for the oxidation of naphthalene, but they used a kinetic network different from that assumed above and based upon Drott's analyses[2] of the data of D'Alessandro and Farkas.[3]

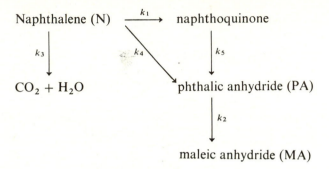

Since k_4 is small and $k_5 \gg k_1$, this reduces to

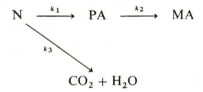

Since the fixed-bed results stand in general agreement with those of White, discussed earlier, we focus upon the PTWR model.

The model Species concentration in the steady state ($C \equiv N$, PA, etc.) is

$$\frac{\partial C}{\partial z} - \frac{\mathscr{A}}{\mathrm{Pe}_{r,m}} \left(\frac{\partial^2 C}{\partial \bar{r}^2} + \frac{1}{\bar{r}} \frac{\partial C}{\partial \bar{r}} \right) = 0 \qquad (10\text{-}48)$$

since there is no generation in the packed phase; i.e., reaction is a boundary condition. For gas temperature

$$\frac{\partial t}{\partial z} - \frac{\mathscr{A}}{\overline{\mathrm{Pe}_{r,m}}} \left(\frac{\partial^2 t}{\partial \bar{r}^2} + \frac{1}{\bar{r}} \frac{\partial t}{\partial \bar{r}} \right) = 0 \qquad (10\text{-}49)$$

The boundary conditions are

At $z = 0$:　　$N = 1$　　$PA = 0$　　$t = 1$

At $\bar{r} = 0$:　　$\dfrac{\partial t}{\partial \bar{r}} = 0$　　$\dfrac{\partial C}{\partial \bar{r}} = 0$

[1] T. G. Smith and J. J. Carberry, *Adv. Chem.*, no. 133, p. 362, 1962; *Chem. Eng. Sci.*, **30:** 221 (1975).
[2] D. W. Drott, Ph.D. thesis, University of Minnesota, Minneapolis, 1972.
[3] A. F. D'Alessandro and A. Farkas, *J. Colloid Sci.*, **11:** 653 (1956).

and at the catalytic wall ($\bar{r} = 1$) for mass

$$\mathscr{D}_r \frac{\partial C}{\partial \bar{r}} = k_g R_0 (C - C_w)$$

$$\frac{\partial C}{\partial \bar{r}} = j_D \, Sc^{-2/3} \, Pe_r \, m(C - C_w) \qquad (10\text{-}50)$$

and for heat

$$\frac{\partial t}{\partial \bar{r}} = St \, \overline{Pe_r} \, m(t - t_w)$$

$$j_D = \frac{k_g}{u} \, Sc^{2/3} \qquad m = \frac{R_0}{d_p}$$

We must now express the surface concentration of N and PA, that is, C_w in terms of mass transport followed by reaction at the wall

$$k_g(N - N_w) = (k_1 + k_3)N$$

$$k_g(PA - PA_w) = k_1 N_w - k_2 PA_w$$

Thus

$$N_w = \frac{N}{1 + Da_3} \qquad Da_3 = \frac{k_1 + k_3}{k_g}$$

$$PA_w = \frac{PA + Da_1 N_w}{1 + Da_2} \qquad Da_1 = \frac{k_1}{k_g} \qquad Da_2 = \frac{k_2}{k_g}$$

The catalytic wall temperature which dictates reaction velocities is established by a heat balance in terms of the process-side heat-transfer coefficient h and that $\bar{h}$ on the shell side

$$\bar{h}a(T_c - T_w) = ha(T_w - T) = \Sigma - \Delta H \, \mathscr{R}$$

or

$$t_w = \frac{T_w}{T_0} = \frac{hat + \bar{h}at_c + k_1' N_w(-\Delta H_1) + k_3' N_w(-\Delta H_3) + k_2' PA_w(-\Delta H_2)}{(h + \bar{h})a}$$

Given this more or less general model of the PTWR, the appropriate transport coefficient correlations are for interphase, at the wall

$$\frac{k_g}{u_0} \, Sc^{2/3} = \frac{0.2}{(d_p u_0/v)^{0.2}} = \frac{h}{\rho u_0 \rho} \, Pr^{2/3} \qquad (10\text{-}51)$$

Given a value of Pe_r (mass dispersion), the thermal radial-dispersion Peclet number $\overline{Pe_r}$ is that recommended by Beek's modification of the Argo-Smith model[1]

---

[1] J. Beek, *Adv. Chem. Eng.*, **3**: 229 (1962).

[Eq. (10-31)], which accounts for conductive and radiative contributions to radial heat dispersed by mass.

Note that in the simulation the specific rate constants are expressed in terms of reaction velocity coefficients, that is, k is in centimeters per second. In consequence for the fixed bed of spheres

$$\bar{k} = k(S + a)$$

where $S + a$ is the total of external-internal area per unit volume, while for the PTWR of tube diameter d_t

$$\bar{k} = k\frac{4}{d_t} = ka_t$$

if it is assumed that the wall of the PTWR is nonporous. Obviously $a_t > 4/d_t$ if the wall catalyst is porous.

The drastic reduction in catalyst area associated with wall deposition in the PTWR might be expected to be taxing relative to the great area offered by a fixed bed of high-area catalyst pellets. However, the BET area of a tube wall wash-coated with Al_2O_3 upon which catalyst is deposited can be several square meters per gram. This value is surely equivalent to a fixed bed of α-Al_2O_3 pellets or η- or γ-aluminas, which by reason of several diffusional intrusions, actually give only a small fraction of their total BET area to reaction. That is to say, if one is confronted with an exothermic reaction of such vigor that intraphase diffusion drastically reduces catalyst utilization, the PTWR suggests itself insofar as core thermal resistance is eliminated and interparticulate interphase-intraphase heat transport become localized at the wall.

In Figs. 10-17 and 10-18 computed gas and wall temperatures and conversion-yield profiles are shown for the PTWR. From Table 10-10, where the PTWR is

Table 10-10   COMPUTED COMPARISON OF OPTIMUM PTWR WITH A FIXED-BED REACTOR FOR NAPHTHA-LENE OXIDATION†

	Catalytic bed	Catalytic wall
T_{feed}, °C	470	450
C_{feed}, %	0.54	9.0
$T_{coolant}$, °C	460	450
Tube diameter, cm	2.85	1.0
Pellet diameter, cm	0.15	0.3
Maximum radial ΔT, °C	175	155
Yield, %	77	81
Productivity‡	1000	2000

† T. G. Smith and J. J. Carberry, *Chem. Eng. Sci.*, **30:** 221 (1975).

‡ Productivity $= \dfrac{\text{g mol anhydride produced}}{\text{(s)(cm}^2 \text{ tube cross section)}}$

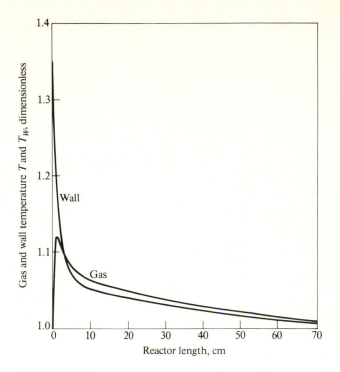

FIGURE 10-17
Tubular-wall-reactor profiles.   [*T. G. Smith and J. J. Carberry,*
Chem. Eng. Sci., **30**: *221 (1975).*]

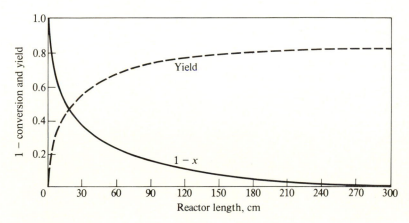

FIGURE 10-18
Tubular-wall-reactor profiles.   [*T. G. Smith and J. J. Carberry, Chem. Eng. Sci.,*
30: *221 (1975).*]

contrasted with the fixed-bed reactor, it is evident that for particularly rapid exothermic catalytic reactions such as are involved in naphthalene oxidation the
PTWR provides higher yield of PA, allows the use of higher reactant-feed concentration, and hence is characterized by greater productivity P. While there is
little doubt that the PTWR unit will be larger due to its comparatively small
catalytic-surface area, in terms of yield or productivity, its merits are noteworthy.

10-4 THE FLUIDIZED-BED CATALYTIC REACTOR

If one passes a gas upward through a vertical fixed bed of fine particles in the
absence of a retaining grid or screen at the top of the pack bed, one observes an
expected increase in pressure drop with gas velocity while the bed remains fixed
(static), as illustrated in Fig. 10-19a. At some point (a particular gas velocity)
particle movement will be observed. At higher velocities, considerable agitation
and bubbling of the bed will be noted (Fig. 10-19b). Pressure drop and agitated
bed height will become constant at velocity values beyond that at which initiation
of particle movement is observed

Point of incipient fluidization, $u = u_{mf}$

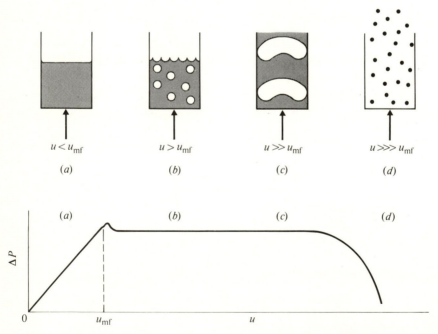

FIGURE 10-19
Schematic of (a) fixed, (b) fluid, (c) slug-flow, and (d) transport-line reactors.

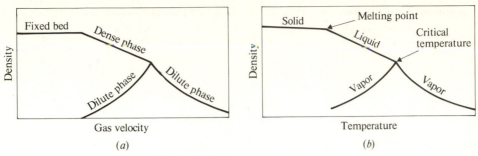

FIGURE 10-20
Analogy between (*a*) fluidization and (*b*) vapor-liquid-solid system.

A realm of free bubbling behavior (fluidization) exists over a wide range of velocity. When bubbles grow to tube size, slug flow exists (Fig. 10-19*c*). Ultimately fluid velocity will be great enough to carry or literally transport the entire contents of the fluid bed out of the unit [region of dilute phase transport (Fig. 10-19*d*)].

The fluidization phenomenon has been fruitfully compared to boiling of a liquid (Fig. 10-20). Indeed, where the fluidizing fluid-solid density difference is great (as is usually the case in gas-solid systems up to modest pressures) gas bubbles relatively free of particles will be observed surrounded by a dense, particle-rich emulsion phase. Hence an analogy with vapor bubbles in a boiling-liquid system.

General Character of Fluidization

The velocity at which fluidization appears is readily computed by a simple force balance. The difference between gravitational and buoyant forces at initiation of fluidization is manifested as pressure drop. For particles of density ρ_s, gas density ρ, and void volume e in a bed of cross-sectional area A and length L

$$(\rho_s - \rho)AL(1 - e)g = \Delta P \, A \qquad (10\text{-}52)$$

The Ergun equation for pressure drop in a packed bed is, at u_{mf},

$$\frac{\Delta P}{L} = \frac{150(1 - e)}{d_p{}^2 e^3} \mu u_{mf} + 1.75 \frac{\rho u_{mf}{}^2}{d_p e^3} \qquad (10\text{-}53)$$

where d_p = particle diameter
μ = fluid viscosity
g = the gravitational constant [Eq. (10-52)]

The minimum fluidization velocity is implicitly given by

$$(1 - e)(\rho_s - \rho)g = \frac{150(1 - e)}{d_p{}^2 e^3} \mu u_{mf} + 1.75 \frac{\rho u_{mf}{}^2}{d_p e^3} \qquad (10\text{-}54)$$

Equation (10-52) indicates the force necessary to move the bed vertically, and Eq. (10-53) relates that force to the fixed-bed pressure-drop functionality.

Quality of Fluidization

Fluidization quality, i.e., whether smooth (homogeneous) or aggregative (inhomogeneous or bubbling), will depend, as Wilhelm has shown, upon the Froude number $Fr = u^2/gd_p$. When $Fr < 1$, one has homogeneous, bubble-free, fluidization, while bubble–emulsion-phase inhomogeneous fluidization is found when $Fr > 1$. As the fluidizing fluid and particle densities approach each other in magnitude, homogeneous fluidization occurs. For gas-solid fluidized systems at modest pressure levels, inhomogeneity of fluidization (bubbling) will be found, while a liquid-solid fluidized bed will tend to be homogeneous.

Fluid-Bed Entrainment

As schematized in Fig. 10-19d, at a particular velocity, the solids are entrained or carried out of the bed, an event which occurs at the terminal velocity of a particle of diameter d_p. As a first approximation (ignoring particle interaction, etc.), Stokes' law is invoked:

$$(\rho_s - \rho)gV_p = 3\pi\mu d_p u_t \qquad (10\text{-}55)$$

where for a spherical particle $V_p = \frac{4}{3}\pi(d_p/2)^3$, and so we obtain

$$u_t = d_p{}^2 \frac{(\rho_s - \rho)g}{18\mu} \qquad (10\text{-}56)$$

the velocity at which one can anticipate entrainment. In general fluid beds are operated at about one-half the entrainment velocity.

Depending, then, upon the density difference between fluidizing fluid and the fluidized particles, we may be witnesses to a uniform, bubble-free system or, as is common in the commonest gas-solid system, we shall encounter a multiphase system consisting of (1) a relatively particle-free bubble phase of gas, (2) a relatively particle-rich emulsion phase, and (3) a partially particle-endowed cloud phase existing between bubble and emulsion phases. Characteristics 1 and 2 follow upon Wilhelm's definitions, while phenomenon 3 is the product of detailed observations of bubbles and their environs in fluidization experiments.

Solids movement in the emulsion phase has been studied in detail. In essence, bubble size and rise velocity dictate the magnitude and direction of solids movement. At low gas velocities, roughly when $u_0 < 2u_{mf}$, small, slowly rising bubbles exist. Gas percolates through the emulsion phase at a rate greater than the bubble-rise velocity. Little net movement of solids is observed.

In the range of practical interest ($u_0 > 2u_{mf}$) solids are observed to move downward, a phenomenon explained by the fact that the rapidly rising bubbles carry solids in their wake. Thus a given bubble draws particles into its wake upon formation in the bottom of the bed and carries them upward. When the bubble emerges from the top of the bed, the particles are released and recirculated downward. With increasing bubble size and velocity, this solid-recirculation rate

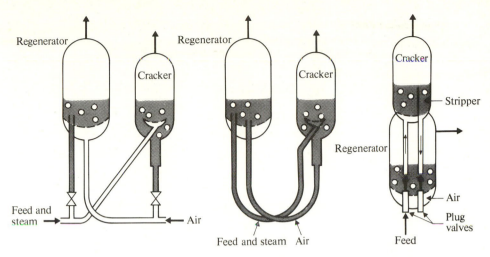

FIGURE 10-21
Diverse vessel arrangements for fluidized-solids systems.

can become great enough to cause flow reversal in the emulsion-phase gas.   The bubbles act as an emulsion-phase pumping device.

We see, then, that a fluidized bed is a complex multiphase network involving exchange of mass and/or heat between the constituent phases.   From this brief sketch of fluidization the advantages and disadvantages of a fluidized-bed catalytic reactor become apparent.   Consider a simplified sketch of a fluid catalytic cracking unit (Fig. 10-21).   The advantages are:

1   Very small catalyst sizes can be employed in a fluid bed with consequent advantages in terms of catalytic effectiveness.
2   As shown in Fig. 10-21, spent or coked catalyst can be continuously circulated between the reactor and regenerator.   Also fresh catalyst is easily injected.
3   By reason of agitation induced by the bubbles, the catalyst-rich emulsion phase is rather well mixed both radially and axially; therefore conditions approaching isothermality characterize the fluid-bed reactor.
4   Pressure drop is independent of flow rate.

The disadvantages are:

1   Catalyst losses by entrainment and attrition can be significant.
2   Mixing of the emulsion phase creates backmixing conditions somewhere between plug and total backmixed flow.
3   Since the bubble phase is a bypassing mechanism, a portion of feed gas may escape without contact with the catalyst, which is largely found in the emulsion and wake phases.
4   The fluid mechanics of the multiphase fluid bed is indeed complex, and our understanding of these phenomena is not as clear as it is for fixed-bed fluid mechanics.

The fluidized-bed catalytic reactor becomes a serious candidate in the process of reactor selection when isothermal operation is necessary, catalyst life is short and regeneration is practicable, or the non-plug-flow and bubble-bypassing features can be tolerated.

Thus it is not surprising that fluidized catalytic cracking was the first demonstration of the fluid-bed reactor. For in catalytic cracking catalyst life is brief, the fouled (coked) catalyst is readily regenerated, and bypassing or imperfect mixing must be tolerated since fixed-bed cracking is clearly impractical.

10-5 FLUID-BED-REACTOR MODELING

A number of models of the fluidized bed have been invoked, and detailed discussions and literature citations can be found in the text of Davidson and Harrison[1] and the later treatise of Kunii and Levenspiel.[2] Two models of the fluid-bed reactor will be cited here, as each contributes an insight into the complexities of these systems, and, most importantly, bases for analysis and design are set forth by the Davidson-Harrison model and the Kunii-Levenspiel bubbling-bed model.

All fluid-bed models assume that an isothermal environment prevails throughout the reaction phases, an assumption justified in beds well agitated by vigorous bubbling, although some evidence of modest axial and radial temperature gradients has been reported.

Davidson-Harrison Model

In Fig. 10-22 the basic features of the two-phase fluidization model are set forth. In addition to isothermality, the following assumptions characterize the Davidson-Harrison model:

1 All gas entering the bed at velocity u_0 is partitioned between the emulsion phase, where the velocity is that of minimum fluidization u_{mf}, and the bubble phase, where the velocity is $u_0 - u_{mf}$.

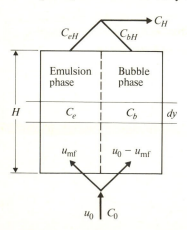

FIGURE 10-22
Two-phase model of a fluidized bed.

[1] J. F. Davidson and D. Harrison, "Fluidized Solids," Cambridge University Press, London, 1963.
[2] D. Kunii and O. Levenspiel, "Fluidization Engineering," Wiley, New York, 1968.

2   The increase in bed height $H - H_0$ relative to the static value H_0 is due to the total volume of bubbles in the bed.

3   The bubble phase is to be particle-free and moving upward in plug flow. Catalytic reaction cannot occur in the solid-free bubble phase.

4   As the emulsion phase contains all the solid-catalyst particles, reaction occurs only in the emulsion phase.

5   The reacting emulsion phase can be considered to be either totally backmixed or in plug flow.

6   Exchange between bubble and emulsion phases occurs, as the bubbles ascend, by cross-flow of mass q and diffusion through the bubble cap of area S; hence, the exchange coefficient is

$$Q = q + k_g S \qquad \text{vol/s} \qquad (10\text{-}57)$$

7   Bed height H at conditions of bubbling fluidization is related to static height H_0 by the number of bubbles per bed volume N and average volume per bubble by

$$H = \frac{H_0}{1 - NV} \qquad (10\text{-}58)$$

Referring to Fig. 10-22, a material balance for the bubble phase in the differential height dy yields (assuming no reaction in the bubble)

$$(q + k_g S)(C_e - C_b) = u_b V \frac{dC_b}{dy} \qquad (10\text{-}59)$$

An overall balance for both phases gives, for first-order reaction,

$$u_{\text{mf}} \frac{dC_e}{dy} + (u_0 - u_{\text{mf}}) \frac{dC_b}{dy} + k C_e (1 - NV) = 0 \qquad (10\text{-}60)$$

Note that the emulsion phase is assumed to be in plug flow.   If total backmixing of the emulsion phase is assumed, then

$$N V u_b (C_0 - C_e) \left[1 - \exp\left(-\frac{QH}{u_b V} \right) \right] + u_0 (C_0 - C_e) = k H C_e (1 - NV) \qquad (10\text{-}61)$$

Following the Davidson-Harrison development, let

$$\beta = 1 - \frac{u_{\text{mf}}}{u_0} \qquad k' = \frac{k H_0}{u_0} \qquad X = \frac{QH}{u_b V}$$

Recall that $u_b = u_0 - u_{\text{mf}}$.   For plug flow in the emulsion phase, Eqs. (10-59) and (10-60) become

$$\frac{dC_b}{dy} + \frac{X}{H} (C_b - C_e) = 0 \qquad (10\text{-}62)$$

and

$$(1 - \beta) \frac{dC_e}{dy} + \beta \frac{dC_b}{dy} + \frac{k'}{H} C_e = 0 \qquad (10\text{-}63)$$

with boundary conditions

At $y = 0$: $C_b = C_0$

At $y = 0$: $\dfrac{dC_b}{dy} = 0$ (10-64)

In operator form

$$\left(D + \frac{X}{H}\right)C_b - \frac{X}{H}\,C_e = 0 \quad (10\text{-}62a)$$

$$\left[D + \frac{k'}{H(1-\beta)}\right]C_e + \frac{\beta}{1-\beta}\,DC_b = 0 \quad (10\text{-}63a)$$

We multiply (10-63a) by X/H and (10-62a) by $D + k'/H(1-\beta)$ and then add the equations to secure a second-order differential equation in C_b:

$$H^2(1-\beta)\frac{d^2C_b}{dy^2} + H(X + k')\frac{dC_b}{dy} + k'XC_b = 0 \quad (10\text{-}65)$$

Equation (10-65) is solved for the appropriate boundary condition. However, to secure the exit concentration of the reactant, we must also solve for C_e by the equation

$$C_e = C_b + \frac{H}{X}\frac{dC_b}{dy} \quad (10\text{-}66)$$

Given then, C_e and C_b at $y = H$, the exit concentration C_H is

$$u_0\,C_H = (u_0 - u_{mf})C_{bH} + u_{mf}\,C_{eH}$$

or $$C_H = \beta C_{bH} + (1 - \beta)C_{eH} \quad (10\text{-}67)$$

The resulting reduced exit concentration for plug flow is

$$\frac{C_A}{C_{0,\,pf}} = \frac{1}{m_1' - m_2'}\left[m_1'\left(1 - \frac{u_{mf}}{u_0}\frac{H}{X}\,m_2'\right)\exp(-m_2'H)\right.$$

$$\left. - m_2'\left(1 - \frac{u_{mf}}{u_0}\frac{H}{X}\,m_1'\right)\exp(-m_1'H)\right] \quad (10\text{-}68)$$

where $$m_{1,\,2}' = \frac{(x + k') \pm \sqrt{(x + k')^2 - 4(1-\beta)k'X}}{2H(1-\beta)} \quad (10\text{-}69)$$

In addition to chemical-kinetic parameters, the application of the Davidson-Harrison model requires knowledge of exchange coefficient Q, bubble volume, and velocity. A vast body of research by Davidson and his students and Rowe and colleagues (to cite but two of many schools of inquiry into fluidization mechanics) provides bases for a priori estimation of the required parameters. Davidson and Harrison[1] analyze a number of fluid-bed catalytic studies in terms of their model,

[1] Loc. cit.

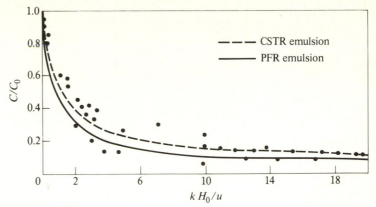

FIGURE 10-23

Comparison of Davidson-Harrison model predictions with ozone-decomposition data.   (*J. F. Davidson and D. Harrison, " Fluidised Solids," Cambridge University Press, London, 1963.*)

and the comparison with one study is shown in Fig. 10-23.   Predictions are set forth for both piston and totally backmixed flow in the emulsion phase.

Kunii-Levenspiel Model

The *bubbling-bed* model of Kunii-Levenspiel,[1] the key features of which are to be found in cited literature as well as in their text cited earlier, views various transport process and catalytic reaction in fluidized beds differently from Davidson and Harrison.   The Kunii-Levenspiel model nevertheless rests upon key fluid- and bubble-mechanic insights provided by the earlier Davidson-Harrison treatise.

Referring to Fig. 10-24, the Kunii-Levenspiel model focuses upon a fluid bed composed of large, fast solids-free bubbles ($u_0 > 2u_{mf}$) surrounded by a cloud and followed by a wake, both rich in solids, passing through an emulsion phase very rich in solids.   Depending upon bubble velocity, the solids in the emulsion phase may have a zero or finite downward velocity.   In emulsion-solids downward flow, emulsion gas may actually flow downward.   In the region of practical interest (bubble velocity much greater than incipient fluidization velocity) Fig. 10-24 reveals the essential features with respect to a single bubble.

In the light of Fig. 10-24 and the accompanying definitions, (1) reactant gas which enters the bed in the bubble can react within the bubble if γ_b is finite and K_r large, (2) unreacted bubble gas can be transported across the bubble-cloud interface (transport coefficient K_{bc}) into the more solids-rich cloud *and* wake phase, where (3) transported reactant can react in the cloud-wake phase at a rate $\gamma_c K_r C_c$. (4) Unreacted species in the cloud can be transported to the very solids-rich

[1] D. Kunii and O. Levenspiel, *Ind. Eng. Chem. Fundam.*, **7**: 446 (1968); *Ind. Eng. Chem. Process Des. Dev.*, **7**: 481 (1968).

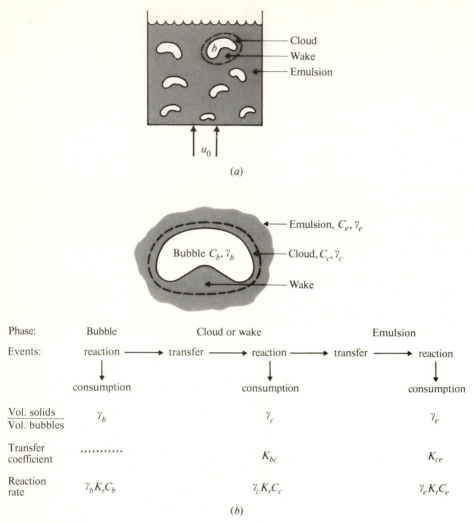

FIGURE 10-24
(a) Bubbling fluid bed; (b) detail of the Kunii-Levenspiel fluid-bed-bubble model.

emulsion phase (transport coefficient K_{ce}) where, as assumed by Kunii and Levenspiel, it is totally consumed by reaction.

The net result of these events is that the observed consumption rate depends upon two transport coefficients and three effective chemical rate coefficients $\gamma_b K_r$, $\gamma_c K_r$, and $\gamma_e K_r$. Hence, we can expect the effective observed chemical rate coefficient for fluidized-bed catalytic reaction (as well as coefficients for physical rate processes) to be less than intrinsic values by reason of interphase-intraphase diffusion, not within catalyst particles necessarily but within the macroscopic bubble, cloud-wake, and emulsion phases which characterize the fluidized-bed reactor.

Thus we can focus analytical attention upon a single bubble as it ascends through the fluid bed as a multiphase batch reactor:

$$-\frac{1}{V_b}\frac{dn}{dt} = -u_b\frac{dC_b}{dy} = \mathscr{K}_f C_b \qquad (10\text{-}70)$$

where $\mathscr{K}_f$ is the effective multiphase-reaction-rate coefficient. Given a definitive relationship between $\mathscr{K}_f$ in Eq. (10-70) and K_r, K_{bc}, K_{ce}, γ_b, γ_c, and γ_e, fluidized-bed conversion is found by

$$1 - x = \exp\left(-\mathscr{K}_f\theta_b\right) \qquad (10\text{-}71)$$

where θ_b is the bubble contact time.

Anatomy of the Overall Rate Coefficient

In general, the Kunii-Levenspiel model for chemical reaction in a bubbling fluid bed states, for a key reactant of concentration C, that

Overall disappearance = reaction in bubble + transfer to cloud and wake

Transfer to cloud and wake $\approx$ reaction in cloud and wake + transfer to emulsion

and   Transfer to emulsion $\approx$ reaction in emulsion

$$-u_b\frac{dC_b}{dy} = \mathscr{K}_f C_b = \gamma_b K_r C_b + K_{bc}(C_b - C_c) \qquad (10\text{-}72)$$

$$K_{bc}(C_b - C_c) \approx \gamma_c K_r C_c + K_{ce}(C_c - C_e) \qquad (10\text{-}73)$$

$$K_{ce}(C_c - C_e) \approx \gamma_e K_r C_e \qquad (10\text{-}74)$$

The Kunii-Levenspiel model specifies $\mathscr{K}_f$ as an overall rate coefficient which accounts for the reaction-transport processes in the fluidized bed. By eliminating concentrations in the above equations, we obtain

$$\mathscr{K}_f = K_r\left(\gamma_b + \cfrac{1}{\cfrac{K_r}{K_{bc}} + \cfrac{1}{\gamma_c + \cfrac{1}{K_r/K_{ce} + 1/\gamma_e}}}\right) \qquad (10\text{-}75)$$

We can take the Kunii-Levenspiel result [Eq. (10-75)] and cast it into a somewhat simpler form by recognizing that the reaction-transport coefficient ratio is a Damköhler number of the type encountered in Chap. 5:

$$\frac{K_r}{K_{bc}} = \mathrm{Da}_c \qquad \text{and} \qquad \frac{K_r}{K_{ce}} = \mathrm{Da}_e \qquad (10\text{-}76)$$

Further, again in the spirit of Chap. 5, let us define a phase effectiveness factor as

$$\eta = \frac{\gamma}{1 + \gamma\mathrm{Da}} \qquad (10\text{-}77)$$

This fluid-bed phase effectiveness has the following properties:

Small Da $\eta \to \gamma$, total utilization of the catalyst in that phase.
Large Da $\eta \to 0$, that phase is not utilized.

Therefore for bubble phase: $\eta_b = \gamma_b$

For cloud-wake phase:

$$\eta_c = \frac{\gamma_c}{1 + \gamma_c \, \mathrm{Da}_c}$$ (10-78)

For emulsion phase:

$$\eta_e = \frac{\gamma_e}{1 + \gamma_e \, \mathrm{Da}_e}$$

Let us also define the overall fluidized-bed efficiency as

$$\mathscr{E}_f = \frac{\mathscr{K}_f}{K_r} \frac{\text{vol bubbles}}{\text{vol solids}} = \frac{\mathscr{K}_f}{K_r} \frac{V_b}{V_s} = \eta_o \frac{V_b}{V_s}$$

In these terms the Kunii-Levenspiel model becomes

$$\mathscr{E}_f = \frac{V_b}{V_s} \left[\gamma_b + \frac{1}{\mathrm{Da}_c + 1/(\gamma_c + \eta_e)} \right]$$ (10-79)

Fluid-bed efficiency has the following limiting properties.   For large Da

$$\eta_e = 0 \qquad \mathscr{E}_f = \frac{V_b}{V_s} (\gamma_b + \eta_c) < 1$$ (a)

$$\eta_e = 0 \qquad \eta_c = 0 \qquad \mathscr{E}_f = \frac{V_b}{V_s} \gamma_b \ll 1$$ (b)

and for small Da

$$\mathscr{E}_f = \frac{V_b}{V_s} (\gamma_b + \gamma_c + \gamma_e) \to 1.0$$ (c)

Recall that $\gamma_e > \gamma_c \gg \gamma_b$.

If we suppose that the catalytic reaction is very slow indeed, Da_c and $\mathrm{Da}_e \to 0$, and the fluid bed exhibits behavior approaching that of the fixed bed [condition (c)]. For very very rapid reaction, condition (b) is approached, and since the solids concentration in the bubble is very low, fluid-bed efficiency will be very poor— poorer in fact than the performance of a CSTR, as will be demonstrated shortly. The intermediate situation, condition (a), is the one most likely to be encountered.

Satisfying as the Kunii-Levenspiel model is, its usefulness rests upon the specification of its parameters, K_{bc}, K_{ce}, and the γ values.   Indeed the great merit of the Kunii-Levenspiel model is the fact that the authors are able to estimate a priori these key parameters solely in terms of the effective bubble diameter in a fluidized bed.

Specification of Fluid-Bed Parameters

The interphase transport coefficients are defined as

$$K_{12} = \frac{\text{vol. of gas exchanged between phases 1 and 2}}{(\text{vol. of bubbles in bed})(\text{time})}$$

Now K_{bc} is found on the assumption that bubble-cloud (or bubble-wake) exchange occurs by flow q and diffusion or by the Davidson-Harrison model [Eq. (10-57)]

$$K_{bc}(C_b - C_c) = \frac{(q + k_g S)(C_b - C_c)}{V_b} \qquad (10\text{-}80)$$

Davidson and Harrison derived expressions for q and k_g in terms of bubble diameter d_b

$$q = \frac{3\pi}{4} u_{\text{mf}} d_b^2 \qquad (10\text{-}81)$$

$$k_g = 0.975 \sqrt{\mathscr{D}} \left(\frac{g}{d_b}\right)^{1/4} \qquad (10\text{-}82)$$

Therefore

$$K_{bc} = 4.5 \frac{u_{\text{mf}}}{d_b} + 5.85 \frac{\mathscr{D}^{1/2} g^{1/4}}{d_b^{5/4}} \qquad (10\text{-}83)$$

Between cloud and emulsion, transport occurs solely by diffusion. Assuming a Higbie penetration model, as employed in the Davidson-Harrison model [Eq. (10-82)], the cloud- or wake-emulsion transport coefficient is

$$K_{ce} \approx 6.78 \left(\frac{e_{\text{mf}} \mathscr{D} u_b}{d_b^3}\right)^{1/2} \qquad (10\text{-}84)$$

Bubble velocity u_b is related to the superficial gas velocity u_0, the velocity at incipient fluidization u_{mf}, and d_b by the Davidson model

$$u_b = u_0 - u_{\text{mf}} + 0.711(gd_b)^{1/2} \qquad (10\text{-}85)$$

The solids distribution is given by

$$\delta(\gamma_b + \gamma_c + \gamma_e) = (1 - e_{\text{mf}})(1 - \delta) \qquad (10\text{-}86)$$

where e_{mf} = void fraction at minimum fluidization
δ = fraction of bed containing bubbles

$$\delta = \frac{u_0 - u_{\text{mf}}}{u_b} \qquad (10\text{-}86a)$$

Experimental data suggest $\gamma_b = 0.001$ to 0.01. For cloud and wake, the Kunii-Levenspiel development suggests

$$\gamma_c = (1 - e_{\text{mf}}) \left[\frac{3 u_{\text{mf}}/e_{\text{mf}}}{0.711(gd_b)^{1/2} - u_{\text{mf}}/e_{\text{mt}}} + \alpha\right] \qquad (10\text{-}87)$$

$$\alpha = \frac{\text{volume of emulsion in bubble wake}}{\text{volume of bubbles}} \approx \frac{V_w}{V_b}$$

Rowe and Partridge[1] present data indicating $\alpha = \frac{1}{3}$ to $\frac{1}{2}$. The value of γ_e is obtained from Eq. (10-86) once δ, γ_b, γ_c, and e_{mf} are specified. Finally V_b/V_s, the volume of bubbles per *total* volume of solids, is

$$\frac{V_b}{V_s} = \frac{\delta}{(1-\delta)(1-e_{mf})} = \frac{1}{\gamma_b + \gamma_c + \gamma_e} \qquad (10\text{-}87a)$$

So our fluidized-bed efficiency (relative to the fixed bed) is

$$\mathscr{E}_f = \frac{\delta}{1-\delta} \frac{1}{1-e_{mf}} \left[\gamma_b + \frac{1}{Da_c + 1/(\gamma_c + \eta_e)} \right] \qquad (10\text{-}88)$$

Once $\mathscr{E}_f$ is computed, the fluid-bed rate coefficient $\mathscr{K}_f$ for a given value of K_r is

$$\mathscr{K}_f = \mathscr{E}_f K_r \frac{V_s}{V_b}$$

and conversion is computed by

$$x = 1 - \exp\left(-\mathscr{K}_f \theta_b\right)$$

The bubble-contact time is H/u_{br}, which can be expressed as

$$\theta_b = \frac{H}{u_{br}} = \frac{H_0(1-e)}{u_{br}(1-e_{mf})} \qquad (10\text{-}89)$$

where H_0 = fixed-unfluidized-bed height
$\quad\ e$ = fixed-bed void fraction
$\quad u_{br}$ = bubble velocity relative to emulsion phase

As derived by Davidson and Harrison, u_{br} is

$$u_{br} = 0.711(gd_b)^{1/2} \qquad (10\text{-}90)$$

Example of Fluid-Bed Conversion

We illustrate the use of the Kunii-Levenspiel model with a modified version of their illustration, namely, an analysis of Kobayashi's ozone-decomposition data:

$$H_0 = 34 \text{ cm} \qquad e_m = 0.45 \qquad e_{mf} = 0.5 \qquad K_r = 2 \text{ s}^{-1}$$
$$u_{mf} = 2.1 \text{ cm/s} \qquad u_0 = 13.2 \text{ cm/s} \qquad \gamma_b \text{ (assumed)} = 0$$
$$\alpha = 0.47 \text{ (Rowe-Partridge)} \qquad \mathscr{D} = 0.204 \text{ cm}^2/\text{s}$$

Assume that $d_b = 3.7$ cm (tube diameter = 20 cm).

SOLUTION

From Eq. (10-83)

$$K_{bc} = 4.5 \frac{2.1}{3.7} + 5.85 \frac{\sqrt{0.204\,980^{1/4}}}{(3.7)^{5/4}} = 5.44 \text{ s}^{-1}$$

[1] P. N. Rowe and B. A. Partridge, *Trans. Inst. Chem., Eng.*, **43T**: 157 (1965).

From Eq. (10-90)

$$u_{br} = 0.711(980 \cdot 3.7)^{1/2} = 42.8 \text{ cm/s}$$

From Eq. (10-85)

$$u_b = 13.2 - 2.1 + 42.8 = 54 \text{ cm/s}$$

From Eq. (10-86a)

$$\delta = \frac{13.2 - 2.1}{54} = 0.206$$

From Eq. (10-84)

$$K_{ce} = 6.78 \left[\frac{0.5(0.204)(54)}{(3.7)^3} \right]^{1/2} = 2.23 \text{ s}^{-1}$$

From Eq. (10-87)

$$\gamma_c = (1 - 0.5) \left\{ \frac{3(2.1/0.5)}{0.711[(3.7)(980)]^{1/2} - 2.1/0.5} + 0.47 \right\} = 0.4$$

From Eq. (10-86)

$$\gamma_e = \frac{(1 - 0.5)(1 - 0.206)}{0.206} = 1.53$$

From Eq. (10-87a)

$$\frac{V_b}{V_s} = \frac{0.206}{0.794(1 - 0.5)} = 0.52$$

For $K_r = 2 \text{ s}^{-1}$, we see that

$$Da_c = 0.368 \qquad Da_e = 0.9$$

$$\eta_e = \frac{1.53}{1 + 1.53(0.9)} = 0.64 \qquad \eta_c = \frac{0.4}{1 + 0.4(0.368)} = 0.35$$

The fluid-bed efficiency from Eq. (10-88) is

$$\mathscr{E} = 0.52 \frac{1}{0.368 + 1/(0.4 + 0.64)} = \frac{V_b}{V_s} \eta_o = 0.52(0.78)$$

$$\%\mathscr{E} = 39\%$$

The fluid-bed conversion is then $\mathscr{K}_f = \eta_o K_r$

$$x_f = 1 - \exp\left[-0.78(2)\theta_b\right]$$

From Eq. (10-89)

$$\theta_b = \frac{34(0.55)}{42.8(0.5)} = 0.87 \text{ s} \qquad x_f = 0.73$$

In the fixed bed (PFR), $\theta_p = \theta_b$

$$x_{PFR} = 82.5\%$$

In a CSTR for some contact time $\theta_c = \theta_b$

$$x_{CSTR} = \frac{K_r \theta}{1 + K_r \theta} = \frac{1.75}{2.75} = 64\%$$

For an intrinsic value of $K_r = 2 \text{ s}^{-1}$ the results are summarized in Table 10-11.
In the above example we have assumed $\theta_b = \theta_{PFR} = \theta_{CSTR}$. Since, in general,

$$\theta = \frac{\text{volume reactor}}{\text{volumetric flow rate}} = \frac{\text{length}}{\text{velocity}}$$

the intrinsically different meaning of θ_b must be appreciated. For in the Kunii-Levenspiel model, it is actual bubble residence time which is determining with respect to conversion. In a sense, the bubble is a batch reactor operating in a reaction-time period given by its passage time, which is much shorter than gas mean residence time because bubble velocity is far greater than reactor-feed velocity u_0. Hence, in our example, based on fixed height,

$$\theta = \frac{L_m}{u_0} = \frac{34}{13.2} = 2.6 \text{ s}$$

in contrast with $\theta_b = 0.87$ s, amply demonstrating the meaning of bubble bypassing. Note also that if the bed is operated as a fixed unit, $\theta_p = 2.6$ s.
We may make the fluid, fixed, and CSTR comparison more realistic in the following fashion:

PFR: $\qquad x = 1 - \exp\left[-K_r(1 - e_0)\frac{H_0}{u_0}\right]$

Fluid bed: $\qquad x = 1 - \exp\left[-\frac{\eta_0 K_r(1 - e_f)H}{u_{br}}\right]$

and for the CSTR, let us assume a bubble-free fluid bed of height H and void volume e_f

CSTR: $\qquad x = \dfrac{K_r(1 - e_f)H/u_0}{1 + K_r(1 - e_f)H/u_0}$

Table 10-11

Reactor	Contact time	Conversion, %
PFR	0.87	82.5
CSTR	0.87	64
Fluid bed	0.87	73

from which we find

PFR: $x = 0.94 = 94\%$ conversion

FBR: $x = 0.73 = 73\%$ conversion

CSTR: $x = 0.74 = 74\%$ conversion

On this basis, the telling effect of bubble bypassing is most forcefully revealed.

Olson's Fluid-Bed-Reactor Analysis

In Chap. 6 the Pavlica-Olson generalized design equations[1] for mass transport (and reaction) in two-phase systems were summarized, and we noted there that these generalizations encompass the fluidized-bed-reactor system, which for a negligible cloud phase involves only two phases, a solids-free bubble phase and a solids-rich emulsion phase. Assuming isothermality, the continuity equation for the dispersed (bubble) phase is, as developed in Chap. 6

Bubble phase:
$$\bar{A}b'' = b' + Bb + \bar{E}(b - C) = 0 \qquad (10\text{-}91)$$

Emulsion phase (countercurrent): $FC'' = -C' + GC + H(C - b) = 0 \qquad (10\text{-}92)$

where $\bar{A}$, $F =$ dispersion (axial-mixing) terms

  B, $G =$ reaction terms

  $\bar{E}$, $H =$ interphase transfer groups

  b'', $b' =$ second and first derivatives with respect to bed height for bubble phase

  C'', $C' =$ derivatives of emulsion-phase concentration

Simplifications are permitted:

 1 No reaction in bubble phase; $B = 0$.
 2 Plug flow of bubbles; $\bar{A} = 0$.
 3 Emulsion-phase fluid velocity is that of minimum fluidization; $u_c = u_{mf}$.
 4 The mass-transfer coefficients contained in $\bar{E}$ and H become simply a bubble-emulsion exchange coefficient Q.

Using the emulsion-phase backmixing (F) and cross-flow (bubble-emulsion exchange) data and their correlation, as assembled by Mireur and Bischoff,[2] Pavlica and Olson reexpress the coefficients of their general equations in terms of fluid-bed-reactor parameters:

$$H = \frac{QL}{u_{mf}} \qquad (10\text{-}93)$$

$$u_b = u - (1 - f_D)u_{mf} \qquad (10\text{-}94)$$

where $f_D =$ dispersed- (bubble-) phase holdup

  $u =$ total superficial velocity

---

[1] R. T. Pavlica and J. H. Olson, *Ind. Eng. Chem.*, **62**(12): 45 (1970).
[2] J. P. Mireur and K. B. Bischoff, *AIChE J.*, **13**: 839 (1967).

F, $\bar{E}$, and H are to be sought in terms of u/u_{mf} and bubble diameter d_b. The key mixing and cross-flow parameters correlated by Mireur and Bischoff are

$$I = \frac{QL}{u} \qquad \text{cross-flow} \qquad (10\text{-}95)$$

$$II = \frac{uL}{f_c \mathscr{D}_c} \qquad \text{emulsion-phase mixing} \qquad (10\text{-}96)$$

where $\mathscr{D}_c$ is the emulsion-phase axial dispersion coefficient. Recall that $uL/\mathscr{D}_c =$ Pe; therefore

$$\bar{E} = I\,\frac{u/u_{mf}}{u/u_{mf} - (1 - f_D)} \qquad (10\text{-}97)$$

$$H = I\,\frac{u/u_{mt}}{1 - f_D} \qquad (10\text{-}98)$$

$$F = \frac{u/u_{mf}}{II(1 - f_D)} \qquad (10\text{-}99)$$

From Eq. (10-94) f_D can be secured in terms of u_b, which in turn, according to Davidson and Harrison, gives [Eq. (10-85)]

$$u_b = u - u_{mf} + 0.711(gd_b)^{1/2}$$

so that
$$f_D = \frac{u/u_{mf} - 1}{u/u_{mf} - 2 + 0.711(gd_b)^{1/2}/u_{mf}} \qquad (10\text{-}100)$$

In consequence $\bar{E}$, H, and F, the dimensionless coefficients governing fluid-bed-reactor behavior, are expressed in terms of u/u_{mf} and $(gd_b)^{1/2}/u_{mf}$.

In Figs. 10-25 to 10-28 emulsion-phase backmixing and bubble-phase and emulsion-phase transfer (F, $\bar{E}$, and H, respectively) are displayed as a function of u/u_{mf} as computed by Pavlica and Olson based upon the mixing and cross-flow correlations of Mireur and Bischoff. Pavlica and Olson conclude that (1) the emulsion-phase Peclet number (Fig. 10-25) is quite small and rather insensitive to d_b over the interesting range of u/u_{mf} values (10 to 30). Thus CSTR emulsion character is a realistic assumption provided bed height to diameter is less than about 10. (2) Bubble-phase mass transfer $\bar{E}$ is extremely insensitive to d_b as a function of u/u_{mf}. Finally (3) emulsion-phase mass transfer H is only moderately affected by d_b over a range of pertinent values of u/u_{mf}. Consequently, fluid-bed-reactor conversion is rather neatly approximated solely in terms of group $G = k_c L/u_{mf}$ and u/u_{mf}, as set forth in Fig. 10-28, which shows that bubble diameter exerts far less influence upon fluid-bed reactor performance than u/u_{mf} and reactivity G. In fact, the value of u/u_{mf} has a telling effect, in agreement with the experimental observations of Heidel,[1] whose data reveal a conversion drop of from 95 to 43 percent upon increasing u/u_{mf} from 3 to 13.

---

[1] K. Heidel et al., *Chem. Eng. Sci.*, **20**: 557 (1965).

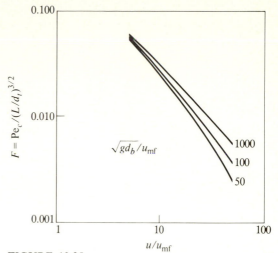

FIGURE 10-25
Emulsion-phase mixing.   [*R .T. Pavlica and J. H. Olson,
Ind. Eng. Chem.*, **62**(*12*): *45 (1970)*.]

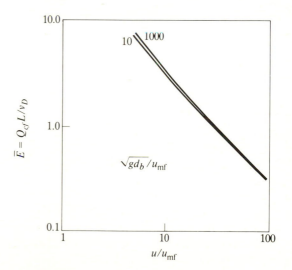

FIGURE 10-26
Bubble-phase mass transfer.   [*R. T. Pavlica and J. H.
Olson, Ind. Eng. Chem.*, **62**(*12*): *45 (1970)*.]

More rigorous analysis of the fluid-bed reactor can be realized by numerical
solution of the dimensionless continuity equations as evolved by Pavlica and
Olson.   Unhappily a paucity of plant-scale operation data still exists, so that
model discrimination is not yet possible.   (See, however, the work of Grace et al
cited in Additional References.)

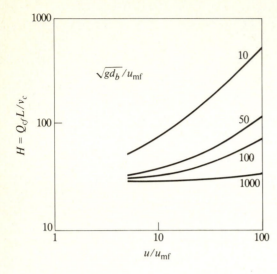

FIGURE 10-27
Emulsion-phase mass transfer. [*R. T.*
Pavlica and J. H. Olson, Ind. Eng. Chem.,
62(*12*): *45 (1970).*]

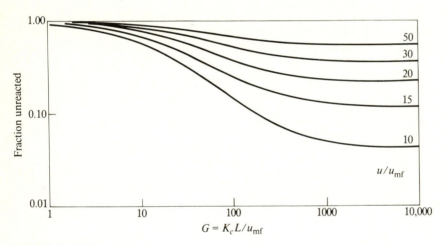

FIGURE 10-28
Design chart for first-order conversion in a fluid bed. [*R. T. Pavlica and J. H. Olson, Ind.*
Eng. Chem., **62**(*12*): *45 (1970).*]

Yield in the Fluidized-Bed Reactor

Poor though the situation may be with respect to the contrast of real conversion data with available models, the situation is even more barren in terms of fluid-bed selectivity predictions and verification. Nevertheless it may prove instructive to formulate a fluid-bed-reactor-yield model for the important consecutive reaction network $A \xrightarrow{k_1} B \xrightarrow{k_2} C$. This problem can be resolved in terms of the Davidson-Harrison model or that of Kunii and Levenspiel. Though neither school has addressed itself to the yield issue,[1] it is evident that the Kunii-Levenspiel model more readily lends itself to the resolution of fluid-bed yield for linear consecutive reaction.

When we use the Kunii-Levenspiel model, a material balance about the rising bubble of effective diameter d_b is found for both the rate of disappearance of A (as realized in the previous section) and the net rate of appearance of B.   For A

$$-\frac{dA_b}{d\theta} = \mathcal{K}_f A_b \quad (10\text{-}101)$$

where $\mathcal{K}$ is given by Eq. (10-75).   For B

$$\frac{dB_b}{d\theta} = \gamma_b(k_2 B_b - k_1 A_b) + K_{bc}(B_b - B_c) \quad (10\text{-}102)$$

where
$$K_{bc}(B_b - B_c) = \gamma_c(k_2 B_c - k_1 A_c) + K_{ce}(B_c - B_e)$$

and
$$K_{ce}(B_c - B_e) = \gamma(k_2 B_e - k_1 A_e) \quad (10\text{-}103)$$

The method of resolution consists of expressing A_e in terms of A_c:

$$A_e = \alpha_1 A_c \qquad \alpha_1 = \frac{K_{ce}}{\gamma_e k_1 + K_{ce}} \quad (10\text{-}104)$$

which gives us B_e in terms of A_c and B_c:

$$B_e = \beta_1 B_c + \beta_2 A_c \qquad \beta_1 = K_{ce}/(\gamma_e k_2 + K_{ce})$$
$$\beta_2 = \gamma_e k_1 \alpha_1 / \gamma k_2 + K_{ce} \quad (10\text{-}105)$$

Cloud-phase concentration is then expressed in terms of bubble-phase concentration

$$A_c = \alpha_2 A_b$$

where
$$\alpha_2 = \frac{K_{bc}}{\gamma_c k_1 + K_{bc} + K_{ce}(1 - \alpha_1)}$$

or
$$\alpha_2 = \cfrac{1}{1 + \cfrac{\gamma_c k_1}{K_{bc}} + \cfrac{1}{1/K_{ce} + 1/\gamma_e k_1}} \quad (10\text{-}106)$$

and
$$B_c = \beta_3 B_b + \beta_4 A_b$$

where
$$\beta_3 = \frac{K_{bc}}{\gamma k_2 + K_{bc} + K_{ce}(1 - \beta_1)} \quad (10\text{-}107)$$

and
$$\beta_4 = \frac{\gamma_c k_1 \alpha_2 + K_c \beta_2 \alpha_2}{\gamma_c k_2 + K_{bc} + K_{ce}(1 - \beta_1)} \quad (10\text{-}108)$$

---

[1] Mea culpa; see the Kunii-Levenspiel text, loc. cit.

The net rate of formation of B in terms of bubble-phase composition is then

$$-\frac{dB}{d\theta} = \beta_5 B_b - \beta_6 A_b$$

where $\qquad \beta_5 = \gamma_b k_2 + K_{bc}(1 - \beta_3) \qquad \beta_6 = K_{bc}\beta_4 + \gamma_b k_1 \qquad$ (10-109)

Recall that $\qquad\qquad\qquad -dA_b/d\theta = \mathscr{K} A_b$

Therefore the point yield of B is

$$\frac{dB_b}{dA_b} = \frac{\beta_5 B_b}{\mathscr{K} A_b} - \frac{\beta_6}{\mathscr{K}} \qquad (10\text{-}110)$$

where A and B are bubble-phase concentrations. Integration of Eq. (10-110) provides the overall yield-conversion profile. Note that the homogeneous case unaffected by diffusive cross-flow is that in which $\beta_1 = 1$ since the cross-flow coefficients K_{ce} and K_{bc} are so great that the rate is diffusion-uninfluenced, i.e.,

$$\beta_1 = \frac{K_{ce}}{\gamma_c k_2 + K_{ce}} \to 1 \qquad \text{since } K_{ce} \gg \gamma_c k_2$$

It then follows that α_1 and α_2 become unity, $\beta_2 \to 0$, $\beta_3 = 1$, $\beta_5 = \gamma_b k_2$, $\beta_4 \to 0$, and

$$\frac{dB}{dA} = \frac{k_2}{k_1}\frac{B}{A} - 1 \qquad (10\text{-}111)$$

the homogeneous batch or PFR point yield.

The integrated yield of B for the fluidized bed, plug-flow homogeneous reactor, and CSTR can now be compared:

FBR: $\qquad \dfrac{B}{A_0} = \dfrac{\beta_6}{\mathscr{K} - \beta_5}(f^{\beta_5/\mathscr{K}} - f) \qquad$ (10-112)

PFR: $\qquad \dfrac{B}{A_0} = \dfrac{1}{1 - K_2}(f^{K_2} - f) \qquad$ (10-113)

CSTR; $\qquad \dfrac{B}{A_0} = \dfrac{1 - f}{1 + K_2(1 - f)/f} \qquad$ (10-114)

where $\qquad\qquad\qquad\qquad f = \dfrac{A}{A_0} \qquad K_2 = \dfrac{k_2}{k_1}$

and β's and $\mathscr{K}$ are as defined earlier.

For an intrinsic value of $K_2 = 0.06$ the comparison is displayed in Fig. 10-29 for diverse bubble diameters and fluidization parameters.

We learn that yield in the fluid bed can be poorer than in a CSTR, thus emphasizing the perils inherent in treating a fluid bed as a homogeneous CSTR. Its two-phase nature invites bypassing, with the result that both conversion and yield can be less than predicted CSTR values. One may well ask: Then why use a fluid-bed reactor?

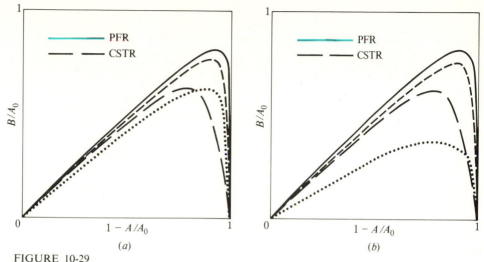

FIGURE 10-29
Yield of B in the reaction $A \xrightarrow{1} B \xrightarrow{2} C$ in PFR, CSTR, and fluid bed (Kunii-Levenspiel model); fluid bed: $————= d_b = 1$ cm, $\cdots = d_b = 5$ cm; $k_1/k_2 = 16$, $u_{mf} = 2.1$ cm/s.   (a) $u_0 = 13.2$ cm/s; (b) $u_0 = 5$ cm/s.

First, virtual isothermality is realized in the fluid bed. In highly sensitive exothermic processes such as naphthalene oxidation, the fluid bed merits considera- tion. Second, where catalyst deactivation proves to be of significance, fluidized systems permit constant circulation of the catalyst between reactor and regenerator as exemplified in modern catalytic-cracking plants. Finally the taxation upon conversion and yield due to bubble bypassing can be reduced (Fig. 10-30) by staging the fluid-bed system, by installing internal baffles or grids to redisperse bubbles, or by fluidizing in tubes of high length-to-diameter ratio so that slug flow results (bubble diameter equals tube diameter), thus eliminating bypassing. A plant reactor may well consist of a parallel array of such slug-flow units immersed in a shell of coolant or heating medium. The most comprehensive treatment of the fluid-bed reactor, which includes interphase diffusion as well as stability and multi- plicity analyses, is that of Amundson.[1]

Example: fluidized-bed regeneration   A telling analysis of continuous-fluid-bed- reactor behavior is found in the work of Pansing,[2] who reported data for regenera- tion of coked cracking catalyst in a pilot-plant unit.   The continuously fed coked catalyst is assumed to be well mixed (therefore average coke level is that of exiting catalyst) while the oxygen-bearing regenerating gas is assumed to be in plug flow. The rate of regeneration is

$$\mathscr{R} = kCp_i \qquad (a)$$

[1] N. R. Amundson et al., *Chem. Eng. Sci.*, **29:** 1173(1974), **30:** 847 (1975).
[2] W. F. Pansing, *AIChE J.*, **2:** 71 (1956).

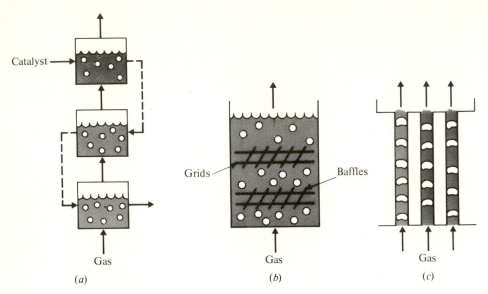

FIGURE 10-30
Remedies for bypassing and backmixing in fluidization: (a) staged fluid beds, (b) baffled fluid-bed reactor, (c) multitubed fluid bed.

where C is the carbon concentration on the regenerated (effluent, well-mixed) emulsified-catalyst phase and p_i is the local oxygen partial pressure.

Anticipating two-phase fluidization models later evolved, Pansing equates O_2 transport from bubble to emulsion (reaction) phase

$$\mathscr{R} = K(p - p_i) = kCp_i \qquad (b)$$

Thus

$$\mathscr{R} = \frac{p}{1/K + 1/kC} \qquad (c)$$

A differential material balance for O_2 in plug flow yields for a feed rate F, O_2 concentration N, and catalyst weight w

$$F\,dN = -\mathscr{R}\,dw$$

On substituting for $\mathscr{R}$, integration gives

$$-\frac{PC}{S \ln f} = \frac{C}{K} + \frac{1}{k} \qquad (d)$$

where $f = N/N_0$, $S = FN_0/w$ is space velocity, and P is total pressure. Obviously the left-hand side is experimentally ascertained, so that in principle k is found if K is known (C does not vary significantly enough to allow determination of K and k). Pansing assumed that

$$K = \frac{1}{\alpha}\frac{G^m}{d_p{}^n} \qquad (e)$$

where G = mass velocity of gas
d_p = particle diameter

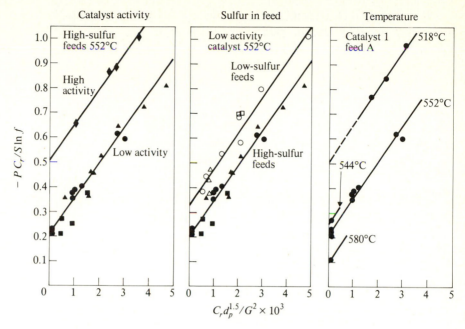

FIGURE 10-31
Correlation of regeneration variables.   For significance of symbols with respect to catalyst and
feed, see the original paper.   [*W. F. Pansing, AIChE J.*, **2:** *71 (1956).*]

So that

$$-\frac{PC}{S \ln f} = \frac{\alpha C d_p{}^n}{G^m} + \frac{1}{k} \qquad (f)$$

Utilizing data secured in 42 pilot-plant runs at various G and d_p values, α, n,
and m were determined with the result

$$-\frac{PC}{S \ln f} = 146 \frac{C d_p{}^{1.5}}{G^2} + \frac{1}{k} \qquad (g)$$

In Fig. 10-31 the fidelity of the correlation is demonstrated for diverse conditions.
Values of k so obtained at four temperature levels give an activation energy of 35
kcal/g mol, in quite good agreement with a value of 36 kcal/g mol secured in fixed-
bed regeneration units devoid of bubble-emulsion exchange resistance.   Pansing's
analysis rather neatly separates the fluid-bed transport-reaction events.

It is of interest to note that in treating the fluidized catalyst phase in simple
CSTR terms, a potent correlation emerges in spite of the acknowledged fact that
these solids should be treated as segregated.   Doubtless the accord is due to the
fact that the intrinsic chemical kinetics dictates this gas-solid (O_2-coke) reaction since
d_p is so small that interphase-intraphase (but not interparticulate) diffusion proves
to have no influence upon local and thus overall rate.

It is evident that the fluid-bed models treated here are rooted in the notion of transfer *followed* by reaction rather than the more realistic postulate that transfer through the cloud (or wake) and emulsion phases actually occurs simultaneously with reaction. The mathematical formulation becomes somewhat more complex if simultaneous reaction-diffusion is assumed to take place in the solids-laden cloud (or wake) and emulsion phases. On the other hand, as noted in Chap. 6, for comparatively slow reactions the series transport-reaction model should suffice.

Transport-Line and Raining-Solids Reactors

If the velocity of the fluidizing gas exceeds that necessary for solid-catalyst entrainment, the transport-line reactor (TLR) is born. Under such circumstances the assumption of near plug flow in both gas and catalyst phases becomes a realistic postulate. Obviously the TLR is gainfully employed only when rapid catalyst deactivation requires a steady-state addition of fresh catalyst with removal of the spent agent at a product-catalyst separation zone at the reactor exit.

If, on the other hand, the solid catalyst is rained downward in countercurrent fashion relative to the reactant-laden gas, we have the raining-solids reactor (RSR). As with the TLR, the RSR is utilized when rapid catalyst decay requires its steady-state replenishment, and, again both the TLR and the RSR can be approximated as a PFR.

Qualitatively, it is apparent that the TLR represents a fixed bed of quite high void fraction, while the RSR is an instance of a moving bed of quite high void fraction. Alternatively the TLR may be visualized as a dilute-phase cocurrent moving bed while the RSR is a dilute-phase countercurrent moving bed. In each case steady state prevails.

The essential distinction between these cocurrent and countercurrent gas-solid catalyst systems and their moving-bed counterparts lies in the fact that gas-solid catalyst velocity, which, of course, dictates interphase heat and mass transport, is less securely defined in these dilute transport reactors. However, given the quite small values of particle size associated with dilute-phase TLRs, one can anticipate a negligible interphase resistance to heat and mass except for quite high intrinsic reactivity. In principle, TLR and RSR design procedures are merely extensions of one-dimensional moving-bed models in which high porosity and low catalyst loading are to be anticipated.

10-6 SLURRY REACTORS

The slurry reactor, cited earlier (Fig. 10-1c) may assume various forms (stirred tank, bubble column, sieve-tray staged reactor) but whatever the reactor geometry, the reaction phases are gas, solid, and liquid, i.e., solid catalyst is suspended in a liquid medium within which gas is dispersed. Figure 10-32 is a schematic of the potential gradients in a slurry system. Slurry reactors offer several advantages not enjoyed in alternative modes of operation:

1 The high heat capacity of the slurry is conducive to isothermality.
2 Heat exchange and therefore recovery is good.

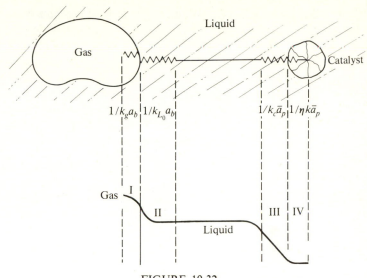

FIGURE 10-32
Gas-liquid-solid resistances in a slurry reactor.

3 Very small catalyst particles can be accommodated with a consequent intra-particle effectiveness benefit.
4 The reactor can be simple, e.g., a stirred autoclave.

The chief disadvantage of the slurry reactor may be the problem of catalyst-product separation, particularly in a continuous system.

Global Reactions in a Slurry

Referring to Fig. 10-32, for the steady state, we can write

$$
\begin{aligned}
\mathcal{R} &= k_g a_b (C_0 - C_i) \qquad \text{mol/(vol)(time)} \\
&= k_L a_b (C_{ii} - C_L) \\
&= k_c a_p (C_L - C_s) \\
&= \eta k a_p C_s \qquad \text{for first-order reaction}
\end{aligned}
\tag{10-115}
$$

where k_g = mass-transfer coefficient for the gas film within gas bubbles
k_L = mass-transfer coefficient for liquid film surrounding bubble
k_c = mass-transfer coefficient for liquid film surrounding catalyst particles
k = specific catalytic rate constant, cm/s
η = intraparticle effectiveness
a_b = bubble area, cm^2/cm^3 expanded slurry[1]
a_p = particle external area, cm^2/cm^3 expanded slurry[1]

---

[1] C. N. Satterfield, "Mass Transfer in Heterogeneous Catalysis," MIT Press, Cambridge, Mass., 1970.

We eliminate concentrations in Eq. (10-115) to obtain

$$\frac{C_0}{\mathscr{R}} = \frac{1}{K_0} = \underbrace{\left(\frac{1}{\eta k} + \frac{1}{k_c}\right)\frac{1}{a_p}}_{\text{Catalyst-liquid}} + \underbrace{\left(\frac{1}{k_L} + \frac{1}{k_g M}\right)\frac{1}{a_b}}_{\text{Liquid-gas}} \quad (10\text{-}116)$$

where M is the Henry's law constant. We define

$$\eta_b = \frac{1}{1 + \mathrm{Da}_b} \qquad \mathrm{Da}_b = \frac{k_L}{k_g M} \quad (10\text{-}117)$$

$$\eta_L = \frac{1}{1 + \mathrm{Da}_L} \qquad \mathrm{Da}_L = \frac{\eta k}{k_c}$$

where η = intraparticle effectiveness = $f(\phi)$
   η_b = bubble effectiveness
   η_L = liquid-particle effectiveness
all in terms of the familiar Damköhler numbers. Equation (10-116) becomes

$$\frac{1}{K_0} = \frac{1}{\eta_L \eta k a_p} + \frac{1}{\eta_b k_L a_b} \quad (10\text{-}118)$$

or the overall observed rate coefficient K_0 is, by rearrangement,

$$K_0 = \frac{\eta_L \eta k a_p}{1 + \eta_L \eta k a_p / \eta_b k_L a_b} = \frac{\eta_L \eta k a_p}{1 + \mathrm{Da}_0} \quad (10\text{-}119)$$

In the spirit of the development in Chap. 5, we note that $1/(1 + \mathrm{Da}_0) = \eta_o$ is the overall effectiveness, and so

$$\eta_o = 1 - \eta_o \mathrm{Da}_0 \quad (10\text{-}119a)$$

If now we divide Eq. (10-119) by $\eta_b k_L a_b$,

$$\frac{\mathrm{Da}_0}{1 + \mathrm{Da}_0} = \eta_o \mathrm{Da}_0 = \frac{K_0}{\eta_b k_L a_b} \quad (10\text{-}120)$$

Hence by Eq. (10-119a) η is obtained from

$$K_0 = \frac{C_0}{\mathscr{R}} = \frac{\text{gas-phase concentration}}{\text{observed global rate}}$$

and a computable $\eta_b k_L a_b$. With this value of η_o secured in terms of observables, the usually unobservable $\eta_L \eta k a_p$ is obtained. This procedure assumes the existence of correlations for estimating k_g, k_L, k_c, a_b, and a_p to permit estimation of ηk.

Coefficient and Area Correlations

Satterfield[1] recommends the following relationships; note that film resistance within bubbles is usually very small. In terms of our development, this suggests that

[1] Ibid.

$\mathrm{Da}_b = 0$ and $\eta_b = 1$.   For gas-liquid transport from bubbles

$$k_L \mathrm{Sc}^{1/2} = 0.42 \left(\frac{\mu g\, \Delta \rho}{L^2} \right)^{1/3} \qquad (10\text{-}121)$$

$$a_b = \frac{\mathrm{cm}^2}{\mathrm{vol\ slurry}} = \frac{6H}{d_b} \qquad (10\text{-}122)$$

where H = gas holdup, cm³/cm³ expanded slurry
$\quad d_b$ = average bubble diameter
$\quad \Delta \rho$ = liquid-gas density difference
For liquid-solid transport

$$\left(\frac{k_c d_p}{\mathcal{D}} \right)^2 = 4 + 1.21 \left(\frac{g d_p^3\, \Delta \rho}{18 \mu \mathcal{D}} \right)^{2/3} \qquad (10\text{-}123)$$

$$a_p = \frac{6m}{\rho_p d_p} \qquad (10\text{-}124)$$

where m = catalyst loading, g/cm³ expanded slurry
$\quad d_p$ = particle diameter
$\quad \rho_p$ = particle density
Equation (10-116) can now be expressed as

$$\frac{C_0}{\mathcal{R}} = \frac{1}{K_0} = \frac{\rho_p d_p}{6 \eta_L \eta k} \frac{1}{m} + \frac{d_b}{\eta_b k_L 6H} \qquad (10\text{-}125)$$

Analysis of First-Order Slurry Reaction Systems

Method 1   Given a series of experimental results which provide global rate at several levels of catalyst loading, according to Eq. (10-125), a plot of the reciprocal of $\mathcal{R}$ versus $1/m$ should (at constant temperature) yield a straight line of slope equal to $\rho_p d_p / 6 \eta_L \eta k$, which is the slope of Eq. (10-125), and intercept $d_b / \eta_b k_L 6H$.   This is to say that at large values of catalyst loading ($1/m \to 0$), the rate of gas absorption becomes controlling.   On the other hand, at low catalyst activity and/or loading, the fluid phase approaches saturation, and reaction-diffusion about (and possibly within) the catalyst becomes rate-controlling.   To be sure, a considerable number of experiments at various values of m are required to find linear relationships from which one can confidently extract accurate parameters from slope and intercept. But in $\mathcal{R}$-versus-m experiments one must be alert to the possibility of catalyst drop-out at high values of loading.   Analysis by method 1 is illustrated in Fig. 10-33.

Method 2   Estimates of key parameters can be extracted from one experiment by use of Eqs. (10-116) to (10-120).   The steps involved, given one value of $\mathcal{R}$, are (for $\eta_b = 1$):

1   Estimate $k_L a_b$.
2   Compute $(\mathcal{R}/C_0)(1/k_L a_b) = \eta_o \mathrm{Da}_0$.

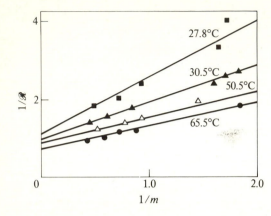

FIGURE 10-33
Slurry-reactor-data correlation [Eq. (10-125)]. [*T. K. Sherwood and E. J. Farkas, Chem. Eng. Sci.*, **21**: *573 (1966).*]

3 From Eq. (10-119a) find η_o.

4 From η_o compute $\eta_L \eta k a_p$ from Da_0, and estimated $k_L a_b$.

5 If necessary, ηk can then be estimated from results (4) and a calculated value of k_c and a_p.

Note that in each step involving an effectiveness factor, the effectiveness factor is easily ascertained in terms of observables, as shown in Chaps. 5 and 9.

General Comments

Certainly more reliable estimates of the physicochemical parameters can be secured by method 1 of analysis. Should time and monies prove to be in short supply, method 2 will at least provide insights into which of the several steps involved in slurry reactions is rate- (and selectivity-) determining.

Selectivity in Slurry Reactors

So long as linearity of kinetics is presumed, yield or selectivity predictions for the slurry reactor can be rooted in an analysis quite similar to that illustrated for the Kunii-Levenspiel bubbling-fluid-bed model. In such formulations, care must be exercised in formulating the boundary conditions; e.g., in a consecutive reaction, *B*, the intermediate, may or may not be volatile. See, for example, Chap. 6.

10-7 THE CATALYTIC-GAUZE REACTOR

Two reactions of signal commercial interest are realized by contacting the reactants over a small but effective quantity of catalyst in the form of a gauze or an assembly of metal screens (see Fig. 10-1*d*). The oxidation of NH_3 over a Pt-alloy gauze network is a classic case of gauze-catalyzed reaction

$$2NH_3 + \tfrac{5}{2}O_2 \xrightarrow{\text{Pt}} 2NO + 3H_2O \quad (10\text{-}126)$$

The catalytic efficiency of this process is so great that virtually 100 percent conversion of NH_3 is achieved in a contact time of about 1 ms. This nonporous-surface-catalyzed reaction is obviously controlled by bulk mass transport. Therefore the multilayer screen or gauze reactor admits an analysis rooted in interphase mass transfer. Several studies have been devoted to such a system, e.g., those of Oele,[1] Satterfield and Cortez,[2] and Roberts and Gillespie.[3]

In j-factor form, mass transfer between a gas and a nonporous catalytic gauze may profitably be expressed as

$$\frac{k_g}{u} \, \mathrm{Sc}^{2/3} = \frac{K_1}{\mathrm{Re}^a} \qquad \mathrm{Re} = \frac{d_{\mathrm{wire}} \, u}{v} \qquad (10\text{-}127)$$

Assuming that the rate-determining step is that of external, fluid film, mass transport, conversion for plug-flow conditions is

$$x = 1 - \exp\left(-k_g a \theta\right) \qquad (10\text{-}128)$$

HCN is produced catalytically by reaction over Pt gauze catalyst: $NH_2 + \frac{3}{2}O_2 + CH_4 \rightarrow HCN + 3H_2O$. As in NH_3 oxidation, bulk mass transport largely governs the global rate; so too in methanol oxidation over metallic silver to produce formaldehyde.

Yield in gauze catalysis is of some importance: in NH_3 oxidation, N_2 is an undesirable by-product. While it has been assumed that N_2 is formed via decomposition of NO

$$NH_3 \xrightarrow[O_2]{1} NO \xrightarrow{2} N_2$$

it is apparent that since bulk mass transport controls NH_3 conversion, yield of intermediate NO should be negligible if consecutive reaction is responsible for N_2 formation. Further, the yield of NO varies inversely with total pressure, thus suggesting an alternative mechanism to that noted above.

In Chap. 5 it was shown that in the consecutive network subject to bulk-mass-transport control, the selectivity B/C (or in this case NO/N_2) is determined by

$$\frac{NO}{N_2} \equiv \frac{B}{C} \approx \frac{k_g a}{k_2} \qquad (10\text{-}129)$$

Except when k_2 is small relative to $k_g a$, the yield of NO should be small and a function of Reynolds number. An alternative rationale[4] which accounts for N_2 formation during NH_3 oxidation suggests that reactant NH_3 can react homogeneously with product NO to produce N_2: $3NO + 2NH_3 \rightarrow \frac{5}{2}N_2 + 3H_2O$, a reaction which, being homogeneous and irreversible, is surely pressure-dependent. NH_3 oxidation over Pt-alloy catalyst can then be visualized in terms of the network

$$NH_3 \xrightarrow{1} NO \qquad \text{heterogeneous}$$

$$NO + NH_3 \xrightarrow{2} N_2 \qquad \text{homogeneous}$$

[1] A. P. Oele, "Chemical Reaction Engineering," Pergamon, Amsterdam, 1957.
[2] C. N. Satterfield and D. H. Cortez, *Ind. Eng. Chem. Fundam.*, **9**: 613 (1970).
[3] D. Roberts and G. P. Gillespie, *Adv. Chem.*, **133**: 600 (1974).
[4] J. J. Carberry, *Ind Eng. Chem.*, **58**: 40 (1966).

which is generalized as

$$A \xrightarrow{\ 1\ } B$$

$$A + B \xrightarrow{\ 2\ } C$$

Yield $NO/(NH_3)_0 = Y$ or selectivity is, assuming linear kinetics in each species [see Eq. (2-123)],

$$Y + (K + 1) \ln \frac{K - Y}{K} = \frac{NH_3}{(NH_3)_0} - 1 \qquad (10\text{-}130)$$

where $K = k_1/k_2 y_0 \pi$.   Hence K varies with total pressure π for a fixed NH_3 mole fraction y_0 in the feed (about 0.09 to 0.11, in practice).

Few data are to be found in the literature relating yield to total pressure; however, Fauser[1] provides data as follows:

$$Y = \begin{cases} 0.96 \text{ to } 0.97 & 1 \text{ atm} \\ 0.91 & 4 \text{ atm} \end{cases}$$

Utilizing the 1-atm value of Y in Eq. (10-130), we find $K = 40$.   Therefore at 4 atm, K should be about 10, which gives a predicted yield at 4 atm of 0.91, in excellent accord with the Fauser datum.

Granted the validity of the empirical model which suggests the yield loss due to homogeneous reactant (NH_3)–product (NO) reaction, the issue of gauze-reactor design becomes one of specifying the global mass-transport coefficient which dictates conversion.

The question of temperature gradients in this highly exothermic reaction is a valid one.   After all, the gas-feed temperature is but 400°C and that of the effluent 900 to 1000°C.   In consequence—there is a 500 to 600°C axial temperature gradient over an axial distance of about several millimeters.   For this nonporous-solid-catalyzed reaction under conditions of virtual total-bulk mass-transport control, the solid metallic phase can be essentially isothermal.

As an exercise, prove that the above statement is true if the Lewis number is unity.   What is the value of Le?   See *AIChE J.*, **20:** 571 (1974).

The Satterfield-Cortez study of bulk-mass-transport-controlled oxidation of hexane and toluene suggests

$$j = C\text{Re}^{-m} = 0.94 \left(\frac{d_w u/e}{v} \right)^{-0.7} \qquad (10\text{-}131)$$

where d_w = wire diameter

u/e = average interstitial fluid velocity

The range of Reynolds numbers embraced by the above correlation is from 10 to 10^4.   The j factor is

$$j = e \frac{k_g}{u} \text{Sc}^{2/3} = \frac{\epsilon h}{\rho u C_p} \text{Pr}^{2/3} \qquad (10\text{-}132)$$

The conversion of NH_3 to produce NO and N_2 is then given by Eq. (10-128).

[1] G. Fauser, *Chem. Met. Eng.*, **37:** 604 (1930).

In terms of the j factor for wire-mesh gauzes, where $a \geqq 4/d_w$,

$$k_g a = \frac{0.94(u/e)(4)}{d_w(d_w u/ve)^{0.7} Sc^{2/3}} \qquad (10\text{-}133)$$

The value of θ, contact time in the gauze, is determined by gauze depth and interstitial velocity

$$\theta = \frac{d_w e}{u} n$$

Therefore

$$k_g a\theta = \frac{4n}{Re^{0.7}}$$

We illustrate the validity of this analysis by computing conversion in terms of $k_g a\theta$. In a high-pressure ammonia-oxidation-process burner of 30 layers of gauze, the Reynolds number based on superficial velocity, according to Roberts and Gillespie, is about 30. Taking the average void function to be about 0.5 and $n = 30$, we find, from Eq. (10-128),

$$x = 1 - \exp\left(-\frac{120}{60^{0.7}}\right)$$

$$= 0.999$$

in excellent accord with plant observations. Backmixing is assumed to be negligible in this model, and there doubtless exists some degree of axial dispersion. However the j-factor correlations were established using a log mean driving force so that actual backmixing effects are hidden in the mass-transfer correlation.

10-8 THE TRICKLE-BED REACTOR

We have noted that certain process circumstances gave birth to the fluidized-bed reactor (catalyst decay, the need for isothermality) and the three-phase slurry reactor (small catalyst particles suspended in a fluid of relatively high heat capacity). Thus circumstances created the trickle-bed catalytic reactor, i.e., the need to remove nitrogen and sulfur from complex petroleum compounds found in that litany of species which constitute petroleum stocks.

Denitrogenation and hydrodesulfurization are accomplished by contacting high-pressure hydrogen with petroleum stocks of broad boiling-point range over a suitable catalyst. As a consequence of the high pressure and high-boiling-point spectrum, a three-phase system (gas-liquid-solid catalyst) is encountered; accordingly, the trickle bed developed, a unit consisting of a fixed bed of catalyst which the gas-liquid mixture traverses. Hydrocracking is of the same nature.

Early Studies

Paradoxically, one of the earliest systematic studies of trickle-bed kinetics issued from the academy, a study of the hydrogenation of α-methylstyrene in a trickle bed of supported noble-metal catalysts. In contrast to commercial practice, the H_2 passed countercurrently to the trickling liquid flow, yet in sound anticipation of the potency of intraphase diffusional intrusions, the catalytic agents were superficially imposed upon the support (partial impregnation).[1] This and earlier studies, conducted at modest temperature levels ($< 100°C$), exhibit diffusion-unaffected kinetic behavior. Because of the paucity of organized reaction-rate data for chemical reactions in trickle beds, the more plentiful studies of hydrodesulfurization will be cited.

Hydrodesulfurization Kinetics

Basically this process involves reaction of gaseous H_2 with a gas-liquid phase containing "bound" sulfur to produce H_2S and an oil product relatively devoid of sulfur. Detailed discussions of hydrodesulfurization are provided by Schuit and Gates.[2]

In the hydrodesulfurization of light cycle oil, the sulfur-laden coreactant passes downward through a fixed bed of catalyst, for example, Co-Mo on Al_2O_3, the feed being partially vaporized in the presence of cocurrently flowing reactant H_2 (usually at superatmospheric pressures). The analysis-design problem is then one of describing this three-phase fixed-bed system, duly recognizing that gas-liquid flow patterns, distribution, and in situ vaporization due to net exothermicity all conspire to challenge one's modeling talents.

The kinetics of hydrodesulfurization of a given sulfur-containing compound conforms generally to first-order kinetics, with some evidence of inhibition due to product H_2S and other adsorbing species which are found in the liquid feed stream. Insofar as we are attempting to focus upon the trickle-bed catalytic reactor rather than the desulfurization process per se, a discussion of the kinetics of that process may seem inappropriate. However, without sacrifice of some generality with respect to the trickle bed, a lesson is conveyed by a respect for the desulfurization process and its intrinsic and phenomenological kinetics.

Figure 10-34 shows the first-order kinetic correlation secured by Frye and Mosby[3] for three individual typical sulfur-containing compounds. Since, in fact, a sulfur-containing oil feed contains a rather broad spectrum of sulfur compounds, each of which is hydrodesulfurized at its distinct rate, it is not surprising that overall (as to opposed to intrinsic individual-compound) kinetics will differ. In terms of total sulfur removal, the overall (global) kinetic display set forth in Fig. 10-35 demonstrates this disparity between intrinsic and overall kinetics. Intrinsic kinetics is linear; observed, overall, kinetics based upon total sulfur is second order. As in catalytic cracking we are witness to mechanistically meaningless second-order kinetics, which is a consequence of a wide distribution of reactivities. Lumped kinetics will exhibit overall order greater than unity.[4]

[1] B. D. Babcock, G. T. Mejdell, and O. A. Hougen, *AIChE J.*, **3**: 366 (1957).
[2] G. C. A. Schuit and B. C. Gates, *AIChE J.*, **19**: 417 (1973).
[3] C. G. Frye and J. F. Mosby, *Chem. Eng. Prog.*, **63**: 66 (1967).
[4] D. Luss and P. Hutchinson, *Chem. Eng. J.*, **1**: 129 (1970), **2**: 172 (1971).

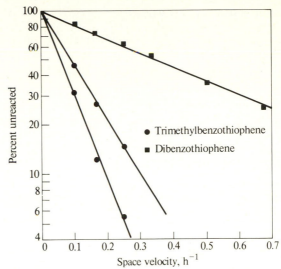

FIGURE 10-34
Desulfurization rate of selected sulfur compounds; tem
perature 288°C, pressure 15.3 atm.   [*C. G. Frye and J. F.
Mosby, Chem. Eng. Prog.*, **63**: *66 (1967)*.]

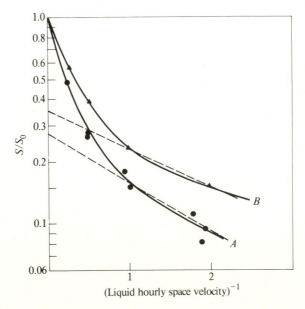

FIGURE 10-35
Desulfurization of Venezuelan Vacuum gas oil at two activ-
ity levels.   Curve *A* is 14°C higher than curve *B*.
(*R. R. Cecil, F. X. Mayer, and E. N. Cart, AIChE Meet.,
Los Angeles, 1968.*)

Trickle-Bed Kinetic Model

Cognizant of the severe problems of liquid-vapor distribution and nonisothermality as well as residence time distributions of both phases, we can fashion a simple yet instructive model of the plug-flow trickle bed within which linear reaction prevails, in the light of the work of Frye et al., cited above. Though fashioned to correlate hydrodesulfurization trickle-bed-reactor performance, the approach may well prove relevant in analyses of quite different systems. Consider the case of a component in the liquid phase which reacts with a gaseous component (sparingly soluble) at a rate $\mathscr{R}$. We fashion a differential material balance over a segment dh for molar flow rates of liquid L and gas G, in which the fraction of liquid vaporized is v and mole fractions in liquid and vapor of reactant are x and y, respectively. Let the vapor pressure of liquid reactant be $\bar{p}$ and total pressures π; hence if p is reactant vapor-phase partial pressure, then, assuming simple solution theory,

$$y = \frac{p}{\pi} \qquad dy = \frac{dp}{\pi}$$

$$(10\text{-}134)$$

but

$$p = x\bar{p} \qquad dy = \frac{\bar{p}}{\pi} dx$$

When we strike a differential material balance and invoke the above relations for kinetics of the form

$$\mathscr{R} = \frac{kpp_g}{(1 + \Sigma I)^m} \qquad (10\text{-}135)$$

where p_g is the gas-phase coreactant partial pressure and p, as noted, that of the liquid-phase reactant; the denominator represents diverse inhibitor I functionalities. The differential balance yields

$$\frac{dx}{dh} = \frac{kp_g \bar{p} x}{L[1 - v + (v + G/L)](\bar{p}/\pi)(1 + \Sigma I)^m} \qquad (10\text{-}136)$$

which is simply integrated with the result, in terms of bed height h

$$\ln \frac{x_0}{x} = \frac{k\bar{p}p_g h}{L[1 - v + (v + G/L)](\bar{p}/\pi)(1 + \Sigma I)^m} \qquad (10\text{-}137)$$

Verification of the Model

Frye and colleagues measured hydrodesulfurization conversion in two reactors (for a single sulfur compound): (1) 50 cm^3 of catalyst in a bed of 22 mm diameter and 150 mm length; (2) 3000 cm^3 of catalyst in a bed of 28 mm diameter and 6000 mm length. The data secured in both reactors were nicely assembled by Eq. (10-137), including the temperature dependencies of inhibition coefficients for H_2S and aromatics.

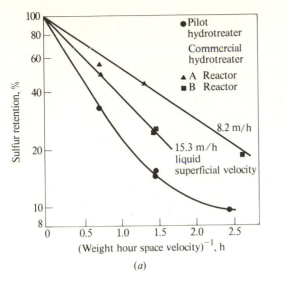
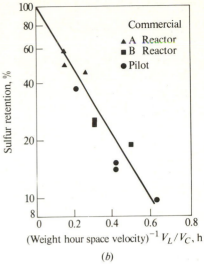

FIGURE 10-36
(*a*) Sulfur removal in commercial and pilot hydrotreaters and (*b*) at constant catalyst utilization; ● = pilot; commercial: ▲ = A reactor, ■ = B reactor. [*L. D. Ross, Chem. Eng. Prog.*, **61**(*10*): *77 (1965)*.]

Trickle-Bed Scale-up

An instructive reference dealing with trickle-bed performance on model (air-water system), pilot-plant, and commercial-plant (hydrodesulfurization) scales is that of Ross,[1] in which residence-time distributions, static and dynamic holdup, and conversion data are presented for each level of scale-up. Ross concludes that effective liquid distribution is achieved with increasing difficulty in moving from the model to commercial-reactor level. His residence-time distribution data also suggest that axial dispersion may be of signal importance in commercial units. If high conversion is the design goal, even modest backmixing can be taxing. The reactor efficiencies for the pilot and two plant scale units are displayed in Fig. 10-36*a*, from which it is quite apparent that trickle-bed scale-up is a hazardous undertaking. However, recognizing that liquid holdup differences with scale of operation account for these efficiency discrepancies, Ross replots the data vs. contact time multiplied by liquid holdup per volume of catalyst, with encouraging results (Fig. 10-36*b*).

The key role played by axial dispersion in trickle beds is examined by Mears.[2] His data (Table 10-12) are clearly unlike those for gas flow through fixed beds, where the number of perfect mixers is $n = Z_0/d_p$ for a pellet Peclet number of 2.

[1] L. D. Ross, *Chem. Eng. Prog.*, **61**: (10) 77 (1965).
[2] D. E. Mears, *Chem. Eng. Sci.*, **26**: 1361 (1971).

Mass Transfer in Trickle Beds

In principle mass-transport resistances in a trickle bed may exist in gas film, trickling-liquid film, or within the porous catalyst. As in the case of slurry reactors, the first is usually negligible while, of course, the importance of liquid-film and pore diffusion will depend upon reaction vigor. Satterfield[1] recommends an observable criterion

$$\frac{3.3 d_p}{C^*} \mathscr{R} > k_L \qquad (10\text{-}138)$$

for a significant mass-transport contribution to observed rate $\mathscr{R}$. In Eq. (10-138) d_p is pellet diameter, C^* the saturation concentration of the gas in the liquid, and k_L a trickle-film mass-transfer coefficient composed of

$$\frac{1}{k_L} = \frac{1}{k_{L_0}} + \frac{1}{k_s} \qquad (10\text{-}139)$$

where k_{L_0} is the usual physical liquid-film mass-transfer coefficient (Chap. 6) and k_s is that coefficient for mass transfer between a flowing liquid and a solid.

By reason of the small values characteristic of liquid diffusivities (10^{-5} to 10^{-6} cm^2/s), the Thiele modulus will be large indeed for trickle-bed commercial-size pellets catalyzing a rapid reaction. For example, in analyzing the data of Satterfield et al.[2] by the one-point method described in Sec. 10-6, Sylvester et al.[3] computed an overall trickle-reactor effectiveness (interphase and intraphase) of 0.16 and intraparticle value of 7×10^{-3} for the hydrogenation of α-methylstyrene on Pd/Al$_2$O$_3$. The intrinsic rate constant for this reaction is 17 s^{-1}, in contrast with values of about 10^{-2} to 10^{-3} s^{-1} typical of trickle-bed hydrodesulfurization.

Finally one must anticipate the possibility of both liquid- and vapor-phase catalysis in an incompletely wetted trickle bed, as the benzene-hydrogenation study of Ware[4] reveals.

Table 10-12

$\dfrac{L}{dp}$	n	Re$_{\text{liquid}}$
150	4	2
260	11	2
580	42	8

[1] C. N. Satterfield, "The Role of Mass Transfer in Heterogeneous Catalysis," MIT Press, Cambridge, Mass., 1970.
[2] C. N. Satterfield, A. A. Pelossoff, and T. K. Sherwood, AIChE J., 15: 226 (1969).
[3] N. D. Sylvester A. A. Kulkarni, and J. J. Carberry, Can. J. Chem. Eng., 53: 313 (1975).
[4] C. H. Ware, Jr., Ph.D. thesis, University of Pennsylvania, Philadelphia, 1959.

10-9 REACTORS SUFFERING CATALYST DEACTIVATION

The fixed-bed catalytic reactor which is subject to catalyst decay can surely be described in terms of the general continuity equations (10-1) and (10-2) in which the transient terms are retained and a suitable deactivation function is incorporated into local species-generation term $\mathscr{R}$. If, however, the transient function must be retained, i.e., the pseudo-steady-state treatment cannot be justified, then decay is so rapid that one must entertain an alternative to the fixed bed. Reality has inspired the development of these alternatives, namely, catalytic cracking of large hydro-carbon molecules to more valuable lower-molecular-weight entities in the moving-bed reactor (MBR) or the continuous-fluid-bed reactor (CFBR). In both cases catalyst is added and removed continuously, the effluent catalyst being regenerated externally and then returned to the reactor. Under such circumstances the steady state prevails, as noted in Chap. 7.

It will prove instructive to consider conversion (and later yield) in isothermal fixed, moving, and fluid-bed catalytic reactors which suffer catalyst decay during use. Catalytic cracking is an excellent example in which (by reason of rapid deactivation due to coking), alternatives to the fixed bed must be entertained, as shown by Weekman's analysis[1] presented here.

Assuming isothermal diffusion-unaffected reaction to occur in a PFR, the transient relationship (due to catalyst decay) is in terms of fraction unconverted f and vapor density ρ

$$\frac{1}{\rho}\frac{\partial \rho f}{\partial t} + u\frac{\partial f}{\partial z} = -\mathscr{R}(f,t) \quad (10\text{-}140)$$

We recast this equation into dimensionless form. First, since a change in moles occurs in catalytic cracking

$$\rho = \frac{\rho_0}{f + a(1-f)}$$

where $a = \dfrac{\text{reactant molecular weight}}{\text{product molecular weight}}$

In terms of liquid hourly space velocity

$$S = \frac{F_0}{V_r \bar{\rho}}$$

where F_0 = mass reactant feed/h

V_r = reactor volume, void fraction e

$\bar{\rho}$ = liquid-reactant density at STP

Then
$$B\frac{\partial f}{\partial \theta} + \frac{\partial f}{\partial x} = -\frac{e\rho}{\bar{\rho}S}\mathscr{R}(y,\theta) \quad (10\text{-}141)$$

[1] V. W. Weekman, Jr., *Ind. Eng. Chem. Process Des. Dev.*, **7**: 90 (1968).

where $x = z/z_0$ and

$$B = \frac{e\rho_0 a}{\bar{\rho} \mathrm{St}_m [f + a(1-f)]^2}$$

B represents the ratio of oil transit time to the time of catalyst decay t_m. Since oil-vapor contact time is quite small compared with the time of catalyst decay, $B = 0$ unless deactivation is virtually instantaneous, and so

$$\frac{df}{dx} = -\frac{e\rho}{\bar{\rho}S} \mathscr{R}(f, \theta) \quad (10\text{-}142)$$

Fixed Bed

Catalytic-cracking kinetics is *phenomenologically* described, in terms of conversion, quite simply by second-order kinetics. Deactivation conforms conveniently to an exponential decay model; thus

$$\mathscr{R}(f, \theta) = k_0 f^2 \exp(-\lambda\theta) \quad \text{where } \lambda = \alpha t_m \quad (10\text{-}143)$$

(The observed second order with respect to cracking conversion rate is a consequence of the fact that with diversity of feed species, each cracks in a linear fashion at widely differing intrinsic rates; this fact, together with a volume change due to extent of reaction, gives rise to the observed higher-order behavior.)

The dimensionless group λ is a length-of-decay parameter. Fixed-bed behavior is thus given by

$$\frac{df}{dx} = -A f^2 \exp(-\lambda\theta) \quad (10\text{-}144)$$

where the severity or extent of reaction is

$$A = \frac{e\rho k_0}{\bar{\rho}S} = \frac{K}{S} \quad (10\text{-}145)$$

As we have assumed that $d\theta/dx = B = 0$, integration yields

$$f = \frac{1}{1 + Ax \exp(-\lambda\theta)} \quad (10\text{-}146)$$

The time-averaged conversion $\bar{e}$ is

$$\bar{e} = 1 - \bar{f} = 1 - \int_0^1 f \, d\theta$$

or

$$\bar{e} = \frac{1}{\lambda} \ln \frac{1 + A}{1 + A \exp(-\lambda)} \quad (10\text{-}147)$$

this result is set forth in Fig. 10-37*a* as $\bar{e}$ versus A for diverse values of λ.

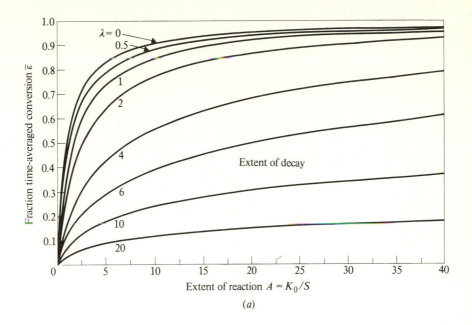

(a)

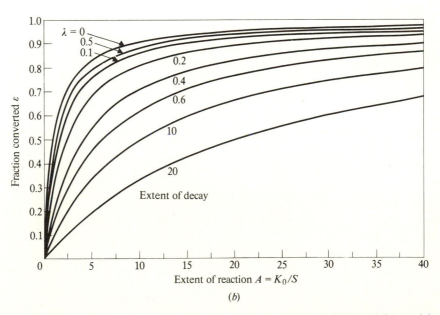

(b)

FIGURE 10-37a and b

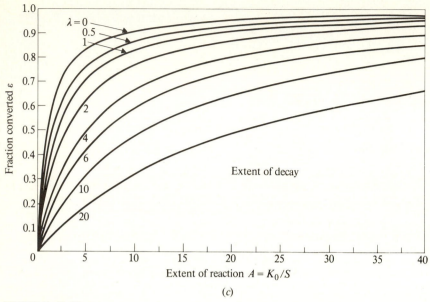

FIGURE 10-37

Time-averaged conversion (a) for fixed beds (piston gas flow), (b) for moving beds (piston gas flow, piston solid flow), and (c) for fluids beds (piston gas flow, backmix solid flow). [V. W. Weekman, Jr., Ind. Eng. Chem. Process Des. Dev., 7: 90 (1968).]

Batch Fluid Bed

If the gas phase is in plug flow and the batch of fluidized solids is subject to deactivation according to the exponential model $\exp(-\lambda\theta)$, Eq. (10-147) aptly describes time-averaged conversion for the batch fluid-bed catalytic reactor. This equivalence (PFR and CST batch fluid reactor) is a consequence of the fact that in the deactivating fixed-bed model, the decay process is assumed to be uniform throughout the bed in time; i.e., decay is not a function of distance within the bed but only of time on-stream.

Moving Bed

For the steady-state moving-bed plug-flow reactor, Eq. (10-140) becomes

$$\frac{df}{dx} = -Af^2 \exp(-\lambda x) \quad (10\text{-}148)$$

where the argument in the decay function is now the product of the dimensionless decay parameter λ and fractional length x. Note that t_m in the product $\alpha t_m = \lambda$ is now the catalyst residence time in the moving bed.

The moving-bed reaction-decay function [Eq. (10-148)] can be integrated to provide the conversion $\epsilon = 1 - f$ at bed exit ($x = 1$)

$$\epsilon = \frac{A[1 - \exp(-\lambda)]}{\lambda + A[1 - \exp(-\lambda)]} \quad (10\text{-}149)$$

a result displayed in Fig. 10-37b.

Continuous Fluid Bed

The deactivation-prone catalyst is fed and withdrawn at a constant rate in the continuous fluid-bed reactor.   Steady state prevails; however, the gas- and catalyst-phase residence-time distribution must be defined.   Weekman assumes plug flow of gas and a CSTR behavior of the catalyst phase.   The CSTR internal age distribution must be invoked to establish an average reaction velocity coefficient $\bar{k}$

$$\bar{k} = k_0 \int_0^\infty I(\theta) \exp(-\lambda\theta)\, d\theta \quad (10\text{-}150)$$

where for CSTR solids $I(\theta) = \exp(-\theta)$; therefore

$$\bar{k} = \frac{k_0}{1 + \lambda} \quad (10\text{-}151)$$

so that Eq. (10-148) becomes

$$\frac{df}{dx} = \frac{Af^2}{1 + \lambda} \quad (10\text{-}152)$$

The terminal conversion is therefore

$$\epsilon = \frac{A}{1 + \lambda + A} \quad (10\text{-}153)$$

This result is shown in Fig. 10-37c.

10-10   COMPARISON OF FIXED, MOVING, AND FLUID BEDS

Conversion comparisons are presented in Fig. 10-38a for the fixed bed relative to the moving bed, while Fig. 10-38b shows the conversion ratio of the fluid bed relative to the moving bed.   The fixed-fluid bed conversion ratio is set forth in Fig. 10-38c. In all cases, the decay "length" λ is a parameter while the reaction "length" A is the variable which dictates the conversion ratio behavior.   These results are, of course, secured by Eqs. (10-147), (10-149), and (10-153) for the fixed, moving, and continuous fluid beds.

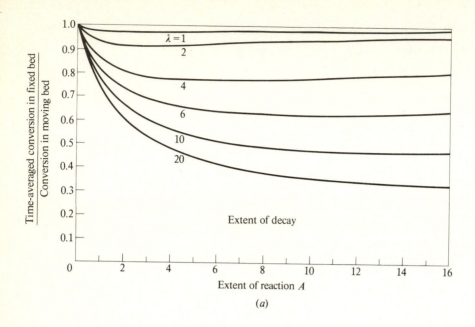

(a)

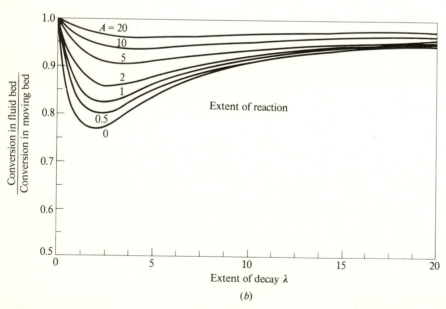

(b)

FIGURE 10-38a and b

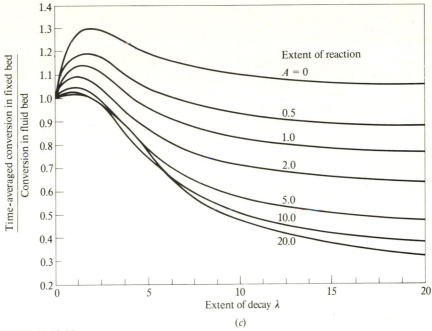

FIGURE 10-38
(*a*) Ratio of fixed-bed to moving-bed conversion; (*b*) ratio of fluid-bed to moving-bed conversion; (*c*) ratio of fixed-bed to fluid-bed conversion. [*V. W. Weekman, Jr., Ind. Eng. Chem. Process Des. Dev.*, **7**: *90* (*1968*).]

Validation of the Model

In Fig. 10-39 experimental data are contrasted with theory for catalytic cracking in fixed beds, while in Table 10-13 both moving- and fixed-bed cracking-conversion data are quite favorably compared with theory [Eqs. (10-147) and (10-149)].

Implications

Weekman goes on to note that reactor comparisons are quite simply realized in terms of the dimensionless decay λ and reaction numbers A. Thus one can compare reactors at a given conversion at:

1 A fixed catalyst residence time (constant λ) and reactant throughput

$$\text{Reactor volume} = \frac{V_1}{V_2} = \frac{A_1}{A_2}$$

2 A given space velocity (constant A)

$$\text{Catalyst-feed ratio} = \frac{\lambda_2}{\lambda_1}$$

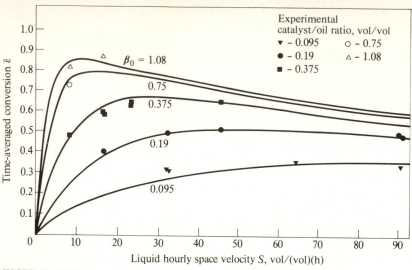

FIGURE 10-39

Comparison of fixed-bed model and fixed-bed zeolite catalyst data; charge Mid- Continent gas oil; catalyst temperature 482°C; ——— Eq. (10-147); model constants: $k_0 = 143.2 \text{ h}^{-1}$, $\alpha = 18.8 \text{ h}^{-1}$.[*V. W. Weekman, Jr., Ind. Eng. Chem. Process Des. Dev.*, **7**: *90 (1968)*.]

Table 10-13 **COMPARISON OF MODELS TO EXPERIMENTAL DATA**†

Mid-Continent gas oil, commercial TCC catalyst
$\alpha = 2.96 \text{ h}^{-1}$
$K_0 = 1069 e^{-q/RT} = 2.9$ at 482°C
$q = 9000$ cal/g mol

Liquid hourly space velocity, vol/(vol)(h)	Cat/Oil, vol/vol	Catalyst residence time, min	Temp, °C	ε Model	Exp
		Moving bed			
2	4	7.5	482	0.53	0.55
	4	7.5	510	0.58	0.58
	2	15	510	0.53	0.50
	1	30	510	0.47	0.50
	1	30	485	0.42	0.41
	2	15	484	0.48	0.45
	4	7.5	454	0.47	0.49
	1	30	510	0.47	0.47
		Fixed bed			
4	0.75	20	482	0.32§	0.33§
	0.375	40	482	0.25§	0.27§
	1.5	10	482	0.36§	0.39§
2	1.5	20	482	0.47§	0.43§

† V. W. Weekman, Jr., *Ind. Eng. Chem. Process Des. Dev.*, **7**: 90 (1968).
§ Time-averaged.

3 A given catalyst residence time (constant λ); space velocity can be compared

$$\frac{S_1}{S_2} = \frac{A_2}{A_1}$$

4 A given space velocity; catalyst residence times can be contrasted

$$\frac{(\text{Catalyst residence time})_1}{(\text{Catalyst residence time})_2} = \frac{\lambda_1}{\lambda_2}$$

To the extent that catalysts generally are mortal, it is apparent that an evaluation of a catalyst based solely upon its activity can be hazardous indeed, as Weekman notes, particularly if fixed-bed data are to be applied to a moving- or fluid-bed unit. He illustrates this point by a comparison of the performance of two hypothetical catalysts in Table 10-14. Extremes of reaction-decay character are evident in the assigned values of K and α. Note, however, the significant differences in computed conversions for the fixed-, moving-, and fluid-bed models. From a comparison of the two catalysts as tested in fixed bed, one would select catalyst 2, while in a moving- or fluid-bed reactor, catalyst 1 proves superior.

10-11 DECAY-AFFECTED SELECTIVITY

Weekman has extended the conversion analysis to the crucial issue of selectivity in decay-affected fixed-, moving-, and fluid-bed reactors,[1] again applied to catalytic cracking. As has been noted, cracking feed stocks are complex; so too will be the reaction network which defines the products and diverse reaction paths. However, as cracking conversion admits to simple rate modeling in terms of second-order behavior, yield or selectivity in catalytic cracking can be fruitfully described in a simple fashion.

Weekman defines F, catalytic cracking feed stock; G, gasoline product; C, dry

Table 10-14 HYPOTHETICAL COMPARISON OF TWO CATALYSTS IN FIXED, MOVING, AND FLUID-BED REACTORS[a]

Catalyst no.	S	t_m	A	λ	Fixed[b]	Moving	Fluid
1[c]	1	0.5	30	10	0.34	0.75	0.73
2[d]	1	0.5	2.5	1	0.59	0.61	0.55

Fraction conversion heads the last three columns (Fixed, Moving, Fluid).

[a] V. W. Weekman, Jr., *Ind. Eng. Chem. Process Des. Dev.*, **7**: 90 (1968).
[b] Time-averaged.
[c] Catalyst 1: $K_0 = 30$ h^{-1}, $\alpha = 20$ h^{-1}.
[d] Catalyst 2: $K_a = 2.5$ h^{-1}, $\alpha = 2$ h^{-1}.

V. W. Weekman, Jr., *AIChE J.*, **16**: 397 (1970).

gas, for example, CH_4, and coke.   C appears to be the product of both precursor A and B; thus the network in its simplest form is

As indicated earlier, conversion of F, in terms of f, the fraction remaining is

$$\mathscr{R}(f, \theta) = k_0 f^2 \exp(-\lambda\theta) \quad \text{where } k_0 = k_1 + k_3 \quad (10\text{-}143)$$

Proceeding as we did in the previous section, we find the dimensionless continuity equations for second-order cracking of feed F and first-order cracking of the gasoline G

$$B\frac{\partial f}{\partial \theta} + \frac{df}{\partial x} = -\frac{\rho}{\bar{\rho}S} k_0 f^2 \phi \quad (10\text{-}154)$$

$$B\frac{\partial g}{\partial \theta} + \frac{\partial g}{\partial x} = \frac{\rho_0}{\bar{\rho}S}(k_1 f^2 \phi - k_2 g\bar{\phi}) \quad (10\text{-}155)$$

where $\phi, \bar{\phi}$ = decay functionalities
   g = reduced gasoline concentration
   B = ratio of vapor to catalyst residence time

For reasons set forth in an earlier discussion, B is equal to zero.

As an exercise, discuss and justify the assumption of first-order gasoline cracking in contrast with second-order charge-stock cracking.

Under the restraint of $B = 0$, the continuity equations which apply to the isothermal fixed, moving, and both batch and continuous fluid beds become

$$\frac{df}{dx} = -\frac{K_0}{S}f^2\phi \quad (10\text{-}156)$$

$$\frac{dg}{dx} = \frac{K_1 f^2}{S}\phi - \frac{K_2}{S}g\phi \quad \text{where } K_i = \frac{\rho k_i}{\bar{\rho}} \quad (10\text{-}157)$$

Solution of Eqs. (10-156) and (10-157) relies upon specification of ϕ and $\bar{\phi}$. As Weekman observed, it is reasonable to suppose that if those sites which poison charge-stock cracking also poison gasoline cracking, $\phi = \bar{\phi}$, thus simplifying subsequent analyses.

Weekman shows that the now uniform deactivation function ϕ, space velocity S, and distance can be concisely reexpressed by defining a stretched reaction time u; that is,

$$du = \frac{\phi}{S} dx \quad (10\text{-}158)$$

In consequence the continuity equations become

$$\frac{df}{du} = -K_0 f^2 \quad (10\text{-}159)$$

and

$$\frac{dg}{du} = K_1 f^2 - K_2 g \qquad (10\text{-}160)$$

Solution of the total conversion function [Eq. (10-159)] is set forth above. The matter of yield, that is, dg/df or overall yield g/f, is readily resolved by eliminating u by division of (10-160) by (10-159), with the result, for point yield,

$$\frac{dg}{df} = \frac{K_2}{K_0}\frac{g}{f^2} - \frac{K_1}{K_0} \qquad (10\text{-}161)$$

Overall yield g/f or G/F_0 is analytically secured by integration of the point-yield expression

$$g = \frac{K_1 K_2}{K_0}\left(\frac{K_0}{K_1}\exp\frac{K_2}{K_0} - \frac{K_0}{K_2}f\exp\frac{K_2}{K_0 f} - E_{in}\frac{K_2}{K_0} + E_{in}\frac{K_2}{K_0 f}\right)\exp\frac{K_2}{K_0 f} \qquad (10\text{-}162)$$

where

$$E_{in} = \int_{-\infty}^{x}\frac{\exp x}{x}\,dx$$

This is the *instantaneous* yield, i.e., the value of g issuing from the fixed-bed reactor at any moment. As one collects product, the time-averaged yield results

$$\bar{g} = \int_0^1 g\,d\theta \qquad (10\text{-}163)$$

Both instantaneous and time-averaged yields, as computed, are shown in Fig. 10-40. In the moving and continuous fluid-bed units, g and $\bar{g}$ are identical, of

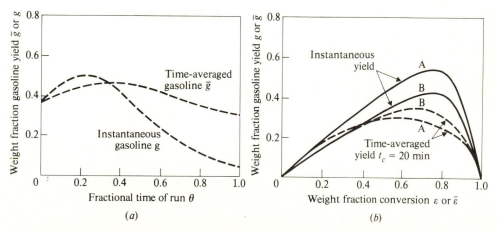

FIGURE 10-40

(a) Comparison of instantaneous and time-averaged gasoline yield in fixed-bed cracking; calculations from Eqs. (10-162) and (10-163); $K_0 = 22.9$, $K_1 = 18.1$, $K_2 = 1.7$, $\alpha = 42.7$, $t_c = 7.5$ min $S = 2.0$ vol/(vol)(h). (b) Effect of time averaging on catalyst selectivity evaluation.

Catalyst	K_1/K_0	K_2/K_0	α
A	0.9	0.1	20
B	0.7	0.1	10

[*V. W. Weekman, Jr., AIChE J.*, **16**: 397 (1970).]

course.   Note, however, that in the continuous fluid-bed model, the rate constant, as shown earlier, is

$$\bar{k} = k \int_0^\infty \exp\left[-\theta(1 + \lambda)\right] d\theta = \frac{k}{1 + \lambda}$$

for first-order decay.

For the fixed-bed, moving-bed, batch and continuous fluid-bed catalytic reactors, all of which suffer catalyst deactivation via a first-order decay process, that is, $\phi = \exp(-\theta)$, both isothermal conversion and yield can be analytically assessed with instructive results, as we have illustrated for a complex catalytic-cracking reaction network.   It may well be asked: How is the accord between model and reality for these complex reaction-reactor networks?

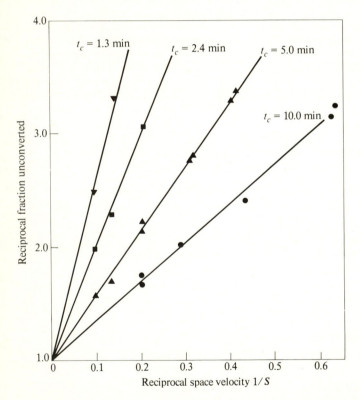

FIGURE 10-41
Second-order test of moving-bed data; t_c is catalyst residence time. Catalyst is Durabead 5, and stock is Mid-Continent gas oil; temperature 482°C.   Slope $= (K_0/\alpha t_c)[1 - \exp(-\alpha t_c)]$.   [*V. W. Weekman, Jr., AIChE J.*, **16**: *397 (1970).*]

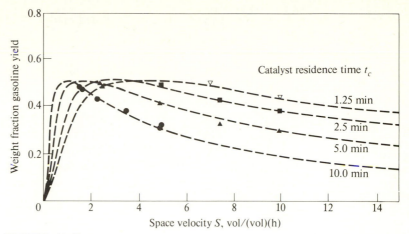

FIGURE 10-42

Comparison of gasoline yield with experimental moving-bed data; first-order decay model, Mid-Continent gas oil, Durabead 5, temperature 482°C; dashed line = theoretical values; experimental t_c: $\triangle = 1.3$ min, $\square = 2.4$ min, $\triangle = 5.0$ min, $\bigcirc = 10.0$ min; $K_0 = 22.9$, $K_1 = 18.1$, $K_2 = 1.7$, $\alpha = 42.7$. [*V. W. Weekman, Jr., AIChE J.*, **16**: *397 (1970)*.]

Verification of the Models

Happily, the cracking conversion and yield data, supplied by Weekman, are in magnificent agreement with the Weekman models for both conversion and yield.

Cracking-conversion kinetics As Fig. 10-41 indicates, Weekman's data for catalytic cracking conform quite nicely to second-order phenomenological kinetics.

Yield of gasoline data This is well organized by the Weekman model, as evident in Fig. 10-42, where yield is plotted vs. space velocity for moving beds.

Yield of gasoline data and model predictions These are set forth in Fig. 10-43 as a function of feed-stock conversion; these results are independent of u, that is, the stretched reaction time (deactivation, space velocity, and reduced bed length).

Other, more sophisticated, aspects of catalytic-cracking conversion and selectivity or yield are discussed in the Weekman reference and references cited therein.

We should emphasize that while the behavior of catalyst-decay-affected catalytic cracking in fixed, moving, and fluids beds has been the topic of illustration, these analyses should surely be relevant to other reactant-catalyst systems in which deactivation plays a signal role.

A note of caution is required with respect to the potential hazards of applying lumped kinetic models secured in, say, a plug-flow system to another reactor of a radically different residence-time distribution. As Luss and his students show,

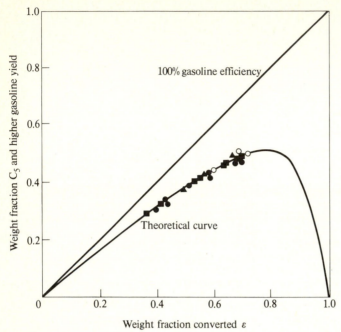

FIGURE 10-43
Comparison of model with moving-bed data, for Mid-Continent gas oil, Durabead 5, temperature 482°C, $K_0 = 22.9$, $K_1 = 18.1$, $K_2 = 1.7$, $\alpha = 42.7$; space velocity 1.5 to 10, catalyst/oil ratio 1.2 to 8.6. Experimental data: catalyst residence time: ○ = 10.0 min, ▼ = 7.6 min, ■ = 5.0 min, ▲ = 2.5 min, ◆ = 1.2 min. [*V. W. Weekman, Jr., AIChE J.*, **16**: *397 (1970)*.]

conversion and yield or selectivity predictions may be in serious error in such lumped-model extrapolations.[1]

10-12 REACTOR POISONING IN TERMS OF SPM

Weekman's model of fixed and batch fluid-bed poisoning assumes that the poisoning or deactivation event is solely a function of time on-stream not of distance throughout the reactor. Furthermore either exponential or nth-order decay is envisioned. In Chap. 8 it was noted that the pseudo-steady-state SPM might well describe poisoning (or regeneration, Chap. 7) for a single pellet. Utilizing the SPM, Olson[2] has addressed himself to the problem of isothermal behavior of a fixed bed subject to poisoning via the SPM dictates. Assuming plug flow and the absence of radial gradients, the continuity equation becomes identical to that of ion exchange

[1] S. V. Golkeri and D. Luss, *Chem. Eng. Sci.*, **29**: 845 (1974); *AIChE J.*, **18**: 277 (1972), **21**: 865 (1975).
[2] J. H. Olson, *Ind. Eng. Chem. Fundam.*, **7**: 185 (1968).

(Chap. 7), and so by a change of variables detailed in Chap. 7, the mass-continuity equation for poisoning becomes

$$\frac{\partial C}{\partial Z} + \frac{\partial Q}{\partial \tau} = 0 \quad (10\text{-}164)$$

where C = dimensionless concentration of poison precursor
Q = volumetric fraction of porous solid phase saturated with poison

For SPM applied to sphere

$$Q = 1 - \left(\frac{R}{R_0}\right)^3 = 1 - r^3$$

Z = reduced bed length, z/L

and

$$\tau = \text{dimensionless time} = \left(\frac{vt}{L} - Z\right) \frac{ebC_0}{(1-e)q_s}$$

where v = fluid velocity
e = bed void fraction
C_0 = inlet poison
q_s = solid-phase saturation concentration

One now expresses Q in terms of the SPM steps, i.e., in terms of τ

$$-\frac{q_s}{bC_0} r^2 \frac{\partial r}{\partial \tau} = \frac{k_g}{R_0} (C_b - C_s) = \frac{\mathscr{D}}{R_0{}^2} (C_s - C_1) \frac{r}{1-r} = \frac{k}{R_0} r^2 C_1$$

Since $dQ = -3r^2\, dr$, we have

$$\frac{\partial Q}{\partial \tau} = \frac{[(1-e)/e]C_b N_s}{\dfrac{1}{\text{Bi}} + \dfrac{1 - (1-Q)^{1/3}}{(1-Q)^{1/3}} + \dfrac{1}{\text{Da}(1-Q)^{2/3}}} \quad (10\text{-}165)$$

where *for the poisoning species*

$$N_s = \frac{3\mathscr{D}L}{R_0{}^2 v} = \text{number of solid diffusion transfer units}$$

$$\text{Bi} = \frac{k_g R_0}{\mathscr{D}} = \text{Biot number}$$

$$\text{Da} = \frac{kR_0}{\mathscr{D}} = \text{Damköhler number}$$

and k is the poisoning rate constant.

What is of interest is the effectiveness for the principal reaction in terms of extent of poisoning. For an irreversible first-order reaction occurring in a porous catalyst subject to SPM poisoning, local effectiveness is

$$\bar{\eta} = \frac{3}{\phi^2 \left\{ \dfrac{1}{\text{Bi}_m} + \dfrac{1-r}{r} + \dfrac{1}{r[r\phi \coth (r\phi) - 1]} \right\}} \quad (10\text{-}166)$$

where r = reduced radius identifying poisoned-unpoisoned boundary

$\phi = R_0\sqrt{k_1/\mathscr{D}_1}$

k_1 = principal reaction-rate coefficient

$\mathscr{D}_1$ = reactant intraphase diffusivity

and
$$\mathrm{Bi}_m = \frac{k_g R_0}{\mathscr{D}_1} \qquad \text{since } \mathscr{D} \neq \mathscr{D}_1$$

As an exercise, derive Eq. (10-166) for effectiveness of first-order reaction when SPM poisoning prevails.

The overall influence of SPM poisoning in a fixed bed is readily assessed by contrasting overall effectiveness with that at zero time of exposure, i.e.,

$$A(\tau) = \frac{\bar{\eta}(\tau)}{\eta(0)} \qquad (10\text{-}167)$$

where
$$\bar{\eta}(\tau) = \int_0^1 \eta(z, \tau)\, dz \qquad (10\text{-}168)$$

which, as Olson notes, is the fractional activity ratio $A(\tau)$. The governing equations are resolved by Olson to yield (1) the fractional activity of the catalytic fixed bed as a function of time, as graphically displayed in Fig. 10-44 (note that at reduced

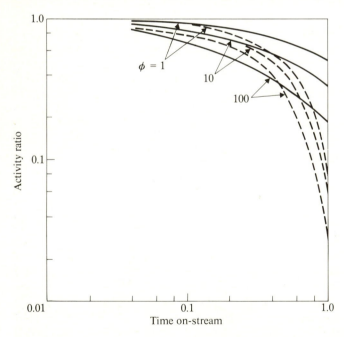

FIGURE 10-44
Fixed-bed fractional activity vs. time on-stream; $\mathrm{Bi} = \mathrm{Da} = 20$;
———— : $N_s = 0.1$, — — — : $N_s = 1.0$. [*J. H. Olson, Ind. Eng. Chem. Fundam.*, **7**: 185 (1968).]

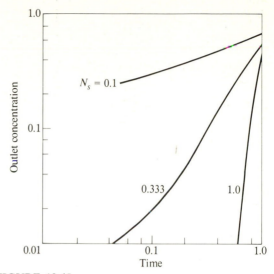

FIGURE 10-45
Breakthrough behavior of a guard reactor; Bi = Da = 20.
[*J. H. Olson, Ind. Eng. Chem. Fundam.*, **7**: *185 (1968)*.]

time of unity, total poisoning of the bed is realized and recall that N_s refers to the poison while ϕ refers to the reactant) and (2) the breakthrough behavior of a guard reactor situated upstream of the catalytic reactor.   The purpose of the guard reactor is to consume poison precursors before they impose themselves upon the principal-reactor catalyst.   Figure 10-45 is a display of guard-reactor break-through behavior in terms of N_s as a parameter.

Olson concludes that both values of N_s and ϕ must be borne in mind in assessing the average activity-time behavior of the fixed-bed catalytic reactor. Thus when ϕ is large, penetration of reactant is limited, so that for large N_s, a narrow band region of deactivation results.   In contrast, if N_s is small, the bed is poisoned more uniformly, resulting in a rapid decline in activity for systems sensitive to pore-mouth poisoning (large ϕ).   Hence in the distributed fixed-bed system rapid poisoning (high N_s) may well result in a greater average activity cycle than for modest decay (N_s small).

As for the guard-bed reactor (designed to remove poison precursors), the higher the value of N_s, the more effective the guard bed.   In sum (1) a guard bed should invite poison deposition (large N_s), (2) the reactor should invite large values of N_s if ϕ for the principal reaction is large, but (3) on the other hand, if ϕ is modest, N_s should be minimized to extend fixed-bed life.

Time-dependent behavior of fixed beds subject to decay under isothermal conditions has been generalized by Bischoff,[1] whose treatment embraces a number of particular cases.

[1] K. B. Bischoff, *Ind. Eng. Chem. Fundam.*, **8**: 665 (1969).

Thermal Waves in Fixed-Bed Regeneration

Again following Olson,[1] the general case of fixed-bed regeneration or for that matter poisoning under conditions of nonisothermality can be analyzed in terms of the SPM.

In this instance of transient fixed-bed behavior, three contact-time parameters emerge:

$$\text{Fluid contact time} = t_f = \frac{L}{v} \quad (10\text{-}169)$$

$$\text{Reactant-gas contact time} = t_m = \frac{1-e}{e}\frac{q_c}{(C)b}t_f \quad (10\text{-}170)$$

where q_c = solid coreactant initial concentration, e.g., coke

   b = gas-solid stoichiometric coefficient

   C = gas-phase concentration

Thermal-wave contact time

$$t_T = \frac{1-e}{e}\frac{(\rho C_p)_s}{(\rho C_p)_g}t_f \quad (10\text{-}171)$$

where subscripts s and g refer to solid and gas.

Finally we have the measure of reactivity contact time as expressed by the number of solid-phase-diffusion transfer units

$$N_s = \frac{1}{v}\frac{3\mathscr{D}_{\text{eff}}L}{R_p^{\,2}} = \frac{3\mathscr{D}_{\text{eff}}}{R_p^{\,2}}t_f = \frac{t_f}{t_D} \quad (10\text{-}172)$$

N_s is, of course, dimensionless and represents the ratio of fluid contact time to that of intraphase diffusion. As Olson suggests, the meaningful ratio of thermal to reactant-mass contact times establishes another dimensionless ratio

$$\bar{v} = \frac{t_T}{t_m} = \frac{C}{q_c}\left[\frac{(\rho C_p)_s}{(\rho C_p)_g}b\right] = \frac{\text{mass velocity}}{\text{thermal velocity}} \quad (10\text{-}173)$$

Thus N_s and $\bar{v}$ define the behavior of the adiabatic fixed-bed regeneration system. Olson defines a dimensionless temperature t as

$$t = \frac{(T-T_0)(\rho C_p)_s b}{q_c(-\Delta H)} \quad (10\text{-}174)$$

which is reexpressed in terms of $\bar{v}$ and the gas-phase reduced adiabatic temperature as

$$t = \frac{\Delta T\,(\rho C_p)_g}{C(-\Delta H)}\bar{v} = \beta\frac{t_T}{t_m} \quad (10\text{-}175)$$

[1] J. H. Olson, Thermal Waves in Fixed Bed Reactor Regeneration, unpublished; see also J. J. van Deemter, *Ind. Eng. Chem.*, **45**: 1227 (1953).

Although t is a limiting value, it is nevertheless evident that t can be greater or less than the adiabatic limit when $\bar{v} = t_T/t_m \gtrless 1.0$, that is, whether the mass reactant wave is slow or fast relative to the thermal wave.

Essentially $\bar{v}$ depends solely upon the gaseous-reactant concentration C for a fixed regenerable concentration q_c and gas-solid-phase volumetric heat-capacity ratio. Obviously the relative mass- and thermal-wave velocities are governed by N_s, as Olson's analyses indicate. Some results are summarized in Table 10-15.

In exothermic regeneration Olson's analyses show that two parameters can be negotiated to limit damaging temperature excursions: (1) fluid velocity, which affects N_s, and (2) reactant concentration, which affects $\bar{v}$.

10-13 OPTIMIZATION

In principle, as we have noted earlier, optimization can be realized with respect to conversion, yield, STC, or STY by a clever allocation of reactor sequences (CSTR-PFR combinations), coreactant distribution (semibatch or side-stream addition), and temperature distribution (as a function of time or distance). These remedies amount to allocation of residence-time distributions, concentration distributions, and temperature distribution.

The principles which guide the designer with respect to one or more of these optimization schemes should now be clear. What may not be evident is the fact that in the distributed-parameter, mixed-rate-controlled catalytic reactor, the simple optimization schemes suggested above cannot be simply invoked. Again, as in the cases of multiplicity and stability for the complex reactor, explorations of appropriate numerical solutions appear to be the most prudent route to optimization. Again we refer to the literature for detailed optimization analyses of such complex reactors as the fixed bed and tubular wall reactor.[1]

If the circumstances of nonisothermal interparticulate, interphase-intraphase heat- and mass-diffusion-affected catalytic reaction are ignored, the consequent isothermal, simple-kinetic-controlled but deactivation-affected reactor performance can be viewed in terms of optimization. Specifically, we address ourselves to the following optimization problem.

Table 10-15 VALUES OF t FOR DIVERSE N_s AND $\bar{v}$

N_s	$\bar{v}$	
	0.5	2
1	<1	<1
10	<1	>1
100	~ 1	$\gg 1$

[1] T. G. Smith and J. J. Carberry, *Adv. Chem.* No. 133, p. 362, 1974; *Chem. Eng. Sci.*, **30**: 221 (1975).

Given a fixed-bed reactor subject to deactivation and a method of regeneration, we have two periods, an operating period and a regeneration period. When operating and regenerating times are of comparable magnitude, a unique operation-regeneration cycle exists which results in optimal efficiency. Catalytic cracking with decay due to coking and cracker regeneration by coke burnoff represents such a optimization problem, as Weekman's work neatly illustrates.[1] His analysis is summarized here.

Optimum Operation-Regeneration Cycles

As noted earlier, the time-averaged conversion for a fixed-bed or batch fluid-bed catalytic cracker is

$$\bar{\varepsilon} = \frac{1}{\lambda} \ln \frac{1+A}{1 + A \exp(-\lambda)} \qquad (10\text{-}176)$$

Weekman defines reactor efficiency as the total productivity over any number of operation-regeneration cycles divided by productivity at 100 percent conversion in the absence of regeneration. For N cycles of operation at time t_0

$$\text{Efficiency} = \mathscr{E} = \frac{N t_0 \bar{\varepsilon}}{t_t} \qquad (10\text{-}177)$$

where t_t is total time equal to $t_t = N(t_0 + t_r)$ and t_r is regeneration time. Noting that $\lambda = \alpha t_0$, we see that efficiency becomes

$$\mathscr{E} = \frac{1}{\alpha(t_0 + t_r)} \ln \frac{1+A}{1 + A \exp(-\lambda t_0)} \qquad (10\text{-}178)$$

We must now relate regeneration time to operating time, for it is evident that coke formation will be a function of t_0; hence coke burnoff (regeneration) time t_r will be a function of t_0.

Coke formation can be described by the Voorhies model

$$\frac{dC}{dt_0} = k_c C^{-n} \qquad (10\text{-}179)$$

Coke-burning kinetics, in a diffusion-uninhibited case, is assumed at constant O_2 concentration to obey the relation

$$\frac{dC}{dt_r} = -k_b C^m \qquad (10\text{-}180)$$

Weekman poses the question: For what values of m and n will an optimum operation-regeneration cycle exist?

Integrating the coking equation ($t_0 = 0$, $C = 0$) gives $C = b t_0^{\gamma}$, where $b = [(1 + n)k_c]^{\gamma}$ and $\gamma = 1/(1 + n)$.

[1] V. W. Weekman, Jr., *Ind. Eng. Chem. Process Des. Dev.*, 7: 252 (1968).

Solving the regeneration relation ($t_r = 0$; $C = bt_0{}^\gamma$), we find

$$t_r = \frac{b^{1-m}t_0{}^{1-m/n+1}}{k_b(1-m)} \qquad m \neq 1 \qquad (10\text{-}181)$$

Inspection of the reactor-efficiency equation (10-178) indicates that

$$\mathscr{E} \to 0 \qquad \begin{array}{c} t_0 \to 0 \\[4pt] t_0 \to \infty \end{array}$$

Therefore an optimum must exist in terms of coking and regeneration kinetic orders n and m.

As an exercise, show that an optimum exists for $m \geq 0$ and values of n between 0 and 1.

Regeneration time t_r becomes essentially independent of t_0 when the time required for mechanical changeover is great compared to t_r. In this case Eq. (10-178) is simply displayed in Fig. 10-46, for example, where $t_r = 0.1$ h for the indicated values of K_0 and α and space velocity is the parameter. Optimum profiles, t_0 versus t_r are set forth in Fig. 10-46b.

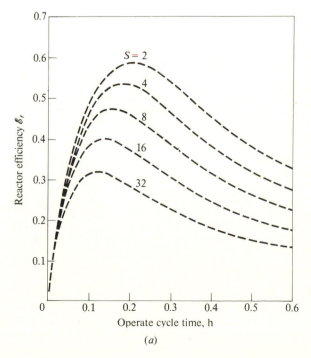

(a)

FIGURE 10-46a

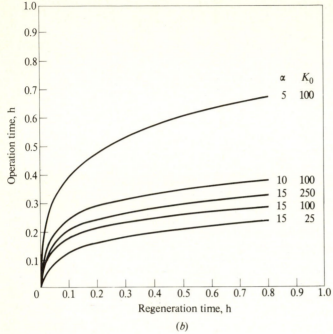

FIGURE 10-46
(a) Reactor efficiencies at various space velocities for fixed regeneration time of 0.1 h; $K_0 = 143.0 \text{ h}^{-1}$, $\alpha = 18.8 \text{ h}^{-1}$. (b) Locus of operation-regeneration times for optimum reactor efficiency; space velocity $= 8$ vol/(vol)(h). [*V. W. Weekman, Jr., Ind. Eng. Chem. Process Des. Dev.*, **7**: *252 (1968)*.]

Comments on the Model

Weekman notes that his treatment rests upon the assumption of isothermality, whereas in fact the crackers operate adiabatically. Since the activation energy is only 10 kcal, the isothermal postulate is not too severe a compromise with reality. It is also to be noted (see Chap. 7) that regeneration of coke tends to be near zero order in the intrinsic-burning regime. However regeneration in other catalyst systems may approach first order in poison-residue concentration, in which case Eq. (10-181) cannot apply since $m = 1$.

As an exercise, derive the relationship for operation-regeneration efficiency $\mathscr{E}$ in terms of A, λ, and t_0 for first-order regeneration. Assume that the final poison-residue concentration at the end of the regeneration cycle is 0.0003 and that the poison-deposition rate conforms to the Voorhies model.

Do not forget that insofar as the chemical reactor is only part of the plant, overall plant optimization may not necessarily require optimum reactor performance. For example, the optimum conditions in terms of fixed-bed yield may call for high Reynolds number and small-particle operation at high air-to-reactant feed ratio, as in naphthalene oxidation. But these optimum conditions may well be uneconomical because the consequent high pressure drop and high airflow requirements impose penalities in terms of the air-compressor operation.

Multiplicity and Stability

Both isothermal and nonisothermal multiplicity of the steady state have been touched upon in earlier chapters; in the course of such discussions the relevant parameters have been identified. It is inappropriate and, in fact, beyond our capabilities to venture into analyses of multiplicity and stability which embrace the multidimensional, mixed-rate controlled-reactor environments cited in this chapter. The seminal works of authorities in this field are cited at the end of this chapter. Suffice it to note that multiplicity and stability will be governed in the distributed-parameter, mixed-rate-controlled heterogeneous catalytic reactor by those same dimensionless groups which dictate such behavior for the CSTR, nonporous-particle, gas-liquid reactor, and porous-solid catalytic pellet, some of which have been delineated in earlier chapters.

Prudence dictates that in the distributed-parameter, mixed-rate-controlled reactor environment the matters of multiplicity and stability are best explored by cautious numerical resolution of those continuity equations judged to best describe the system at hand, as illustrated in the work of Lapidus and his students[1] for the fixed-bed nonisothermal nonadiabatic reactor in which interphase-intraphase gradients of both concentration and temperature are anticipated in both steady and unsteady states.

Parametric sensitivity for fixed-bed o-xylene oxidation is discussed by Froment,[2] while the pioneering analyses of stability of packed adiabatic reactors are those of Amundson and his students.[3] When the reactor is subject to fouling, additional complexities arise, as discussed by Luss,[4] who presents criteria for uniqueness.

Finale

Now that the terrain has been viewed and reviewed, a summation would seem to be in order.

Other treatises, cited earlier, focus in depth upon aspects of chemical and catalytic reaction engineering which we have chosen to but allude to, or prudently ignore.

The key topics of optimization and stability, for example, were lightly treated herein, since the complexities of these topics are judged to be of such gravity as to be best left to those of acknowledged expertise in these areas.

[1] H. Deans and L. Lapidus, *AIChE J.*, **6:** 656 (1960); M. L. McGuire and L. Lapidus, *AIChE J.*, **11:** 85 (1965).
[2] G. Froment, *Ind. Eng. Chem.*, **59**(2): 18 (1967).
[3] N. R. Amundson et al., *Ind. Eng. Chem. Fundam.*, **1:** 200 (1962), **2:** 12 (1963); *AIChE J.*, **14:** 636 (1968).
[4] M. A. Ervin and D. Luss, *AIChE J.*, **16:** 980 (1970).

Recognizing these limitations in the scope embraced by this text, it is to be hoped that certain concepts, central to chemical and catalytic reaction engineering, have been conveyed.

Experts in one or more of the diverse aspects of this cosmic subject will understandably recognize that this work was created primarily for students who do not as yet know the nature of the terrain created by the experts.

Details have been specified when it was judged that such specification would advance understanding—the ultimate goal of any teaching enterprise.

After all, even Virgil served as a guide.

My indebtedness to my many Virgils is, I trust, evident.

Each error is mine.

ADDITIONAL REFERENCES

The anatomy of heat transport in the fixed bed is analyzed by E. Singer and R. H. Wilhelm, *Chem. Eng. Prog.*, **46**: 343 (1950). Design procedures are found in G. Damköhler, "Der Chemie-Ingenieur," A. Euken and M. Jakob (eds.), vol. 3, pt. 1, p. 448, Leipzig (1947); L. V. Grossman, *Trans. AIChE*, **42**: 535 (1946); D. U. von Rosenberg, R. Durrill, and F. Spencer, *Br. Chem. Eng.*, **7**(3): 186 (1962); D. Hurt, *Ind. Eng. Chem.*, **35**: 522 (1943); G Froment, *Adv. Chem.*, **109**: 1 (1972); J. J. Carberry and D. White, *Ind. Eng. Chem.*, **61**(7): 27 (1969); T. G. Smith and J. J. Carberry, *Adv. Chem.*, **133**: 362 (1974); and J. Beek, *Adv. Chem. Eng.*, **3**: 259 (1962). V. Hlavacek, *Ind. Eng. Chem.*, **62**: 8 (1970), presents an exhaustive review of the mathematical aspects of fixed-bed modeling; see also N. G. Karnath and R. Hughes, *Catal. Rev.*, **9**: 169 (1974). Axial-dispersion effects in fixed-bed modeling are analyzed in J. J. Carberry and M. M. Wendel, *AIChE J.*, **9**: 129 (1963), and B. Finlayson, *Catal. Rev.*, **10**: 69 (1974), a review of collocation techniques.

The status of catalytic reaction engineering is reviewed in J. J. Carberry and J. B. Butt, *Catal. Rev.*, **10**: 221 (1974). Elements of catalytic-reactor design are nicely set forth in T. E. Corrigan and W. C. Mills, *Chem. Eng.*, April to June, 1956, which is but a fraction of Corrigan's comprehensive refresher series on kinetics and reactor design. Important design and operating aspects of fixed-bed adiabatic reactors for a number of industrially significant reactions are discussed in P. Mars, "Chimica e Dinamica della Catalisi," Consiglio Nazionale della Ricerche, Rome, 1964; see also P. Mars, *Chem. Eng. Sci. 2d Eur. Symp. Chem. React. Eng.*, vol. K-3, p. 1, which deals with the adiabatic water-gas shift reactor.

Catalytic-reactor design problems associated with ortho-para H_2 conversion in a hydrogen-liquefaction process are discussed and a simplified design procedure illustrated in M. S. Lipman, H. Cheung, and O. P. Roberts, *Chem. Eng. Prog.*, **59**(8): 49 (1963), while a dynamic model of a catalytic reformer is detailed in W. G. Welter, *Chem. Eng. Prog.*, **59**(2): 78 (1963). Comparisons of two- and one-dimensional fixed-bed models are set forth in J. Beek and E. Singer, *Chem. Eng. Prog.*, **47**: 534 (1951), and in J. E. Crider and A. Foss, *AIChE J.*, **11**: 1012 (1965).

A comprehensive review of fluidization with respect to the transport phenomena encountered there is found in the three-part survey in J. F. Frantz, *Chem. Eng.*, 1962,

Sept. 17, p. 161; Oct. 1, p. 89; Oct. 29, p. 103.   These articles are rich in data, correlation, and references.   Model and data for backmixing in fluid beds are presented in R. Latham, C. Hamilton, and O. E. Potter, *Br. Chem. Eng.*, **13**(5): 666 (1968).   A critique of diverse fluid-bed reactor models and their conformity with data is given in C. Chavarie and J. R. Grace, *Ind. Eng. Chem. Fundam.*, **14**: 75, 79, 86 (1975).   Fluid-bed catalytic-reaction data are contrasted with fixed-bed results in M. Goldman, L. N. Canjar, and R. B. Beckman, *J. Appl. Chem.*, **7**: 274 (1957) for SO_2 oxidation and in A. Kuloor and associates, *Ind. Eng. Chem. Process Des. Dev.*, **9**: 293 (1970), for dehydrogenation of propanol.   Finally, fluid-bed bubble-diameter correlation is provided in S. Mori and C. Y. Wen, *AIChE J.*, **21**: 209 (1975).

Effectiveness factors in a three-phase slurry reactor are reported in C. N. Kenny and W. Sedriks, *Chem. Eng. Sci.*, **27**: 2029 (1972).   D. R. Cova, *Ind. Eng. Chem. Process Des. Dev.*, **5**: 20 (1966), treats catalyst suspensions in gas-agitated reactors.

G. R. Gillespie and R. Kenson, *Chem. Tech.*, **1**: 627 (1971), describe the catalytic-gauze system employed in the commercial oxidation of NH_3.   The phenomenon of temperature fluctuations (flickering) of catalytic wires and gauzes is treated theoretically and experimentally by D. Luss and his students, *AIChE J.*, **20**: 571 (1974).

Trickle-bed data and modeling are reviewed in C. N. Satterfield, *AIChE J.*, **21**: 209 (1975).   An interesting study of partial wetting in trickle beds is that of W. Sedricks and C. N. Kenny, *Chem. Eng. Sci.*, **28**: 559 (1973), while the influence of transport processes upon trickle-bed conversion is treated in N. Sylvester, *AIChE J.*, **19**: 640 (1973). Fluid-flow and heat-transfer characteristics in concurrent gas-liquid flow in packed beds are detailed in V. W. Weekman and J. Myers, *AIChE J.*, **10**: 951 (1964), **11**: 13 (1965).

Hybrid-computer simulation of a moving-bed regenerator is reported in V. W Weekman, M. D. Harter, and G. E. Marr, *Ind. Eng. Chem.*, **59**: 84 (1967), while B. L. Schulman, *Ind. Eng. Chem.*, **55**(12): 44 (1963), discusses fixed-bed regeneration.

A classic example of catalytic-reactor optimization is that of O. A. Hougen and students, *Chem. Eng. Prog.*, **57**: 51 (1961); *AIChE J.*, **8**: 5 (1962).   Stability and multiplicity for the nonisothermal, nonadiabatic fixed bed are treated by M. L. McGuire and L. Lapidus, *AIChE J.*, **2**: 82 (1965), who utilize the cell model of H. A. Deans and L. Lapidus.

PROBLEMS

10-1   A new catalyst proposed for commercial use is evaluated in the laboratory at 370°C.   A conversion of 71 percent is achieved with a 100-g sample.   At the same conditions 40 g of the present commercially employed but far more expensive catalyst gives the same conversion.   If the activation energy of the proposed catalyst is 30 kcal/mol, by how many degrees must the plant-reactor temperature be increased to match the performance of the present commercial catalyst? Repeat for an activation energy of 10 kcal/mol.

10-2   At 300°C, 100 g of $Pt/\alpha\text{-}Al_2O_3$ oxidizes 95 percent of 100 ppm of CO to CO_2 in air.   A somewhat sintered sample of this catalyst oxidizes 92.2 percent of CO at identical conditions.   How much of this latter catalyst is required to match the performance of 1 kg of the former?

10-3 A dual-function isomerization catalyst-reaction system is governed by the rate, at constant temperature,

$$r = \frac{k(A - 0.2B)}{1 + 4B} \qquad \text{kg mol/(kg cat)(min)}$$

where A and B are partial pressures (in atmospheres) of reactant and product, respectively, and $k = 0.04$. How deep an isothermal catalyst bed of 1 m² cross-sectional area is required to produce 75 percent conversion of 2.5 kg mol/min of feed A of 2 atm partial pressure? Catalyst bulk density is 0.8 g/cm³.

10-4 A catalyst bed 1.2 m in diameter and 2.4 m depth is packed with 1.25-cm-diameter spherical catalyst particles. Mass velocity of the gas, 90 percent air, is 0.44 kg/(m²)(s), and gas temperature is 400°C. Use the Kozeny equation to determine the diameter and bed depth of 0.6-cm-diameter spheres of the same weight which will provide the identical pressure drop. Repeat using the Ergun equation and Talmadge's modification of it [*AIChE J.*, **16:** 1092 (1970)].

10-5 In a plant installation, an adiabatic bed of fresh catalyst provides the design conversion goal at an inlet temperature of 460°C and an exit value of 437°C. After several months of operation, to maintain design conversion, it is necessary to raise the inlet temperature to 470°, the exit value then being 448°C. If the energy of activation is known to be 20 kcal, what percentage of fresh catalyst activity is lost?

10-6 The rate of the catalytic hydrogenation of CO_2 to produce methane is [*Ind. Eng. Chem.*, **47:** 140 (1955)]

$$r = \frac{\text{kg mol } CH_4}{(\text{kg cat})(\text{h})} = \frac{kCH^4}{(1 + K_1H + K_2C)^5}$$

where C and H are the partial pressures of CO_2 and H_2 respectively. At a total pressure of 30 atm and 314°C, $k = 7$, $K_1 = 1.73$, and $K_2 = 0.3$. For a feed of 100 kg mol/h of CO_2 and stoichiometric H_2/CO_2 ratio, compute the weight of catalyst required for 20 percent isothermal conversion of the CO_2. Repeat, ignoring mole change due to reaction.

10-7 Weller [*AIChE J.*, **2:** 59 (1956)] observed, as mentioned in Chap. 8, that a simple power law [Eq. (8-80)] adequately describes CO hydrogenation kinetics. Examine the data referred to in Prob. 10-6 and attempt to fashion a simpler rate model. If successful, use the model in repeating Prob. 10-6.

10-8 Pasquon and Dente [*J. Catal.*, **1:** 508 (1962)] report detailed diffusion-affected rate data for a methanol-synthesis plant reactor. For the conditions tabulated on p. 509 of their paper, use the information displayed in their figure 1 to determine catalyst weight required for 30 percent conversion in an adiabatic industrial reactor. Repeat assuming catalytic effectiveness to be unity.

10-9 It has been shown that if the aspect ratio of bed length to catalyst diameter is greater than 150, axial mixing of heat and mass may be considered negligible. However, some catalytic reactors are of quite low aspect ratio since reaction vigor is such as to require a narrow zone of reaction, for example, NH_3 oxidation over Pt/Rh gauze, HCN synthesis, and methanol oxidation over Ag. In each of the above cases the rate is largely determined by bulk mass transport. Derive the mass and energy continuity equations, including axial dispersion of heat and mass, for a totally external mass-transport-governed reaction. Solve for and plot the reactant concentration-temperature profiles assuming:

(*a*) Pe for heat and mass equals 2.

(*b*) Pe for heat equals 0.5; for mass, 2; $d_p = 6$ mm: gas properties are those of air at 400°C; gas velocity = 10 cm/s (= 0.2355 mi/h); $Z/d_p = 2, 5, 10$, and 100.

10-10 Repeat the illustration for naphthalene oxidation assuming a one-dimensional fixed-bed model where the overall effective heat-transfer coefficient of Beek and Singer [*Chem. Eng. Prog.*, **47**: 534 (1951)] is assumed to apply.

10-11 Compute the isothermal yield-conversion profile for naphthalene oxidation in terms of the simple consecutive model at 320, 370, 400, and 450°C assuming catalyst-pellet sizes of 1, 3, 5, and 10 mm. Intraphase diffusivity is 0.1 cm²/s. The mass Biot number may be considered infinite. Repeat for $\mathscr{D} = 0.01$ cm²/s.

10-12 If the tubular catalytic wall reactor discussed in Sec. 10-3 is unpacked, how would the rate processes at the wall be affected for the same linear superficial gas velocity?

10-13 Compute the minimum fluidization and entrainment velocities, $\Delta P/L$, at the fluidization point and the quality of fluidization for the following systems:

	Fluid	T, °C	Pressure, atm	d_p, μm
(*a*)	Air	400	1	10
				50
				100
				200
	H_2	400	1	10
				50
				100
				200
(*b*)	H_2O	70	1	10
				100
				1000
(*c*)	Benzene	50	1	1000
(*d*)	Repeat part (a) for 10, 100, and 300 atm			

10-14 Martinez (Ph.D. thesis, University of Notre Dame, Notre Dame, Ind., 1972) reports kinetic data for Ag/Al_2O_3-catalyzed oxidation of ethylene

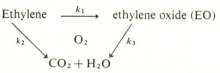

Use the data below to establish ethylene oxide yield EO/E_0 versus ethylene conversion profiles at each temperature for a fluidized-bed reactor in terms of the Kunii-Levenspiel model. At 230°C, Martinez' data are

$$k_1 = 0.57 \ \text{s}^{-1} \qquad E_1 = 23.4 \ \text{kcal}$$
$$k_2 = 0.585 \ \text{s}^{-1} \qquad E_2 = 31.2 \ \text{kcal}$$
$$k_3 = 0.24 \ \text{s}^{-1} \qquad E_3 = 5.8 \ \text{kcal}$$

The reactions are essentially first order in ethylene and zero order in excess O_2. Feed is 5% ethylene. Temperature is 230, 250, 275, and 300°C. Pressure: 1 and 10 atm. $d_b = 2, 5$, and 10 cm. $u_{mf} = 2.1$ cm/s. $u_0 = 13.2$ and 25 cm/s.

10-15   For the conditions cited in Prob. 10-14 compute ethylene conversion vs. bubble-contact time using (*a*) the Kunii-Levenspiel, (*b*) the Davidson-Harrison, and (*c*) the Olson model.

10-16   A complex antibiotic precursor is to be prepared in a batch slurry reactor via Raney nickel–catalyzed hydrogenation. Overall stoichiometry is

$$A + 3H_2 \underset{k_{-1}}{\overset{k_1}{\longleftrightarrow}} B$$

$$k_1 = 0.1\sqrt{p} \quad \text{and} \quad k_{-1} = 0.1 \text{ h}^{-1}$$

$$p = H_2 \text{ partial pressure, atm}$$

The autoclave is to be charged with 1 kg mol of A plus catalyst of 50 μm diameter. Free gas space above the solution is 5 liters. Compute the conversion vs. batch time for

(*a*)   100 atm of pure H_2 initial pressure
(*b*)   100 atm of cracked NH_3 initial pressure
(*c*)   100 atm of H_2 recirculated with make-up addition

Note: In cases (*a*) and (*b*) gas phase is batch, as is liquid, while in (*c*) the batch liquid is contacted with a continuous gas phase.

10-17   Derive relationships for conversion-deactivation for isothermal fixed-, moving- and fluid-bed reactors in which a first-order reaction occurs.

10-18   Develop the isothermal decay-affected yield-of-B relation for fixed-, moving-, and fluid-bed reactors assuming linear kinetics for the reaction

Assume $\phi = \bar{\phi}$, as did Weekman.

10-19   Consider the data of Table 10-5. Assuming $-\Delta H_1 = 430$ kcal and $-\Delta H_2 = 870$ kcal, explore the stability issue for each catalyst at each temperature level for an adiabatic fluid bed at naphthalene feed compositions of $\frac{1}{2}$, 1, and 5 percent. For purposes of simple stability analysis, treat the reactor as a simple CSTR. Assume an exit conversion of 90 percent and 100 percent yield of anhydride, i.e., the reaction is $N \overset{k_1}{\rightarrow} A$.

10-20   With the data set forth in Table 10-5 construct the dimensionless heat-generation–temperature curves for $\frac{1}{2}$, 1, and 5% naphthalene in air for a CSTR environment. Assume $\theta = 10$ s and $T_0 = 320°C$ for each catalyst.

10-21   Use the kinetic data presented by Callahan, Grasselli, Miberger, and Strecker [*Ind. Eng. Chem. Prod. Des. Dev.*, **9**: 134 (1970)] to compare isothermal fixed-bed (PFR) yield of acrylonitrile (in the ammoxydation of propylene over Bi/Mo catalyst) with that in a fluidized bed for diverse bubble diameters and gas velocity. Assume a small Thiele modulus.

10-22   Given the rate coefficients in figure 9 of the paper cited in Prob. 10-21, what maximum pellet diameter could be utilized in a fixed bed for an η value of unity if (*a*) $\mathscr{D} = 10^{-2}$ cm²/s, (*b*) $\mathscr{D} = 10^{-3}$ cm²/s? Should an isothermal diffusional gradient be avoided or encouraged for ammoxydation as schematized in figure 8 of the paper?

10-23 Wilson [*Trans. Inst. Chem. Eng.*, **24:** 77 (1946)] presents experimental axial temperature-profile data for fixed-bed catalyzed hydrogenation of nitrobenzene. Attempts to predict these data with use of a one-dimensional homogeneous model have failed. Use the one-dimensional model but include interphase and intraphase mass diffusion and interphase heat transport and first-order kinetics (at E equal twice that reported by Wilson) in an attempt to match the data.

10-24 Walsh and Katzer [*Ind. Eng. Chem. Process Des. Dev.*, **12:** 477 (1973)] studied the kinetics of phenol oxidation in air over copper oxide. Consult that work and:

    (*a*) From the data of their table I, prove the absence of interphase and intraphase diffusion influences in terms of the observables and other criteria cited by Mears [*J. Catal.*, **20:** 127 (1971); *Ind. Eng. Chem. Process Des. Dev.*, **10:** 541 (1971)].

    (*b*) Design an adiabatic fixed-bed reactor to achieve 99 percent removal of 1% phenol in air (phenol feed rate is 1 kg mol/min). To minimize pressure drop, 8-mm-diameter catalyst spheres are specified; intraphase diffusivity is 2×10^{-3}. Assume the pellets to be isothermal at a temperature dictated by interphase heat transport. Reactor feed temperature can be varied between 150 and 200°C.

10-25 Mars (*2d Eur. Symp. Chem. React. Eng. 19*, vol. K-3, p. 1, Pergamon Press) describes the use of laboratory catalytic data and transport factors in predicting temperature profiles in a plant water-gas shift reactor. Analyze this paper and attempt to fashion a more detailed model which more faithfully predicts plant measurements.

INDEX

INDEX

Page numbers in **boldface** indicate references are to either illustrations or tables.